SALAMFESTSCHRIFT

World Scientific Series in 20th Century Physics

Published

Vol. 44 *Symmetries in Nature — The Scientific Heritage of Louis Michel* edited by T. Damour, I. Todorov and B. Zhilinksii

Vol. 43 *Nambu: A Foreteller of Modern Physics* edited by T. Eguchi and M. Y. Han

Vol. 42 *Selected Papers of K. C. Chou* edited by Y.-L. Wu

Vol. 41 *Many-Body Theory of Molecules, Clusters, and Condensed Phases* edited by N. H. March and G. G. N. Angilella

Vol. 40 *Murray Gell-Mann — Selected Papers* edited by H. Fritzsch

Vol. 39 *Searching for the Superworld — A Volume in Honour of Antonino Zichichi on the Occasion of the Sixth Centenary Celebrations of the University of Turin, Italy* edited by S. Ferrara and R. M. Mössbauer

Vol. 38 *Matter Particled — Patterns, Structure and Dynamics — Selected Research Papers of Yuval Ne'eman* edited by R. Ruffini and Y. Verbin

Vol. 37 *Adventures in Theoretical Physics — Selected Papers with Commentaries* by Stephen L. Adler

Vol. 36 *Selected Papers (1945–1980) with Commentary* by Chen Ning Yang

Vol. 35 *A Career in Theoretical Physics (2nd Edition)* by P. W. Anderson

Vol. 34 *Quantum Mechanics in Phase Space: An Overview with Selected Papers* edited by C. K. Zachos, D. B. Fairlie and T. L. Curtright

Vol. 33 *Formation and Evolution of Black Holes in the Galaxy — Selected Papers with Commentary* edited by H. A. Bethe, G. E. Brown and C.-H. Lee

Vol. 32 *In Conclusion — A Collection of Summary Talks in High Energy Physics* edited by J. D. Bjorken

Vol. 31 *From the Preshower to the New Technologies for Supercolliders — In Honour of Antonino Zichichi* edited by B. H. Wiik, A. Wagner and H. Wenninger

Vol. 30 *Selected Papers of J. Robert Schrieffer — In Celebration of His 70th Birthday* edited by N. E. Bonesteel and L. P. Gor'kov

Vol. 29 *Selected Works of Emil Wolf (with Commentary)* edited by E. Wolf

Vol. 28 *The Legacy of Léon Van Hove* edited by A. Giovannini

Vol. 27 *Selected Papers of Richard Feynman (with Commentary)* edited by L. M. Brown

For information on Vols. 1–26, please visit http://www.worldscientific.com/series/wsscp

World Scientific Series in 20th Century Physics – Vol. 4

SALAMFESTSCHRIFT

A Collection of Talks from the Conference on Highlights of Particle and Condensed Matter Physics

ICTP, Trieste, Italy   8 – 12 March 1993

Editors

A Ali *(DESY)*

J Ellis *(CERN)*

S Randjbar-Daemi *(ICTP)*

World Scientific
Singapore • New Jersey • London • Hong Kong

Published by

World Scientific Publishing Co. Pte. Ltd.
5 Toh Tuck Link, Singapore 596224
USA office: 27 Warren Street, Suite 401-402, Hackensack, NJ 07601
UK office: 57 Shelton Street, Covent Garden, London WC2H 9HE

British Library Cataloguing-in-Publication Data
A catalogue record for this book is available from the British Library.

World Scientific Series in 20th Century Physics — Vol. 4
SALAMFESTSCHRIFT

ISBN-13 978-981-02-1421-0
ISBN-10 981-02-1421-9
ISBN-13 978-981-02-1422-7 (pbk)
ISBN-10 981-02-1422-7 (pbk)

PREFACE

During the week 8 to 12 March 1993, a Conference was held at the International Centre for Theoretical Physics in Trieste to honour Professor Abdus Salam on the occasion of his retirement as a professor of theoretical physics at the Imperial College of Science and Technology, London University. The Conference, which was called "*Highlights of Particle and Condensed Matter Physics*", brought together physicists from 32 countries to honour a man who has devoted all of his active life to the advancement of science around the world.

It was, therefore, a pleasure to see the friends, colleagues, collaborators and former students of Professor Salam who attended the Salamfest. Their mere presence was an acknowledgement of their deep admiration for the man being honoured.

Many of the friends and the former students of Professor Salam presented colourful recollections of their earlier meetings and work with him. It became apparent as the week went by, and memories surfaced more readily, that everyone had always enjoyed a productive relationship with Salam — he became a close friend, as well as a strong critic, of those he supervised and researched with.

This book is a collection of the articles presented at the meeting. It also contains papers by some our colleagues who kindly accepted our invitation to contribute to the volume.

Although there were several distinguished condensed matter physicists who contributed to the Conference, this volume contains mainly papers on particle physics.

We would like to express our sincere gratitude to our co-organizers, Professors D. Amati, S. Weinberg and Yu Lu, as well as to all of Professor Salam's friends, collaborators and students who, by their participation in this memorable meeting, or by sending us their papers, have enriched the Conference and have helped us to put this *festschrift* together.

We would also like to thank the Deputy Director of the Centre, Professor L. Bertocchi, and his administrative staff for all the logistic support which was put at our disposal so that the meeting could proceed in a very cordial and festive atmosphere.

Last, but not least, we should like to express our sincere thanks to the secretary of the Conference, Ms. S. Bencich, for her diligent and professional help.

A. Ali
J. Ellis
S. Randjbar-Daemi

OPENING ADDRESS

by S. Randjbar-Daemi

Good Morning, Ladies and Gentlemen.

On behalf of the Organizers and the staff of the ICTP, I should like to welcome you to the International Centre for Theoretical Physics.

As you know already, this Conference has been organized to honour Professor Abdus Salam, who has been an active scientist for over forty years. In this remarkably long period, his interests have covered a wide spectrum of topics, ranging from the intricacies of renormalization theory in the early fifties, to the problem of the origin of life with which he has been recently involved.

Professor Salam is one of the main architects of modern fundamental physics and, in particular, the Standard Model of Particle Physics. Many contributions to this meeting will address the importance of this model in our understanding of the structure and behaviour of matter in its most fundamental form.

Professor Abdus Salam founded the International Centre for Theoretical Physics almost thirty years ago, and has been its Director ever since.

The motivation behind the creation of the International Centre for Theoretical Physics was the idealistic conviction of a distinguished scientist from a developing country who believed, and who still believes today, that the scientific heritage belongs to all humanity and it cannot, or rather it should not, be monopolised. The Centre was created to help the scientists of developing countries to fight their isolation and to have an easier access to the international scientific community.

Over the past three decades the International Centre for Theoretical Physics has grown into a Research Institute, and has hosted many scientists from the develping, as well as the advanced, countries.

Despite the remarkable success of the ICTP in attaining many of its original objectives, sadly many of the unfavourable socio-political conditions which make it impossible to create a meaningful scientific culture in developing countries are still with us. For this reason, the Centre is expected to respond to the same needs as those of thirty years ago, only maybe at a higher level.

The generous support of the Italian Government over the years, as well as the hospitality of the city of Trieste, has been of prime importance in the achievements of the Centre. Equally important, however, has been the support of the international scientific community. We needed this support in the past, and we also need it now.

I thank you for accepting our invitation to be with us today, and I hope you will enjoy this week of overview of a large part of theoretical physics of our time, to which Professor Salam has contributed so much.

CONTENTS

III. Astro-Particle Physics and Cosmology

IV. Strings and Quantum Gravity

V. Mathematical Physics and Condensed Matter

VI. Salam's Collaborators and Students

I. Standard Model

Unbreaking Symmetries*

Steven Weinberg†
Theory Group, University of Texas
Austin, Texas 78712

†Research supported in part by the Robert A. Welch Foundation and NSF Grant PHY 9009850.

I first met Abdus Salam in 1960, when he knocked on the door of the little cubbyhole that I had for an office at the Rad Lab in Berkeley. I had spent a fair fraction of my time as a graduate student in mastering Salam's analysis of overlapping ultraviolet divergences, and then as a post-doc at Columbia in using Salam's result[1] , that renormalization makes all integrals and all subintegrals however chosen superficially convergent, to show that the integrals actually converge.[2] (I have always been proud of this proof, because it is the only place I know in physics where one actually needs to use the Heine-Borel theorem. Arthur Wightman had done me the great favor of reading through the manuscript, thus enabling me to claim that my paper satisfied a criterion that had been laid down earlier by Salam, that a notation should be understandable to at least two persons, of whom one may be the author. Understandable it may have been, but not elegant; Wightman told me that he could see blood on every lemma.) Knowing as I did the importance of Salam's contributions to renormalization theory, I was pleased and flattered that he had stopped by to say hello during his visit to Berkeley. It began a friendship that for thirty years has been a valuable part of my life in physics.

I think it was then that I proposed that I use a Sloan Fellowship to make a visit to London for several months the following year. Salam was good enough to offer the hospitality of his department at Imperial College, and I arrived in October 1961. By then both Salam and I had independently become interested in a new topic, spontaneously broken symmetries. I can't speak for Salam, but I know that I was at first confused about the role that broken symmetries might play in physics; I had some sort of idea that they would appear as approximate symmetries, something like the new $SU(3)$ of Gell-Mann and Ne'eman. Of course, it was not long before I learned better, through the work that Salam and I did that year at Imperial College (together with Jeffrey Goldstone as a long-distance collaborator), on the massless particles required by symmetry breaking[3]. As everyone knows now, broken global symmetries in general don't look at all like approximate ordinary symmetries, but show up instead as low energy theorems for the interactions of these massless Goldstone bosons.

Given this history, it seems appropriate that I return in this festschrift to the old topic of broken symmetries. I want to discuss here certain circumstances under which the Goldstone bosons together with certain massless partners do form a complete representation of the broken symmetry group.

Suppose we vary some parameter, such as a mass or the temperature, so that a theory with broken symmetries goes through a second-order phase transition in which these symmetries become unbroken. On the broken symmetry side of the transition, there are massless particles, but in general no degenerate multiplets forming representations of the symmetry group. On the unbroken symmetry side of the transition, there are degenerate multiplets forming full representations of the symmetry group, but in general no massless particles. Continuity therefore requires that *at* the phase transition, there must be a multiplet of massless particles, containing not only nascent Goldstone bosons, but other massless particles needed to fill out a full representation of the symmetry group. These massless Goldstone partners are very important physically, because unlike the Goldstone bosons they participate in non-derivative interactions, and can therefore produce infrared divergences that radically change various scaling laws at the phase transition.

The question then arises, to what sort of multiplet do the Goldstone bosons belong at a second-order phase transition? This is potentially of importance not only in statistical mechanics, but also in ordinary particle physics, for it may be that in some broken symmetry theories we are actually not so far from a second-order phase transition. For instance, a recent paper of Georgi[4] can be interpreted as proposing that quantum chromodynamics at zero temperature is near a second-order phase transition, at which the broken chiral $SU(3) \otimes SU(3)$ symmetry has a $(8,1) \oplus (1,8)$ representation, consisting of the octet of pseudoscalar Goldstone bosons plus an octet of massless scalars, that on the broken symmetry side of the phase transition become the helicity-zero states of the massive vector meson octet. (This interpretation raises the question of the nature of the $SU(3) \otimes SU(3)$ multiplet that contains the helicity ± 1 states of the vector meson octet, which of course must also be massless at the phase transition.)

It is important not jump to conclusions about the nature of

the multiplet that contains the Goldstone bosons at a second-order phase transition. For instance, consider the time-honored example of a theory with broken symmetries, the linear σ-model. It has a Lagrangian density

$$\mathcal{L} = -\frac{1}{2}\partial_\mu\phi_n\partial^\mu\phi_n - \frac{1}{2}m^2\phi_n\phi_n - \frac{\lambda}{4}(\phi_n\phi_n)^2 \tag{1}$$

with the index n running over values 1, 2, ..., N. It is natural on inspecting Eq.(1) to conclude immediately that at the second-order phase transition where m^2 passes through 0, the $N-1$ Goldstone bosons join up with the component of ϕ_n in the direction of $< (\phi)_n >_{vac}$ to form an N-vector representation of $O(N)$. This is indeed true if the coupling λ is sufficiently weak. However if λ is strong, it is perfectly possible that the particles that become massless at a phase transition are bound states belonging to some other tensor representation of $O(N)$. In that case, (1) would not be a useful effective Lagrangian for studying physics near the phase Lagrangian, even if it is the Lagrangian of the underlying theory.

So how are we to decide to what sort of multiplet the Goldstone bosons belong at second order phase transitions in strongly-coupled theories like quantum chromodynamics? Fortunately, it is not impossible to find clues to help us with this question. Let us first recall some old results[5] about the coupling of Goldstone bosons. Consider a theory with an arbitrary broken symmetry group $\mathcal{G}$ associated with a set of Goldstone bosons Π_a. The S-matrix for emission of a Goldstone boson in a transition $I \to F + \Pi_a$ in Lorentz frames where the transition is *colinear*, can be written in terms of a 'reduced matrix element' X_a, as:

$$< F, \mathbf{p}', \lambda'; \Pi_a, \mathbf{q}|S|I, \mathbf{p}, \lambda > \equiv \frac{2(m_I^2 - m_F^2)(F, \lambda'|X_a|I, \lambda)\delta^4(p' + q - p)}{(2\pi)^{9/2}(8q^0 p^0 p'^0)^{1/2} f_a}. \tag{2}$$

[Here p',λ' and p,λ are the four-momenta and helicities of particles F and I, respectively, and f_a is the Goldstone boson 'decay constant', like $f_\pi \simeq 190$ MeV.] The complete set of all sum rules of the Adler-Weisberger type in the narrow resonance approximation tell us that the X_a together with unbroken symmetry generators T_i satisfy the commutation relations of the Lie algebra of $\mathcal{G}$. Physical particle states can then be expressed as superpositions of elements of various representations of this Lie algebra.

In the broken symmetry phase of a theory the X_a are not really symmetry generators. For one thing, they don't commute with the mass matrix, so a given particle state must in general be expressed as a sum of a large number of elements of various different irreducible representations of the Lie algebra composed of X_a and T_i. Also, the S-matrix does not commute with the X_a. However, it is not hard to see that the X_a do become true symmetry generators at a second-order phase transition at which $\mathcal{G}$ is restored.

For one thing, Eq. (2) shows that particles I and F for which $(F, \lambda'|X_a|I, \lambda)$ doesn't vanish must have masses that approach each other at a phase transition where $f_a \to 0$, to avoid the matrix element for Goldstone boson emission from blowing up in this limit. Also, the collinear matrix elements of the conserved current J_a^μ associated with one of the broken symmetries may be put in the form:

$$< F, \lambda', p'|J_a^\mu(x)|I, \lambda, p > =$$
$$\left((p+p')^\mu + \frac{(p-p')^\mu(m_F^2 - m_I^2)}{(p-p')^2}\right) e^{i(p-p')\cdot x} \frac{(F, \lambda'|X_a|I, \lambda)}{(2\pi)^3(4p^0p'^0)^{1/2}}$$
$$+ \ transverse \ terms \tag{3}$$

(The dependence of (3) on p and p' is dictated by current conservation. The numerical factor in (3) follows, as in the derivation of the Goldberger-Treiman relation, from the interpretation of the term proportional to $(p-p')^\mu$ as being due to exchange of a Goldstone boson.) As one approaches the second-order phase transition where the symmetry is restored, we have $m_F \to m_I$, and the $(p-p')^\mu$ term can be dropped. Then setting $\mu = 0$ and integrating over $\vec{x}$, we see that at the phase transition

$$< F, \lambda', p'| \int d^3x \ J^0(x)|I, \lambda, p >= \delta^3(\vec{p} - \vec{p}')(F, \lambda'|X_a|I, \lambda) \tag{4}$$

Thus at the phase transition, X_a is the ordinary conserved 'charge' associated with the current J_a^μ. (Another way to see that X_a commutes with the S-matrix at the phase transition is to note that otherwise the S-matrix for bremstrahlung[6] of a soft Goldstone boson from external lines of general processes would blow up in the limit $f_a \to 0$.)

It follows that if we can determine experimentally that the Goldstone boson states are sums of elements of a set of representations $\mathcal{D}_j$ of the Lie algebra of X_a and T_i in the broken symmetry phase, and if we have some reason to believe that this representation assignment does not change qualitatively as we approach the phase transition, then at the phase transition there must be sets of massless particles forming all of the representations $\mathcal{D}_j$.

In quantum chromodynamics with 2 massless quarks, it is possible under certain assumptions (the 'large $\mathcal{N}_{color}$' approximation, and the neglect of exclusive inelastic scattering at high energy and small momentum transfer) to show[7] that the pion is a superposition of the **4** and **6** representations of chiral $SU(2)\otimes SU(2)$, each with equal strength, with the other elements of these representations supplied by the helicity zero states of the ρ, ϵ, and a_1 mesons. This quartet structure compares well with experiment; in fact, it was found long ago by Gilman and Harari[8] as a successful saturation schemes for sum rules of the Adler-Weisberger and superconvergence type. (The equal strength of the **4** and **6** components of the pion is borne out by a successful prediction[7] of the ρ width.) We conclude that under the assumptions indicated above, the ρ, ϵ, and a_1 mesons as well as the pion all become massless at a second-order phase transition in quantum chromodynamics at which the chiral symmetry becomes unbroken. It is intriguing and mysterious that at the phase transition at which the chiral $SU(2)\otimes SU(2)$ symmetry of massless quantum chromodynamics becomes unbroken, this symmetry may become *local*, with ρ and a_1 as massless gauge bosons.

References

1. A. Salam, Phys. Rev. **82**, 217 (1951); **84**, 426 (1951).
2. S. Weinberg, Phys. Rev. **118**, 838 (1960).
3. J. Goldstone, A. Salam, and S.Weinberg, Phys. Rev. **127**, 965 (1962).
4. H. Georgi, Nuclear Physics **B331**, 311 (1990).
5. S. Weinberg, Phys. Rev. **177**, 2604 (1969).
6. S. Weinberg, Phys. Rev. **D2**, 674 (1970).
7. S. Weinberg, Phys. Rev. Letters **65**, 1177 (1990).
8. F. Gilman and H. Harari, Phys. Rev. **165**, 1803 (1969).

PRECISION TESTS OF THE STANDARD MODEL

G. Altarelli
Theory Division, CERN, Geneva, Switzerland

Contents

The electroweak theory originally developed by Glashow, Salam and Weinberg [1], together with QCD, is the Standard Model of fundamental particle interactions that provides us with the basic framework for the description of all observed particle properties. I think that presenting the updated evidence that sofar completely supports the intuition of the inventors of the Standard Model is a very appropriate way to honour the great achievements of Abdus Salam in particle physics. In the following I shall discuss the analysis of the available data in the Standard Model and beyond. In writing this article I decided to include all the present information, that is, also those data that became available after the Meeting in honour of Abdus Salam.

1. Precision Electroweak Data and the Standard Model

For the analysis of electroweak data in the Standard Model [1-3] one starts from the input parameters: some of them, α, G_F and m_Z, are very well measured, some other ones, m_{flight} and $\alpha_s(m_Z)$ are only approximately determined while m_t and m_H are largely unknown. Then one computes the radiative corrections [4,5] to a sufficient precision to match the experimental capabilities, compares the theoretical predictions for the numerous observables which have been measured, checks the consistency of the theory and derives constraints on m_t and hopefully also on m_H. The light quark masses m_{flight} enter in the large logs that determine the running of α from the very small momenta where it is defined up to m_Z. By using a dispersion relation, the relevant contribution to the vacuum polarisation function of the photon can be directly obtained [6] from the experimental data on the hadronic cross-section in e^+e^- annihilation at energies below m_Z. The resulting value [7] of $\alpha(m_Z)$, in a specified definition for this quantity, is $\alpha(m_Z) = (128.87 \pm 0.12)^{-1}$. The error quantifies the theoretical ambiguity associated to m_{flight} for precision tests of the Standard Model (the largest source of error apart from our relative ignorance on m_t, m_H and $\alpha_s(m_Z)$).

Recently there has been some additional progress [8] in the control of radiative corrections by new computations of some potentially dominant higher-loop effects: terms of order $(G_F m_t^2)^2$ in the $Z \to b\bar{b}$ vertex and, for all values of m_H, in $\Delta\rho$ [9], terms of order $\alpha_s G_F m_t^2$ in the $Z \to b\bar{b}$ vertex [10] and, for some refinements, in Δr and $\Delta\rho$ [11], terms of order $(\alpha_s m_b/m_Z)^2$ in $\Gamma(Z \to \text{hadrons})$ [12]. The impact of these new additions is a shift upward of 2-3 GeV in the fitted central value of m_t.

The experimental values, presented at the Marseille Conference [13-15], of the most relevant observables are collected in table 1, together with, for comparison, their values at the '93 winter meetings [19]. Not all LEP measurements listed in table 1 are independent. From resonance scanning m_Z and Γ_T are obtained. At the peak one measures the ratios σ_h, R_h, R_{bh} ... and the asymmetries. Thus, for example, Γ_l and Γ_h are derived quantities.

In figs.1-6 we present a direct comparison of the data and the Standard Model predictions (as functions of m_t, for m_H=50-1000 GeV and α_s= 0.111-0.125) for the six most relevant observables, assuming lepton universality: m_W/m_Z, Γ_T, σ_h, R_h, R_{bh} and g_V/g_A, the last quantity, which refers to charged leptons, being derived from all the asymmetries listed in table 1.

Table 1
The results presented at this Conference compared with the situation last winter.

	LATHUILE-MORIOND [19]	MARSEILLE
m_Z(GeV)	91.187 ± 0.007	
Γ_T(MeV)	2488 ± 7	2489 ± 7
$R_h=\Gamma_h/\Gamma_l$	20.83 ± 0.06	20.77± 0.05
$\sigma_h=12\pi\Gamma_e\Gamma_h/m_Z^2\Gamma_T^2$ (nb)	41.44 ± 0.17	41.55± 0.14
Γ_l(MeV)	83.52 ± 0.28	83.79 ± 0.28
Γ_h(MeV)	1739.7 ± 6.3	1740± 6
Γ_b(MeV)	383 ± 6	383± 5
$R_{bh}=\Gamma_b/\Gamma_h$	0.220 ±0.003	0.220±0.0027
A_{FB}^l	0.0164 ± 0.0021	0.0162 ± 0.0020
A_{pol}^τ	0.142 ± 0.017	0.138 ± 0.014
A^e	0.130 ± 0.025	
A_{FB}^b	0.098 ± 0.009	0.098 ± 0.006
A_{FB}^c	0.090 ± 0.019	0.075 ± 0.015
A_{LR} (SLD)	0.100±0.044	
g_V/g_A (all asymmetries)	0.0725 ±0.0033	0.0712 ±0.0028
m_W/m_Z [16]	0.8798 ± 0.0028	
$R_{neutrino}=\sigma_{NC}/\sigma_{CC}$ [17]	0.312 ±0.003	
APV(Cs) Q_W [18]	-71.04 ± 1.81	
$\alpha_s(m_Z)$ [21,22]	0.118± 0.007	

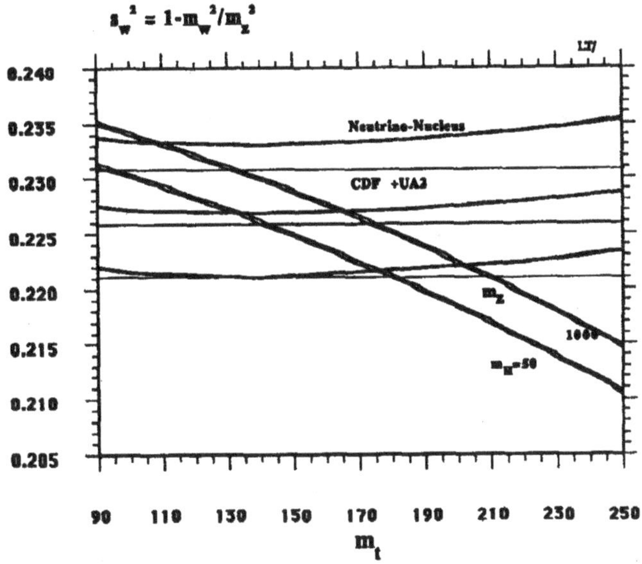

Figure 1: Labelled by m_Z is the prediction for $s_W^2 = 1 - m_W^2/m_Z^2$ obtained in the Standard Model as a function of m_t for m_H=50-1000 GeV. The 1σ error bands implied by the CDF and UA2 measurements of m_W/m_Z [16] and by the data on R_ν [17] are also shown.

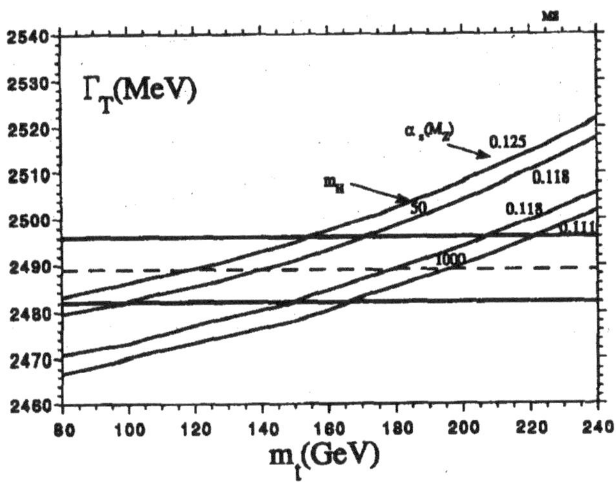

Figure 2: The Standard Model prediction for Γ_T as function of m_t for m_H=50-1000 GeV and $\alpha_s(m_Z)$ = 0.111-0.125 is compared with the LEP result.

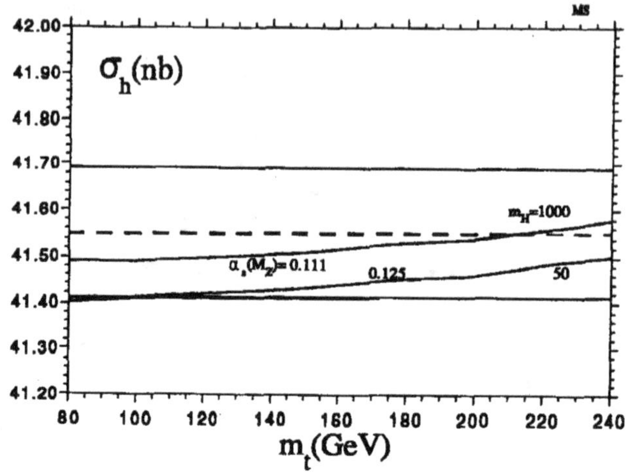

Figure 3: The Standard Model prediction for σ_H as function of m_t for m_H=50-1000 GeV and $\alpha_s(m_Z) = 0.111\text{-}0.125$ is compared with the LEP result.

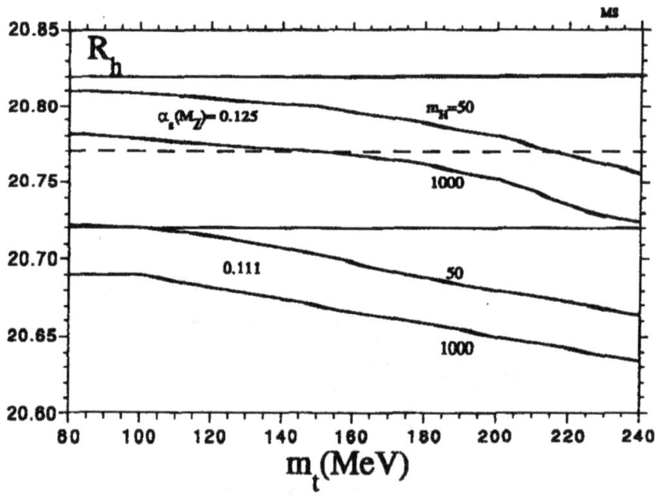

Figure 4: The Standard Model prediction for R_h as function of m_t for m_H=50-1000 GeV and $\alpha_s(m_Z) = 0.111\text{-}0.125$ is compared with the LEP result.

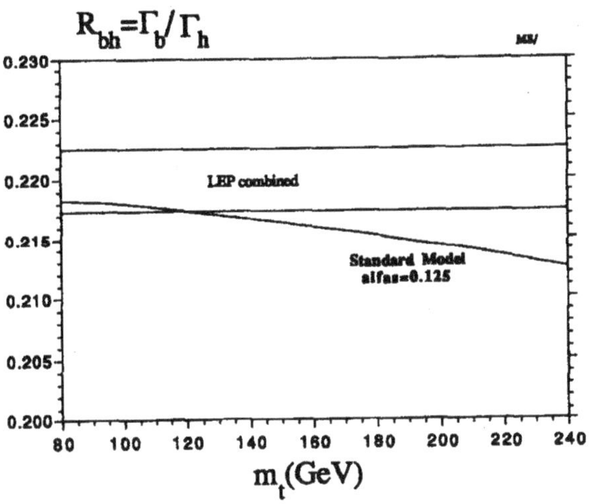

Figure 5: The Standard Model prediction for R_{bh} as function of m_t for m_H=50-1000 GeV and $\alpha_s(m_Z)$ = 0.125 (that leads to the maximum prediction in the range 0.111-0.125) is compared with the LEP result.

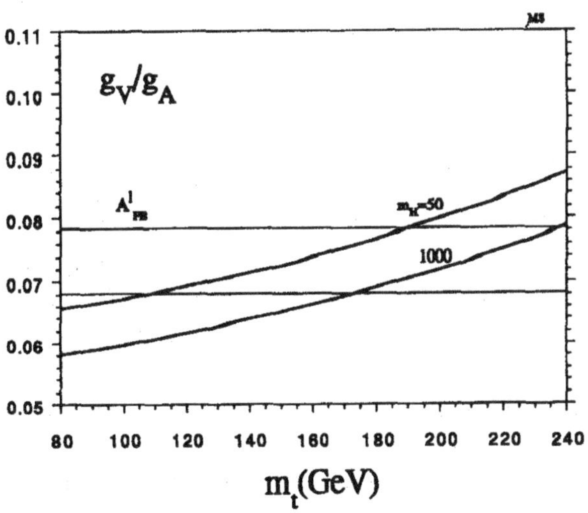

Figure 6: The Standard Model prediction for g_V/g_A as function of m_t for m_H=50-1000 GeV is compared with the LEP result obtained by combining all the measured asymmetries (see table 1).

Complete Standard Model fits are discussed in ref.14. Here I present the results of a simplified fit [20] based on the six most important quantities discussed above. Fitting at fixed m_H in terms of m_t and $\alpha_s(m_Z)$ one obtains:

$$\begin{aligned} m_t &= 141 \pm 20 \text{ GeV},\ \alpha_s(m_Z) = 0.118 \pm 0.006 \text{ for } m_H = 60 \text{ GeV} \\ m_t &= 162 \pm 18 \text{ GeV},\ \alpha_s(m_Z) = 0.120 \pm 0.006 \text{ for } m_H = 300 \text{ GeV} \\ m_t &= 180 \pm 18 \text{ GeV},\ \alpha_s(m_Z) = 0.123 \pm 0.006 \text{ for } m_H = 1 \text{ TeV} \end{aligned} \qquad (1)$$

No important differences are found if $\alpha_s(m_Z)$ is fixed in the range shown in table 1. Comparing these results with the complete fits of ref.14 one indeed realises that the above six observables completely dominate the fit. The previous results imply significant upper and lower limits on m_t: at 1.64σ one has:

$$106 \text{ GeV} < m_t < 208 \text{ GeV} \qquad (2)$$

The fitted value of $\alpha_s(m_Z)$, also including the uncertainty from m_H is given by $\alpha_s(m_Z) = 0.120 \pm 0.007$, in perfect agreement with the value obtained from jet studies: $\alpha_s(m_Z) = 0.123 \pm 0.006$ [21] and with the world average of table 1 [21,22].The χ^2 prefers smaller values of m_H (fig.7), but within 2σ no significant upper bound on m_H is obtained [20] (i.e. one below 1 TeV which is already too large according to the bound from avoiding the Landau pole [23]). Actually a 2-variable fit (fig.8), i.e. using $\chi^2(m_t,m_H)$, indicates a value for m_H below [20,24] the direct experimental lower limit $m_H>63.5$ GeV [25]. From the fit one can also obtain the best estimates for any other quantity. For example, one gets $s_W^2 = 1-m_W^2/m_Z^2 = 0.2259 \pm 0.0019$, which, given m_Z from LEP, corresponds to $m_W=80230 \pm 100$ MeV. The effective $\sin^2\theta_W$ defined from the $Z \to l\bar{l}$ vertex via g_V/g_A [15] is given by $\sin^2\theta_W=\bar{s}_W^2 = 0.2325 \pm 0.00055$ (while from g_V/g_A given in table 1 we have $\bar{s}_W^2 = 0.2322 \pm 0.0007$).

The Standard Model fits the data better now than at the winter conferences: the χ^2/degrees of freedom was 4.2/5 and is now 2.8/5 [20]. The major change was that Γ_l went up by about 1σ (fig.9), because of comparable shifts in σ_h and R_h (see table 1). As a consequence, the value of $\alpha_s(m_Z)$, which used to be large, went down. Similarly, we shall see that the value of ε_3, which used to be low, went up.

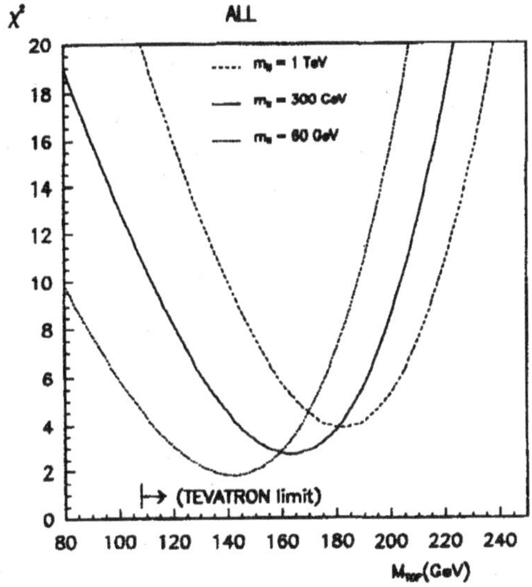

Figure 7:The χ^2 vs m_t of the Standard Model fit for different fixed values of m_H [20].

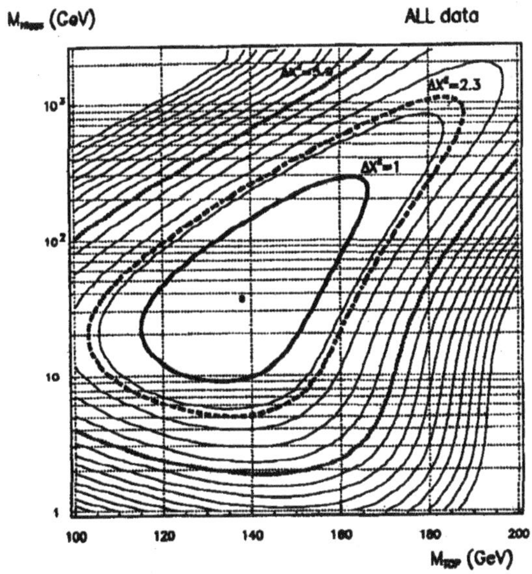

Figure 8: Contour plot of the function $\chi^2(m_t,m_H)$ for the Standard Model fit [20].

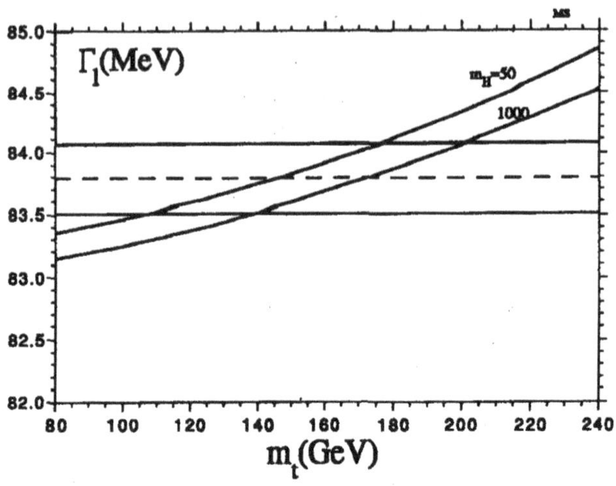

Figure 9: The Standard Model prediction for Γ_l as function of m_t for m_H=50-1000 GeV is compared with the LEP result.

In a sense my presentation could well end here. The LEP experiments have already reached a remarkable accuracy of about or below the 0.5% level, yet the Standard Model passes all precision tests. Concerning the search for new physics, that we discuss in the following, the consequence is that, by now, only those extensions of the Standard Model can survive that very delicately perturb its basic framework.

2. Strategy for a (More) Model Independent Analysis.

Recently we have proposed [26,27] a general strategy for the analysis of precision electroweak tests in view of the search for new physics beyond the Standard Model. Our method is more complete and less model dependent than the similar approach based on the variables S, T and U, [28-31] which, from the start, necessarily assumes dominance of vacuum polarisation diagrams from new physics and truncation of the q^2 expansion of the corresponding amplitudes. In a completely model independent way we define [27] four variables, called[f1] ε_1, ε_2, ε_3 and ε_b, that are precisely measured and can be compared with the predictions of different theories. The quantities ε_1, ε_2, ε_3 and ε_b are defined in ref.27 in one to one correspondence with the set of observables m_W/m_Z, Γ_l, A^l_{FB} and Γ_b. The four epsilons are defined without need of specifying m_t (and m_H). In the Standard Model, for all

[f1] Here we resume the notation ε_i for exactly the same quantities as defined in ref.27, where they were denoted ε_{Ni} (the index N, for "new", had been inserted to signal some small differences with respect to the original definitions in refs.26,29.

observables at the Z pole, the whole dependence on m_t arising from one-loop diagrams only enters through the epsilons. The same is true for any extension of the Standard Model such that all possible deviations only occur through vacuum polarisation diagrams and/or the $Z\to b\bar{b}$ vertex. As discussed in detail in ref.27, for such a model one can compare the theoretical predictions with the experimental determination of the epsilons as obtained from the whole set of LEP data. If a model does not satisfy this requirement then the comparison is to be made with the epsilons determined from the defining variables only, or with some more limited enlargement of the same set, depending on the particular case. For example, if lepton universality is maintained, then the data on A^l_{FB} can be replaced by the combined result on g_V/g_A from all lepton asymmetries (see, e.g., the following section on extended gauge models).

The epsilons represent an efficient parametrisation of the small deviations from what is solidly established in a way that is unaffected by our ignorance of m_t. The variables S, T, U depend on m_t because they are defined as deviations from the complete Standard Model prediction for specified m_t (and m_H). Instead the epsilons are defined with respect to a reference approximation which does not depend on m_t. In fact the epsilons are defined in such a way that they are exactly zero in the Standard Model in the limit of neglecting all pure weak loop-corrections (i.e. when only the predictions from the tree level Standard Model plus pure QED and pure QCD corrections are taken into account). This very simple version of improved Born approximation is amazingly well supported by the data [27,32,33]. In table 2 we compare the Born predictions, computed for $\alpha(m_Z)=1/128.87$ and $\alpha_s(m_Z)=0.118$ (the values used in the definitions of the epsilons) with the experimental data. We see that there is agreement to better than 1-1.5 σ. This does not mean that LEP experiments were badly designed and their accuracy is not sufficient to detect the pure weak corrections. Rather it reflects a very important a-posteriori dynamical result that the experimental values of the radiative corrections are smaller than their a-priori estimate, signalling the presence of important cancellations in the key quantities like $\Delta\rho$, Δr, $\Delta k'$ [3,4] that determine the whole pattern of radiative corrections. This information is very constraining both within and beyond the Standard Model. For example, in the Standard Model the lower limit on m_t in eq.2 arises from the necessity of cancelling in $\Delta\rho$ the negative residual corrections with the positive contribution of m_t.

Table 2

The QED and QCD improved Born approximation, obtained with $\alpha(m_Z)$ =1/128.87 and $\alpha_s(m_Z)$ = 0.118, is compared to the data.

	Born[27]	Experiment
m_W/m_Z	0.8768	0.8798 ± 0.0028
Γ_T(MeV)	2488	2489 ± 7
$R_h=\Gamma_h/\Gamma_l$	20.81	20.77 ± 0.05
σ_h (nb)	41.425	41.55 ± 0.14
$R_{bh}=\Gamma_b/\Gamma_h$	0.218	0.220 ± 0.0027
g_V/g_A (all asym.s)	0.0753	0.0712 ± 0.0028

By combining the value of m_W/m_Z [16] with the LEP results on the charged lepton partial width and the forward-backward asymmetry, all given in table 1, and following the definitions of ref.27, one obtains:

$$g_A^2 = 0.2508 \pm 0.00085$$
$$x = g_V/g_A = 0.0738 \pm 0.0046 \quad \text{or} \quad \bar{s}_W^2 = 0.2316 \pm 0.0012 \tag{3}$$

and:

$$\begin{aligned} \varepsilon_1 &= \Delta\rho = (3.0 \pm 3.4)\ 10^{-3} \\ \varepsilon_2 &= (-5.3 \pm 7.6)\ 10^{-3} \\ \varepsilon_3 &= (3.2 \pm 4.1)\ 10^{-3} \end{aligned} \tag{4}$$

Finally, by adding the value of Γ_b listed in table 1 and using the definition of ε_b given in ref.27 one finds:

$$\varepsilon_b = (3.0 \pm 5.8)\ 10^{-3} \tag{5}$$

In fig.10 the experimental 1σ ellipse in the ε_1-ε_3 plane is compared with the Standard Model predictions for different m_t and m_H values. We recall that ε_1 and ε_3 are completely determined by Γ_l and A_{FB}^l. We see that the value of ε_3 went up in the latest data and is by now well compatible within 1σ with the Standard Model. This is because ε_3 fine-tunes the value of Γ_l once $\bar{s}_W^2$ is given from A_{FB}^l [27]. In the data presented at Marseille Γ_l went up with A_{FB}^l essentially unchanged (see table 1). In fig.11 the experimental value of ε_2 is compared with the Standard Model prediction as a function of m_t and there is consistency at all practical values of m_t. Note that ε_2 also depends on m_W/m_Z and better measurements of this quantity are needed in order to make this test more stringent. Finally, in fig.12 we compare the experimental value of ε_b with the Standard Model prediction. Here we see that ε_b prefers small values of m_t. This result is a simple and direct consequence of the fact that the measured value of R_{bh} is a bit high (Γ_b went up over the last year and the error considerably decreased).

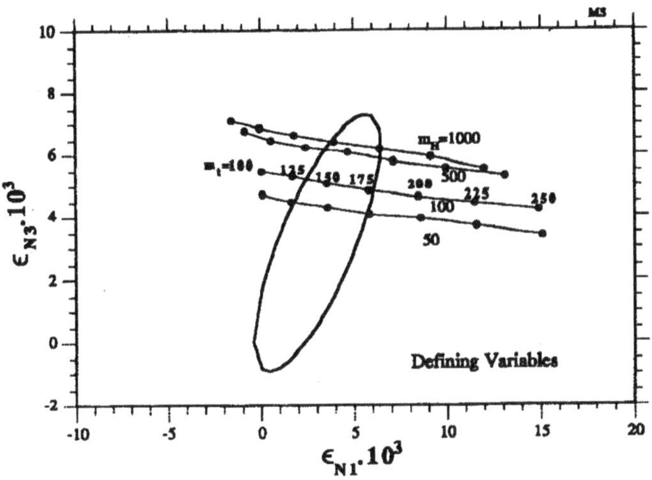

Figure 10: The 1σ ellipse in the plane ε_1-ε_3 obtained from the data on the defining variables Γ_l and A^l_{FB} compared with the Standard Model predictions for the indicated values of m_t and m_H.

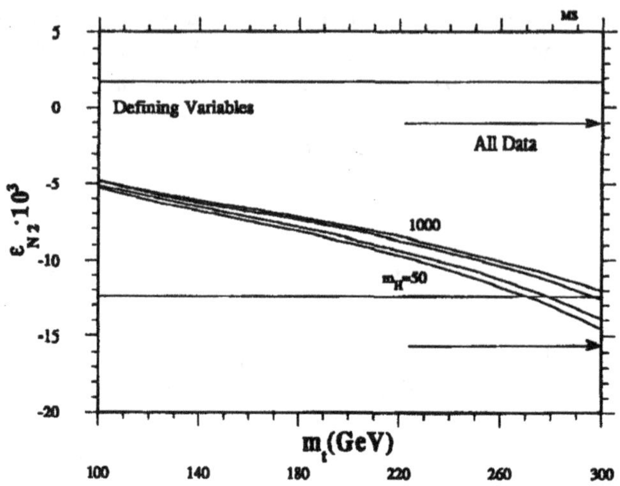

Figure 11: The 1σ data on ε_2 obtained from the data on the defining variables Γ_l, A^l_{FB} and m_W/m_Z compared with the Standard Model predictions as functions of m_t for the indicated values of m_H. The arrows indicate the experimental 1σ band from the fit in eq.8 to all electroweak data.

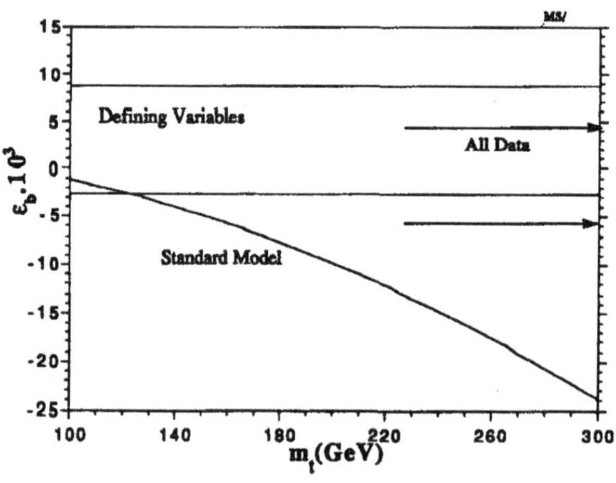

Figure 12: The 1σ data on ε_b obtained from the data on the defining variables Γ_l, A^l_{FB}, m_W/m_Z and Γ_b compared with the Standard Model predictions as functions of m_t for $\alpha_s(m_Z)$=0.118. The arrows indicate the experimental 1σ band from the fit in eq.8 to all electroweak data.

To proceed further, and include other measured observables in the analysis we need to make some dynamical assumptions. The minimum amount of model dependence is introduced by including other purely leptonic quantities at the Z pole such as A^{τ}_{pol} , A_e (measured [15] from the angular dependence of the τ polarisation) and A_{LR} (measured by SLD [34]). At this stage, one is simply relying on lepton universality. With essentially the same assumptions one can also include the data on the b-quark forward backward asymmetry A^b_{FB}. In fact it turns out that A^b_{FB} is almost unaffected by the $Z\rightarrow b\bar{b}$ vertex correction.

As a result of the previous discussion, we can combine the values of $x = g_V/g_A$ obtained from the whole set of asymmetries measured at LEP and we get the value reported in table 1. At this stage the best values of ε_1, ε_2 and ε_3 are modified according to

$$
\begin{aligned}
\varepsilon_1 &= \Delta\rho = (3.4 \pm 3.4)\ 10^{-3} \\
\varepsilon_2 &= (-6.3 \pm 7.0)\ 10^{-3} \\
\varepsilon_3 &= (5.0 \pm 3.2)\ 10^{-3} \\
\varepsilon_b &= (3.2 \pm 5.8)\ 10^{-3}
\end{aligned} \tag{6}
$$

From fig.13 we see that the agreement with the Standard Model is even improved by this step: the center of the 1σ ellipse of the data in the ε_1-ε_3 plane is very close to the point m_t=150 GeV, m_H=100 GeV.

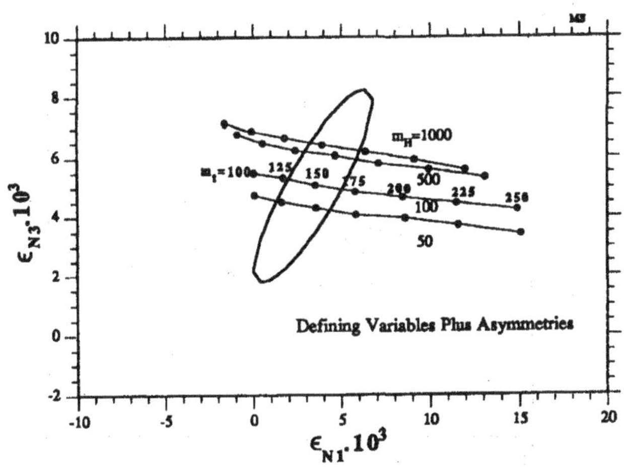

Figure 13: The 1σ ellipse in the plane ε_1-ε_3 obtained from the data on Γ_l and g_V/g_A derived from all the asymmetries (see table 1) compared with the Standard Model predictions for the indicated values of m_t and m_H.

All observables measured on the Z peak at LEP can be included in the analysis provided that we assume that all deviations from the Standard Model are only contained in vacuum polarisation diagrams (without demanding a truncation of the q^2 dependence of the corresponding funtions) and/or the $Z \to b\bar{b}$ vertex. Note that this is true for whatever partition of the new effect between g_{bV} and g_{bA}, because only one combination of them is measured in Γ_b, while, as already mentioned, A^b_{FB} is nearly independent of the $Z \to b\bar{b}$ vertex.

From a global fit of the data on m_W/m_Z, Γ_T, R_h, σ_h, R_{bh} and $x = g_V/g_A$ given in table 1 (for LEP data, we have taken the correlation matrix for Γ_T, R_h and σ_h given in ref.15, while we have considered the additional information on R_{bh} and x as independent) we obtain :

$$\begin{aligned} \varepsilon_1 &= \Delta\rho = (1.8 \pm 3.1)\ 10^{-3} \\ \varepsilon_2 &= (-7.7 \pm 6.9)\ 10^{-3} \\ \varepsilon_3 &= (4.0 \pm 3.1)\ 10^{-3} \\ \varepsilon_b &= (-0.5 \pm 5.1) 10^{-3} \end{aligned} \tag{7}$$

The comparison of theory and experiment in the planes ε_1-ε_3, ε_b-ε_3 and ε_b-ε_1 is shown in figs. 14, 15 and 16, respectively. We see that, except for ε_b which noticeably moves in the direction of the Standard Model prediction, the inclusion of all LEP quantities does not change the epsilons very much. This is because Γ_T, σ_h and R_h (or equivalently the ratios of Γ_T, Γ_h and Γ_l) are normal.

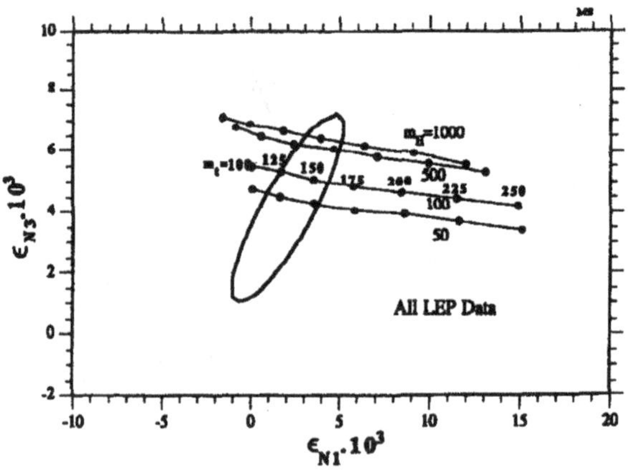

Figure 14: The 1σ ellipse in the plane ε_1-ε_3 obtained from the data on m_W/m_Z, Γ_T, σ_h, R_h, R_{bh} and g_V/g_A derived from all the asymmetries (see table 1) compared with the Standard Model predictions for the indicated values of m_t and m_H.

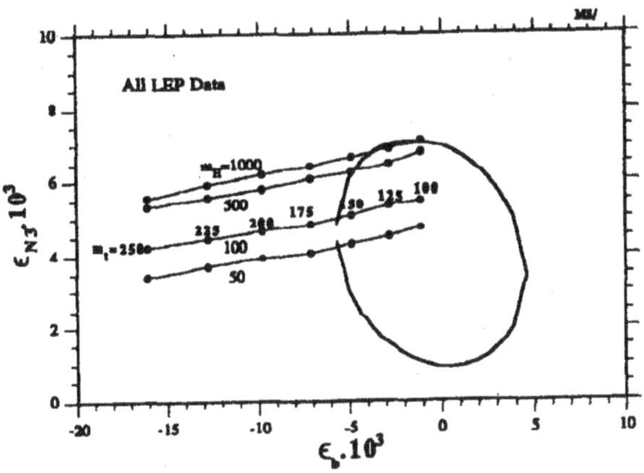

Figure 15: The 1σ ellipse in the plane ε_b-ε_3 obtained from the data on m_W/m_Z, Γ_T, σ_h, R_h, R_{bh} and g_V/g_A derived from all the asymmetries (see table 1) compared with the Standard Model predictions for the indicated values of m_t and m_H.

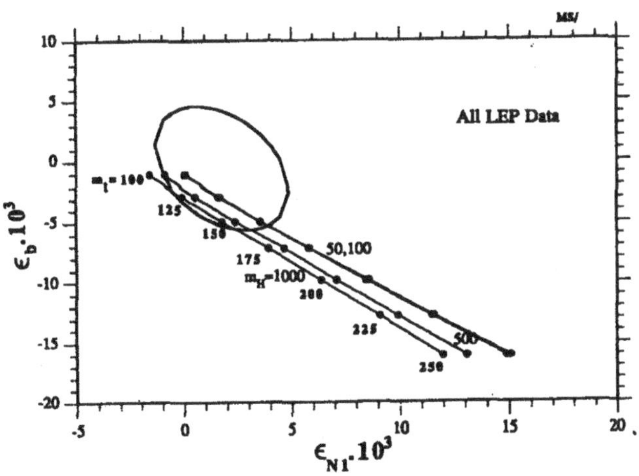

Figure 16: The 1σ ellipse in the plane ε_1-ε_b obtained from the data on m_W/m_Z, Γ_T, σ_h, R_h, R_{bh} and g_V/g_A derived from all the asymmetries (see table 1) compared with the Standard Model predictions for the indicated values of m_t and m_H.

To include in our analysis lower energy observables as well, a stronger hypothesis needs to be made: only vacuum polarization diagrams are allowed to vary from the Standard Model ones and only in their constant and first derivative terms in a q^2-expansion

[28-30], a likely picture, e.g., in technicolour theories [35,36]. In such a case, one can, for example, add to the analysis the ratio R_ν of neutral to charged current processes in deep inelastic neutrino scattering on nuclei [17],the "weak charge" Q_W measured in atomic parity violation experiments on Cs [18] and the measurement of g_V/g_A from $\nu_\mu e$ scattering [38] (the final result of CHARM-II corresponds to $\bar{s}_W^2$ =0.2324 ± 0.0086). In this way one obtains the global fit:

$$\begin{aligned} \varepsilon_1 &= \Delta\rho = (1.5 \pm 2.6)\ 10^{-3} \\ \varepsilon_2 &= (-7.8 \pm 6.8)\ 10^{-3} \\ \varepsilon_3 &= (3.5 \pm 2.8)\ 10^{-3} \\ \varepsilon_b &= (-0.5 \pm 5.0)\ 10^{-3} \end{aligned} \tag{8}$$

With the progress of LEP the low energy data, while important as a check that no deviations from the expected q^2 dependence arise, play a lesser role in the global fit. The ε_1-ε_3 plot for all data is shown in fig.17. Note that the present ambiguity on the value of $\alpha(m_Z) = (128.87 \pm 0.12)^{-1}$ [7] corresponds to an uncertainty on ε_3 (the other epsilons are not much affected) given by $\Delta\varepsilon_3 \cdot 10^3 = \pm 0.6$ [27]. Thus the theoretical error is still confortably less than the experimental error but the two will become close at the end of the LEP1 phase. The values of ε_2 and ε_b in eq.8 are compared with the Standard Model predictions in figs. 11 and 12. The fitted value of ε_b analysed in the Standard Model implies m_t<190 GeV at 1.64σ if the CDF limit m_t>113 GeV [39] is not taken into account. Note that the limit from ε_b is independent of $\varepsilon_1 = \Delta\rho$, i.e. of a large class of possible large vacuum polarisation effects. It however depends on the assumed value of $\alpha_s(m_Z)$ that we have taken as in table 1. If we repeat the fit with $\alpha_s(m_Z)$ free, ε_b moves up to $\varepsilon_b \cdot 10^3 = 3.1 \pm 7.1$ (to fix R_{bh} which is nearly independent of $\alpha_s(m_Z)$) while $\alpha_s(m_Z)$ goes down to $\alpha_s(m_Z) = 0.105 \pm 0.013$. Finally, ε_1, ε_2 and ε_3 are quite insensitive to $\alpha_s(m_Z)$ and closely keep their values in eq.8.

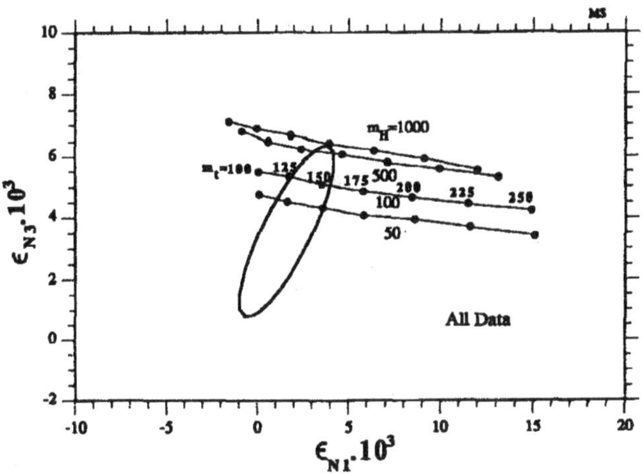

Figure 17: The 1σ ellipse in the plane ε_1-ε_3 obtained from all the data also including the low energy data compared with the Standard Model predictions for the indicated values of m_t and m_H.

3. Specific Examples.

We now concentrate on a number of well known extensions of the Standard Model which not only are particularly important per se but also are interesting in that they clearly demonstrate the constraining power of the present level of precision tests.

3.1 Technicolour. It is well known that technicolour models [35-37] tend to produce large and positive corrections to ε_3. As the central value of ε_3 went up one might imagine that the experimental problems of technicolour with respect to electroweak tests are now solved. This is not the case. First, by glancing at fig.18 [37], where the data on ε_3 and ε_1 are compared with the predictions of a class of simple versions of technicolour models, one realises, that the experimental errors on ε_3 are by now small enough that these models remain clearly disfavoured with respect to the Standard Model.

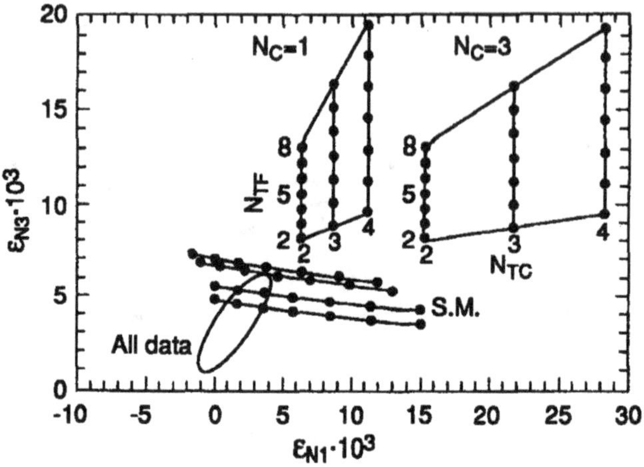

Figure 18: The same as fig.17 but with the predictions of simple technicolour models also shown [37] , where N_C, N_{TC} and N_{TF} are the numbers of colours, technicolours and techniflavours, respectively.

Second, recently it has been shown [40] that the data on ε_b also produce evidence against technicolour models. The same mechanism that in extended technicolour generates the top quark mass also leads to large corrections to the $Z \rightarrow b\bar{b}$ vertex that have the wrong sign. For example, in a simple model with two technidoublets, (N_{TC}=2), the Standard Model prediction is decreased by the amount [40]:

$$\Delta\varepsilon_b = -\ 16\cdot10^{-3} \left| \frac{\xi}{\xi'} \left(\frac{m_t}{100\ \mathrm{GeV}}\right) \right| \tag{9}$$

where ξ and ξ' are Clebsch-like coefficients, expected to be of order 1. The effect is even larger for larger N_{TC}. In a more sophisticated version of the theory, the so called "walking" technicolour [41], where the relevant coupling constants walk (i.e. they evolve slowly) instead of running, the result is somewhat smaller [42] but still gigantic (fig.19).

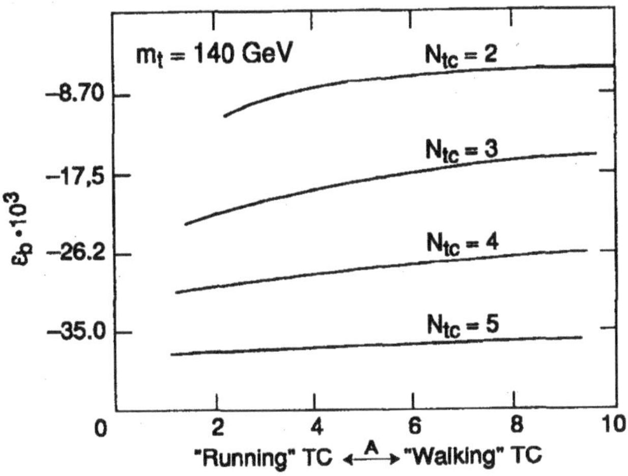

Figure 19: Evolution of the contributions to $\Delta\varepsilon_b$ in moving from running (A small) towards walking (A large) tecnicolour, for m_t=140 GeV and different values of N_{TC} (from ref.42).

In conclusion, it is difficult to really exclude technicolour because it is not a completely defined theory and no realistic model could be built sofar out of this idea. Yet, it is interesting that the most direct realisations tend to produce $\Delta\varepsilon_3 \gg 0$ and $\Delta\varepsilon_b \ll 0$ which are both disfavoured by experiment.

3.2 Minimal Supersymmetric Standard Model (MSSM). Contrary to technicolour the MSSM [43-45] is a completely specified, consistent and computable theory. There are too many parameters to attempt a direct fit of the data to the most general framework. So in ref.46 we restricted ourselves to two significant limiting cases: the "heavy" and the "light" MSSM.

The "heavy" limit correspond to all s-particles being sufficiently massive, still within the limits of a natural explanation of the weak scale of mass. In this limit a very important result holds [47]: for what concerns the precision electroweak tests, the MSSM predictions tend to reproduce the results of the Standard Model with a light Higgs, say $m_H \lesssim 100$ GeV.

In the "light" MSSM option some of the superpartners have a relatively small mass, close to their experimental lower bounds. In this case the pattern of radiative corrections may sizeably deviate from that of the Standard Model. The most interesting effects occur in vacuum polarisation amplitudes and/or the $Z \to b\bar{b}$ vertex and therefore are particularly suitable for a description in terms of the epsilons (because in such a case, as explained in ref.27, the predictions can be compared to the experimental determination of the epsilons from the whole set of LEP data). They are:

i) a threshold effect in the Z wave function renormalisation [47] mostly due to the vector coupling of charginos and (off-diagonal) neutralinos to the Z itself. Defining the vacuum polarisation functions by $\Pi_{\mu\nu}(q^2)$=-$ig_{\mu\nu}[A(0)+q^2F(q^2)]+q_\mu q_\nu$ terms, this is a positive contribution to $e_5=m_Z^2 F'_{ZZ}(m_Z^2)$ [47],the prime denoting a derivative with respect to q^2 (i.e. a contribution to a higher derivative term not included in the usual S,T,U formalism [28-31]). The e_5 correction shifts ε_1, ε_2, ε_3 by $-e_5$, $-c^2e_5$ and $-c^2e_5$ respectively, where $c^2=\cos^2\theta_W$, so that all of them are reduced by a comparable amount. Correspondingly all the Z widths are reduced without affecting the asymmetries. This effect can be significant but requires the lightest chargino to have a mass close to the experimental lower limit of 45 GeV.

ii) a positive contribution to ε_1 from the virtual exchange of the scalar top and bottom superpartners [48], analogous to the contribution of the top-bottom quark doublet. The needed isospin splitting requires one of the two scalars (in the MSSM the s-top) to be light.

iii) a negative contribution to ε_b due to the virtual exchange of a charged Higgs [49]. If one defines, as customary, $tg\beta=v_2/v_1$ (v_1 and v_2 being the vacuum expectation values of the Higgs doublets giving masses to the down and up quarks, respectively), then, for negligible bottom Yukawa coupling or $tg\beta \ll m_t/m_b$, this contribution is proportional to $m_t^2/tg^2\beta$.

iv) a positive contribution to ε_b due to virtual chargino--s-top exchange [50] which in this case is proportional to $m_t^2/\sin^2\beta$. This effect again requires the chargino and the s-top to be light in order to be sizeable.

As an example, in fig. 20-22, we give the pair correlations ε_3- ε_1, ε_3- ε_b and ε_b- ε_1 in the MSSM in the form of scatter plots [46]. The ellipses are the 1σ contours obtained from the present combined LEP experimental data (with the addition of the measurements of m_W/m_Z) [16]. The theoretical points in each plot are for fixed m_t=130 GeV. The Standard Model prediction as a function of m_H is shown by two black stars corresponding to m_H=50 GeV and m_H=1000 GeV connected by a line. The MSSM scatter plot are obtained for $tg\beta>1$ (which is relevant to the effects described in iii) and iv) above), the charged Higgs mass m_{H^+}>100 GeV (relevant for iii)) the lightest s-top and the approximately degenerate s-bottom masses (relevant to ii) and iv)), m_{stop}>50 GeV and $m_{sbottom}$>150 GeV. This last constraint requires an additional qualification. In the mass matrix of the two top superpartners we limit the off diagonal term by taking $m_A<3m_{sbottom}$. This constrains the relative s-top--s-bottom splitting when $m_{sbottom}$ gets large and consequently the size of the effect on ε_1. A particularly important role is played by the lightest chargino mass m_{χ^+}, relevant to i) and iv). The white bullets refer to very light charginos: 60 >m_{χ^+}>48 GeV, while the white stars, that by superposition form a dark area, refer to m_{χ^+}>60 GeV. In figs.20-22 the dark area leans towards the heavy MSSM while the white bulletts are from the (very) light MSSM. As apparent from figs.20-22, the differences between the theoretical predictions of the Standard Model and of the MSSM are not large in comparison with the present accuracy of the experiments, but they will become more important with the increase of statistics at LEP.

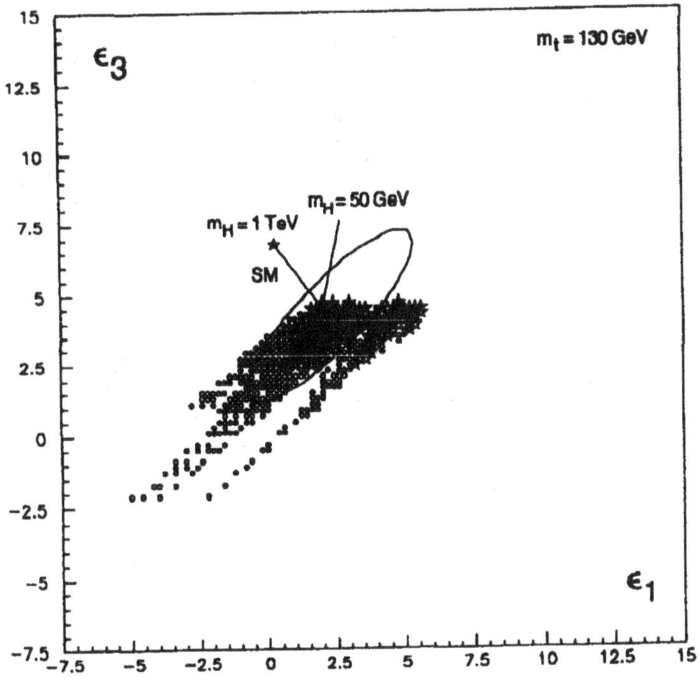

Figure 20: ε3 vs ε1 at fixed m_t: m_t=130 GeV (from ref.46). The ellipse is the 1σ contour from all LEP data.The Standard Model prediction as a function of m_H is shown by two black stars corresponding to m_H=50 GeV and m_H=1000 GeV connected by a line. The MSSM scatter plot are obtained for tgβ>1, the charged Higgs mass m_{H^+}>100 GeV, the lightest s-top and the approximately degenerate s-bottom masses m_{stop}>50 GeV and $m_{sbottom}$>150 GeV.The white bullets refer to very light charginos: 60 >m_{χ^+}>48 GeV, while the white stars, that by superposition form a dark area, refer to m_{χ^+}>60 GeV. The figures are actually realised by taking tgβ=1 and 4, m_{H^+}=100 GeV or ∞ (i.e. its contributions set to zero), m_{stop}=50 or 200 GeV, $m_{sbottom}$>150 GeV and varying the chargino masses in the mentioned ranges.

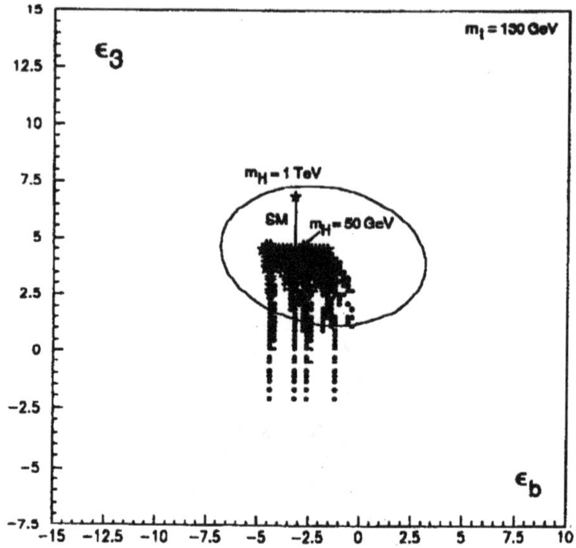

Figure 21: The same as fig.20, but for ε_3 vs ε_b.

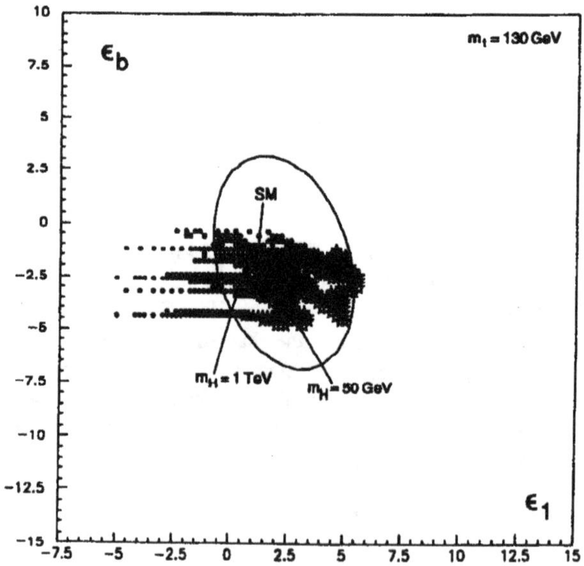

Figure 22: The same as fig.20, but for ε_b vs ε_1.

In conclusion, the present electroweak data are well consistent with the MSSM. As the present data are very constraining this statement is highly non trivial. In its heavy version, the pattern of radiative corrections predicted by the MSSM is practically indistinguishable from that obtained from the Standard Model with a light Higgs. Sizeable departures from the Standard Model arise in the MSSM if a light gaugino and a light s-top exist, with masses close to their experimental bounds and such as to make them visible at LEP200. The resulting effects are still consistent with the data on ε_3 and ε_b.

3.3 Models with an Extended Gauge Group. In the simplest models with an extra U(1), for a specified definition of that U(1), i.e. for given couplings to fermions of the associated neutral vector boson Z_N, irrespective of assumptions on the new Higgs sector (namely on the heavy Z_H mass) two new parameters are introduced, tgξ and $\delta\rho_M$ [51]. The angle ξ defines the mixtures of the standard (Z_S) and the new (Z_N) vector boson that make up the light (Z_L), observed at LEP, and the heavy (Z_H) mass eigenstates, e.g. $Z_L = \cos\xi Z_S + \sin\xi Z_N$. $\delta\rho_M$ describes the shift induced at tree level in the ρ parameter by the mixing: $\cos^2\theta_W = m_W^2/(m_{Z_L}^2\rho)$ with $\rho = 1 + \delta\rho_M$ (we assume that the breaking of the ordinary SU(2) is induced by Higgs doublets). One has $\delta\rho_M > 0$ because the mixing pushes the lowest state down. If the Z_H mass was known $\delta\rho_M$ and $\sin^2\xi$ would be related: $\delta\rho_M = (m_{Z_H}^2/m_{Z_L}^2 - 1)\sin^2\xi$. More details can be found in ref.[52]. Here for lack of space we go directly to the results for the class of models based on E6 [53] (with arbitrary orientation of the extra U(1) in the group space, determined by an angle θ_2) and for the left-right symmetric model (assuming heavy W_R) [54].The constraints on ξ and $\delta\rho_M$ are very strong [52]. In fig.23 the 90% c.l. bounds on ξ are plotted for m_t=100-200 GeV, m_H=100 GeV and $\alpha_s(m_Z)$=0.118. The dependence on m_t of the limits on ξ_0 is very moderate, except in the region near θ_2=0. One finds that in most cases $|\xi| < 0.5\%$ is implied for all values of m_t, m_H and $\alpha_s(m_Z)$, with $|\xi| < 1\%$ in the least favourable instances. Similarly the 90% c.l. bounds on the quantity $\Delta\rho = (\varepsilon_1)_{SM} + \Delta\rho_M$ are displayed in fig.24, for the same values of m_t, m_H and $\alpha_s(m_Z)$. Note that considering $\Delta\rho$ is useful because the corresponding bounds are nearly independent of m_t. On the contrary the allowed range for $\Delta\rho_M$ rapidly decreases with increasing m_t because of the quadratic m_t dependence of $(\varepsilon_1)_{SM}$ and the fact that $\Delta\rho_M > 0$. A typical value for the upper bound on $\Delta\rho$ (at $\alpha_s = 0.118$) of about 0.006 is already saturated by the contribution from $\Delta\rho_t$ with $m_t \sim 175$ GeV. We have also studied the dependence on α_s in the range $0.111 < \alpha_s < 0.125$. It turns out that, for $m_H = 100$ GeV, m_t=150 GeV, by increasing α_s of 0.01 the allowed region for $\Delta\rho_M$ is shifted down by about 0.0015.

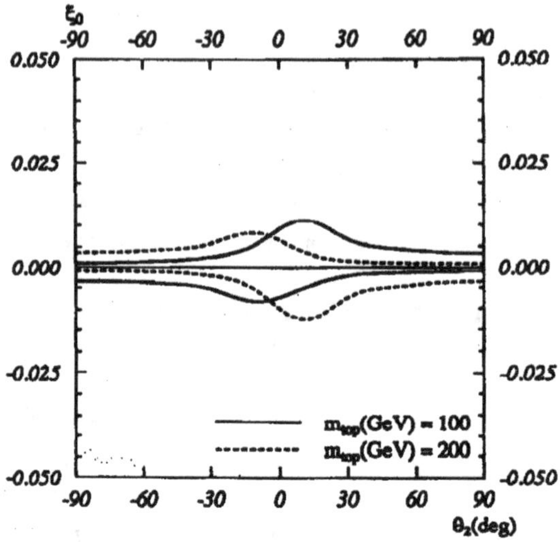

Figure 23: Allowed range (90% c.l.) for ξ versus θ_2, for m_H =100 GeV, α_s=0.118 (from ref.52).

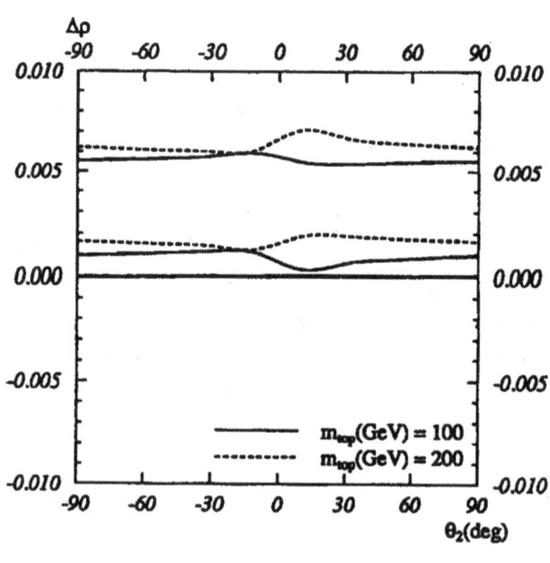

Figure 24: Allowed range (90% c.l.) for $\Delta\rho$ versus θ_2, for m_H =100 GeV, α_s=0.118 (from ref.52).

One can obtain a pictorial impression of the comparison with the Standard Model by using the epsilon variables . As discussed in ref.[26], in extended gauge models ε_1, ε_2 and ε_3 receive the additional contributions (with respect to the Standard Model). In figs.25 we plot for m_t=m_H=100 GeV and α_s =0.118 a set of dots in the plane ε_3 vs ε_1 that correspond to the best fit for each θ_2 value obtained [52] by considering the whole set of LEP data, plus m_W/m_Z, given in table 1. The solid lines, for each dot, show the 1σ variation due to $\Delta\rho$, while the dashed lines span 1σ in ξ. The result for the LR model also appears in the figures, marked by a box symbol. The corresponding Standard Model point is also shown, together with the ellipse of the experimental data (with 1σ projections on both axis). The data correspond to a fit to the defining variables for ε_1, ε_2 and ε_3, i.e. m_W/m_Z, Γ_l and the ratio g_V/g_A obtained by combining all the asymmetries. In general, it is at small values of m_t that there is a substancial improvement in going from the Standard Model to the extended gauge models, mostly because $\Delta\rho_M$ increases $\Delta\rho$. Finally, the variable ε_b is not particularly relevant in the present context because, in all extended gauge models we consider, all the couplings are family independent. Thus, whatever correction to the $Z\to b\bar{b}$ vertex would also affect all down quark vertices by the same amount . Since the accuracy of the data on Γ_b is relatively modest, we cannot adjust the value of Γ_b without at the same time spoil the agreement with the other hadronic observables.

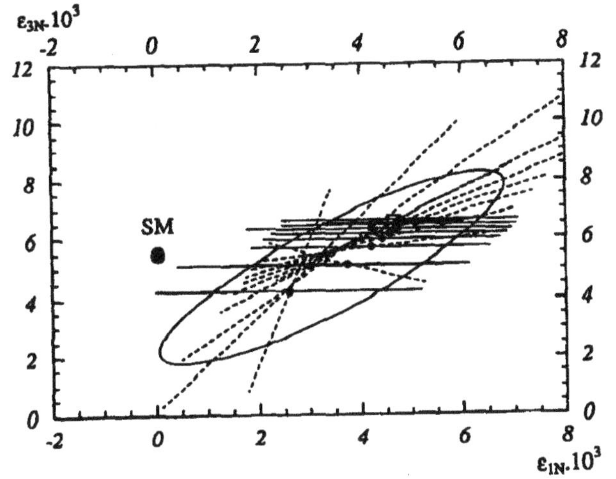

Figure 25: In the ε_{N1}, ε_{N3} plane the 1σ ellipse corresponding to the experimental data on m_W/m_Z, Γ_l and the ratio g_V/g_A obtained by combining all the asymmetries is shown together with the Standard Model point . The set of dots correspond to the best fit for different θ_2 values. The box symbol corresponds to the LR model.The solid lines, for each dot, show the 1σ variation due to $\Delta\rho$, while the dashed lines span 1σ in ξ. Here we have assumed m_H=100 GeV, α_s =0.118, with m_t=100 GeV (from ref.52).

In summary, the data collected at LEP on the Z peak impose very severe constraints on the mixings of a non standard Z'. At present, the Standard Model is in such good agreement with the data that no indication for additional corrections is found, at least for unspecified m_t. In fact, only a very small amount of mixing is allowed, with ξ always less than 1%.

4.Conclusion

With the progress of the LEP programme the electroweak precision tests have become extremely constraining. There is at the moment perfect agreement with the Standard Model. A quite restricted range of m_t is indicated by the data: for example, if we take the central value obtained in sect.1 for m=300 GeV, and we enlarge the error so as to include the uncertainty on m_H (a procedure of merely indicative value), we get

$$m_t = 162 \pm 27 \text{ GeV} \qquad (10)$$

Similarly, tight bounds on all conceivable forms of new physics are also obtained. We have considered a few particularly important examples. Technicolour theories are not in better shape now in spite of the fact that the ε_3 deficit has disappeared from the data, because of the smaller errors and the problem with ε_b. The MSSM is the most hard to beat model of new physics because it only makes the Standard Model more stable and robust, without disrupting its structure. It predicts corrections that are in most cases smaller than present errors, unless some of the spartners are very light. In models with extended gauge structure the amount of allowed mixing of the observed Z with the non standard component must be below 1%.

In the '93 and '94 runs of LEP one aims at collecting ~50 pb^{-1} of additional integrated luminosity per experiment with resonance scanning. The errors on the epsilons will then go down by about a factor of two. Another big step in probing the Standard Model is ahead of us !

Acknowledgements. It is pleasure for me to thank R. Casalbuoni, F. Caravaglios, S. De Curtis, M. Martinez, M. Pepe-Altarelli for their qualified help in prepapring the material for this talk, and R. Barbieri, D. Bardin, L. Okun , G. Passarino for important interactions.

References

1. S.L. Glashow, *Nucl. Phys.* **22** (1961) 579;
S. Weinberg, *Phys. Rev. Lett.* **19** (1967) 1264;
A. Salam, Proc. 8th Nobel Symposium, Aspenägården, 1968, ed. N. Svartholm (Almqvist and Wiksell, Stockholm, 1968) p. 367.

2. See, for example, the following textbooks:
I. Aitchison and A. Hey, *Gauge theories in particle physics: a practical introduction* (A. Hilger, Bristol, 1982);
D. Bailin, *Introduction to gauge field theory* (A. Hilger, Bristol, 1986);
E.D. Commins and P.H. Bucksbaum, *Weak interactions of leptons and quarks*, (Cambridge Univ. Press, Cambridge, 1983);
L.B. Okun, *Leptons and quarks* (North Holland Publ. Co., Amsterdam, 1982).

3. G. Altarelli, Proceedings of the 32nd Int. Winter School, Schladming,Austria, CERN preprint TH.6867/93 (1993); Proceedings of the EPS Conference on High Energy Physics, Marseille, France, 1993.

4. G. Altarelli, R. Kleiss and C. Verzegnassi (eds.), Z Physics at LEP 1 (CERN 89-08, Geneva, 1989), Vols. 13.

5. V. Novikov, L. Okun and M. Vysotsky, *Nucl. Phys.* **B397** (1993) 35; CERN preprint CERN-TH.6696/92, CERN-TH.6855/93 (1993); *Phys. Lett.* **299** (1993) 329; ITEP 104-92 (1992).

6. H. Burkhardt, F. Jegerlehner, G. Penso and C. Verzegnassi, *Z.Phys.* **C43** (1989) 497.

7. F. Jegerlehner, in Proceedings of the 1990 Theoretical Advanced Study Institute in Elementary Particle Physics, ed. by P. Langacker and M. Cvetic (World Scientific, Singapore, 1991), p. 476.

8. B. Kniehl, Proceedings of the EPS Conference on High Energy Physics, Marseille, France, 1993.

9. R. Barbieri, M. Beccaria, P. Ciafaloni, G. Curci and A. Viceré, *Phys. Lett.* **B288** (1992) 95 and CERN preprint CERN-TH.6713/92 (1992);
A. Denner, W. Hollik and B. Lampe, CERN preprint CERN-TH.6874/93 (1993).

10. J. Fleischer, F. Jegerlehner, P. Raczka and O.V. Tarasov, *Phys.Lett.* **B293** (1992) 437;
G. Buchalla and A.J. Buras, Munich preprint MPI-PTh/111-92 (1992), TUM-T31-36/92 (1992);
G. Degrassi, Padua preprint DFPD 93/TH/03 (1993).

11. S. Fanchiotti, B. Kniehl and A. Sirlin, CERN preprint CERN-TH.6749/92 (1992).

12. K.G. Chetyrkin and J.H. Kühn, *Phys. Lett.* **B248** (1990) 359;
K.G. Chetyrkin, J.H. Kühn and A. Kwiatkowski, *Phys. Lett.* **B282** (1992) 221.

13. G. Quast, Proceedings of the EPS Conference on High Energy Physics, Marseille, France, 1993.

14. J. Lefrançois, Proceedings of the EPS Conference on High Energy Physics, Marseille, France, 1993.

15. The LEP Electroweak Working Group, CERN preprint CERN/PPE/93-157 (1993).

16. CDF Collaboration, F. Abe et al., *Phys. Rev. Lett.* **65** (1990) 2243; *Phys. Rev.* **D43** (1991) 2070;
UA2 Collaboration, *Phys. Lett.* **B276** (1992) 354; the value shown in Table 1 is from the Particle Data Group 1992: *Phys. Rev.* **D45** (1992).

17. CHARM Collaboration, J.V. Allaby et al., *Phys. Lett.* **B177** (1986) 446; *Z. Phys.* **C36** (1987) 611;
CDHS Collaboration, H. Abramowicz et al., *Phys. Rev. Lett.* **57** (1986) 298;
A. Blondel et al., *Z. Phys.* **C45** (1990) 361;

CCFR Collaboration, A. Bodek, Proceedings of the EPS Conference on High Energy Physics, Marseille, France, 1993.

18. M.C. Noecker et al., *Phys. Rev. Lett.* **61** (1988) 310;
M. Bouchiat, M.A. Proceedings of the 12th International Atomic Physics Conference (1990).

19. M. Pepe-Altarelli, Proceedings of the Rencontres de la Vallée D'Aoste, La Thuile, Italy, 1993;
C. De Clerq, V. Innocente and R. Tenchini, Proceedings of the Rencontres de Moriond, Les Arcs, France, 1993.

20. I thank M. Martinez for providing me with the results of this simplified fit from exactly the same code used for the complete fit presented in Ref. [14].

21. S. Catani, Proceedings of the EPS Conference on High Energy Physics, Marseille, France, 1993.

22. G. Altarelli, in QCD-20 years later, ed. by P.M. Zerwas and H.A. Kastrup, World Scientific 1993;
S. Bethke, Proceedings of the XXVI International Conference on High Energy Physics, Dallas, 1992.

23. M.A.B. Bég et al., *Phys. Rev.* **52** (1984) 883;
D.J. Callaway, *Nucl. Phys.* **B233** (1984) 189;
R. Dashen and H. Neuberger, *Phys. Rev. Lett.* **50** (1983) 189;
K.J. Babu and E. Ma, *Phys. Rev.* **D31** (1984) 2861;
E. Ma, *Phys. Rev.* D31 (1985) 322;
M. Lindner, *Z. Phys.* **C31** (1986) 295;
P. Hasenfratz, *Nucl. Phys.* (Proc. Suppl.) **B9** (1989) 3;
J. Kuti, ibid., p. 55;
H. Neuberger, Proc. Symposium on Lattice field theory, Capri, 1989 (to appear as Nucl. Phys. B. Proc. Suppl. 17, 1990);
M. Lüscher and P. Weisz, *Nucl. Phys.* **B290** (1987) 5; **B295** (1988) 65 and **B318** (1989) 705.

24. M. Martinez, private communication;
V. Novikov et al., *Phys. Lett.* **B308** (1993) 123.

25. D. Treille, Proceedings of the EPS Conference on High Energy Physics, Marseille, France, 1993.

26. G. Altarelli, R. Barbieri and S. Jadach, *Nucl. Phys.* **B369** (1992) 3.

27. G. Altarelli, R. Barbieri and F. Caravaglios, *Nucl. Phys.* **B405** (1993) 3; CERN preprint CERN-TH.6859/93 (1993).

28. M.E. Peskin and T. Takeuchi, *Phys. Rev. Lett.* **65** (1990) 964 and *Phys. Rev.* **D46** (1991) 381.

29. G. Altarelli and R. Barbieri, *Phys. Lett.* **B253** (1990) 161;
B.W. Lynn, M.E. Peskin and R.G. Stuart, SLAC-PUB-3725 (1985); in Physics at LEP, Yellow Book CERN 86-02, Vol. I, p. 90.

30. B. Holdom and J. Terning, *Phys. Lett.* B247 (1990) 88;
D.C. Kennedy and P. Langacker, *Phys. Rev. Lett.* **65** (1990) 2967 and preprint UPR-0467T;
B. Holdom, Fermilab 90/263-T (1990);

W.J. Marciano, as in Ref. [73];
A. Ali and G. Degrassi, DESY preprint DESY 91-035 (1991);
E. Gates and J. Terning, *Phys. Rev. Lett.* **67** (1991) 1840;
E. Ma and P. Roy, *Phys. Rev. Lett.* **68** (1992) 2879;
G. Bhattacharyya, S. Banerjee and P. Roy, *Phys. Rev.* **D45** (1992) 729.

31. M. Golden and L. Randall, *Nucl. Phys.* **B361** (1991) 3;
M. Dugan and L. Randall, *Phys. Lett.* **B264** (1991) 154;
A. Dobado et al., *Phys. Lett.* **B255** (1991) 405;
J. Layssac, F.M. Renard and C. Verzegnassi, Preprint UCLA/93/TEP/16 (1993).

32. V. Novikov, L. Okun and M. Vysotksy, *Mod. Phys. Lett.* **A8** (1953) 2529.

33. M. Bilenky et al., preprint BI-TP93/46 (1993);
D. Schildknecht, Proceedings of the EPS Conference on High Energy Physics, Marseille, France, 1993.

34. The SLD Collaboration, SLAC-PUB-6030 (1993);
W. Ash, Proceedings of the EPS Conference on High Energy Physics, Marseille, France, 1993.

35. S. Weinberg, *Phys. Rev.* **D13** (1976) 974 and *Phys. Rev.* **D19** (1979) 1277;
L. Susskind, *Phys. Rev.* **D20** (1979) 2619;
E. Farhi and L. Susskind, *Phys. Rep.* **74** (1981) 277.

36. R. Casalbuoni et al., *Phys. Lett.* **B258** (1991) 161;
R.N. Cahn and M. Suzuki, LBL-30351 (1991);
C. Roisnel and Tran N. Truong, *Phys. Lett.* **B253** (1991) 439;
T. Appelquist and G. Riantaphylou, Yale Univ. preprint YCTP-p49-91 (1991);
T. Appelquist, Proceedings of the Rencontres de la Vallée d'Aoste, La Thuile, Italy, 1993.

37. J. Ellis, G.L. Fogli and E. Lisi, CERN preprint CERN-TH.6383/92 (1992).

38. The CHARM II Collaboration, R. Berger, Proceedings of the EPS Conference on High Energy Physics, Marseille, France, 1993.

39. The CDF Collaboration, A. Barbaro-Galtieri, Proceedings of the EPS Conference on High Energy Physics, Marseille, France, 1993.

40. R.S. Chivukula, S.B. Selipsky and E.H. Simmons, *Phys. Rev. Lett.* **69** (1992) 575.

41. B. Holdom, *Phys. Lett.* **105** (1985) 301;
K. Yamawaki, M. Bando and K. Matumoto, *Phys. Rev. Lett.* **56** (1986) 1335;
V.A. Miransky, *Nuov. Cim.* **90A** (1985);
T. Appelquist, D. Karabali and L.C.R. Wijewardhana, *Phys. Rev.* **D35** (1987) 389; 149;
T. Appelquist and L.C.R. Wijewardhana, *Phys. Rev.* **D35** (1987) 774;
Phys. Rev .**D36** (1987) 568.

42. R.S. Chivukula et al., Preprint BUHEP-93-11 (1993).

43. H.P. Nilles, *Phys. Rep.* **C110** (1984) 1;
H.E. Haber and G.L. Kane, *Phys. Rep.* **C117** (1985) 75;
R. Barbieri, *Riv. N. Cim.* **11** (1988) 1.

44. E. Eliasson, *Phys. Lett.* **147** (1984) 67;
S. Lim et al., *Phys. Rev.* **D29** (1984) 1488;
J. Grifols and J. Sola, *Nucl. Phys* **B253** (1985) 47;
B. Lynn et al., in Physics at LEP, ed. by J. Ellis and R. Peccei, CERN86-02, Vol. 1 (1986);
R. Barbieri et al., *Nucl. Phys.* **B341** (1990) 309;
P. Gosdzinsky and J. Sola, preprint UAB-FT-247/90 (1990);
M. Drees, K. Hagiwara and A. Yamada, DESY preprint DTP/91/34 (1991);
J. Ellis, G. Fogli and E. Lisi, CERN preprint CERN-TH.6642/92 (1992);
J. Lopez et al., Texas Univ. preprint CTP.TAMU/19/93 (1993).

45. R. Barbieri, S. Ferrara and C. Savoy, *Phys. Lett.* **119B** (1982) 343;
P. Nath, R. Arnowitt and A. Chamseddine, *Phys. Rev. Lett.* **49** (1982) 970.

46. G. Altarelli, R. Barbieri and F. Caravaglios, *Phys. Lett.* **B314** (1993) 357.

47. R. Barbieri, F. Caravaglios and M. Frigeni, *Phys. Lett.* **B279** (1992) 169.

48. R. Barbieri and L. Maiani, Nucl. Phys. B224 (1983) 32;
L. Alvarez-Gaumé, J. Polchinski and M. Wise, Nucl. Phys. B221 (1983) 495.

49. W. Hollik, *Mod. Phys. Lett.* **A5** (1990) 1909.

50. A. Djouadi et al., *Nucl. Phys.* **B349** (1991) 48;
M. Boulware and D. Finell, *Phys. Rev.* **D44** (1991) 2054. The sign discrepancy between these two papers appears now to be solved in favour of the second one.

51. G. Altarelli, R. Casalbuoni, S. De Curtis, F. Feruglio and R. Gatto, *Mod. Phys. Lett.* **A5** (1990) 495; *Nucl. Phys.* **B342** (1990) 15;
G. Altarelli, R. Casalbuoni, S. De Curtis, F. Feruglio and R. Gatto, *Phys.Lett.* **B235** (1990) 669;
G. Altarelli, R. Casalbuoni. S. De Curtis, N. Di Bartolomeo, F. Feruglio and R. Gatto, *Phys. Lett.* **B263** (1991) 459;
A. Chiappinelli, *Phys. Lett.* **B263** (1991) 287;
M.C. Gonzalez-Garcia and J.W.F. Valle, *Phys. Lett.* **B259** (1991) 365;
F. Del Aguila, W. Hollik, J.M. Moreno and M. Quiros, *Nucl. Phys.* **B372** (1992) 3;
P. Langacker and M. Luo, *Phys. Rev.* **D45** (1992) 278;
J. Layssac, F.M. Renard and C. Verzagnassi, *Z. Phys.* **C53** (1992) 114;
A. Leike, S. Riemann and T. Riemann, *Phys. Lett.* **B291** (1992) 187;
E. Nardi, E. Roulet and D. Tommasini, *Phys. Rev.* **D46** (1992) 3040.

52. G. Altarelli, R. Casalbuoni, S. De Curtis, N. Di Bartolomeo, R. Gatto and F. Feruglio, CERN preprint CERN-TH.5947/93 (1993).

53. R.W. Robinet and J.L. Rosner, *Phys. Rev.* **D26** (1982) 2396;
E. Witten, Nucl. Phys. B258 (1985) 75;
M. Dine, V. Kaplunovsky, M. Mangano, C. Nappi and N. Seiberg, Nucl. Phys. B259 (1985) 519;
S. Cecotti, J.-P. Derendinger, S. Ferrara, L. Girardello and M. Roncadelli, *Phys. Lett.* **156B** (1985) 318;
J.D. Breit, B.A. Ovrut and G. Segré, *Phys. Lett.* **158B** (1985) 33;
E. Cohen, J. Ellis, K. Enqvist and D.V. Nanopoulos, *Phys. Lett.* **165B** (1985) 76;
J. Ellis, K. Enqvist, D.V. Nanopoulos and F. Zwirner, *Nucl. Phys.* **B276** (1986) 14; *Mod. Phys. Lett.* **A1** (1986) 57;

F. Del. Aguila, G. Blair, M. Daniel and G.G. Ross, *Nucl. Phys.* **B272** (1986) 413;
D. London and J.L. Rosner, *Phys. Rep.* **34** (1986) 1530;
G. Belanger and F. Godfrey, *Phys. Rev.* **D35** (1987) 378;
L. Ibànez and J. Mas, *Nucl. Phys.* **B286** (1987) 107.

54. J.C. Pati and A. Salam, *Phys. Rev.* **D10** (1974) 275;
R.N. Mohapatra and J.C. Pati, *Phys. Rev.* **D11** (1975) 566 and ibid. 2559;
G. Senjanovic and R.N. Mohapatra, *Phys. Rev.* **D12** (1975) 152;
G. Senjanovic, *Nucl. Phys.* **B153** (1979) 334;
R.W. Robinet and J.L. Rosner, *Phys. Rev.* **D25** (1982) 3035;
C.N. Leung and J.L. Rosner, *Phys. Rev.* **D29** (1982) 2132.

Standard Model of Flavour Mixing and CP Violation - Status Report and Perspectives

A. Ali
Theory Division, CERN
CH-1211 Geneva 23, Switzerland[1]

1 Introduction

In the electroweak theory of Glashow, Salam and Weinberg, popularly known as the Standard Model (SM) [1], fermions and gauge bosons ($W^{\pm}, Z^0$) get their masses through the Higgs mechanism. For the quarks, the gauge- and mass-eigenstates are not the same, leading to intergenerational couplings in the weak charged-current interactions. These flavour non-diagonal couplings are described by the unitary Cabibbo-Kobayashi-Maskawa (CKM) matrix [2]. The CKM matrix is one of the few aspects of the SM that remains to be fully tested. The complex phase residing in the CKM matrix is the only source of CP violation in the SM. So far CP violation has been observed only in the decays of the neutral kaons in laboratory experiments. However, its macroscopic implication is all too obvious, as it is regarded as the key ingredient responsible for the baryon asymmetry that we observe today in the universe. The CKM matrix elements are thus fundamental constants of nature. In their understanding lies the clue to our understanding of quark masses, mixings, CP violation, and eventually of such existential questions as the baryon asymmetry of the universe.

The primary aim of experiments in flavour physics is to determine the CKM matrix elements precisely. This is done by measurements of the charged current (CC) and the so-called flavour-changing neutral-current (FCNC) processes. The latter involve quark transitions such as $s \to d$, $c \to u$, $b \to s$. FCNC transitions, being forbidden in the SM Lagrangian [3], are induced by higher-order CC weak interactions and hence they are intrinsically sensitive to much larger mass scales than m_W. In the SM, FCNC transitions among down-type quarks involve, as virtual state, the top quark, and these transitions are at present the only source of information for the CKM matrix elements involving the t-quark, V_{tj}, $j = d, s, b$. Precise knowledge of these matrix elements would determine the CKM parameters completely, hence also the CKM phase and with it all the CP asymmetries in the quark sector.

In this article, written for the *Festschrift* in honour of Professor Abdus Salam, I review the present status of the CKM matrix, which in the SM governs the flavour transitions of the left-chiral matter fields and CP violation. In theoretical discussions of these matters,

[1] On leave of absence from DESY, Hamburg, FRG.

the constraints that unitarity imposes on the parameters of the CKM matrix play a central role. Such constraints are often expressed as triangle relations in the complex plane of the CKM parameter space, and there are altogether six of them. I present a state-of-the-art profile of perhaps the most interesting of the CKM triangles following from the constraint $\sum_i V_{id}V_{ib}^* = 0$ [4, 5]. The constraints that follow from the recent measurement of the electromagnetic penguin decay $B \to K^* + \gamma$ [6, 7] and the existing CC data [8] on another triangle following from the unitarity relation $\sum_i V_{is}V_{ib}^* = 0$ are also quantified. As discussed by Altarelli [9], the SM has passed rather excruciating experimental tests in the electroweak sector. At the end of this review, the same message will also emerge from the flavour physics, although here the arguments in favour of the SM are not yet completely quantitative.

The profile of the CKM unitarity triangles will become increasingly sharp in future as precision experiments in B and K physics bring to bear the full force of their measurements. In anticipation of a major experimental surge foreseen for the rest of this and the next decade(s) in precision studies of B physics, I review a number of rare FCNC B decays and CP violation involving B hadrons. These measurements will provide a quantitative test of the SM in the flavour sector, in particular, its short-distance physics. Hopefully, these experiments will also serve as windows on non-SM physics, which inevitably lies there and which hopefully will bring us closer to an understanding of the large number of parameters in the SM, such as fermion masses and mixing angles.

2 Charged Weak Currents in the SM

We retrace here some of the essential steps leading to the introduction of the CKM matrix in the CC weak interactions. The SM particle content and their (weak) isospin and chiral properties of the fields may be seen in any standard textbook on particle physics. We shall concentrate here on the interaction term $\mathcal{L}(f, \phi)$ involving the fermions and the Higgs fields in SM, which has the Yukawa form:

$$\begin{aligned} \mathcal{L}(f,\Phi) &= \sum_{j=1}^{3} \left\{ (h_l)_j \, \bar{l}_L^j \Phi \bar{l}_R^j \right\} \\ &+ \sum_{j,k=1}^{3} \left\{ \left(h'_q\right)_{jk} \bar{q}_L^j \Phi u_R^k + (hq)_{jk} \bar{q}_L^j \Phi^c d_R^k \right\}, \end{aligned} \tag{1}$$

where

$$\Phi^c = i\sigma_2 \phi^* = \begin{pmatrix} \phi^{0*} \\ -\phi^- \end{pmatrix},$$

and both Φ and Φ^c transform as a (weak) isospin doublet. Here $q_L^j(q_R^j)$ and $l_L^j(l_R^j)$ are, respectively, the $SU(2)_L(U(1)_R)$ quark and lepton fields, respectively, and we have assumed three families of fermions. The Yukawa coupling constants $(h_l)_j$, $(hq)_{jk}$ and $(h'q)_{jk}$ are arbitrary complex numbers and each term above is independently $SU(2)_L \otimes U(1)_Y$-invariant due to the fact that the $SU(2)_L$ acts only on $\bar{l}_L$ and $\bar{q}_L$ and on the Higgs doublet Φ and Φ^c, whose products are $SU(2)$ scalars. After spontaneous symmetry breaking

$SU(2)_L \otimes U(1)_Y \longrightarrow U(1)_{EM}$, which preserves the symmetry under $U(1)$ electromagnetism, the gauge bosons, fermions and the neutral scalar field ϕ, acquire non-zero masses through the Higgs mechanism. The resulting Lagrangian involving the fermion fields and ϕ (the physical Higgs scalar) reads as:

$$\begin{aligned}\mathcal{L}(f,\phi)^{SSB} &= \sum_{j=1}^{3}(m_j)_l \bar{l}_L^j l_R^j \left(1+\frac{1}{v}\phi\right) \\ &\quad - \sum_{j,k=1}^{3}\left\{(m_{jk})_U\, \bar{u}_L^j u_R^k + (m_{jk})_D\, \bar{d}_L^j d_R^k\right\}\left(1+\frac{1}{v}\phi\right) + \text{h.c.},\end{aligned} \tag{2}$$

where

$$\begin{aligned}(m_j)_l &= (h_j)_l \frac{v}{\sqrt{2}}, \\ (m_{jk})_U &= -(hq)_{jk}\frac{v}{\sqrt{2}}, \\ (m_{jk})_D &= -(h'q)_{jk}\frac{v}{\sqrt{2}},\end{aligned} \tag{3}$$

and v is the neutral-Higgs vacuum expectation value. Since in the SM there are no right-handed fields ν_R^i $(i = 1,2,3)$, the neutrinos ν^i remain massless and the charged lepton mass matrix $(m_j)_l$ is diagonal. Hence, there are no family-changing leptonic interactions. This aspect of the SM is now considered as vulnerable. In particular, as data on the solar neutrino flux are consolidating, suggestions that neutrinos may have a tiny mass and that they mix with each other are gaining currency. Likewise, astrophysical data from COBE and elsewhere having a bearing on the dark matter problem of the universe are more comfortably accommodated if there exists a significant component of the so-called hot dark matter, which may be identified with massive neutrinos. These aspects are being discussed by J. Ellis [10].

In (3), $(m_{jk})_U$ and $(m_{jk})_D$ are the (3×3) quark mass matrices for the up- and down-type quarks, respectively. In order to write the Lagrangian in terms of the quark mass eigenstates one has to diagonalize the mass matrices. This can be done with the help of two unitary matrices for each of the two mass matrices. We denote them for the up-type quarks by $V_L^{\mathrm{u}}$and $V_R^{\mathrm{u}\dagger}$ (likewise the ones for the down-type quarks are denoted by $V_L^{\mathrm{d}}$and $V_R^{\mathrm{d}\dagger}$):

$$\begin{aligned}V_L^{\mathrm{u}} m_U V_R^{\mathrm{u}\dagger} &\equiv (m_{diag.})_U \equiv Diag.(m_u\ , m_c\ , m_t) \\ V_L^{\mathrm{d}} m_D V_R^{\mathrm{d}\dagger} &\equiv (m_{diag.})_D \equiv Diag.(m_d\ , m_s\ , m_b)\end{aligned} \tag{4}$$

with $V_L^{\mathrm{u}\dagger} V_L^{\mathrm{u}} = 1$, etc. The physical quark fields (mass eigenstates) are related to the corresponding fields in the Lagrangian by the transformation:

$$\hat{u}_L = V_L^{\mathrm{u}} u_L = V_L^{\mathrm{u}} \begin{pmatrix} u_L \\ c_L \\ t_L \end{pmatrix}$$

$$\hat{d}_L = V_L^{\rm d} d_L = V_L^{\rm d} \begin{pmatrix} d_L \\ s_L \\ b_L \end{pmatrix} . \tag{5}$$

Likewise, $\hat{u}_R = V_R^{\rm u} u_R$, $\hat{d}_R = V_R^{\rm d} d_R$. One can now rewrite the term $\mathcal{L}(f, \Phi)$ in the SM Lagrangian in terms of the physical fields, obtaining

$$\mathcal{L}(f, \Phi)^{SSB} = - \left(1 + \frac{\phi}{v}\right) \left\{ \sum_{i=1}^{6} m_{q_i} \bar{q}_i q_i + \sum_{j=1}^{3} m_{l_j} \bar{l}_j l_j \right\} \tag{6}$$

where now it is understood that $q_1 = \frac{1}{2}({u^1}_L + {u^1}_R) = u$, etc., and we have dropped the hat over the quark fields. The identification of the parameters m_{l_i}, m_{q_i} with the lepton and quark masses is now evident. Written in terms of the physical boson ($W_\mu^\pm$, Z_μ^0, A_μ) and fermion fields, it is easy to show that the neutral current (NC) part of $\mathcal{L}(f, W, B)$ is manifestly flavour-diagonal. Thus, all flavour-changing transitions in the SM are confined to the charged current (CC) sector. This is an important feature of the SM, which we will confront with experiments. The absence of FCNC interactions in the SM Lagrangian is due to the choice of a single Higgs doublet and to the unitarity of the mixing matrix [3].

The charged current J_μ^{CC} is derived from the SM Lagrangian using the mass eigenstates. The CC couplings in the SM involve only the left-handed fermions q_L^i and l_L^i and the charged current J_μ^{CC}, which couples to the $W^\pm$, is:

$$J_\mu^{CC} = (\bar{u}, \bar{c}, \bar{t})_L \gamma_\mu V_{CKM} \begin{pmatrix} d \\ s \\ b \end{pmatrix}_L , \tag{7}$$

where we have again dropped the hats and $V_{CKM} \equiv V_L^{\rm u} V_L^{\rm d\dagger}$ is a (3×3) unitary matrix in the flavour space, first written down by Kobayashi and Maskawa in 1973 [2]. It is a generalization of the Cabibbo rotation [2] for the three-quark-flavour (u, d, s) case, invented to keep the universality of weak interactions, which took the form of a (2×2) matrix by the inclusion of a c-quark with the GIM construction [3], and is called the Cabibbo-Kobayashi-Maskawa (CKM) matrix. In the SM, CC have a $(V - A)$ structure, they violate P and C maximally, conserve the electric charge and the lepton- and baryon-number separately, but otherwise there are no other known restrictions on them except that $V_{CKM}^\dagger V_{CKM} = 1$. In general, $\mathcal{L}^{CC}$ violates CP due to the appearance of a non-trivial phase in V_{CKM}.

3 The CKM Matrix and Unitarity Triangles

Symbolically, the CKM matrix can be written as:

$$V_{CKM} \equiv \begin{pmatrix} V_{ud} & V_{us} & V_{ub} \\ V_{cd} & V_{cs} & V_{cb} \\ V_{td} & V_{ts} & V_{tb} \end{pmatrix} . \tag{8}$$

The nine quantities V_{ij} have to be determined experimentally via the couplings $W^{\pm}q\bar{q}'$. The CKM matrix elements obey unitarity constraints, which state that any pair of rows, or any pair of columns, of the CKM matrix are orthogonal. This leads to six orthogonality conditions. The three involving the orthogonality of columns are listed below, with the quark pair in the parenthesis (jk) representing the product of the j'th and k'th columns:

$$\begin{aligned} (ds): \quad \sum_i V_{id}V_{is}^* &= 0, \\ (sb): \quad \sum_i V_{is}V_{ib}^* &= 0, \\ (db): \quad \sum_i V_{id}V_{ib}^* &= 0. \end{aligned} \tag{9}$$

Similarly, there are three more such orthogonality conditions on the rows:

$$\begin{aligned} (uc): \quad \sum_j V_{uj}V_{cj}^* &= 0, \\ (ct): \quad \sum_j V_{cj}V_{tj}^* &= 0, \\ (ut): \quad \sum_j V_{uj}V_{tj}^* &= 0. \end{aligned} \tag{10}$$

The six orthogonality conditions can be depicted as six triangles in the complex plane. This is shown in fig. 1, in which each triangle is marked by the pair of quarks, representing the pair of columns, or of rows, whose orthogonality this triangle represents. Another way to view them is through the FCNC transitions in which the pair of quarks depicted participate. The triangle labelled (ds) represents the unitarity constraints on the transition $s \to d$, which, for example, one encounters in the K^0–$\overline{K^0}$ transition. The two others in this group, namely (sb) and (db), are encountered in the FCNC transitions in B decays, such as particle-antiparticle mixings B^0–$\overline{B^0}$ and rare B decays $B \to K^* + \gamma$ and $B \to \omega + \gamma$. These unitarity relations will, of course, be tested in CP violation experiments as well.

The (uc), (ct) and (ut) triangles can be measured in FCNC transitions involving the $Q = +2/3$ quarks, i.e. $c \to u$, $t \to c$ and $t \to u$. The analogous FCNC processes in this case are D^0–$\overline{D^0}$ and T^0–$\overline{T^0}$ mixings, and decays such as $D \to \rho + \gamma$, $T \to D^* + \gamma$ and $T \to (D, D^*, \rho) + Z$ etc. However, due to the fast decays of the charmed hadrons, which do not have any CKM suppression, unlike the K and B decays, and the even faster decays expected for the top quark, such FCNC transitions will be very difficult to measure. Of course, the sides of all of these triangles can be measured in CC decays of the quarks, as indicated in fig. 1. However, for the (ct) and (ut) triangles this would require the CKM matrix elements V_{td} and V_{ts}, once again a dismal proposition in direct top quark decays due to the anticipated tiny magnitudes of these matrix elements (see below). The (uc) triangle involves CKM matrix elements that have all been measured, but getting a sharp profile of this triangle imposes enormous demands on the accuracy of the matrix elements involved. It should be noted that not all the sides and angles in these triangles are independent. It can be shown that only four of the eighteen angles in the six CKM unitarity triangles are independent [5].

It is obvious that measuring all these triangles independently will keep the experimental flavour physics community busy for some time to come! The unitarity triangle that has, in our opinion, the best chance to be measured precisely is the triangle (db) following from the constraint:

$$V_{ud}V_{ub}^* + V_{cd}V_{cb}^* + V_{td}V_{tb}^* = 0 \ . \tag{11}$$

This is so because the three terms here are expected to be of comparable magnitude. This can be readily seen in the Wolfenstein parametrization of the CKM matrix. It was noticed some time ago by Wolfenstein [11] that the elements of this matrix exhibited a hierarchy in terms of λ, the Cabibbo angle. In this parametrization the CKM matrix can be written approximately as

$$V_{CKM} \simeq \begin{pmatrix} 1-\frac{1}{2}\lambda^2 & \lambda & A\lambda^3(\rho - i\eta) \\ -\lambda & 1-\frac{1}{2}\lambda^2 - iA^2\lambda^4\eta & A\lambda^2 \\ A\lambda^3(1-\rho-i\eta) & -A\lambda^2 & 1 \end{pmatrix} . \tag{12}$$

Using this representation of the CKM matrix, the (db) unitarity relation can be written as:

$$\frac{V_{ub}^*}{\lambda V_{cb}} + \frac{V_{td}}{\lambda V_{cb}} = 1 \ , \tag{13}$$

which is a triangle relation in the complex plane (i.e. ρ–η space). This is illustrated in fig. 2.

4 An Update of the CKM Matrix

In what follows we give an update of the Wolfenstein parameters A, λ, ρ and η. Having done this, we will use these updates to get the profile of the CKM unitarity triangle shown in fig. 2. We recall that $|V_{us}|$ has been extracted with good accuracy from $K \to \pi e\nu$ and hyperon decays [8] to be

$$|V_{us}| = \lambda = 0.2205 \pm 0.0018 \ . \tag{14}$$

This agrees quite well with the determination of $V_{ud} \simeq 1 - \frac{1}{2}\lambda^2$ from β-decay:

$$|V_{ud}| = 0.9744 \pm 0.0010 \ . \tag{15}$$

The parameter A is related to the CKM matrix element V_{cb}, which can be obtained from semileptonic decays of B mesons. We shall restrict ourselves to the methods based on the heavy-quark effective theory (HQET). In the heavy quark limit it has been observed that all hadronic form factors describing the dominant semileptonic transitions $B \to D\ell\nu$ and $B \to D^*\ell\nu$ can be expressed in terms of a single function, the Isgur-Wise function [12]. It has been shown that the HQET-based method works best for $B \to D^*\ell\nu$ decays, since these are unaffected by $1/m_b$ corrections [13, 14, 15]. Since $O(1/m_b)$ corrections to the Isgur-Wise function $\xi_{A_1}(v \cdot v' = 1)$ determining the rate for the decay $B \to D^*\ell\nu_\ell$ at the symmetry point are absent, deviation from the relation $\xi_{A_1}(v \cdot v' = 1) = 1$ is dominantly of perturbative QCD origin (here v and v' denote 4-velocities of the B and D^*, respectively). The differential decay rate can be expressed as [15]:

$$\lim_{v\cdot v' \to 1} \frac{1}{\sqrt{(v\cdot v')^2 - 1}} \frac{d\Gamma(B \to D^*\ell\nu_\ell)}{d(v\cdot v')} = \frac{G_F{}^2}{4\pi^3} M_{D^*}^3 (M_B - M_{D^*}) |V_{cb}|^2 \xi_{A_1}^2 \cdot$$

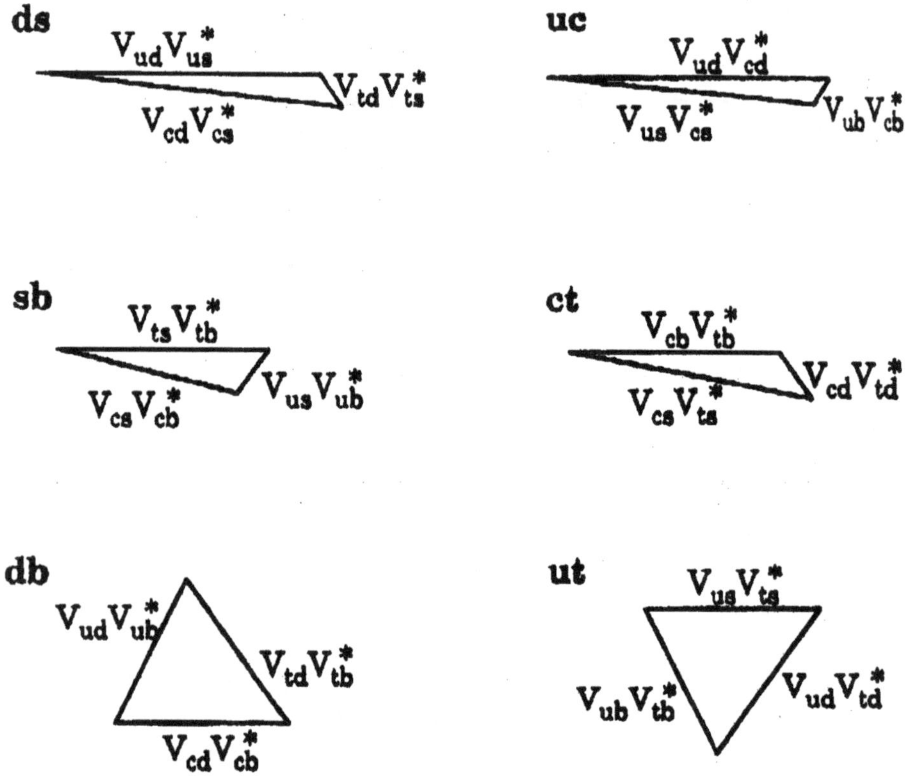

Figure 1: The six CKM unitarity triangles following from the orthogonality of the indicated pairs of columns, or of rows. The sides are drawn to the presently estimated lengths.

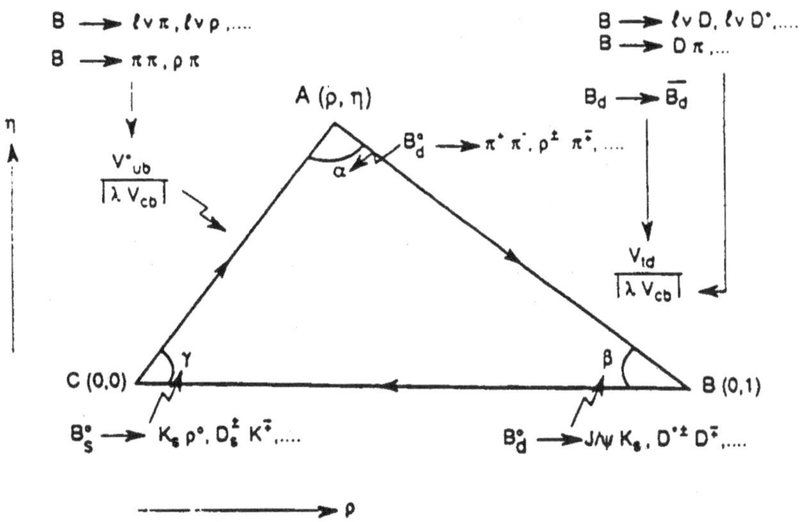

Figure 2: The (db) unitarity triangle. The sides can be measured in CC and FCNC B transitions. The angles α, β and γ can be measured via CP violation in the B system.

The QCD corrected function $\xi_{A_1}(v \cdot v' = 1)$ turns out to be essentially 1, namely $\xi_{A_1}(v \cdot v' = 1) \simeq 0.99$. There also exist theoretical estimates of the $O(1/m_b^2)$ power corrections [16]. Although these corrections are model-dependent and the present estimates of their contribution at the level of $O(\pm 3\%)$ may or may not be entirely realistic, in all likelihood, at the symmetry point $v \cdot v' = 1$, the semileptonic decays $B \to D^* \ell \nu$ are under theoretical control.

However, there are no data at the point $v \cdot v' = 1$. Recently, lattice-QCD calculations of the Isgur-Wise function have been reported in the $v \cdot v'$ range where data are available [17]. The experimental data and the lattice version of the Isgur-Wise function are in reasonable agreement [18]. One could also use a parametrization of the Isgur-Wise function $\xi_{A_1}(y)$ to extrapolate the data to the point $y = v \cdot v' = 1$, where the heavy quark symmetry holds. Four such parametrizations have been used in the experimental analysis. A fit of the data on $B \to D^* \ell \nu_\ell$ determines the slope and $|V_{cb}|$. With the branching ratios measured by the ARGUS and CLEO experiments [19, 20], and using $\tau_B = 1.49 \pm 0.04$ ps, the present world average of the B-hadron lifetime [18], the results of a recent update of $|V_{cb}|$ based on HQET are shown in fig. 3. The parametrizations used are also given here. While there is some spread in the extracted value, typical values from the ARGUS data are centred around $|V_{cb}| = 0.045$, which incidentally is also the central value quoted by the lattice-QCD analysis in [17], and those of the CLEO data around $|V_{cb}| = 0.037$ [20]. The errors are still significant so that all these determinations are consistent with each other. Eventually, data will become quite precise so that the required extrapolation to the symmetry point will be much reduced and with it the attendant theoretical uncertainties.

In the fits for the CKM unitarity triangle shown later, the following value for $|V_{cb}|$ has been used:

$$|V_{cb}| = 0.042 \pm 0.005 \,. \tag{16}$$

This gives for the CKM parameter A:

$$A = 0.86 \pm 0.10 \; . \tag{17}$$

The other two CKM parameters ρ and η are constrained at present by the measurements of $|V_{ub}/V_{cb}|$, $|\epsilon|$ (the CP-violating parameter in the kaon system), x_d (the B_d^0–$\overline{B_d^0}$ mixing ratio) and (in principle) ϵ'/ϵ ($\Delta S = 1$ CP violation in the kaon system). We shall not discuss the constraints from ϵ'/ϵ, due to the various experimental and theoretical uncertainties surrounding it at present and refer to [21] for details, but take up the rest in turn and present fits in which the allowed region of ρ and η is shown.

First of all, $|V_{ub}/V_{cb}|$ can be obtained by looking at the endpoint of the inclusive lepton spectrum in semileptonic B decays [22]. The leptons from the semileptonic transition $b \to c\ell\nu_\ell$ range out earlier than the ones from the transition $b \to u\ell\nu_\ell$. The extraction of $|V_{ub}/V_{cb}|$, however, requires a model for the inclusive lepton spectra. The result of a recent analysis of the available data (from ARGUS [23], CLEO 1.5 (old data) [24] and CLEO II (new data) [25]), for four different models [27, 28, 29, 30], is shown in fig. 4. The CLEO II data lie systematically below the CLEO 1.5 and ARGUS data, and the spread in the values of the CKM ratio $|V_{ub}/V_{cb}|$ due to models is noteworthy. Clearly, there is much room for improvement here. Unlike the semileptonic decay $B \to D^* + \ell\nu_\ell$, the $b \to u$ transitions are not determined by the HQET rules alone and some model building is inevitable. The present average by consensus (barring the ISGW model) is [18, 26, 31]:

$$\left|\frac{V_{ub}}{V_{cb}}\right| = 0.08 \pm 0.02 \; . \tag{18}$$

This gives for the Wolfenstein parameters:

$$\sqrt{\rho^2 + \eta^2} = 0.36 \pm 0.09 \; . \tag{19}$$

The value of $|V_{ub}/V_{cb}|$ has considerable effect on the allowed domains for the parameters (ρ, η), and in determining the CP asymmetries in a number of B decays. Further constraints on these parameters are obtained from $|\epsilon|$. The experimental value of $|\epsilon|$ is [8]:

$$|\epsilon| = (2.26 \pm 0.02) \times 10^{-3} \; . \tag{20}$$

Theoretically, $|\epsilon|$ is essentially proportional to the imaginary part of the box diagram for K^0–$\overline{K^0}$ mixing and is given by [32]

$$\begin{aligned} |\epsilon| &= \frac{G_F^2 f_K^2 M_K M_W^2}{6\sqrt{2}\pi^2 \Delta M_K} B_K \left(A^2\lambda^6\eta\right) \Big(y_c \left\{\eta_{ct} f_3(y_c, y_t) - \eta_{cc}\right\} \\ &\quad + \eta_{tt} y_t f_2(y_t) A^2\lambda^4(1-\rho)\Big). \end{aligned} \tag{21}$$

Here, the η_i are QCD correction factors, $\eta_{cc} \simeq 0.82$, $\eta_{tt} \simeq 0.62$, $\eta_{ct} \simeq 0.35$ for $\Lambda_{QCD} =$ 200 MeV [33], $y_i \equiv m_i^2/M_W^2$, and the functions f_2 and f_3 are given by

$$\begin{aligned} f_2(x) &= \frac{1}{4} + \frac{9}{4}\frac{1}{(1-x)} - \frac{3}{2}\frac{1}{(1-x)^2} - \frac{3}{2}\frac{x^2 \ln x}{(1-x)^3}, \\ f_3(x, y) &= \ln\frac{y}{x} - \frac{3y}{4(1-y)}\left(1 + \frac{y}{1-y}\ln y\right). \end{aligned} \tag{22}$$

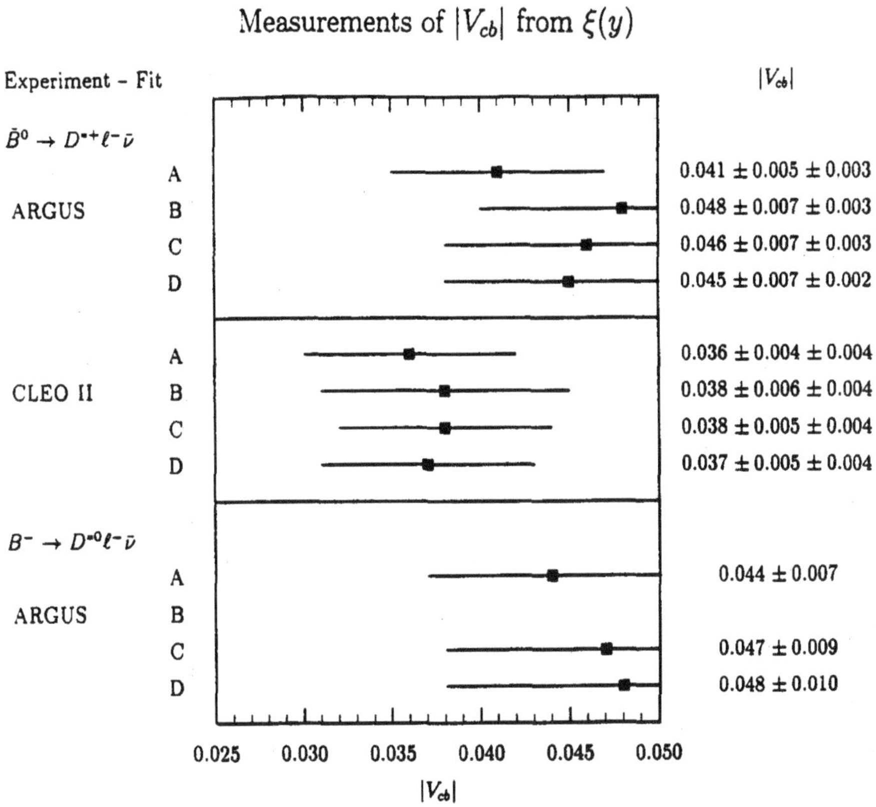

Fit	Function
A	$1-\rho^2(y-1)$
B	$\frac{2}{y+1}\exp\left[-(2\rho^2-1)\frac{y-1}{y+1}\right]$
C	$\left(\frac{2}{y+1}\right)^{2\rho^2}$
D	$\exp[-\rho^2(y-1)]$

Figure 3: Measurements of the CKM angle $|V_{cb}|$ from data on the semileptonic decay $B^0 \to D^*\ell^-\bar{\nu}$ from ARGUS [19] and CLEO [20] and $B^- \to D^*\ell^-\bar{\nu}$ from ARGUS [19], using the Isgur-Wise function ξ_{A_1}. The parametrizations used are also indicated (from [26]).

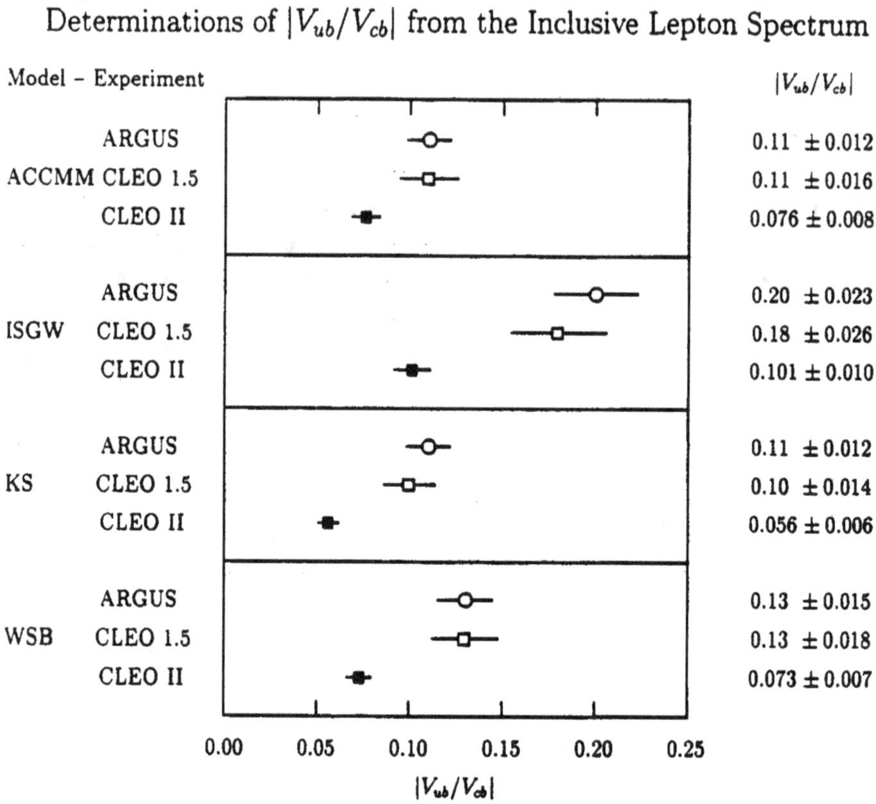

Figure 4: Determinations of $|V_{ub}|/|V_{cb}|$ from the inclusive lepton energy spectrum in B decays. The data are from ARGUS [23], CLEO 1.5 [24] and CLEO II [25]. The extracted values are shown on the r.h.s., and the models used ACCMM[27], ISGW [28], KS [29] and WSB [30] are indicated on the l.h.s. (from [26]).

(The above form for $f_3(x,y)$ is an approximation, obtained in the limit $x \ll y$. For the exact expression, see ref. [34].) One of the unknowns in eq. (21) is the top-quark mass. The present CDF bound from direct top searches gives $m_t > 113$ GeV [35]. While the exact range of m_t from electroweak analysis depends on a number of details, in particular on the value of α_s [9], a recent analysis based on the combined LEP data gives [36]:

$$m_t = 164 \pm 27 \text{ GeV}. \tag{23}$$

The final parameter in the expression for $|\epsilon|$ is B_K, which involves the matrix element $\langle K^0|(\bar{d}\gamma^\mu(1-\gamma_5)s)^2|\overline{K^0}\rangle$. The evaluation of this matrix element has been the subject of much work, reviewed in [4]. Although the entire range of B_K is $1/3 \le B_K \le 1$, the $1/N$ and lattice approaches are generally considered more reliable. They give estimates, which can be stated as

$$B_K = 0.8 \pm 0.2\ . \tag{24}$$

We now turn to B_d^0-$\overline{B_d^0}$ mixing. The value of x_d, which we use in our analysis, is

$$x_d = 0.68 \pm 0.09\ . \tag{25}$$

This is slightly different from the value $x_d = 0.71 \pm 0.07$ quoted in [18], but certainly consistent with it. The mixing parameter x_d is calculated from the B_d^0-$\overline{B_d^0}$ box diagram. Unlike the kaon system, where the contributions of both the c- and the t-quarks in the loop are important, this diagram is dominated by t-quark exchange:

$$x_d \equiv \frac{(\Delta M)_B}{\Gamma} = \tau_B \frac{G_F^2}{6\pi^2} M_W^2 M_B \left(f_{B_d}^2 B_{B_d}\right) \eta_B y_t f_2(y_t) |V_{td}^* V_{tb}|^2\ , \tag{26}$$

where, in the Wolfenstein parametrization, $|V_{td}^* V_{tb}|^2 = A^2\lambda^6\left[(1-\rho)^2 + \eta^2\right]$. Here, η_B is the QCD correction. In ref. [37], this correction is analysed in great detail, including the effects of a heavy t-quark. They find that η_B depends sensitively on the definition of the t-quark mass, and that, strictly speaking, only the product $\eta_B(y_t)f_2(y_t)$ is free of this dependence. In the fits presented here we use the value $\eta_B = 0.55$, following ref. [37].

For the B system, the hadronic uncertainty is given by $f_{B_d}^2 B_{B_d}$, analogous to B_K in the kaon system, except that in this case, also f_{B_d} is not measured. And, just like B_K, the evaluation of $f_{B_d}^2 B_{B_d}$ has been the subject of much work, reviewed in [4]. Until very recently the scaling law

$$f_B(m_B)\sqrt{m_B} = \text{constant}, \tag{27}$$

was thought to be valid for the B system. This led to rather small values for $f_{B_d}^2 B_{B_d}$ in the range

$$100 \text{ MeV} \le f_{B_d}\sqrt{B_{B_d}} \le 170 \text{ MeV}. \tag{28}$$

However, recent lattice calculations have indicated that there are significant scaling violations. These have led to larger estimates for $f_{B_d}^2 B_{B_d}$, although the final number is still in a state of flux. We will use the following values in our analysis

$$\begin{aligned} f_{B_d} &= 180 \pm 50 \text{ MeV}, \\ B_{B_d} &= 1.0 \pm 0.2\ . \end{aligned} \tag{29}$$

The errors are large enough to cover most present lattice-based numbers.

Table 1: Parameters used in the CKM fits.

Quantity	Input Value
$\|V_{cb}\|$	0.042 ± 0.005
$\|V_{ub}\|/\|V_{cb}\|$	0.08 ± 0.02
$\tau(B)$	1.49 ± 0.04 (ps)
$\|\epsilon_K\|$	$(2.26 \pm 0.02) \times 10^{-3}$
x_d	0.68 ± 0.09
m_t	164 ± 27 GeV
$\hat{B}_K$	0.8 ± 0.2
$\hat{B}_B$	1.0 ± 0.2
$f(B_d)$	180 ± 50 MeV

5 A Profile of the (db) Unitarity Triangle

We shall concentrate on the (db) triangle, shown in fig. 2, in this section. In order to find the allowed unitarity triangle, the computer programme MINUIT was used in ref. [4] to fit the CKM parameters A, ρ and η to the experimental values of $|V_{cb}|$, $|V_{ub}/V_{cb}|$, $|\epsilon|$ and x_d. These fits have since then been updated and the results are being included here. The parameters used in this analysis have been discussed in the previous section and their input values from experimental measurements or theoretical estimates are listed in table 1. The results of the fit themselves are shown in fig. 5. In all these graphs, the solid line has $\chi^2 = \chi^2_{min} + 1$, which corresponds to 39% confidence level region [8]. For comparison, we include the dashed line, which is the 90% C.L. region ($\chi^2 = \chi^2_{min} + 4.6$). (Note that, strictly speaking, these fits do not represent the true sense of confidence levels since some of the errors shown in table 1 are theoretical, and theoretical "errors" are not Gaussian! However, this method is not unreasonable, being quite similar to the practice of adding experimental statistical and systematic errors in quadrature.) It is clear that, as we pass from fig. 5(a) to fig. 5(c), the "most likely" unitarity triangle becomes more and more acute. However, it is also clear that there is a substantial overlap between the 90% C.L. regions. The "best values" of the parameters (ρ, η), corresponding to the triangles drawn in the figures, together with their χ^2, are given in table 2.

We remark that the range of η is well constrained by the very accurate measurement of $|\epsilon|$, and it depends only modestly on x_d (and hence f_{B_d}) or, as we are going to argue in the next section, on the ratio x_s/x_d. It should also be remarked that if the constraint from $|\epsilon|$ is removed from this analysis, then the present data are not enough to prove that indeed one has a triangle in the $\rho-\eta$ plane. The value of ρ is strongly affected by the x_d measurement, which in turn depends strongly on both m_t and f_{B_d}. It is therefore qualitatively clear that a good measurement of x_s (and hence x_s/x_d) will be very effective in narrowing down the allowed range for ρ. Likewise, rare B decays $B \to \rho + \gamma$, $B \to \omega + \gamma$, $B \to \rho + \ell^+\ell^-$ etc. would also be very helpful. We will return to a quantitative discussion of these points in the next sections.

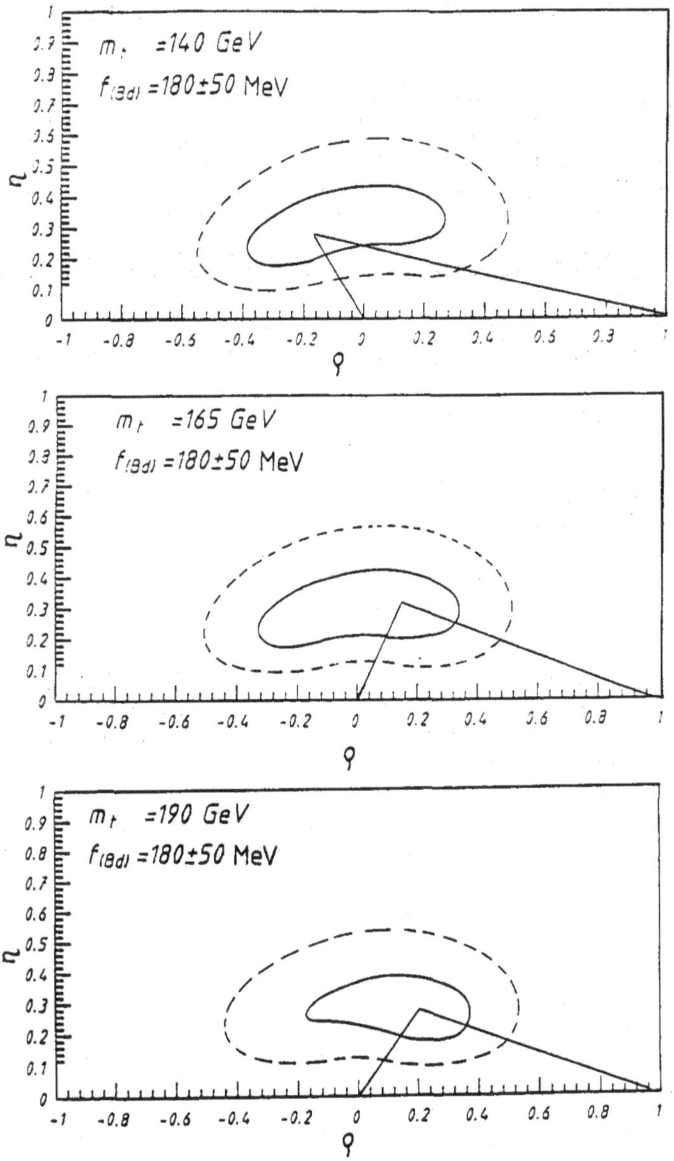

Figure 5: Allowed region in ρ-η space for different values of the Standard Model parameters given in table 1. The three figures correspond to m_t = 140, 165 and 190 GeV, respectively. The solid line represents the region with $\chi^2 = \chi^2_{min} + 1$; the dashed line denotes the 90% C.L. region. The triangles show the best fit (updated from [4].)

Table 2: "Best Values" for the Wolfenstein parameters ρ and η following from the analysis based on fig. 5.

m_t (GeV)	f_{B_d} (MeV)	B_{B_d}	(ρ, η)	χ^2_{min}
140	180 ± 50	1.0 ± 0.2	(−0.19, 0.29)	0.04
165	180 ± 50	1.0 ± 0.2	(0.14, 0.31)	0.19
190	180 ± 50	1.0 ± 0.2	(0.21, 0.27)	0.13

6 B_s^0–$\overline{B_s^0}$ Mixing and the CKM Unitarity Triangle

Mixing in the B_s^0–$\overline{B_s^0}$ system follows quite closely that of the B_d^0–$\overline{B_d^0}$ system. The B_s^0–$\overline{B_s^0}$ box diagram is again dominated by t-quark exchange, and the mixing parameter x_s is given by a formula analogous to that of eq. (26):

$$x_s \equiv \frac{(\Delta M)_{B_s}}{\Gamma_{B_s}} = \tau_{B_s} \frac{G_F^2}{6\pi^2} M_W^2 M_{B_s} \left(f_{B_s}^2 B_{B_s}\right) \eta_{B_s} y_t f_2(y_t) |V_{ts}^* V_{tb}|^2 \, . \tag{30}$$

Using the fact that $|V_{cb}| = |V_{ts}|$ (eq. (12)), it is clear that one of the sides of the unitarity triangle, $|V_{td}/\lambda V_{cb}|$, can be obtained from the ratio of x_d and x_s:

$$\frac{x_d}{x_s} = \frac{\tau_{B_d} \eta_{B_d} M_{B_d} \left(f_{B_d}^2 B_{B_d}\right)}{\tau_{B_s} \eta_{B_s} M_{B_s} \left(f_{B_s}^2 B_{B_s}\right)} \left|\frac{V_{td}}{V_{ts}}\right|^2 . \tag{31}$$

All dependence on the t-quark mass drops out, and we are left with the square of the ratio of CKM matrix elements, multiplied by a factor that reflects $SU(3)_{flavour}$-breaking effects. The only real uncertainty in this factor is the ratio of hadronic uncertainties – the other quantities will be either calculated or measured. Whether or not x_s can be used to help constrain the unitarity triangle will depend crucially on the theoretical status of the ratio $f_{B_d}^2 B_{B_d} / f_{B_s}^2 B_{B_s}$.

Lattice calculations indicate that the ratio f_{B_d}/f_{B_s} could be calculated much more accurately than either f_{B_d} or f_{B_s}, due to the cancellation of some systematic uncertainties. For example, ref. [38] gives

$$\begin{aligned} f_{B_d} &= 188\text{–}246 \text{ MeV}, \\ f_{B_s} &= 204\text{–}241 \text{ MeV}. \end{aligned} \tag{32}$$

Along the same lines, Abada et al. [39] quote $f_{B_s}/f_{B_d} = 1.06 \pm 0.04$; a recent average over a number of lattice calculations gives $f_{B_s}/f_{B_d} = 1.08 \pm 0.06$ [40]. Likewise, a recent calculation of the pseudoscalar coupling constant ratios in the QCD sum rule approach by Narison gives [41]:

$$\frac{f_{B_s}}{f_{B_d}} = 1.16 \pm 0.05 \tag{33}$$

These calculations are all very encouraging and though there is still some spread in theoretical predictions, the results are clearly consistent with each other within ±10%. Having motivated the necessity for measuring x_s, we will now turn to an estimate of its size in the Standard Model.

6.1 Standard model prediction for x_s

The SM expression for x_s is given in eq. (30). Using eq. (12) to substitute for $V_{ts}^* V_{tb}$, we obtain

$$x_s = \tau_{B_s} \frac{G_F^2}{6\pi^2} M_W^2 M_{B_s} \left(f_{B_s}^2 B_{B_s} \right) \eta_{B_s} A^2 \lambda^4 y_t f_2(y_t) \, . \tag{34}$$

The main uncertainties in this equation are again the t-quark mass and $f_{B_s}^2 B_{B_s}$. The lifetime of the B_s meson has now been measured at LEP. We give below the LEP average for the B_s^0 and B_d^0 lifetimes [18]:

$$\begin{aligned} \tau_{B_s^0} &= (1.26^{+0.22}_{-0.17}) \times 10^{-12} \text{ sec} \\ \tau_{B_d^0} &= (1.48 \pm 0.10) \times 10^{-12} \text{ sec.} \end{aligned} \tag{35}$$

Within the measurement errors, both these values are consistent with the averaged value of τ_B used in the previous section. The mass of the B_s^0 mesons has also now been measured and its present world average is $\langle M_{B_s} \rangle = 5373.2 \pm 4.2$ MeV [18], thus removing one of the unknowns. We will also take the QCD correction η_{B_s} to be equal to its B_d counterpart, i.e. $\eta_{B_s} = 0.55$. Using the central value for A in eq. (17), we obtain

$$x_s = (173) \frac{f_{B_s}^2 B_{B_s}}{(1 \text{ GeV})^2} y_t f_2(y_t). \tag{36}$$

The function $y_t f_2(y_t)$ is numerically equal to 2.03 for $m_t = 150$ GeV. As for $f_{B_s}^2 B_{B_s}$, as mentioned earlier, there is some controversy regarding its value. We use a value with a sufficiently large error:

$$f_{B_s} \sqrt{B_{B_s}} = 210 \pm 50 \text{ MeV}. \tag{37}$$

This leads to

$$x_s = (7.6 \pm 3.5) \, y_t f_2(y_t), \tag{38}$$

The "central values" (taking $m_t = 150$ GeV) lies in the range:

$$x_s = 15.4 \pm 7.1, \tag{39}$$

giving $x_s \simeq$ 8–24. The Standard Model therefore predicts very large values for x_s. This is to be expected since, from eq. (31), one has $x_s/x_d \simeq |V_{ts}/V_{td}|^2$, which in our estimates lies between 11 and 30, apart from $SU(3)_{flavour}$-breaking effects. Using the central value for $x_d = 0.68$, used above, this gives $x_s \simeq$ 8–21. Due to these large values, time-dependent measurements are necessary to obtain x_s. Present measurements at LEP and elsewhere yield a theoretically uninteresting lower bound: $x_s > 0.8$ at 90% C.L. [42]. Time-dependent measurements on x_s at LEP will be exploring the lower fringe of the expected domain of x_s. Luckily, several proposals for doing dedicated B physics in hadronic collisions, such as will be carried out at the LHC and HERA-B experiments, are capable of reaching the entire x_s range we have presented.

Table 3: The "best values" of the CKM parameters (ρ, η) and the ratio $|V_{td}/V_{ts}|^2$, obtained by a minimum χ^2 fit of the experimental data discussed in the text, for 3 different values of m_t. Note that $x_s = 15.0 \pm 3.0$ has been assumed for the fits. The resulting minimum χ^2 values from the MINUIT fits are also given.

m_t (GeV)	(ρ, η)	$\lvert V_{td}/V_{ts}\rvert^2$	χ^2_{min}
100	(0.01, 0.37)	0.056	0.03
140	(0.02, 0.34)	0.056	0.24
180	(0.03, 0.33)	0.057	0.93

6.2 CKM fits imposing constraints from x_s

In the previous section we have given our CKM fits, using the measured value of x_d and its SM analysis, which depends on m_t, the coupling constant product $f_{B_d}^2 B_{B_d}$, and the QCD correction factor η_B. These fits show that the allowed region in ρ–η is enormous, reflecting the large uncertainties in some of the input parameters. We have also argued that much of the uncertainty in the allowed region would disappear if, instead of x_d, one used the ratio x_s/x_d to constrain the CKM parameters. The main point is that if this ratio were measured then one could restrict the allowed value of ρ. Unfortunately, x_s has not been measured. However, the point can be illustrated by *assuming a value for* x_s, based on our SM estimates of this quantity in the previous section. We assume, for the sake of the CKM fits in this section:

$$x_s = 15.0 \pm 3.0 \,. \tag{40}$$

We now impose the constraints from this *hypothetical measurement of* x_s, together with the measured value for x_d, and the lattice QCD estimates of the coupling constant product [43]:

$$\frac{f_{B_s}^2 B_{B_s}}{f_{B_d}^2 B_{B_d}} = 1.19 \pm 0.10 \,. \tag{41}$$

The other constraints from the measured values of $|\epsilon|$, $|V_{cb}|$ and $|V_{ub}|$, which we have employed earlier, remain unchanged in these fits. The resulting CKM fits are shown in figs. 6(a)–(c), for the three asumed values of $m_t = 100, 140$ and 180 GeV. Note that the allowed regions in the (ρ, η) for the three choices of m_t are hardly distinguishable. The reason is that using the ratio x_s/x_d, the residual m_t dependence is very mild. Our "best values" for the parameters (ρ, η), together with the value of the CKM matrix element ratio $|V_{td}/V_{ts}|^2$, and the χ^2 are given in table 3. It is clear that the measurement of x_s would greatly help in reducing the uncertainty in the unitarity triangle.

7 Rare B-Decays and the CKM Matrix Elements

FCNC B and K decays also provide valuable information on the CKM matrix. We shall concentrate here on the so-called $|\Delta B| = 1,\ \Delta Q = 0$ decays, leading to transitions such

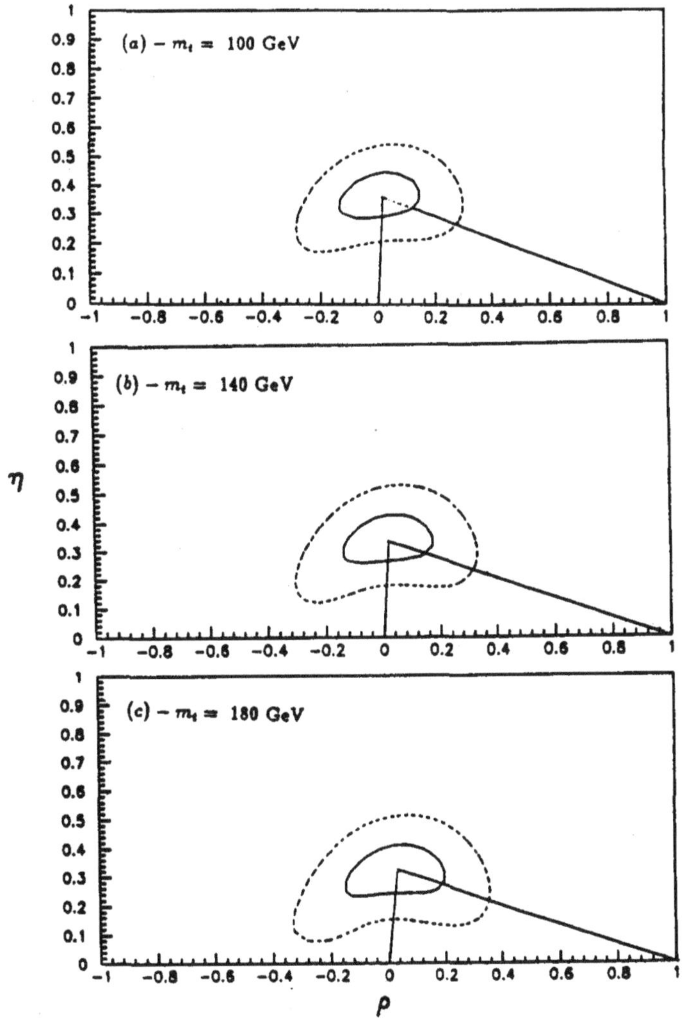

Figure 6: Allowed region in ρ-η space for different values of the Standard Model parameters, assuming $x_s = 15.0 \pm 3.0$. Figs. (a)–(c) have m_t = 100, 140 and 180 GeV, respectively. The solid line represents the region with $\chi^2 = \chi^2_{min} + 1$; the dashed line denotes the 90% C.L. region. The triangles show the best fit (from [4] .)

as $b \to (s,d) + X$, where $X = g,\ \gamma,\ \ell^+\ell^-$, and $\nu\bar{\nu}$, as well as the purely leptonic decays $B_s \to \ell^+\ell^-$ and $B_d \to \ell^+\ell^-$. We shall consider the radiative B decays in some detail since the first such decay has recently been measured by the CLEO collaboration with a branching ratio $BR(B \to K^* + \gamma) = (4.5 \pm 1.0 \pm 0.9) \times 10^{-5}$ [6]. This measurement, together with the upper bound on the inclusive decay rate $BR(B \to \gamma + X) < 5.4 \times 10^{-4}$, also reported by CLEO [7], provide an upper and lower bound on the CKM matrix element ratio $|V_{ts}|/|V_{cb}|$ [44].

We briefly discuss the SM theoretical framework for inclusive and exclusive radiative B decays dominated by the electromagnetic penguins. The inclusive decays are grouped as $B \to X_f + \gamma$, where we use the flavour of the light quark $f = s, d$ in the transition $b \to f$ to characterize the hadronic system recoiling against the photon. Including lowest-order QCD corrections and gluon bremsstrahlung, these decays are described at the parton level by the transitions $b \to f\gamma$ and $b \to f\gamma g$. In calculating the inclusive decay widths, we shall follow the work reported in [44]–[47]. The exclusive decays that are of interest may be grouped in an analogous way into $b \to s$ transitions:

- $B_u \to K^* + \gamma,\ \ B_d \to K^* + \gamma,$
- $B_s \to \phi + \gamma,$

which we shall also term CKM-allowed, according to the dominant CKM matrix element dependence of their decay rates, and $b \to d$ transitions:

- $B_d \to \rho + \gamma,\ \ B_d \to \omega + \gamma,\ \ B_u \to \rho + \gamma,$
- $B_s \to K^* + \gamma,$

which will be called CKM-suppressed.

7.1 Effective Hamiltonian for $B \to X_s + \gamma$ and $B \to X_d + \gamma$

The framework to incorporate (perturbative) short-distance QCD corrections in a systematic way is that of an effective low-energy theory with five quarks. It is obtained by integrating out the heavier degrees of freedom, i.e. the top quark and $W^\pm$ bosons.

Before using the unitarity properties of the CKM matrix, the effective Hamiltonian relevant for the processes $b \to f\gamma$ and $b \to f\gamma g$ has the form[2]

$$\begin{aligned} H_{eff}(b \to f) &= -\frac{4G_F}{\sqrt{2}} \left(V_{tb}V_{tf}^* \sum_{j=3}^{8} C_j(\mu)\, O_j(\mu) \right. \\ &\left. + V_{cb}V_{cf}^* \sum_{j=1}^{8} C'_j(\mu)\, O'_j(\mu) + V_{ub}V_{uf}^* \sum_{j=1}^{8} C''_j(\mu)\, O''_j(\mu) \right), \end{aligned} \tag{42}$$

and j runs through a complete set of operators with dimension up to six; the $C_j(\mu)$ are their Wilson coefficients evaluated at the scale μ.

[2] The treatment here follows closely the one given in [48].

We recall here that $O_1^{\prime(\prime)}$ and $O_2^{\prime(\prime)}$ represent the colour-singlet and colour-octet four-fermion operators, respectively, obtained from the SM charged-current Lagrangian written in the charge retention form:

$$\begin{aligned} O_1' &= (\bar{c}_{L\alpha}\gamma^\mu b_{L\beta})(\bar{f}_{L\alpha}\gamma_\mu c_{L\beta}) \quad \text{and} \quad O_1'' = (\bar{u}_{L\alpha}\gamma^\mu b_{L\beta})(\bar{f}_{L\alpha}\gamma_\mu u_{L\beta}) \,, \\ O_2' &= (\bar{c}_{L\alpha}\gamma^\mu b_{L\alpha})(\bar{f}_{L\beta}\gamma_\mu c_{L\beta}) \quad \text{and} \quad O_2'' = (\bar{u}_{L\alpha}\gamma^\mu b_{L\alpha})(\bar{f}_{L\beta}\gamma_\mu u_{L\beta}) \,. \end{aligned} \tag{43}$$

The remaining operators are the same for all three CKM prefactors, i.e. $O_j' = O_j'' = O_j$ for $j \neq 1, 2$. At tree level, the only contribution to $b \to f\gamma$ comes from the magnetic moment operator

$$O_7 = \frac{e}{16\pi^2}\, \bar{f} \sigma^{\mu\nu} \left(m_b R + m_f L\right) b \; F_{\mu\nu} \,, \tag{44}$$

where $L = (1 - \gamma_5)/2$ and $R = (1 + \gamma_5)/2$, and e is the QED coupling constant. The four-fermion operators $O_3, ..., O_6$ and the QCD magnetic moment operator O_8, which is the gluonic counterpart of O_7, arise from penguin diagrams in the full theory (before integrating out W and t). These operators enter indirectly in $b \to f\gamma$ decays due to operator mixing and through (virtual and bremsstrahlung) gluon corrections.

Taking into account the unitarity of the CKM matrix, only two of the three combinations $\lambda_i \equiv V_{ib}V_{is}^*$ — or $\xi_i \equiv V_{ib}V_{id}^*$ for CKM-suppressed transitions — are independent. In the case of $b \to s$ transitions, one finds $|\lambda_u| \ll |\lambda_c|$, $|\lambda_t|$; therefore, when neglecting terms proportional to λ_u, the effective Hamiltonian (42) can be brought into a form proportional to $\lambda_t \approx -\lambda_c$:

$$H_{eff}(b \to s\gamma) = -\frac{4G_F}{\sqrt{2}}\, \lambda_t \sum_{j=1}^{8} C_j(\mu)\, O_j(\mu) \quad . \tag{45}$$

This makes it clear that measuring the transition $B \to X_s + \gamma$, inclusively or through exclusive decay modes, amounts to measuring the CKM matrix element $\lambda_t = V_{tb}V_{ts}^*$.

For $b \to d$ transitions, where ξ_u, ξ_c and ξ_t are all of the same order of magnitude, it is most convenient to choose ξ_u and ξ_c as independent CKM factors during the matching for the Wilson coefficients at $\mu = m_W$ and during the renormalization group evolution to $\mu \approx m_b$. Since thereby terms of order $O(m_u^2/m_W^2)$ and $O(m_c^2/m_W^2)$ (or $O(m_u^2/m_t^2)$ and $O(m_c^2/m_t^2)$) are neglected, the terms proportional to ξ_u and ξ_c are multiplied by just the same coefficient functions, i.e. $C_j'(\mu) = C_j''(\mu)$ [although the corresponding operators are, of course, different, see (43)]. Finally, exploiting again the CKM unitarity constraint $\xi_u + \xi_c = -\xi_t$, one ends up with

$$H_{eff}(b \to d\gamma) = -\frac{4G_F}{\sqrt{2}} \left(\xi_t \sum_{j=3}^{8} C_j(\mu)\, O_j(\mu) - \sum_{j=1}^{2} C_j(\mu) \left\{ \xi_c O_j'(\mu) + \xi_u O_j''(\mu) \right\} \right) \quad , \tag{46}$$

where the Wilson coefficients $C_j(\mu)$ are precisely the same functions as in (45). Details of the full operator basis, the matching of the Wilson coefficients C_j at $\mu \approx m_W$, and the complete leading-logarithmic renormalization group evolution to $\mu \ll m_W$ can be found in literature [49].

7.2 A determination of $|V_{ts}|/|V_{cb}|$ from electromagnetic penguins

We now discuss the implications of the present measurement of $BR(B \to K^* + \gamma)$ and the eventual measurement of the branching ratio $BR(B \to X_s + \gamma)$ for the SM parameters. A much desired goal at present is to determine m_t either directly or indirectly through the loop corrections in which the top quark contributes. Unfortunately, the measurement of $BR(B \to X_s + \gamma)$ would not be a big help here as the m_t-dependence of $BR(B \to X_s + \gamma)$ is rather mild and the QCD scale dependence more pronounced. (For details see [44].) In the SM context, the best use of the measurement of $BR(B \to X_s + \gamma)$, in our opinion, lies in the determination of the CKM matrix element ratio $\lambda_t/|V_{cb}|^2$. The unitarity constraint states that $(\lambda_t/|V_{cb}|)^2 \simeq (\lambda_c/|V_{cb}|)^2 = |V_{cs}|^2$. Since the CKM matrix element $|V_{cs}|$ is known from independent measurements, such as charmed hadron decays, the electromagnetic penguin decays $B \to X_s + \gamma$ provide a quantitative test of CKM unitarity.

In estimates of the inclusive branching ratio for $BR(B \to X_s + \gamma)$ one has to include the contribution of the QCD bremsstrahlung process $b \to s + \gamma + g$ and the virtual corrections to $b \to s + \gamma$, both calculated in $O(\alpha\alpha_s)$ in ref. [45]. Expressed in terms of the inclusive semileptonic branching ratio $BR(B \to X\ell\nu_\ell)$, one finds:

$$BR(B \to X_s + \gamma) = 6\frac{\alpha}{\pi}\frac{|\lambda_t|^2}{|V_{cb}|^2}\frac{|C_7(x_t, m_b)|^2 K(x_t, m_b)}{g(m_c/m_b)(1 - 2/3\frac{\alpha_s}{\pi}f(m_c/m_b))} \times BR(B \to X\ell\nu_\ell)\ , \quad (47)$$

where $x_t = m_t^2/m_W^2$ indicates the explicit m_t-dependence in $C_7(x_t, m_b) \equiv C_7(m_b)$, and $g(r) = 1 - 8r^2 + 8r^6 - r^8 - 24r^4\ln(r)$ is the phase-space function for $\Gamma(b \to c + \ell\nu_\ell)$. The function $f(m_c/m_b)$ accounts for QCD corrections to the semileptonic decay and can be found, for example, in ref. [50]. It is a slowly varying function of r, and for a typical quark mass ratio of $r = 0.35 \pm 0.05$, it has the value $f(r) = 2.37 \mp 0.13$. The contributions from the decays $b \to u + \ell\ \nu_\ell$ have been neglected in the denominator in (47) since they are numerically inessential ($|V_{ub}| \ll |V_{cb}|$). For the semileptonic branching ratio $BR(B \to X\ell\nu_\ell)$ one can insert the measured value of about 11%. The Wilson coefficient $|C_7(x_t, m_b)|$ and the function $K(x_t, m_b)$, which is a K-factor in the sense of QCD corrections, have been computed in [44]. The resulting branching ratio [3] in the SM is estimated as [44]:

$$BR(B \to X_s + \gamma) = (3.0 \pm 0.5) \times 10^{-4} \quad (48)$$

for m_t in the range 100 GeV $\leq\ m_t\ \leq$ 200 GeV. This is to be contrasted with the present upper limit on the exclusive radiative decay, $BR(B \to X_s + \gamma) < 5.4 \times 10^{-4}$ (at 90% C.L.) [7].

The QCD scale dependence of $C_7(x_t, \mu)$ entering through the renormalization of the Wilson coefficients is very pronounced in the presently available leading-logarithmic approximation, as first emphasized in [44] and also confirmed in a recent paper [51]. This uncertainty has been estimated by varying the scale parameter μ in the range $m_b/2 \leq \mu \leq 2m_b$ and introduces an additional theoretical error of $\pm 1 \times 10^{-4}$ in the above estimates for $BR(B \to X_s + \gamma)$.

[3] obtained by setting the CKM matrix element ratio $|\lambda_t|^2/|V_{cb}|^2$ to 1, according to our present knowledge of the CKM matrix (readily seen e.g. in the Wolfenstein representation [11]).

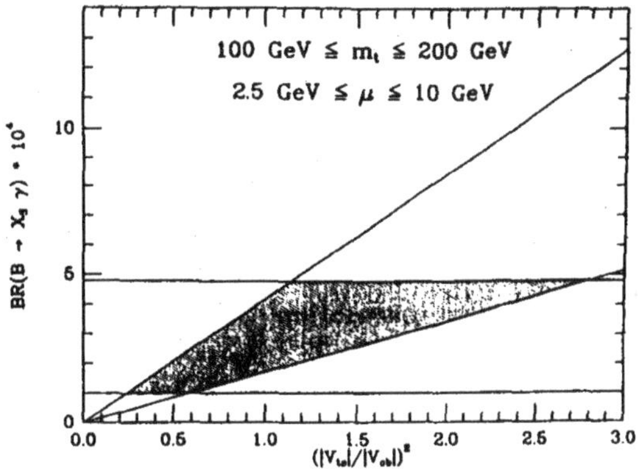

Figure 7: Inclusive branching ratio $BR(B \to X_s + \gamma)$ in the Standard Model as a function of the CKM matrix element ratio $(|V_{ts}|/|V_{cb}|)^2$, with the top-quark mass in the range 100 GeV $\leq m_t \leq$ 200 GeV and the QCD scale parameter μ in the range 2.5 GeV $\leq \mu \leq$ 10 GeV. The derived upper and lower (95% C.L.) bounds from the CLEO measurements [6, 7] are also shown (from [44]).

The dependence of the branching ratio $BR(B \to X_s + \gamma)$ on the CKM matrix element ratio $|V_{ts}|^2/|V_{cb}|^2$ is shown in fig. 7.[4] The two curves starting from the origin delimit the SM expectations with m_t and μ varied in the range 100 GeV $\leq m_t \leq$ 200 GeV and 2.5 GeV $\leq \mu \leq$ 10.0 GeV. An upper bound on the CKM matrix element ratio from the experimental upper limit on the inclusive branching ratio $BR(B \to X_s + \gamma)$ can be readily obtained from fig. 7. For the CLEO upper limit $BR(B \to X_s + \gamma) < 5.4 \times 10^{-4}$ (95% C.L.), the corresponding upper bound is:

$$|V_{ts}|/|V_{cb}| < 1.77. \tag{49}$$

It has been argued in [44] that this bound can be improved theoretically, if the QCD-improved photon energy spectrum in the inclusive decay $B \to X_s + \gamma$ is used in the experimental analysis. The detailed calculations of the photon energy (equivalently hadron mass) spectrum in the decays $B \to X_s + \gamma$ are given in [45, 46]. In [44], where the first analysis for the determination of the CKM parameter λ_t from the electromagnetic penguins was reported, a theoretically improved upper bound $BR(B \to X_s + \gamma) < 4.8 \times 10^{-4}$ (95% C.L.) was derived based on the CLEO upper bound, giving (see fig. 2):

$$|V_{ts}|/|V_{cb}| < 1.67. \tag{50}$$

The CLEO measurements of the exclusive branching ratio $BR(B \to K^* + \gamma) = (4.5 \pm 1.5 \pm 0.9) \times 10^{-5}$ yields, at 95% C.L., the following upper and lower limits [7]:

$$1.6 \times 10^{-5} \leq BR(B \to K^* + \gamma) \leq 7.4 \times 10^{-5}. \tag{51}$$

[4]Since $|V_{tb}| \simeq 1$ to a very good approximation, we have set $\lambda_t = |V_{ts}|$.

To convert these experimental bounds into the ones on the CKM matrix element ratio, one has to calculate the exclusive branching ratio $BR(B \to K^* + \gamma)$. For that purpose we introduce the following exclusive-to-inclusive B-decay width ratio:

$$R(K^*/X_s) \equiv \frac{\Gamma(B \to K^* + \gamma)}{\Gamma(B \to X_s + \gamma)}. \tag{52}$$

The determination of this ratio has been taken up in a number of theoretical papers based on quark model wave-function type estimates [52], QCD sum rules of the older [53] and new-vintage [48, 54, 55], and lattice-QCD [56, 57]. Barring the earlier QCD sum rule results, most other estimates yield (5–15)%. We use the estimates from [44], yielding:

$$R(K^*/X_s) = (13 \pm 3)\%, \tag{53}$$

This can be used to predict the branching ratio for the exclusive decay $BR(B \to K^* + \gamma)$ [44]:

$$BR(B \to K^* + \gamma) = (4.0 \pm 1.6 \pm 1.0) \times 10^{-5}, \tag{54}$$

where the first error reflects both the uncertainty in $R(K^*/X_s)$, given above, and in m_t in the range 100 GeV $\leq m_t \leq$ 200 GeV; the second error is due to the variation in μ in the range 2.5 GeV $\leq \mu \leq$ 10.0 GeV. The estimate (54) is in very comfortable agreement with the CLEO measurements, $BR(B \to K^* + \gamma) = (4.5 \pm 1.0 \pm 0.9) \times 10^{-5}$ [6]. This demonstrates that the SM is describing the electromagnetic penguin decays correctly – no mean achievement since we are discussing rates at the level of one per 20,000 B decays!

It is easy to see that a more stringent upper bound on the ratio $|V_{ts}|/|V_{cb}|$ follows at present from the 95% C.L. upper limit on the inclusive branching ratio $BR(B \to X_s + \gamma)$. The measured branching ratio $BR(B \to K^* + \gamma)$, however, provides at present the lower bound on the matrix element ratio, since the corresponding experimental lower bound on $BR(B \to X_s + \gamma)$ is still being awaited. Using $R(K^*/X_s)^{max} = 16\%$ from the analysis in [44] and the 95% C.L. experimental lower bound from CLEO given in (51), one gets a 95% C.L. lower bound on the inclusive branching ratio:

$$BR(B \to X_s + \gamma) \geq 1.0 \times 10^{-4}. \tag{55}$$

With this lower bound on $BR(B \to X_s + \gamma)$ and the upper limit quoted in (50) above, one derives the following bounds on the CKM ratio [44]:

$$0.50 \leq |V_{ts}|/|V_{cb}| \leq 1.67. \tag{56}$$

This is the first determination of this ratio. The lower bound depends on $R(K^*/X_s)^{max}$, which is a model-dependent but now converging enterprise. While these estimates are not yet completely quantitative, the point that the electromagnetic penguins in B decays provide a determination of the matrix element $|V_{ts}|/|V_{cb}|$, using the CKM unitarity constraints, has been demonstrated. Once the inclusive branching ratio $BR(B \to X_s + \gamma)$ has been measured, this model dependence can be largely removed and the route involving the exclusive decay $BR(B \to K^* + \gamma)$ that we have undertaken here will no longer be necessary. The determination of the CKM matrix element ratio from the electromagnetic penguins in B decays obtained above can be compared with the indirect bounds due to the CKM unitarity that can be worked out from the fits reported in the Particle Data Group [8]:

$$0.55 \leq |V_{ts}|/|V_{cb}| \leq 1.68. \tag{57}$$

The bounds (56) and (57) are quite comparable, although they have been derived by using very different considerations, which can be taken as a consistency check for the SM normalization of the electromagnetic penguins in B decays. With improved data and theory these tests of the CKM unitarity will become much more precise.

7.3 Determining V_{td} from electromagnetic penguins

Whereas in the branching ratio for $B \to X_s + \gamma$ the CKM matrix dependence was contained in an overall factor $|\lambda_t|^2$, the case of inclusive $B \to X_d + \gamma$ decays is somewhat more complicated with respect to the factorization of the CKM parameters. When evaluating the one-loop matrix elements for the parton process $b \to d\gamma g$, at least the four-quark operators O_2' and O_2'' should be kept since their coefficients are of order unity (they contribute in penguin diagrams emitting both a gluon and a photon). These contributions bring to the fore a non-trivial dependence on the CKM matrix elements through terms proportional to ξ_u and ξ_c [see eq. (46)]. Explicit expressions for this dependence on the ρ and η parameter in the Wolfenstein representation of the CKM matrix have been evaluated in ref. [47]. The dependence of the branching ratio can be written explicitly as:

$$BR(B \to X_d + \gamma) = D_1|\xi_t|^2 \left\{1 - \frac{1-\rho}{(1-\rho)^2+\eta^2} D_2 - \frac{\eta}{(1-\rho)^2+\eta^2} D_3 + \frac{D_4}{(1-\rho)^2+\eta^2}\right\}. \tag{58}$$

Here we have expressed ξ_t and ξ_c in terms of the Wolfenstein parameters A, λ, ρ and η:

$$\begin{aligned} \xi_t &= A\lambda^3(1-\rho+i\eta), \\ \xi_c &= -A\lambda^3. \end{aligned} \tag{59}$$

The m_t-dependence of the coefficients D_i is rather weak, as shown in table 4. To get the inclusive branching ratio as a function of m_t, we vary the CKM parameters ρ and η over the presently allowed range of the variables listed in table 1. We have described this procedure in detail earlier. The resulting branching ratio $BR(B \to X_d + \gamma)$ is shown in fig. 8 as a function of m_t, from which one obtains [59]:

$$BR(B \to X_d + \gamma) = (0.8\text{–}3.2) \times 10^{-5} \tag{60}$$

for the top-quark mass in the range 100 GeV $\leq m_t \leq$ 200 GeV. We remark that the μ-dependence of the branching ratio $BR(B \to X_d + \gamma)$ is very significant, which can again be traced to the very significant QCD renormalization of the Wilson coefficients $C_i(m_W) \to C_i(\mu)$, in particular C_7.

A determination of the Wolfenstein parameters ρ and η will follow if one is able to measure the inclusive radiative branching ratios $BR(B \to X_s + \gamma)$ and $BR(B \to X_d + \gamma)$. That a number of unknowns, such as the top-quark mass and the QCD scale parameter, drop out from this ratio is obvious from the expressions given earlier. However, from the experimental point of view measuring these inclusive rates is a formidable proposition. It

Table 4: Values of the coefficients D_i entering in eq. (58) as a function of m_t (from [47]) .

m_t (GeV)	D_1	D_2	D_3	D_4
100	0.16	0.21	0.04	0.14
120	0.17	0.20	0.04	0.13
140	0.19	0.18	0.04	0.12
160	0.20	0.17	0.04	0.11
180	0.21	0.17	0.03	0.10
200	0.22	0.16	0.03	0.10

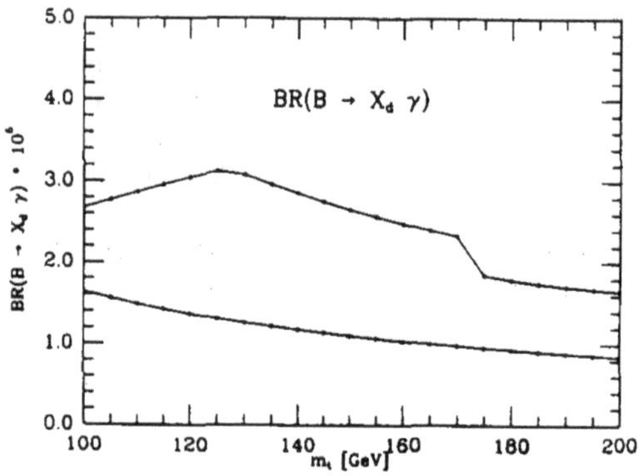

Figure 8: Branching ratio for $B \to X_d + \gamma$ as a function of m_t by varying over the allowed range in the (ρ, η)-plane (from [59]).

is perhaps easier to measure the exclusive radiative decays, as CLEO has demonstrated on the example of the decay $B \to K^* + \gamma$. In the following we shall focus our attention on *exclusive* decays, such as $B_{u,d} \to K^* + \gamma$, $B_s \to \phi + \gamma$, and the corresponding CKM-suppressed modes listed at the beginning of this section, and determine them in a reasonable theoretical framework such as QCD sum rules.

For a generic radiative decay $B \to V + \gamma$ one defines a transition form factor $F_1(q^2)$ as:

$$\langle V, \lambda | \bar{f} \sigma_{\mu\nu} q^\nu b | B \rangle = i\epsilon_{\mu\nu\rho\sigma} e_\nu^{(\lambda)} p_B^\rho p_V^\sigma 2F_1^{B\to V}(q^2) \ , \tag{61}$$

where V is a vector meson ($V = \rho, \omega, K^*$ or ϕ) with the polarization vector $e^{(\lambda)}$; and B is the generic B-meson B_u, B_d or B_s. The vectors p_B, p_V and $q = p_B - p_V$ denote the four-momenta of the initial B-meson and the outgoing vector meson and photon, respectively.

With the above definition, the exclusive decay widths are given by ($B = B_u$ or B_d):

$$\Gamma(B \to K^* + \gamma) = \frac{\alpha}{32\pi^4} G_F^2 |\lambda_t|^2 |F_1^{B\to K^*}(0)|^2 C_7(x_t, m_b)^2 (m_b^2 + m_s^2) \frac{(m_B^2 - m_{K^*}^2)^3}{m_B^3}, \quad (62)$$

and by an analogous expression for $\Gamma(B_s \to \phi + \gamma)$. The branching ratios of these exclusive decays can again be written in terms of the inclusive semileptonic branching ratio:

$$BR(B \to K^* + \gamma) = 6 \frac{\alpha}{\pi} \frac{|\lambda_t|^2}{|V_{cb}|^2} \frac{|C_7(x_t, m_b)|^2 |F_1^{B\to K^*}(0)|^2}{g(m_c/m_b)(1 - 2/3 \frac{\alpha_s}{\pi} f(m_c/m_b))} \frac{(1 - m_{K^*}{}^2/m_B^2)^3}{(1 - m_s^2/m_b^2)^3} \times (0.11), \quad (63)$$

The ratio of the exclusive-to-inclusive radiative decay widths introduced earlier can now be expressed as ($B = B_u$ or B_d):

$$R(K^*/X_s) = \frac{(1 - m_{K^*}{}^2/m_B^2)^3}{(1 - m_s^2/m_b^2)^3} \frac{m_B^3}{m_b^3} \frac{|F_1^{B_i \to K^*}(0)|^2}{K(x_t, m_b)} . \quad (64)$$

The function $K(x_t, m_b)$ is almost independent of m_t and has been estimated as $K(x_t, m_b) \simeq 0.83$, for m_t in the range 100 GeV $\leq m_t \leq$ 200 GeV [44]. In an analogous way, one defines the exclusive-to-inclusive ratio involving B_s decays.

Since the same short-distance-corrected coefficient function $C_7(m_b)$ enters in the Hamiltonian for CKM-allowed (45) and CKM-suppressed (46) modes, the QCD scaling for the two-body decays $b \to s + \gamma$ and $b \to d + \gamma$ is identical and does not effect the ratio of the decay widths $\Gamma(b \to d + \gamma)/\Gamma(b \to s + \gamma)$. The same applies for the exclusive decays such as $B \to \rho + \gamma$ and $B \to \omega + \gamma$, in which case also only the magnetic moment operator O_7 contributes and the exclusive decay rates factorize in the CKM factors. From this observation a number of relations between the exclusive decay rates follow in the Standard Model [47]. This is exemplified by the decay rates for $B_{u,d} \to \rho + \gamma$ and $B_{u,d} \to K^* + \gamma$:

$$\frac{\Gamma(B_{u,d} \to \rho + \gamma)}{\Gamma(B_{u,d} \to K^* + \gamma)} = \frac{|V_{td}|^2}{|V_{ts}|^2} \frac{|F_1^{B\to\rho}(0)|^2}{|F_1^{B\to K^*}(0)|^2} \Phi_{u,d} , \quad (65)$$

where $\Phi_{u,d}$ is a phase-space factor:

$$\Phi_{u,d} = \frac{(m_b^2 + m_d^2)}{(m_b^2 + m_s^2)} \frac{(m_{B_{u,d}}^2 - m_\rho^2)^3}{(m_{B_{u,d}}^2 - m_{K^*}^2)^3} . \quad (66)$$

The ratio (65) depends only on the CKM matrix elements and the ratio of form factors, while it is independent of the top-quark mass (and of the renormalization scale μ).

The transition form factors have been evaluated in the light-cone QCD sum rule approach [48], yielding:

$$\begin{aligned} F_1^{B\to K^*\gamma} &= 0.32 \pm 0.05, \\ F_1^{B\to(\rho,\omega)\gamma} &= 0.24 \pm 0.04, \\ F_1^{B_s\to\phi\gamma} &= 0.29 \pm 0.05, \\ F_1^{B_s\to K^*\gamma} &= 0.20 \pm 0.04. \end{aligned} \quad (67)$$

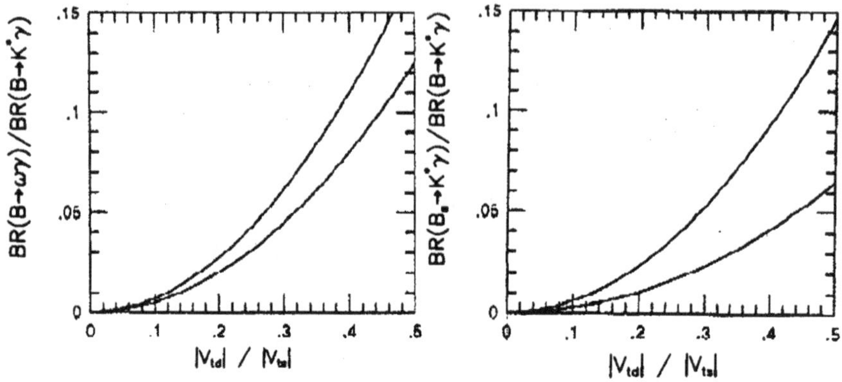

Figure 9: The ratios of the branching ratios $BR(B_d \to \omega + \gamma)/BR(B_d \to K^* + \gamma)$ and $BR(B_s \to K^* + \gamma)/BR(B_d \to K^* + \gamma)$ as a function of the CKM matrix element ratio $|V_{td}|/|V_{ts}|$. The two curves represent theoretical model uncertainties (from [48]).

For the ratios of the form factors, which are needed to determine the ratios of the CKM matrix elements, the following estimates have been reported in [48]:

$$\frac{F_1^{B\to(\rho,\omega)\gamma}}{F_1^{B\to K^*\gamma}} = 0.76 \pm 0.06\,, \tag{68}$$

$$\frac{F_1^{B_s\to K^*+\gamma}}{F_1^{B_s\to\phi+\gamma}} = 0.66 \pm 0.09\,, \tag{69}$$

and

$$\frac{F_1^{B_s\to K^*+\gamma}}{F_1^{B\to K^*+\gamma}} = 0.60 \pm 0.12\,. \tag{70}$$

The eventual measurements of the CKM-suppressed radiative decays $(B_u, B_d) \to \rho + \gamma$, $B_d \to \omega + \gamma$ and $B_s \to K^* + \gamma$ could then be used to determine the CKM parameters. This is illustrated in fig. 9, where the ratio of the branching ratios $BR(B_d \to \omega + \gamma)/BR(B \to K^* + \gamma)$ and $BR(B_s \to K^* + \gamma)/BR(B \to K^* + \gamma)$ are plotted as a function of the CKM matrix element ratio $|V_{td}|/|V_{ts}|$. As a check of this method, we note that it predicts BR$(B \to K^* + \gamma) = (4.8 \pm 1.5) \times 10^{-5}$ [48], which agrees well with the experimental measurement by CLEO. We expect that the estimates of the ratio of the exclusive branching ratio should, in any case, be more stable against variation of the parameters in the QCD sum rule method. So, the SM dependence shown in fig. 9 should be useful to determine the indicated ratio of the CKM matrix elements.

Apart from the radiative B decays, there is an entire class of FCNC semileptonic B decays, $B \to X_f + \ell^+\ell^-$, $B \to X_f + \nu\bar{\nu}$, where $f = d, s$, the purely leptonic decays $(B_d, B_s) \to \ell^+\ell^-$ and the $\gamma\gamma$ decays, $(B_d, B_s) \to \gamma\gamma$, which in the SM provide information on the CKM matrix elements and top-quark mass. Once again, these transitions can be classified as CKM-allowed and CKM-suppressed. Also, the ratios of the exclusive decay branching ratios, such as those involving the decays $B \to (\rho,\omega)+\ell^+\ell^-$ and $B \to K^*+\ell^+\ell^-$, should have dependence on the CKM matrix elements very similar to what we have shown for the corresponding decays in fig. 9. All of these remarks apply to the short-distance contribution. As is well known, for the FCNC semileptonic decays involving charged leptons, the long-distance contribution is quite important. So, an analysis where the long- and short-distance contributions have been treated coherently is required. The present best limit on the FCNC semileptonic decay is from the UA1 collaboration, giving $BR(B \to X+\mu^+\mu^-) < 5.0\times10^{-5}$ at 90 % C.L. [58], which lies about an order of magnitude away from the SM estimates. Since these and related matters have been reviewed recently in [59], we shall not discuss them any further. The expected rates for a number of FCNC B decays that have been estimated in [59], together with the present experimental bounds and the measured rate for $BR(B \to K^* + \gamma)$, are given in table 5. We expect that some of these SM predictions will soon be tested in present experiments at CLEO and the Tevatron, but many more will be scrutinized in dedicated B facilities expected to operate in the later part of this decade. With these measurements, the structure of the FCNC processes will be completely pinned down. There are already stringent constraints on models having enhanced FCNC couplings such as motivated by technicolor models.

8 CP Violation in the B System

Mixing and CP violation in the B system can be described in much the same way as in the kaon system. In the B system, the states $|B^0\rangle$ and $|\overline{B^0}\rangle$ are eigenstates of the strong and electromagnetic interactions, but not of the weak interactions, which are responsible for their decay. Taking into account the weak interactions, one writes the 2×2 Hamiltonian (in the B^0–$\overline{B^0}$ basis)

$$H = M - \frac{i}{2}\Gamma \,, \tag{71}$$

where the mass matrix M and the decay matrix Γ are Hermitian. (Since neutral B's do decay, H itself is not Hermitian.) CPT invariance implies that the diagonal components of H are equal, and if CP is conserved, M and Γ are real. Allowing for the possibility of CP violation, diagonalizing the Hamiltonian

$$H = \begin{bmatrix} m & M_{12} \\ M_{12}^* & m \end{bmatrix} - \frac{i}{2}\begin{bmatrix} \gamma & \Gamma_{12} \\ \Gamma_{12}^* & \gamma \end{bmatrix} \tag{72}$$

yields the eigenstates B_L and B_H (L and H denote the light and heavy states, respectively):

$$\begin{aligned} |B_L\rangle &= p|B^0\rangle + q|\overline{B^0}\rangle, \\ |B_H\rangle &= p|B^0\rangle - q|\overline{B^0}\rangle, \end{aligned} \tag{73}$$

where

$$\frac{p}{q} = \frac{1+\epsilon_B}{1-\epsilon_B} = \sqrt{\frac{M_{12} - \frac{i}{2}\Gamma_{12}}{M_{12}^* - \frac{i}{2}\Gamma_{12}^*}} \,. \tag{74}$$

Table 5: Estimates of the branching fractions for FCNC B decays in the Standard Model for m_t = 150 GeV, μ = 5 GeV and f_{B_d} = 200 MeV. Note that the CKM-suppressed decays depend quadratically on $|V_{td}|$, and the numbers correspond to $|V_{td}| = 9.2 \times 10^{-3}$. Numbers shown in the third column are present experimental upper limits except for $(B_d, B_u) \to K^* + \gamma$, which has been measured (updated from [59]).

Decay modes	BR	Experimental upper limits (90 % C.L.)
$(B_d, B_u) \to X_s\gamma$	3.5×10^{-4}	5.4×10^{-4} [CLEO][7]
$(B_d, B_u) \to K^*\gamma$	$(3.5 - 6.5) \times 10^{-5}$	$4.5 \pm 1.0 \pm 0.9 \times 10^{-5}$ [CLEO][6]
$(B_d, B_u) \to X_d\gamma$	1.5×10^{-5}	-
$(B_d, B_u) \to \rho + \gamma$	$(2.8 - 4.5) \times 10^{-6}$	-
$(B_d, B_u) \to X_s e^+e^-$	1.5×10^{-5}	-
$(B_d, B_u) \to X_d e^+e^-$	6.5×10^{-7}	-
$(B_d, B_u) \to X_s\mu^+\mu^-$	8.5×10^{-6}	5.0×10^{-5} [UA1][58]
$(B_d, B_u) \to X_d\mu^+\mu^-$	3.7×10^{-7}	-
$(B_d, B_u) \to K e^+e^-$	6.0×10^{-7}	5.0×10^{-5} [PDG][8]
$(B_d, B_u) \to K\mu^+\mu^-$	6.0×10^{-7}	1.5×10^{-4} [PDG][8]
$(B_d, B_u) \to K^* e^+e^-$	5.6×10^{-6}	-
$(B_d, B_u) \to K^*\mu^+\mu^-$	2.9×10^{-6}	2.3×10^{-5} [UA1][58]
$(B_d, B_u) \to X_s\ \nu\bar{\nu}$	8.0×10^{-5}	-
$(B_d, B_u) \to X_d\ \nu\bar{\nu}$	3.5×10^{-6}	-
$(B_d, B_u) \to K\ \nu\bar{\nu}$	6.4×10^{-6}	-
$(B_d, B_u) \to K^*\ \nu\bar{\nu}$	2.3×10^{-5}	-
$B_s \to \gamma\gamma$	1.5×10^{-8}	-
$B_d \to \gamma\gamma$	6.5×10^{-10}	-
$B_s \to \tau^+\tau^-$	3.8×10^{-7}	-
$B_d \to \tau^+\tau^-$	1.7×10^{-8}	-
$B_s \to \mu^+\mu^-$	1.8×10^{-9}	-
$B_d \to \mu^+\mu^-$	8.0×10^{-11}	-
$B_s \to e^+e^-$	4.2×10^{-14}	-
$B_d \to e^+e^-$	1.9×10^{-15}	-

In eq. (74), the symbol ϵ_B has been introduced to indicate the relation between this notation and that used normally in the kaon system. Analogously to the kaon system, a non-zero value of ϵ_B indicates CP violation in B^0–$\overline{B^0}$ mixing. However, there is a big difference in the B system – here it is found that $\Gamma_{12} \ll M_{12}$ (recall that $\Gamma_{12} \sim 2M_{12}$ for kaons). This implies that ϵ_B is much smaller than ϵ_K, so that prospects for the observation of CP violation in the $\Delta B = 2$ transitions of B^0–$\overline{B^0}$ mixing are virtually hopeless. Luckily for us, CP violation in B decays ($\Delta B = 1$) can be large [60, 61]. (This is another difference between the K and B systems; in the kaon system, CP violation in K decays (ϵ'/ϵ) is small.)

The most promising signal of CP violation in B decays comes by considering a CP eigenstate final state f to which both B^0 and $\overline{B^0}$ can decay. CP violation is manifested in a non-zero value of the time-dependent asymmetry [62]:

$$A_f(t) = \frac{\Gamma(B^0(t) \to f) - \Gamma(\overline{B^0}(t) \to \overline{f})}{\Gamma(B^0(t) \to f) + \Gamma(\overline{B^0}(t) \to \overline{f})} \, . \tag{75}$$

The expression for this asymmetry is

$$A_f(t) = -\mathrm{Im}\,\alpha_f \, \sin x_q \frac{t}{\tau_B} \, , \tag{76}$$

in which x_q, $q = d, s$ is the B_q^0–$\overline{B_q^0}$ mixing parameter (eqs. (26), (34)), τ_B is the B lifetime, and

$$\alpha_f = \frac{q}{p}\rho_f, \tag{77}$$

where ρ_f is the ratio of the amplitudes for $\overline{B_q^0}$ and B_q^0 to decay to f,

$$\rho_f = \frac{\langle f|H|\overline{B_q^0}\rangle}{\langle f|H|B_q^0\rangle}. \tag{78}$$

If all amplitudes that contribute to the decay $B_q^0 \to f$ have the same CKM phase ϕ_D, then ρ_f is a pure phase:

$$\rho_f = e^{-2i\phi_D} \, . \tag{79}$$

This is indeed the case, up to corrections from penguin diagrams [63], for all CP eigenstate final states. In addition, since $\Gamma_{12} \ll M_{12}$, q/p is also a pure phase:

$$\frac{q}{p} = \sqrt{\frac{M_{12}^*}{M_{12}}} = e^{-2i\phi_M} \, . \tag{80}$$

Therefore

$$\alpha_f = e^{-2i(\phi_M+\phi_D)} \Longrightarrow \mathrm{Im}\,\alpha_f = -\sin 2(\phi_M + \phi_D). \tag{81}$$

Note that, although ϕ_M and ϕ_D each depend on the form chosen for the CKM matrix, the sum $\phi_M + \phi_D$ is convention-independent.

In the Wolfenstein parametrization (eq. (12)), only the matrix elements V_{ub} and V_{td} have large phases. In the following we will ignore all small CKM phases. It is not difficult to relate ϕ_M and ϕ_D to the CKM matrix elements. Like x_q, M_{12} is calculable from the box diagram for B_q^0–$\overline{B_q^0}$ mixing (in fact, $\Delta M = 2|M_{12}|$). Thus, $M_{12} \sim (V_{tb}V_{td}^*)^2$, and therefore the phase information from the mixing is

$$\left(\frac{q}{p}\right)_{B_d} = \frac{V_{td}}{V_{td}^*}, \quad \left(\frac{q}{p}\right)_{B_s} = 1. \tag{82}$$

The phase information in the decay depends on whether the b-quark decays into a c- or a u-quark:

$$(\rho_f)_{b\to u} = \frac{V_{ub}}{V_{ub}^*}, \quad (\rho_f)_{b\to c} = 1. \tag{83}$$

There are therefore three classes of B decays into CP eigenstates that can have sizeable CP-violating asymmetries. These are related to the three angles α, β, γ of the unitary triangle (fig. 2) as follows.

- B_d decays with $b \to c$: For this class of decays, we have

$$\alpha_f = \frac{V_{td}}{V_{td}^*}, \quad \mathrm{Im}\,\alpha_f = -\sin 2\beta . \tag{84}$$

  The classic example here is $B_d \to J/\Psi K_S$. For this particular final state, since $J/\Psi K_S$ is a CP-odd state, there is an additional minus sign in the asymmetry.

- B_d decays with $b \to u$: Here we have

$$\alpha_f = \frac{V_{ub}V_{td}}{V_{ub}^*V_{td}^*}, \quad \mathrm{Im}\,\alpha_f = \sin 2\alpha . \tag{85}$$

  One example of such a decay is $B_d \to \pi^+\pi^-$.

- B_s decays with $b \to u$: In this case

$$\alpha_f = \frac{V_{ub}}{V_{ub}^*}, \quad \mathrm{Im}\,\alpha_f = -\sin 2\gamma . \tag{86}$$

  This angle is the most difficult to measure. One possible decay mode is $B_s \to \rho K_S$. Like the final state $J/\Psi K_S$, ρK_S is CP-odd, which changes the sign of the asymmetry. For other suggestions on measuring the angle γ and further references, see [5].

The asymmetries in eqs. (84), (85) and (86) can be expressed straightforwardly in terms of the CKM parameters ρ and η. The 90% C.L. constraints on ρ and η found previously can be used to predict the ranges of $\sin 2\alpha$, $\sin 2\beta$ and $\sin 2\gamma$ allowed in the Standard Model. The allowed ranges that correspond to each of the figures in figs. 5(a)–(c) are found in table 6. In this table we have assumed that the angle β is measured in $B_d \to J/\Psi K_S$, and have therefore included the extra minus sign due to the CP of the final state.

Since the CP asymmetries all depend on ρ and η, the ranges for $\sin 2\alpha$, $\sin 2\beta$ and $\sin 2\gamma$ shown in table 6 are correlated. That is, not all values in the ranges are allowed simultaneously. We illustrate this in figs. 10(a)–(c) by showing the region in $\sin 2\alpha$–$\sin 2\beta$

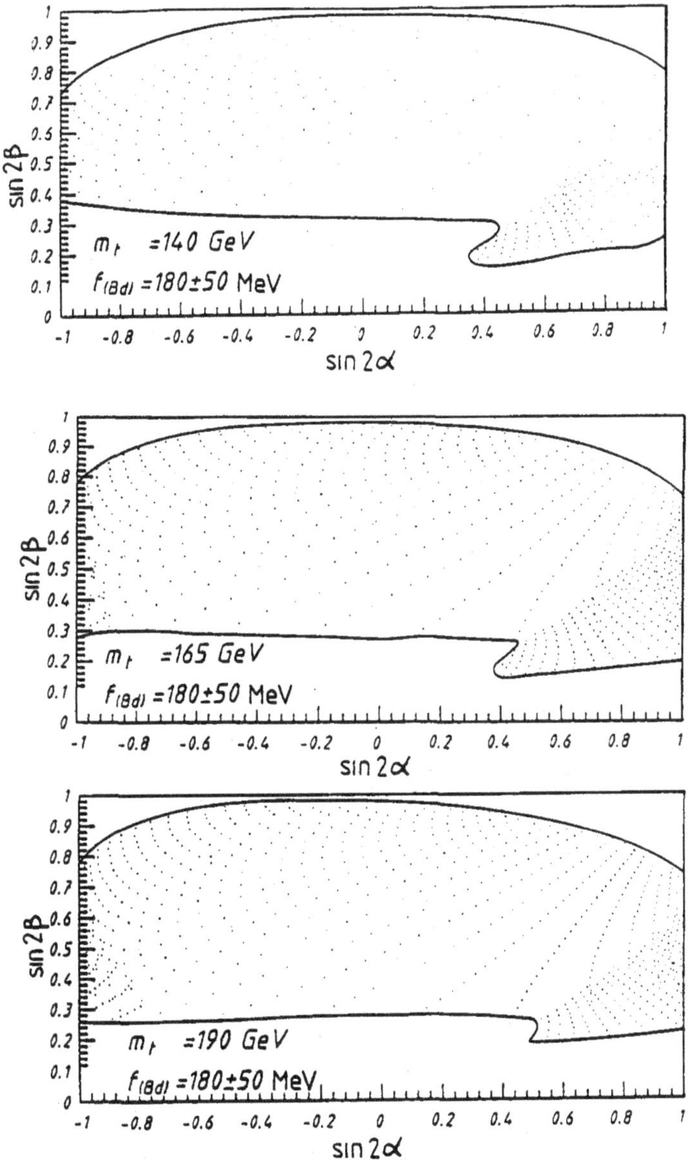

Figure 10: Allowed values of the CP asymmetries $\sin 2\alpha$ and $\sin 2\beta$ for different values of the Standard Model parameters given in table 1. Figs. (a)-(c) have m_t = 140, 165 and 190 GeV, respectively (updated from [4].)

Table 6: The allowed ranges for the CP asymmetries $\sin 2\alpha$, $\sin 2\beta$ and $\sin 2\gamma$, corresponding to the constraints on ρ and η shown in figs. 5. Values of m_t and the coupling constant $f_{B_d}\sqrt{B_{B_d}}$ are stated. The range for $\sin 2\beta$ includes an additional minus sign due to the CP of the final state $J/\Psi K_S$ (updated from [4]).

m_t (GeV)	$f_{B_d}\sqrt{B_{B_d}}$ (MeV)	$\sin 2\alpha$	$\sin 2\beta$	$\sin 2\gamma$
140	180 ± 50	-1 - 1	0.15 - 0.97	-1 - 1
165	180 ± 50	-1 - 1	0.16 - 0.98	-1 - 1
190	180 ± 50	-1 - 1	0.18 - 0.97	-1 - 1

space allowed by the data, for various values of m_t and $f_{B_d}^2 B_{B_d}$. From these figures one sees that the smallest values of $\sin 2\beta$ occur only in a small region of parameter space around $\sin 2\alpha \simeq 0.2$–0.6. Excluding this small tail, one expects the CP asymmetry in $B_d \to J/\Psi K_S$ to be at least 30%, which is good news for the B-physics community interested in CP violation.

9 Summary

One of the most important remaining tests of the Standard Model involves the precise determination of the CKM matrix parameters. The knowledge of the values of these parameters might give us some insight into the mechanism of spontaneous symmetry breaking, particularly as regards the fermion mass matrices. For example, it is known that SM and some unorthodox models, such as Extended Technicolor, fulfil their charge in this respect very differently. New and crucial experiments are needed to settle many of these grey areas. Furthermore, this information would allow us to ascertain whether the CKM matrix is the sole source of CP violation, or if new physics is needed. The CKM triangles provide a convenient framework to interpret present and planned experiments. Presence of new physics, contributing to the various rates and asymmetries discussed in this report, will show itself in these triangles through deviations from the CKM expectations. In view of this we discussed most of the flavour physics in terms of these triangles, equivalently the CKM matrix elements.

We have given here an update of the CKM matrix taking into account CC and FCNC processes, with the latter coming from K and B decays and neutral meson mixings. The results for the CKM-Wolfenstein parameters ρ and η, shown in terms of the unitarity triangles, are summarized in fig. 5. The present status of the CKM matrix elements themselves is summarized in table 7. This table also includes the constraints from the electromagnetic B penguins measured recently. Estimates of the CKM angles α, β and γ are summarized through the correlations shown in fig. 10. The picture that emerges from this analysis is rather rosy for the Standard Model! It is, however, also clear that while the present measurements are entirely consistent with the SM, there are enough uncertainties which inhibit drawing this conclusion at a completely quantitative level. We hope that the next round of experiments in flavour physics will greatly help in improving the precision

Table 7: Present status of the CKM matrix elements.

CKM matrix element	Present value
$\|V_{ud}\|$	0.9734 ± 0.0007
$\|V_{us}\|$	0.2205 ± 0.0018
$\|V_{cd}\|$	0.215 ± 0.016
$\|V_{cs}\|$	0.98 ± 0.02
$\|V_{cb}\|$	0.042 ± 0.005
$\|V_{ub}\|/\|V_{cb}\|$	0.08 ± 0.02
$\|V_{td}\|$	$0.005 - 0.016$
$\|V_{ts}\|/\|V_{cb}\|$	$0.50 - 1.67$
$\|V_{tb}\|$	$0.998 - 0.9995$

in a number of crucial parameters. In the meanwhile, we have also to learn the mundane aspects - how to calculate coupling constants and form factors from first principles - and the sublime aspects - how to relate fermion masses and mixing parameters in a well-defined theoretical concept. Whether the SM will pass the impending experimental hurdles with flying colours and with its feathers intact can only be answered when the experimental jury has given its verdict. While this is being awaited, the romance between the Standard Model and experiments, which started in 1973 with the neutral current discovery in the Gargamelle bubble chamber, continues!

Acknowledgements

I owe my scientific career largely to the generosity of Professor Abdus Salam. He helped me at a crucial stage in my academic life, without which I would have definitely not been able to pursue a scientific career. I thank him sincerely for his timely help in rather difficult circumstances. I also take this opportunity to thank him for his scientific guidance and advice, which I have had the good fortune to enjoy in all these years.

The work reported here is based mostly on collaborations with Volodya Braun, Christoph Greub, David London, Thomas Mannel and Hubert Simma. I thank them for many valuable inputs and discussions. I also acknowledge the help of David Cassel and David London in making available updates of the CKM matrix elements presented here. Helpful discussions with Michael Danilov, John Ellis, Klaus Honscheid, Boris Kayser, Stephan Narison, Peter Schlein, Henning Schröder and Sheldon Stone are also thankfully acknowledged.

References

[1] S.L. Glashow, Nucl. Phys. **22** (1961) 579; A. Salam in *Elementary Particle Theory*, ed. N. Svartholm (Almqvist and Wiksell, Stockholm, 1968); S. Weinberg, Phys. Rev. Lett. **19** (1967) 1264.

[2] N. Cabibbo, Phys. Rev. Lett. **10** (1963) 531; M. Kobayashi and T. Maskawa, Prog. Theor. Phys. **49** (1973) 652.

[3] S.L. Glashow, J. Iliopoulos and L. Maiani, Phys. Rev. **D2** (1970) 1285.

[4] A. Ali and D. London, J. Phys. G: Nucl. Part. Phys. **19** (1993) 1069; and in *B-Physics Working Group Report, ECFA Workshop on a European B-Meson Factory*, editors: R. Aleksan and A. Ali, ECFA 93/151, DESY 93-053 (1993).

[5] R. Aleksan, B. Kayser and D. London, McGill Univ. Report, entitled: *In Pursuit of Gamma*, (1993) (to be published).

[6] R. Ammar et al. (CLEO), Phys. Rev. Lett. **71** (1993) 674.

[7] E. Thorndike (CLEO), in Proc. of the American Physical Society Meeting, Washington (DC), April 1993; T. Browder, K. Honscheid and S. Playfer, CLNS 93/1261, UH-511-778-93, OHSTPY-HEP-E-93-018, HEPSY 93-10 (1993).

[8] M. Aguilar-Benitez et al. (Particle Data Group), Phys. Lett. **B239** (1990) 1.

[9] G. Altarelli (these Proceedings).

[10] J. Ellis (these Proceedings).

[11] L. Wolfenstein, Phys. Rev. Lett. **51** (1983) 1945.

[12] N. Isgur and M. Wise, Phys. Lett. **B232** (1989) 113; **B237** (1990) 527.

[13] M.E. Luke, Phys. Lett. **B252** (1990) 447.

[14] C.G. Boyd and D.E. Brahm, Phys. Lett. **B257** (1991) 393.

[15] M. Neubert and V. Rieckert, Nucl. Phys. **B382** (1992) 97; M. Neubert, Phys. Lett. **B264** (1991) 455, Phys. Rev. **D46** (1992) 2212.

[16] A.F. Falk and M. Neubert, Phys. Rev. **D47** (1993) 2965; **D47** (1993) 2982.

[17] S.P. Booth et al. (UKQCD), Edinburgh Univ. Report, EDINBURGH-93-525(1993).

[18] M. Danilov, in Proc. of the Europhysics Conference on High Energy Physics, Marseille (1993).

[19] H. Albrecht et al. (ARGUS), Phys. Lett. **B275** (1992) 195; Z. Phys. **C57** (1993) 533.

[20] G. Crawford et al. (CLEO), CLEO CONF 93-30, submitted to XVI Int. Symp. on Lepton-Photon Interactions, Ithaca, NY (1993).

[21] B. Winstein (these Proceedings).

[22] H. Schröder (these Proceedings).

[23] H. Albrecht et al. (ARGUS), Phys. Lett. **B234** (1991) 297.

[24] R. Fulton et al. (CLEO), Phys. Rev. Lett. **64** (1990) 16.

[25] J. Bartelt et al. (CLEO), Phys. Rev. Lett. **71** (1993) 4111.

[26] These figures are updates made by Dave Cassel and are based on the ARGUS results reported in [19, 23] and the CLEO results reported in [20, 24, 25].

[27] G. Altarelli et al., Nucl. Phys. **B208** (1982) 365.

[28] N. Isgur et al., Phys. Rev. **D39** (1989) 799.

[29] J. Körner and G.A. Schuler, Z. Phys. **C38** (1988) 511.

[30] M. Bauer, B. Stech and M. Wirbel, Z. Phys. **C34** (1987) 103.

[31] S. Stone, in *B Decays* (Second edition); editor: S. Stone, World Scientific, Singapore (1994).

[32] A.J. Buras, W. Słominski and H. Steger, Nucl. Phys. **B238** (1984) 529; **B245** (1984) 369.

[33] J. Flynn, Mod. Phys. Lett. **A5** (1990) 877.

[34] T. Inami and C.S. Lim, Prog. Theor. Phys. **65** (1981) 297.

[35] F. Abe et al. (CDF), Phys. Rev. **D45** (1992) 3921.

[36] M. Swartz, in same Proc. as ref. [20].

[37] A.J. Buras, M. Jamin and P.H. Weisz, Nucl. Phys. **B347** (1990) 491.

[38] C. Alexandrou et al., Nucl. Phys. **B374** (1992) 263.

[39] A. Abada et al., Nucl. Phys. **B376** (1992) 172.

[40] C. Sachrajda, in *Proceedings of the Conference on QCD–20 Years Later*, editors: H. Kastrup and P. Zerwas, Aachen, FRG (1993).

[41] S. Narison, Preprint CERN-TH.7091/93 (1993).

[42] H. G. Moser, Preprint CERN-PPE/93-164 (1993) (to appear in Proc. of the 5th International Symposium on Heavy Flavour Physics, Montreal, Canada, 6-10 July 1993).

[43] O. Pène, talk at the ECFA *B* Workshop, DESY, October 1992.

[44] A. Ali and C. Greub, Z. Phys. **C60** (1993) 433.

[45] A. Ali and C. Greub, Z. Phys. **C49** (1991) 431; Phys. Lett. **B259** (1991) 182.

[46] A. Ali and C. Greub, Phys. Lett. **B287** (1992) 191.

[47] A. Ali and C. Greub, Phys. Lett. **B293** (1992) 226.

[48] A. Ali, V. M. Braun and H. Simma, Preprint CERN-TH.7118/93 (1993).

[49] S. Bertolini, F. Borzumati and A. Masiero, Phys. Rev. Lett. **59** (1987) 180;
R. Grigjanis et al., Phys. Lett. **B213** (1988) 355;
B. Grinstein, M.J. Savage and M.B. Wise, Nucl. Phys. **B319** (1989) 271;
B. Grinstein, R. Springer and M. B. Wise, Nucl. Phys. **B339** (1990) 269;
G. Cella et al., Phys. Lett. **B248** (1990) 181;
M. Misiak, Phys. Lett. **B269** (1991) 161; Nucl. Phys. **B393** (1993) 23;
M. Ciuchini et al., Phys. Lett. **B316** (1993) 127 and Preprint ROMA 973/93 (1993).

[50] A. Ali and E. Pietarinen, Nucl. Phys. **B154** (1979) 519.

[51] A.J. Buras et al., Tech. Univ. Munich Report TUM-T31-50/93 (1993).

[52] T. Altomari, Phys. Rev. **D37** (1988) 677;
N.G. Deshpande et al., Z. Phys. **C40** (1988) 369;
P.J. O'Donnell and H.K.K. Tung, Phys. Rev. **D44** (1991) 741;
A. Ali, T. Ohl and T. Mannel, Phys. Lett. **B298** (1993) 195;
T. Hayashi, M. Matsuda and M. Tanimoto, Kogakkan Univ. Report AUE-01-93 (1993).

[53] T.M. Aliev et al., Phys. Lett. **B237** (1990) 569;
C.A. Dominguez et al., Phys. Rev. Lett. **B214** (1988) 459;
N. Paver and Riazuddin, Phys. Rev. **D45** (1992) 978.

[54] P. Ball, TU-München Report TUM-T31-43/93 (1993).

[55] P. Colangelo et al., Univ. of Bari Report BARI-TH/93-150 (1993).

[56] C. Bernard, P. Hsieh and A. Soni, Univ. Washington Report Wash. U. HEP/93-35 (1993).

[57] K.C. Bowler et al. (UKQCD Coll.), Edinburgh preprint 93/258 (1993).

[58] C. Albajar et al. (UA1), Phys. Lett. **B262** (1991) 163.

[59] A. Ali, C. Greub and T. Mannel, DESY Report 93-016 (1993), and in same Proc. as in ref. [4].

[60] For a review, see I. Bigi, V. Khoze, N. Uraltsev and A. Sanda, in *CP Violation*, ed. C. Jarlskog (World Scientific)(1989) 175, and references quoted therein.

[61] R. Aleksan et al., in same Proc. as in ref. [4].

[62] A. Carter and A. I. Sanda, Phys. Rev. Lett. **45** (1980) 952; Phys. Rev. **D23** (1981) 1567; I. I. Bigi and A. I. Sanda, Nucl. Phys. **B193** (1981) 85; **B281** (1987) 41; Y. Azimov, V. Khoze and M. Uraltsev, Yad. Fiz. **45** (1987) 1412; D. Du, I. Dunietz and D. Wu, Phys. Rev. **D34** (1986) 3414.

[63] M. Gronau, Phys. Rev. Lett. **63** (1989) 1451; D. London and R.D. Peccei, Phys. Lett. **B223** (1989) 257; B. Grinstein, Phys. Lett. **B229** (1989) 280.

Experimental Results on b Quarks

Henning Schröder
DESY, Hamburg, Germany

Abstract

Experimental results on B hadrons are discussed in the framework of the Standard Model. From the data the fundamental parameters of the Standard Model are extracted, namely the quark mixing angles and the phase responsible for CP vioaltion.

1. Introduction

The study of B mesons at e^+e^- colliders and also at hadron machines has allowed us to determine the properties and behaviour of the b quark. The b quark is of particular importance for the determination of Standard Model parameters and allows us to test it. This can be demonstrated by the quark spectrum (Figure 1).

The b quark the heaviest quark which we can study since the t quark has not yet been discovered. The b quark has the virtue that it couples to all lighter quarks, either directly to c and u quarks, or indirectly via a virtual t quark also to s and d quarks. The arrows indicate the allowed transitions between the quarks which are mediated by the weak interactons. We can collect the transition amplitudes in the CKM matrix V_{CKM}

$$V_{CKM} = \begin{pmatrix} V_{ud} & V_{us} & V_{ub} \\ V_{cd} & V_{cs} & V_{cb} \\ V_{td} & V_{ts} & V_{tb} \end{pmatrix}$$

which is a unitary matrix since only the charged current transitions between the quarks are allowed in the Standard Model. This implies that the CKM matrix is

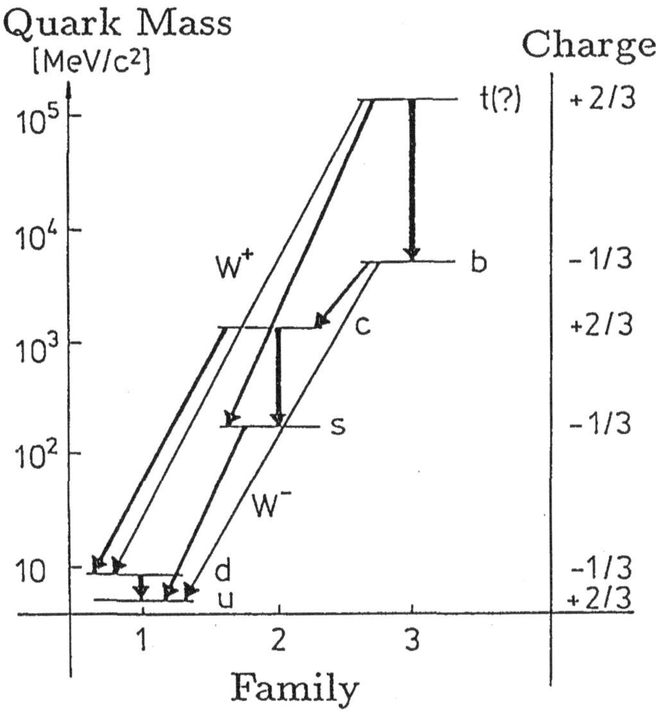

Figure 1: Quark spectrum and allowed transitions between quarks .

given by three Euler angles θ_{12} θ_{23} and θ_{13} and a phase δ which allows for complex matrix elements and therefore for CP violation in weak decays. The standard parametrization [1] is given by:

$$= \begin{pmatrix} c_{\theta_{12}} c_{\theta_{13}} & c_{\theta_{13}} s_{\theta_{12}} & s_{\theta_{13}} e^{i\delta} \\ -s_{\theta_{23}} c_{\theta_{12}} s_{\theta_{13}} e^{i\delta} - s_{\theta_{12}} c_{\theta_{23}} & c_{\theta_{12}} c_{\theta_{23}} - s_{\theta_{12}} s_{\theta_{23}} s_{\theta_{13}} e^{i\delta} & s_{\theta_{23}} c_{\theta_{13}} \\ s_{\theta_{12}} s_{\theta_{23}} - c_{\theta_{12}} c_{\theta_{23}} s_{\theta_{13}} e^{-i\delta} & -s_{\theta_{23}} c_{\theta_{12}} - c_{\theta_{23}} s_{\theta_{12}} s_{\theta_{13}} e^{-i\delta} & c_{\theta_{23}} c_{\theta_{13}} \end{pmatrix}$$

These three angles and the one phase are fundamental parameters in the Standard Model. The study of B mesons allow to determine the angle θ_{23} by a measurement of $|V_{cb}|$ in the decay $\overline{B}^0 \to D^{*+} \ell^- \overline{\nu}$, the angle θ_{13} by a measurement of $|V_{ub}|$ in the decay $\overline{B} \to X_u \ell^- \overline{\nu}$, and the phase δ by a measurement of $|V_{td}|$ in $B^0\overline{B}^0$ mixing. These three measurements together with the well known Cabibbo angle θ_{12} gives us the possibility to test the standard model since they are related by the unitarity of the CKM matrix. A further test of the Standard model can be provided by a measurement of of $|V_{ts}|$ to which $B_s\overline{B}_s$ mixing and penguin decays like $B \to K^*\gamma$ are sensitive.

The study of B hadrons is carried out at the low energy e^+e^- storage rings DORIS

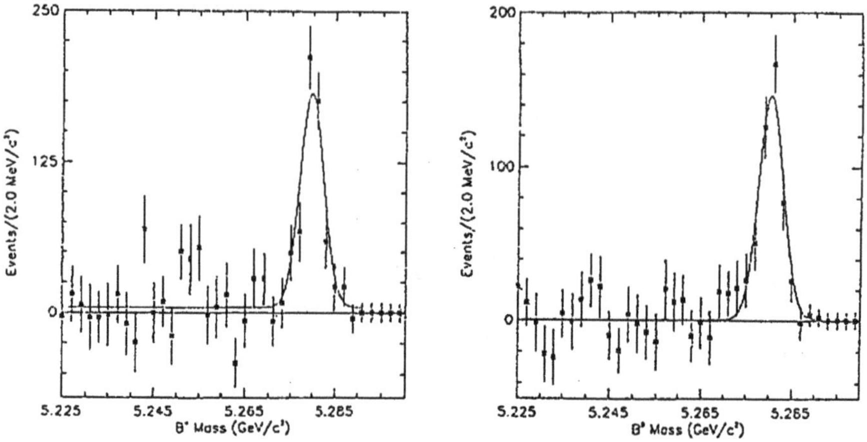

Figure 2: Mass distribution of B candidates with background subtracted

II at DESY and CESR at Cornell where B mesons are decay products of the $\Upsilon(4S)$: $e^+e^- \to \Upsilon(4S) \to B^0\overline{B}^0$ or B^+B^-. The experiments ARGUS and CLEO have collected about 400 000 respectively 3 000 000 B mesons by now. The measurements at the $\Upsilon(4S)$ profit from the virtue that only two B mesons are produced in the decay of the $\Upsilon(4S)$ which are nearly at rest. The study of these B mesons has resulted in the measurements of the masses of B^0 and B^+ [1] mesons as well as their branching ratios. Also, the measurements of $|V_{cb}|$, $|V_{ub}|$,$|V_{td}|$ and $|V_{ts}|$ have been done by these experiments. The study of B hadrons which result from the decay of the Z^0 have been carried out by the four LEP experiments, where each experiment has collected about 300 000 B hadrons. The LEP measurements have resulted in better determination of the B lifetimes and the first evidences for B_s mesons and Λ_b baryons. They have measured the rate for $B\overline{B}$ mixing which is a linear combination of $B^0\overline{B}^0$ and $B_s\overline{B}_s$ mixing. The experiments at hadron colliders have also performed such measurements and have reconstructed some B mesons.

2. Properties of B mesons

B mesons have been reconstructed successfully for the first time in 1986 [2]. The new CLEOII data [3] are displayed in Figure 2.

The masses of charged and neutral B mesons are equal within $0.5 MeV/c^2$ [3] and have a value of $M_{B^0} \approx M_{B^+} \approx M_B = (5279 \pm 2) MeV/c^2$. This implies that B mesons from $\Upsilon(4S)$ decays have an energy equal to half the $\Upsilon(4S)$ mass and that the $\Upsilon(4S)$

[1] Unless otherwise stated references in this paper to a specific charged state are to be interpreted as implying the charge-conjugate state as well.

Mode	Branching Ratio (%)	Mode	Branching Ratio (%)
$\overline{B}^0 \to D^+\pi^-$	0.29 ± 0.05	$B^- \to D^0\pi^-$	0.37 ± 0.06
$\overline{B}^0 \to D^+\rho^-$	0.73 ± 0.21	$B^- \to D^0\rho^-$	1.08 ± 0.27
$\overline{B}^0 \to D^0\rho^0$	< 0.06		
$\overline{B}^0 \to D^+\pi^-\pi^-\pi^+$	0.6 ± 0.2	$B^- \to D^0\pi^-\pi^-\pi^+$	1.1 ± 0.4
$\overline{B}^0 \to D^+D_s^-$	0.8 ± 0.4	$B^- \to D^0D_s^-$	2.0 ± 0.8
$\overline{B}^0 \to D^+D_s^{*-}$	2.4 ± 1.7	$B^- \to D^0D_s^{*-}$	1.4 ± 1.6
$\overline{B}^0 \to D^{*+}\pi^-$	0.29 ± 0.07	$B^- \to D^{*0}\pi^-$	0.41 ± 0.10
$\overline{B}^0 \to D^{*+}\pi^-\pi^0$	1.8 ± 0.6	$B^- \to D^{*+}\pi^-\pi^-$	0.33 ± 0.07
$\overline{B}^0 \to D^{*+}\rho^-$	0.74 ± 0.18	$B^- \to D^{*0}\rho^-$	1.1 ± 0.4
$\overline{B}^0 \to D^{*+}\pi^-\pi^-\pi^+$	1.2 ± 0.5	$B^- \to D^{*+}\pi^-\pi^-\pi^0$	1.8 ± 0.9
$\overline{B}^0 \to D^{*+}a_1^-$	1.39 ± 0.42		
$\overline{B}^0 \to D^{*+}\pi^-\pi^-\pi^+\pi^0$	3.8 ± 1.4	$B^- \to D^{*+}\pi^-\pi^-\pi^-\pi^+$	< 1.0
$\overline{B}^0 \to D^{*+}D_s^-$	1.3 ± 0.7	$B^- \to D^{*0}D_s^-$	1.3 ± 0.9
$\overline{B}^0 \to D^{*+}D_s^{*-}$	2.2 ± 1.3	$B^- \to D^{*0}D_s^{*-}$	3.1 ± 1.7
$\overline{B}^0 \to J/\psi\overline{K}^0$	0.09 ± 0.03	$B^- \to J/\psi\overline{K}^-$	0.09 ± 0.02
$\overline{B}^0 \to J/\psi\overline{K}^{*0}$	0.13 ± 0.03	$B^- \to J/\psi\overline{K}^{*-}$	0.17 ± 0.05
$\overline{B}^0 \to J/\psi K^-\pi^+$	0.10 ± 0.05	$B^- \to J/\psi K^-\pi^-\pi^+$	0.12 ± 0.07
$\overline{B}^0 \to \psi\overline{K}^0$	< 0.15	$B^- \to \psi K^-$	0.18 ± 0.09
$\overline{B}^0 \to \psi\overline{K}^{*0}$	0.14 ± 0.09	$B^- \to \psi K^{*-}$	< 0.35
$\overline{B}^0 \to \psi K^-\pi^+$	< 0.10	$B^- \to \psi K^-\pi^-\pi^+$	0.19 ± 0.12
		$B^- \to \Lambda_{c1}K^-$	0.19 ± 0.14

Table 1: Branching ratios of B mesons

decays with a branching ratio of 50% into a $B^0\overline{B}^0$ pair and a B^+B^- pair [4]. About 25% of all hadronic B decays have been observed [5][6] (Table 1).

The average B lifetime has been measured at LEP with a 5% precision. The b hadron sample in Z^0 decays is selected by choosing events with a hard ($p > 3$ GeV/c) and high p_t ($p_t > 1$ GeV/c) lepton. The measured impact parameter of the lepton with respect to the jet axis in the event is a measure of the b hadron lifetime [7] :

$$\tau_b = (1.40 \pm 0.05)ps.$$

Attempts to determine lifetimes of individual b hadron lifetimes use $X_c\ell^-$ correlations (Table 2). The resulting lifetimes do not differ within the errors. This implies that non-spectator effects do not contribute significantly. . The $D_s\ell^-$ and the $\Lambda_c\ell^-$ correlations [10],[9] show evidence for B_s and Λ_b production in the fragmentation

X_c	B	τ [ps]
D^0	B^-	1.34 ± 0.21
D^{*+}	B^0	1.46 ± 0.19
D_s B_s	1.05 ± 0.32	
Λ_c	Λ_b	0.98 ± 0.23
	$\tau_{<b>}$ (LEP)	1.40 ± 0.045

Table 2: Exclusive and inclusive b hadron lifetimes from LEP

Decay	Branching Ratio [%] all	Branching Ratio [%] semilept.
B $\rightarrow$ D^0 X	$60.0 \pm 4 \pm 4$	$73 \pm 6 \pm 6$
B $\rightarrow$ D^+ X	$29 \pm 4 \pm 4$	$29 \pm 5 \pm 4$
B $\rightarrow$ 'Λ_c'' X	$6.7 \pm 0.7 \pm 1.2$	< 0.4
2 * B $\rightarrow$ J/ψ X	1.7 ± 0.3	
Σ	$97.4 \pm 6 \pm 6$	$101 \pm 7 \pm 4$
B $\rightarrow$ D_s X	$9 \pm 2 \pm 4$	< 1.2

Table 3: Inclusive Branching Ratios of B decays into charmed particles

of b quarks. The first reconstruction of B_s mesons comes from ALEPH [11]. The measured ratio of the lifetimes of charged and neutral B mesons is in agreement with an earlier ARGUS/CLEO measurement of [8]

$$\tau_{B^-}/\tau_{B^0} = 0.96 \pm 0.14$$

Inclusive properties of a 1:1 mixture B^0 and B^- mesons from $\Upsilon(4S)$ decays (Table 3) show the dominance of $b \rightarrow c$ transitions in B decays and in particular also in B semileptonic decays.

3. Measurement of the CKM matrix elements

3.1 Measurement of $|V_{cb}|$

The determination of $|V_{cb}|$, and $|V_{ub}|$ is possible by a measurement of the respective β decay spectra in B decays. Figure 3 displays the expected β decay spectrum of B mesons from $\Upsilon(4S)$ decays. .

The distinction between $b \rightarrow c$ and $b \rightarrow u$ decays arise from the fact that the lightest charm mesons are much heavier than the $\pi, \rho, \omega, \ldots$ mesons which are produced in $b \rightarrow u$ decays. This results in a much harder lepton spectrum for $b \rightarrow u$ β decays. Since these decays are suppressed, as indicated in table 3, an inclusive measurement of $b \rightarrow u$ decays is limited to the momentum range above 2.3 GeV/c.

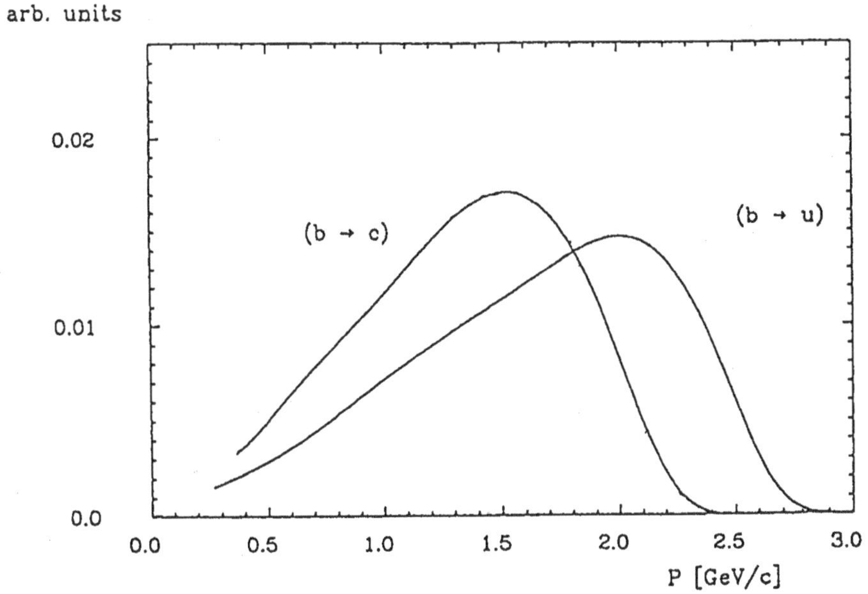

Figure 3: Expected Electron spectra in β decays of B mesons

Analogous to μ decays the total rate is given by

$$\Gamma_{sl} = \frac{BR(b \to (c,u)\ell^-\overline{\nu})}{\tau_b} = \frac{G_F^2 m_b^5}{192\pi^3}(K_u |V_{ub}|^2 + K_c |V_{cb}|^2) \qquad (1)$$

where proportionality factors K_u and K_c are model and quark mass dependent. The measured lepton spectrum (Figure 4) shows again the dominance of the $b \to c$ transitions. Integrating the spectrum a branching ratio $BR(b \to (c,u)\ell^-\overline{\nu}) = (10.7 \pm 0.5)\%$ is obtained. Together with the b lifetime $\tau_b = (1.40 \pm 0.045)ps$ this results in

$$|V_{cb}| = 0.043 \pm 0.008$$

The error on $|V_{cb}|$ is dominated by the uncertainty in the prediction of K_c where we use an average between the ACCMM [12] and ISGW [13] model.

A model independent determination of $|V_{cb}|$ is possible by studying the decay $\overline{B}^0 \to D^{*+}\ell^-\overline{\nu}$ [15]. Over the last few years great progress has been made in our understanding of exclusive semileptonic decays, both in the context of Heavy Quark Effective Theory (HQET) [16, 17] and using QCD sum rules [18]. For the HQET approach it has been shown that the decay $\overline{B}^0 \to D^{*+}\ell^-\overline{\nu}$ is not affected by $O(1/m_Q)$ corrections at the point of zero recoil, i.e. at the maximum momentum transfer $q^2 = q^2_{max}$ [19][20]. The q^2 spectrum for the decay $\overline{B}^0 \to D^{*+}\ell^-\overline{\nu}$ is hard due to

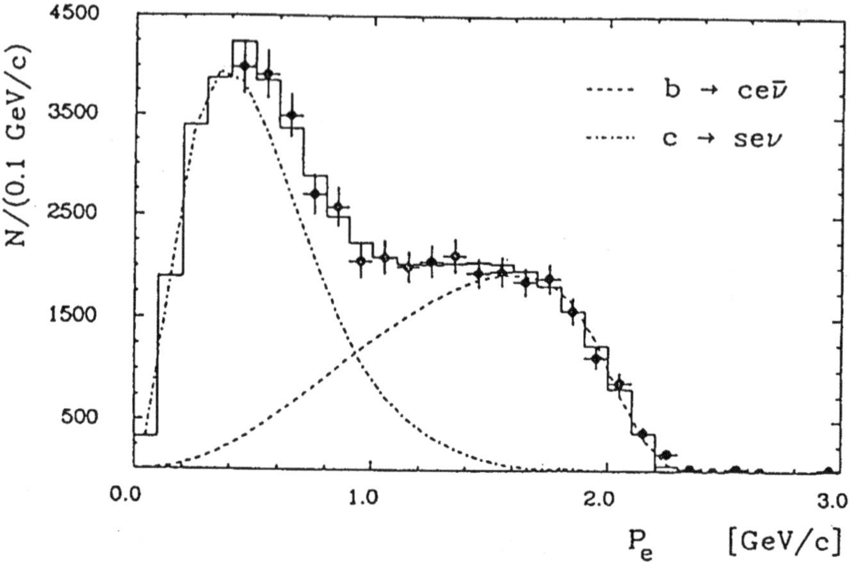

Figure 4: Measured Electron spectra from B decays [14]

the spin $J = 1$ of the D^{*+} meson. In principle, this allows a model independent determination of $|V_{cb}|$ from the study of the q^2 spectrum.

The signal for the reaction $\overline{B}^0 \to D^{*+}\ell^-\overline{\nu}$ is obtained on the $\Upsilon(4S)$ from the distribution of the recoil-mass squared:

$$M^2_{rec} \;=\; (E_{beam} - E_{D^{*+}} - E_{\ell^-})^2 - (\vec{p}_{D^{*+}} + \vec{p}_{\ell^-})^2\,. \tag{2}$$

D^{*+} mesons are reconstructed in the decay $D^{*+} \to D^0\pi^+$.

The M^2_{rec} distribution for the decay $\overline{B}^0 \to D^{*+}\ell^-\overline{\nu}$ peaks at $M^2_{rec} = 0$ since the $\overline{B}^0$ momentum is small, and therefore M^2_{rec} is approximately the neutrino rest mass squared. The contribution from D^{**} resonances which decay into $D^{*+}\pi^-$ is shifted to positive M^2_{rec} values as expected from the non-vanishing $(\overline{\nu}\pi)$ invariant mass.

Fig. 5 shows the measured M^2_{rec} distribution [21] where one observes a prominant signal at $M^2_{rec} = 0$, as expected for the decay $\overline{B}^0 \to D^{*+}\ell^-\overline{\nu}$. The small number of events at $M^2_{rec} < -2\,\mathrm{GeV}^2/c^4$ indicates that there are only minor backgrounds from uncorrelated $D^{*+}\ell^-$ combinations and the continuum. For positive M^2_{rec} values one observes a shoulder which is attributed to the process $\overline{B} \to D^{**}\ell^-\overline{\nu}$.

The fitted number of (206 ± 20) $(D^{*+}\ell^-)$ events for the process $\overline{B}^0 \to D^{*+}\ell^-\overline{\nu}$ lead to, assuming lepton universality, a branching ratio of

$$\mathrm{Br}(\overline{B}^0 \to D^{*+}\ell^-\overline{\nu}) = (5.2 \pm 0.5 \pm 0.6)\%.$$

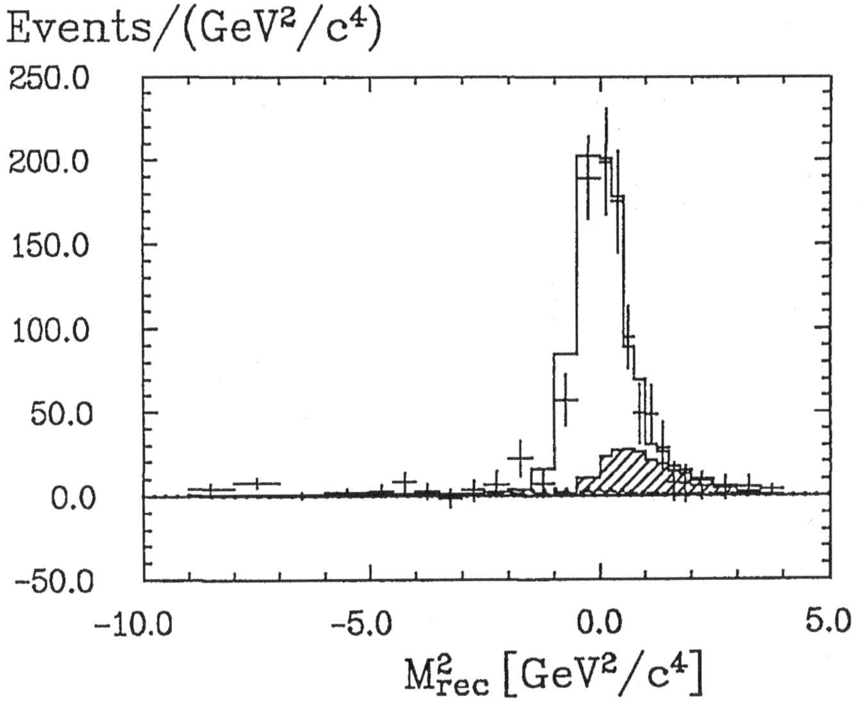

Figure 5: Measured M^2_{rec} distributions (points with error bars) The blank and shaded areas correspond to the rates for the signal process and the feeddown process , respectively. The continuum process is shown as a dashed line [21].

It has been argued in [20] that the observed model dependence in the $|V_{cb}|$ determination can be considerably reduced by a fit to the q^2 spectrum or, equivalently, to the y spectrum with

$$y = \frac{m_B^2 + m_{D^*}^2 - q^2}{2\, m_B\, m_{D^*}}.$$

The prediction from HQET is determined by one universal form factor, the Isgur-Wise function $\xi(y)$ with the normalization $\xi(1) = 1$:

$$A(\overline{B} \to D^{**}\, \ell^-\, \overline{\nu}) = C \cdot V_{cb} \cdot \xi(y),$$

where C is a known, calculable factor. In the case of finite quark masses, corrections of order $1/m_Q$ must be applied. However, it has been shown [20] that, specifically for the decay $B \to D^*\, \ell\, \nu$, these corrections vanish at the point of zero recoil, $y = 1$.

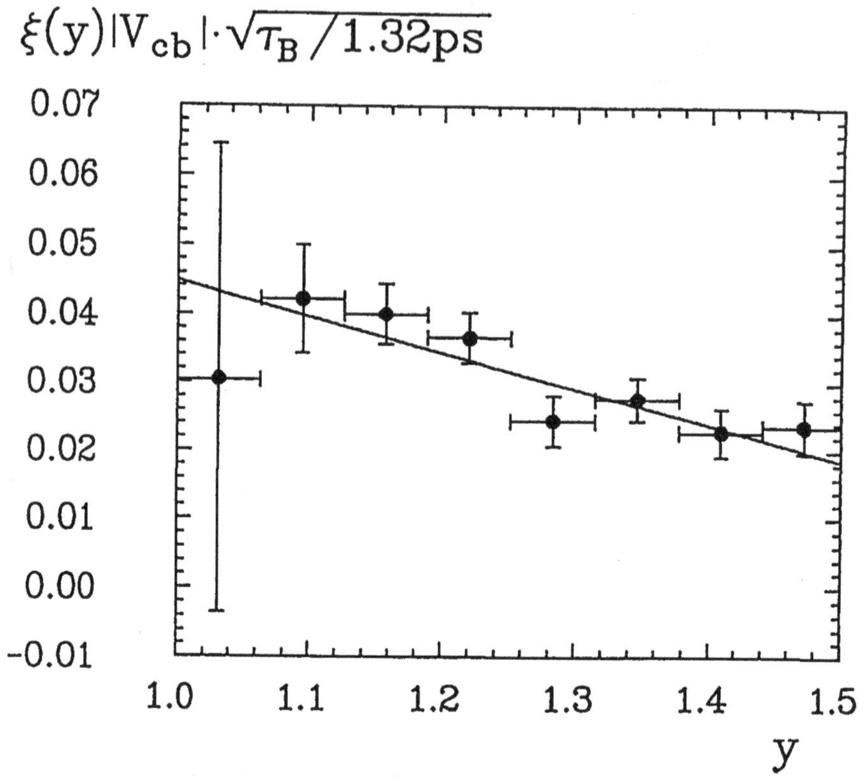

Figure 6: Measured distribution $\Delta \mathrm{Br}(\overline{B}^0 \to D^{*+}\ell^-\overline{\nu})/\Delta y$ transformed to correspond to $|V_{cb}| \cdot \xi(y)$. The dotted line corresponds to $\xi(y) = 1 - \rho^2(y-1)$ for the Isgur-Wise Function where ρ is the charge radius [21].

The measured $|V_{cb}| \cdot \xi(y)$ distribution is shown in Fig. 6. The values of $|V_{cb}|$ are determined by the intersection of the fitted functions $\xi(y)$ with the ordinate since $\xi(1) = 1$.

From the value of $|V_{cb}| \cdot \xi(y)$ at $y = 1$ one obtains for a linear dependence of the Isgur-Function a value of $|V_{cb}| = 0.044 \pm 0.006$, which is very similar to the one from the inclusive β spectrum. However, without further knowledge of this function, there is an additional theoretical uncertainty in the determination of $|V_{cb}|$ of the order ± 0.01 since the $|V_{cb}|$ measurement using HQET is essentially determined by the data points with the lowest y values, where the statistical precision is poor and the exptrapolation uncertain. At B factories, which accumulate more than $20 fb^{-1}$ at the $\Upsilon(4S)$ the Isgur-Wise function could be measured, and in particular the intercept at $y = 1$ could be determined with a 1% accuracy.

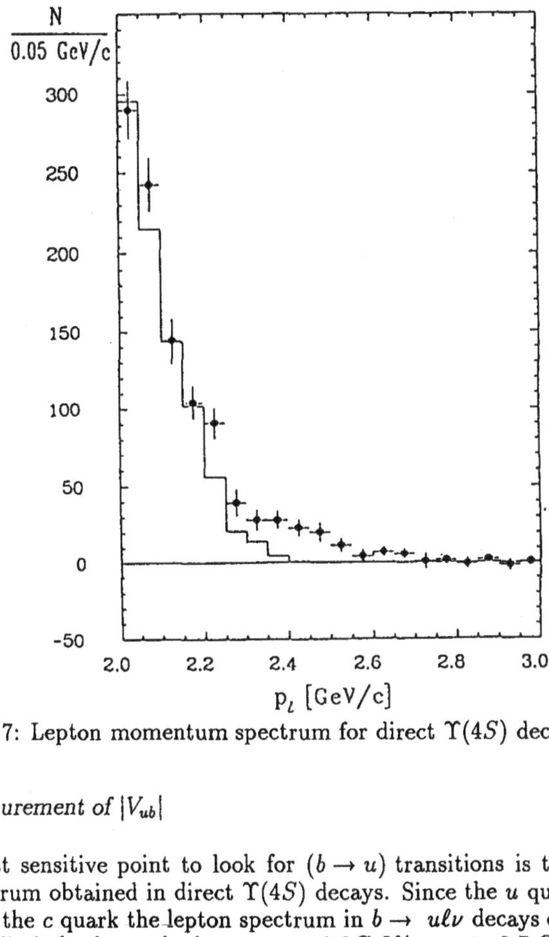

Figure 7: Lepton momentum spectrum for direct $\Upsilon(4S)$ decays (ARGUS)

3.2 Measurement of $|V_{ub}|$

The most sensitive point to look for $(b \to u)$ transitions is the endpoint of the lepton spectrum obtained in direct $\Upsilon(4S)$ decays. Since the u quark is lighter much lighter than the c quark the lepton spectrum in $b \to \; u\ell\nu$ decays extends beyond the kinematical limit for $b \to \; c\ell\nu$ decays, $p_l \approx 2.3 GeV/c$, up to 2.7 GeV/c (Figure 4).

The lepton spectrum in this region, as measured on the $\Upsilon(4S)$ resonance , is dominated by contributions from the continuum below the $\Upsilon(4S)$ resonance (see Figure 4). The strategy to observe $(b \to u)$ transitions in the lepton spectrum consist therefore in suppressing the contribution in this region with respect to the direct $\Upsilon(4S)$ contribution by exploiting the different event toplogies: events from the continuum are jet-like whereas the events from direct $\Upsilon(4S)$ decays are more spherical. Both experiments, ARGUS [22] and CLEO [23], [24] show a significant excess of leptons from charmless B decays in the momentum range above $p_\ell > 2.3 GeV/c$ (Figure 7):

In order to show that this excess of high momentum leptons ($2.3 < p_{\ell^-} < 2.6$

		$\lvert V_{ub}\rvert / \lvert V_{cb}\rvert$		
Model	Reference	CLEO'90	ARGUS'90	CLEO'92
ACM	[12]	0.09 ± 0.01	0.11 ± 0.01	0.062 ± 0.018
WBS	[25]	0.11 ± 0.02	0.13 ± 0.02	0.065 ± 0.019
KS	[26]	0.10 ± 0.01	0.10 ± 0.01	0.053 ± 0.015
ISGW	[27]	0.15 ± 0.03	0.19 ± 0.03	0.095 ± 0.027

Table 4: $\lvert V_{ub}\rvert / \lvert V_{cb}\rvert$ from different models. The errors are not gaussian

GeV/c) observed in the $\Upsilon(4S)$ data do indeed originate from $b \to u\ell^-\overline{\nu}$ transitions, a systematic attempt to completely reconstruct events has been performed by ARGUS. As a result of a search for a hadronic B tag in the data, one fully reconstructed event was found, which is shown in Figure , and consists of a $\Upsilon(4S)$ decay into a pair of $\overline{B}^0$ mesons, indicating that one B^0 meson has oscillated into a $\overline{B}^0$. Thus, the event simultaneously demonstrates the existence of $b \to u$ transitions and $B^0\overline{B}^0$ mixing [28]. The hadronic $\overline{B}^0$ tag was reconstructed in the mode $\overline{B}^0 \to D^{*+}\rho^-$. The second $\overline{B}^0$ meson was seen in the channel $\overline{B}^0 \to \pi^+\mu^-\overline{\nu}$, representing the first direct observation of a $b \to u$ transition.

Having explicitly shown that $b \to u\ell^-\nu$ transitions are responsible for the excess of leptons with momenta above the kinematical limit for $b \to c\ell^-\nu$ transitions, the strength of the $b \to u$ coupling can be obtained from the inclusive spectra. The shape of the spectrum in figure 7 in this interval is compatible with the predictions for semileptonic $b \to u\ell^-\overline{\nu}$ decays [12], [25]-[27]. However, using only this restricted portion of the full momentum spectrum, it is not possible to discriminate between the available models which result in different values for the ratio $\lvert V_{ub}/V_{cb}\rvert$ (Table 4). The table 4 indicates a large spread in the value of $\lvert V_{ub}\rvert$, both from the experimental results as well as from theoretical uncertainties.

Another possibility to measure the strength of the $b \to u$ transition is to search for exclusive hadronic and semileptonic $b \to u$ transitions. So far these searches have only resulted only in upper limits which are a factor of 2-5 above the expectations: [24]

$$BR(B^0 \to \pi^+\pi^-) < 4.8 \cdot 10^{-5}$$
$$BR(B^- \to \omega\ell^-\overline{\nu}) < 3.0 \cdot 10^{-4}$$

3.3 Measurement of $\lvert V_{td}\rvert$

With more data available from ARGUS [30] and in particular CLEO [24] an update on $B^0\overline{B}^0$ and the mixing parameter r can be given, which is defined by

$$r = \frac{Prob(B^0 \to \overline{B}^0)}{Prob(B^0 \to B^0)} = \frac{x^2}{2+x^2}$$

and has been measured by ARGUS in 1987 to be $r = (21 \pm 8)\%$ [28]. The quantity x is given by $x = \Delta M/\Gamma$ where the mass difference between the CP-eigenstates, ΔM,

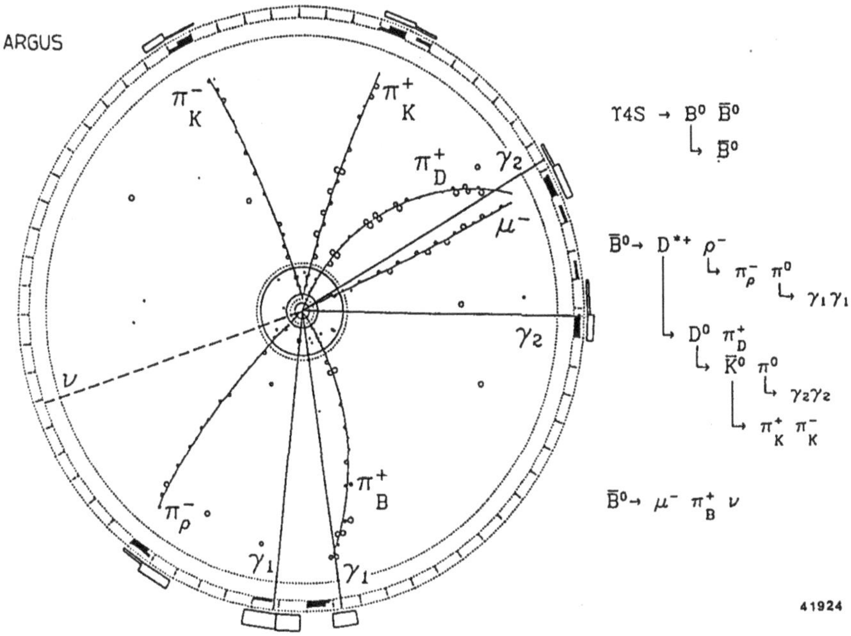

Figure 8: Completely reconstructed event containing a decay $\Upsilon(4S) \rightarrow B^0\overline{B}^0$, $B^0 \rightarrow \overline{B}^0$, where one $\overline{B}^0$ decays into $\pi^+\mu^-\overline{\nu}$ and the other into $D^{*+}\rho^-$

is obtained from the evaluation of the "box"-diagrams [29]. For x one obtains:

$$x = \frac{\Delta M}{\Gamma} = \frac{G_F^2}{6\pi^2} B_B f_B^2 m_b \tau_b \left|V_{tb}^* V_{td}\right|^2 m_t^2 F(\frac{m_t^2}{M_W^2}) \eta_{QCD}$$

The updated results on the $B^0\overline{B}^0$ measurement from $\Upsilon(4S)$ decays were obtained by tagging the b flavour of the B mesons by leptons or partial reconstruction [24]. The average of the different measurements is now

$$r = (19 \pm 3)\%$$

which is in excellent agreement with the original observation [28] so that the general conclusions do not change. For the χ_d value one obtains:

$$\chi_d = \frac{r}{1+r} = 0.162 \pm 0.019$$

and for the mixing parameter x the result is:

$$x = 0.69 \pm 0.06$$

Using $B_B^{\frac{1}{2}} f_B = (220 \pm 40) MeV$ [31] and $m_t = (145 \pm 26) GeV/c^2$ [32] together with the measured value for x, one can infer for the ratio of CKM matrix element the CKM matrix element $|V_{td}|$:

$$|V_{td}| = 0.007 \pm 0.002.$$

3.4 The CKM Matrix

The measurements of V_{cb}, V_{ub} and V_{td} from the investigations of B mesons give, together with the well know Cabibbo angle, constraints on the CKM matrix in terms of the three mixing angles and the complex phase. From a recent analysis [34] one finds that the present measurements are accomadated by the Standard Model which requires a unitary CKM matrix. In the Wolfenstein paramentrization [35] the CKM matrix is approximated by:

$$V_{CKM} \approx \begin{pmatrix} 1 - \frac{1}{2}\lambda^2 & \lambda & A\lambda^3(\rho - i\eta) \\ -\lambda^2 & 1 - \frac{1}{2}\lambda^2 - iA^2\lambda^4\eta & A\lambda^2 \\ A\lambda^3(1 - \rho - i\eta) & -A\lambda^2 & 1 \end{pmatrix}$$

where $\lambda \approx \sin\theta_{12}$. This results in the unitarity condition

$$\frac{V_{ub}^*}{\lambda V_{cb}} + \frac{V_{td}}{\lambda V_{cb}} = 1$$

which represents a triangle in the complex plane (i.e. $\rho - \eta$ plane). Figure shows an example of this triangle using the measured CKM matrix elements as well as specific input parameters on the top mass and the B decay constant [34].

3.5 Tests of the Standard Model

The measurements on B decays allow tests of the standard model. In fact, that the presented data fit to the unitarity triangle can be regarded as a first test, which, however, due to the large experimental errors and theoretical uncertainties, is not too sensitive.

Possible tests of the unitarity of the CKM matrix come from measurements of $B_s - \overline{B}_s$ mixing and $b \to s$ decays which are both sensitive to $|V_{ts}|$ which is predicted to be approximately equal in size to $|V_{cb}|$.

The mixing parameter χ measured at LEP represents an average over $B^0\overline{B}^0$ and $B_s\overline{B}_s$ mixing which are given by χ_d and χ_s: :

$$\chi = (\frac{BR_{sl}^0}{BR_{sl}})p_d\chi_d + (\frac{BR_{sl}^s}{BR_{sl}})p_s\chi_s. \qquad (3)$$

The individual semileptonic branching ratios of B^0 and B_s, BR_{sl}^0 and BR_{sl}^s, are expected to be similar as indicated by the individual lifetime measurements. Table 5 gives the LEP results on the χ measurement in Z^0 decays. The determination of

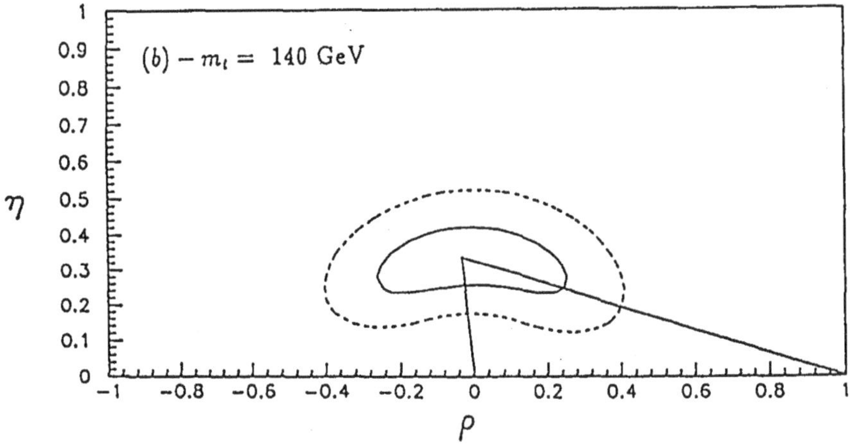

Figure 9: Allowed region for the top of the triangle in the $\rho - \eta$ plane for $m_t = 140 GeV/c^2$ and $f_{B_d}\sqrt{B_{B_d}} = 200 \pm 30MeV$. The solid line is the 1σ contour, the dashed line denotes the 90%C.L. region [34]

the value for χ leads to a straight line in the $\chi_d - \chi_s$ plane where the intercepts with the two axes depend on the choice of the parameters p_d and p_s and the ratio of the semileptonic branching ratios. To compare the different results on $B\overline{B}$ mixing we will use:

$$\chi = 0.375 \cdot \chi_d + 0.15 \cdot \chi_s \tag{4}$$

Comparing the results from the Z^0 and from the continuum with the ARGUS and CLEO result one finds excellent agreement if χ_s is large (Figure 10). With the chosen values for p_d and p_s the measurements imply a value of $\chi_s = 0.47 \pm 0.11$ which gives no sensitivity on $|V_{ts}|$. However, the values of χ_d and χ_s are related from

Table 5: Mixing parameter χ from different experiments [33]

Experiment	χ
DELPHI	$0.121 \pm 0.042 \pm 0.007$
L3	$0.121 \pm 0.017 \pm 0.006$
ALEPH	$0.137 \pm 0.015 \pm 0.007$
OPAL	$0.125 \pm 0.017 \pm 0.015$
$< LEP >$	0.131 ± 0.010

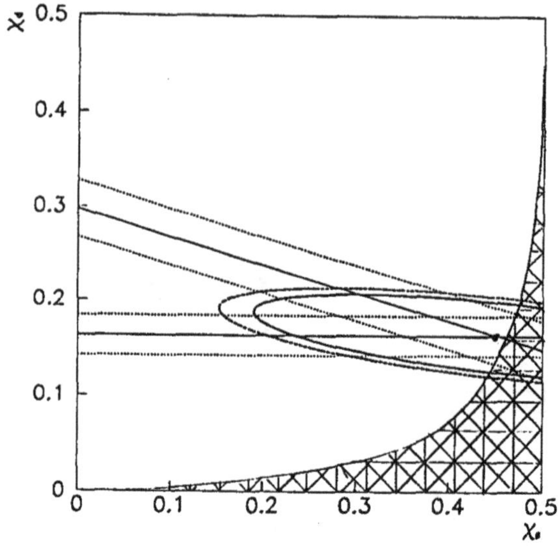

Figure 10: Combined results on χ_d and χ_s.

the calculations of the respective box diagrams:

$$\frac{x_d}{x_s} = \frac{\sqrt{\frac{2\chi_d}{1-2\chi_d}}}{\sqrt{\frac{2\chi_s}{1-2\chi_s}}} \approx \frac{|V_{td}|^2}{|V_{ts}|^2} \tag{5}$$

assuming equal B decay constants. From the unitarity of the KM-matrix [1],an upper limit for $|V_{td}|^2 / |V_{ts}|^2 < 0.21$ is derived (hatched region in Figure Figure 10). The measurement of χ from the high energy experiments coincide with the χ_d measurements from ARGUS and CLEO in the allowed region.

Another test of the Standard model is the measurement of $b \to s$ transistions. The inclusive branching ratio for the $b \to s\gamma$ transition is given by [36]

$$BR(b \to s\gamma) \approx 2 - 5 \cdot 10^{-4} (\frac{V_{ts}}{V_{cb}})^2$$

For this inclusive rate, a search for high energy photons coming from the $b \to s\gamma$ process has yielded a 95% upper limit of [40]:

$$BR(b \to s\gamma) < 5.4 \cdot 10^{-4}$$

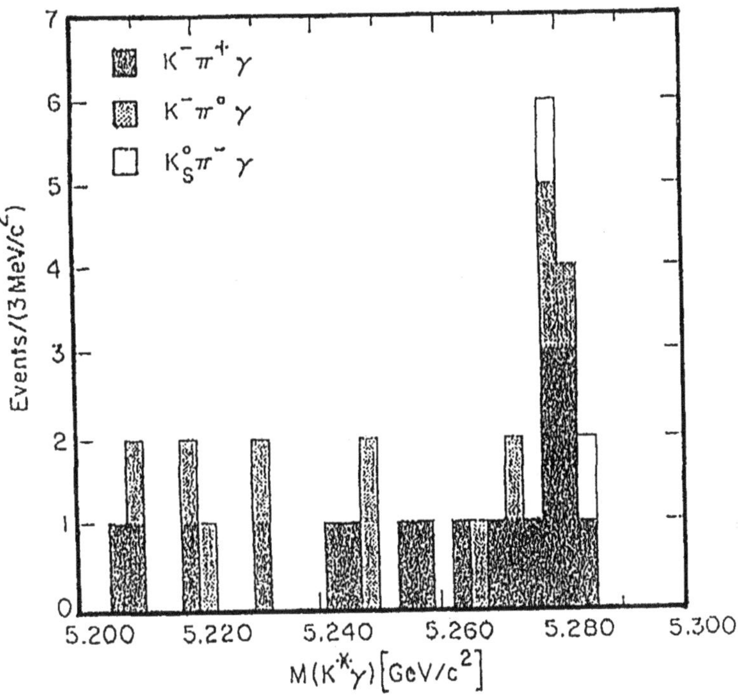

Figure 11: Mass distribution of $K^*\gamma$ candidates from CLEO [40]

The first direct evidence for the penguin type $b \rightarrow s$ decays is obtained from the CLEO observation [40] of the exclusive decays BR($B \rightarrow K^*\gamma$)(Figure 11). With 8 candidates for the decay $B^0 \rightarrow K^{*0}\gamma$ on a background of 1.1 events and 5 candidates for the decay $B^- \rightarrow K^{*-}\gamma$ on a background of 0.1 events, CLEO deduces

$$BR(B \rightarrow K^*\gamma) = (4.5 \pm 1.5 \pm 0.9) \cdot 10^{-5}$$

Theoretical predictions on how much this specific decay contributes to the total inclusive decay rate are rather uncertain and range from about 5% to 40% [37] -[39]. With a value of

$$\frac{BR(B \rightarrow K^*\gamma)}{BR(b \rightarrow s\gamma)} = 30\%$$

one obtains a lower limit of

$$BR(b \rightarrow s\gamma) > 0.6 \cdot 10^{-4} (95\% C.L.)$$

With these bounds one obtains [41]:

$$0.5 < \frac{|V_{ts}|}{|V_{cb}|} < 1.5$$

perfectly consistent with the Standard Model.

4. Conclusions

The results on B decays allow to determine the Standard Model parameters and to test ist. First tests are succesful, however the constraints are also not too stringent. Here we are waiting for high luminosity B factories.

References

[1] Particle Data Group, Phys.Rev. **45D** 1 (1992)

[2] H.Albrecht et al (ARGUS collaboration), Phys.Lett. **185B** (1987)218

[3] D.S.Akerib et al (CLEO collaboration), Contributed paper to the XXVI International Conference on High Energy Physics, Dallas,Texas, August 6-12,1992

[4] G.P.Lepage, Phys.Rev. **D42** (1990) 3251

[5] H.Albrecht et al (ARGUS collaboration), Z.Phys. **C48** (1990) 543.

[6] D.Bortoletto et al (CLEO collaboration), Phys.Rev. **D45** (1992) 21

[7] M.Pohl, Proc. of the XXVI International Conference on High Energy Physics, Dallas,Texas, August 6-12,1992

[8] S.Stone, in S. Stone (editor) "B decays", World Scientific Publishing Co. Pte. Ltd. (Singapore, New Jersey, London, Hong Kong) 1992, p.210.

[9] D.Buskulic et al (ALEPH collaboration), CERN-PPE 92-138, August 1992, Phys.Lett. **294B** (1992) 145

[10] P.Abreu et al (DELPHI collaboration), Phys.Lett. **289B** (1992) 199

[11] D.Buskulic et al (ALEPH collaboration), CERN-PPE 93-42, March 1993,

[12] G.Altarelli et al, Nucl. Phys. **B208** (1982) 365

[13] N.Isgur et al, Phys.Rev. **D39** (1989) 799

[14] H.Albrecht et al (ARGUS collaboration), Phys.Lett. **249B** 359 (1990)

[15] H.Albrecht et al (ARGUS collaboration), Phys.Lett. **197B** 452 (1987)

[16] N. Isgur, Phys. Rev. D43 (1991) 810

[17] S. Balk, F. Hussain, J. G. Körner, G. Thompson, MZ-TH-92-22 (1992)

[18] E.Bagan et al, Phys.Lett. **301B** (1993) 249

[19] M. A. Shifman, M. B. Voloshin, Sov. J. Nucl. Phys. 45 (1987) 292 and 47 (1988) 511

[20] M. Neubert, Phys. Lett. B264 (1991) 455, and SLAC-PUB-5826 (1992)

[21] H.Albrecht et al (ARGUS collaboration), DESY Preprint DESY 92-146, October 1992

[22] H.Albrecht et al (ARGUS collaboration), Phys.Lett. **234B** 409 (1990), Phys.Lett. **255B** 197 (1991)

[23] R.Fulton et al (CLEO collaboration), Phys.Rev.Lett. **64** 16 (1990),

[24] J.Bartelt et al. (CLEO collaboration), CLNS-93-1207, Apr. 1993

[25] M.Wirbel et al., Z.Phys. **C29** (1985) 637.

[26] B.Grinstein et al, Phys.Rev.Lett. **55** (1986) 298,
N.Isgur et al, Phys.Rev. **D39** (1989) 799.

[27] J.G.Körner and G.A.Schuler, Z.Phys. **C38** (1988) 511.

[28] H.Albrecht et al (ARGUS collaboration), Phys.Lett. **192B** 245 (1987)

[29] J.S.Hagelin, Phys.Rev. **D20** 2893 (1979) Nucl.Phys **B193** (1981) 123.

[30] H.Albrecht et al (ARGUS collaboration), Z. Phys. **C55** (1992) 357.

[31] C.T.Sachrajda, in S. Stone (editor) "B decays", World Scientific Publishing Co. Pte. Ltd. (Singapore, New Jersey, London, Hong Kong) 1992, p.444

[32] L.Rolandi, Proc. of the XXVI Int. Conf. on High Energy Physics, Dallas 1992.

[33] P.Drell, Proc. of the XXVI Int. Conf. on High Energy Physics, Dallas 1992.

[34] A.Ali and D.London, DESY Preprint DESY 93-022, February 1993

[35] L.Wolfenstein, Phys.Rev.Lett. **51** 1945 (1983)

[36] A.Ali et al, DESY preprint DESY-93-016, Feb. 1993

[37] N.G.Deshpande et al, Phys.Rev.Lett. **59** 183 (1987).

[38] T.Altomari, Phys.Rev. **D37** 677 (1988).

[39] C.A.Dominguez et al, Phys.Lett. **214B** 459 (1988).

[40] R.Ammar et al (CLEO collaboration), CLNS-93-1212, May 1993

[41] A.Ali, these proceedings .

The Fermilab K Decay Experiments

BRUCE WINSTEIN
The Enrico Fermi Institute and the Department of Physics
The University of Chicago, 5620 South Ellis Avenue
Chicago, Illinois 60637

ABSTRACT

This brief lecture will present some of the activities of the Fermilab K decay experiments. This includes the recently completed Fermilab experiment E731 which concentrated on CP violation studies with the 2π decays and its two follow-on experiments: E773, concentrating on CPT tests, and E799, focused upon rare decays.

CP Violation

The violation of this fundamental symmetry was discovered[1] in 1964. It is most likely a key ingredient in the evolution of the Universe which permits a large matter-antimatter imbalance. It has been the subject of intense study over most of the intervening years. Yet as of today, we still know relatively little about the phenomenon. An interesting conjecture,[2] made by Professor Salam, is that CP symmetry may be restored in very high magnetic fields.

The key parameter measuring the size of the effect is ε. This is a complex number whose imaginary part is convention dependent and whose real part is well measured to be 1.6×10^{-3}. It governs the CP "impurity" in the weak eigenstate K_L.

The parameter ε characterizes what is known as "mass mixing" or the imbalance in the particle-to-antiparticle transition rates. The real part of the K_L-K_S transition gives the mass difference:

$$\Delta m = (0.5286 \pm 0.0028) \times 10^{10}$$

while ε is essentially the much smaller imaginary part. The value for Δm is of the order of 10^{-6} ev and this is why the neutral kaon system has the sensitivity it does to CP violation.

The phenomenon will occur for the following reason: there must be a relative phase between those (virtual) transitions which give the mass difference and those (real) transitions which give kaon decays.

In "superweak" models,[3] all real kaon decay amplitudes are relatively real and the phase enters due to a first order K_L-K_S transition at a very high mass scale. In such a model, essentially the only phenomenon of CP violation would be the asymmetric mass mixing.

A more satisfying approach comes from the realization[4] that CP violation can be incorporated in the 3-generation Standard Model. All three generations are necessary and, if true, then the first experiment that was sensitive to the top quark was the original 1964 discovery of CP Violation.

In this model, asymmetric mass mixing (ε) arises from a second order process with heavy quarks in the intermediate state. The CKM matrix is now imaginary and it can be written[5] so that only the extreme off-diagonal elements are non-real, with only one phase (δ) entering, that of the $b \to u$ transition.

Since we introduce one phase, δ, to explain one complex parameter, ε, this is clearly far from satisfying. As appealing as this model is, and as widely believed as it is, we still need more experimental verification. One manifestation alone is not enough to confirm that the Standard Model indeed correctly describes CP Violation!

Experiments on CP Violation

The Standard Model will allow CP Violation in particle decay, again through heavy intermediate quarks (Penguin Diagrams). In the 2π decays of the neutral kaon, this is parameterized by ε'. The parameter ε' can be found from a measurement of the four $K_{L,S} \to 2\pi$ decay rates:

$$\left|\frac{\eta_{+-}}{\eta_{00}}\right|^2 = \left|\frac{\varepsilon+\varepsilon'}{\varepsilon-2\varepsilon'}\right|^2 \approx \left|1+3\frac{\varepsilon'}{\varepsilon}\right|^2 \approx 1+6\,\frac{\varepsilon'}{\varepsilon}$$

where

$$\eta_{+-} = \frac{\mathrm{amp}(K_L \to \pi^+\pi^-)}{\mathrm{amp}(K_S \to \pi^+\pi^-)} ,$$

and

$$\eta_{00} = \frac{\mathrm{amp}(K_L \to \pi^0\pi^0)}{\mathrm{amp}(K_S \to \pi^0\pi^0)} .$$

There are at present two recent results from experiments attempting to see a non-zero effect among the 2π decay modes. The final results from both NA31 and E731 are now available. These experiments are very difficult and require extreme attention to systematic uncertainty. The final published results of the two are:

$$\varepsilon'/\varepsilon = (7.4 \pm 5.9) \times 10^{-4} \text{ (E731[6])}$$

and

$$\varepsilon'/\varepsilon = (23.0 \pm 6.5) \times 10^{-4} \text{ (NA31[7])}.$$

These results are not in the best agreement. While the NA31 result strongly favors an observation of a new CP violating effect, the E731 result is not far enough from zero to be able to claim an observation. The NA31 experiment ran over a period of 4 years with upgrades incorporated as the experiment advanced; the E731 experiment took all of its data in one 5 month running period.

It is difficult at present to calculate the expected value for ε'/ε. This is because of the unknown value for the top-quark mass (one of the last remaining engineering parameters of the standard model) but also because of the difficulties involved in "soft" QCD. At present, theory[8] favors a value somewhat less than 10^{-3}.

E731 has also reported[9] precise values for other parameters of the neutral kaon system determined from the 2π data samples. The results are:

$$\tau_S = (0.8929 \pm 0.0016) \times 10^{-10} \text{ sec};$$

$$\Delta m = (0.5286 \pm 0.0028) \times 10^{10} \text{ sec}^{-1};$$

$$\Phi_{+-} - \Phi_{00} = (-1.6 \pm 1.2)^0; \text{ and,}$$

$$\Phi_{+-} = (42.2 \pm 1.4)^0.$$

These "auxiliary" results from precise studies of the 2π modes are of considerable interest in their own right. The value for the K_S lifetime is more precise than the current world average; its agreement with the well established world average (4 experiments of comparable precision) is a powerful check on the understanding of the detector acceptance. The value for the mass difference is of comparable precision to the best previous experiment but the value is about two standard deviations lower than the current world average. The value for the phase difference provides the best test of CPT symmetry. And the value for Φ_{+-} is consistent with expectations from unitarity and is also lower than the current world average. However, if the previous best experiments are reanalyzed using the E731 value for Δm, all earlier results nicely fall in line with the expected value of 43.7^0! This is shown in the following Table.

Experiment	Φ+- (degrees)	Internal Error (degrees)	Δm used (10^{10}s)	$\frac{\Delta\Phi+-}{0.01\Delta m}$ (degrees)	Φ+- (E731 Δm)
Geweniger et al.	49.4	1.0	0.5400	3.05	43.0
Carithers et al.	45.5	2.8	0.5348	1.20	44.1
Carosi et al.	46.9	1.6	0.5351	3.10	43.4
E731	42.2	1.4	floated	-	42.2
Expected Value	43.7	0.2	-	-	43.7

Thus the E731 results, taken together with these other experiments, support CPT symmetry. Also, very recently the CPLEAR collaboration has given a new value[10] for Δm which is in agreement with that from E731.

In addition to these results, the same experiment has recently provided physics results on the following topics:

a) A precision measurement of the quadratic slope parameter in $K_L \rightarrow 3\pi^0$ decays. This result[11] represents the first high statistics study of the $3\pi^0$ decay mode. By the high degree of symmetry in the decay, the linear slope parameter of the Dalitz plot is identically zero, affording a good opportunity to study higher order effects. This is the first observation of the effect. The very clean reconstruction of the $3\pi^0$ decay has led to the ability to effectively tag π^0s for rare decay studies as mentioned below.

b) A limit of the branching ratio[12] for the decay $K_L \rightarrow \pi^0\nu\nu$. This mode is obviously extremely difficult to observe. The attraction is that the decay is essentially pure CP violating and, in the standard model, would directly measure the Wolfenstein parameter η which is the height of the Unitarity triangle: its branching ratio is directly proportional to η^2 with a well-known proportionality constant. A long-term goal of the Fermilab group is to pursue this decay mode.

c) The study[13] of the decay $K_L \rightarrow \pi^0\pi e\nu$ (Ke4) . For the first time, a study of five of the form factors of this decay is made. This decay determines with high accuracy one of the important parameters in Chiral Perturbation Theory.

d) High statistics study[14] of both $K_{L,S} \rightarrow \pi\pi\gamma$. The branching ratios for both of these modes were determined with the best available accuracy. In addition, a clean separation between inner bremsstrahlung and direct emission in the K_L decay was made with conclusions concerning the relevant form-factor.

e) Observation[15] of interference in the $\pi\pi\gamma$ time distribution downstream of a regenerator. This constitutes an observation of CP violation in a third channel (adding to those in $\pi\pi$ and $\pi l\nu$). The extracted parameters agree with those of the other two channels, limiting possible new sources of CP violation.

Experiment E773

A set of data was collected over a two month period in late 1991 concentrating on further precise studies of CPT symmetry in the neutral kaon system. The data is in hand to perform a measurement of both Φ_{+-} and of $\Delta\Phi$ to better than in E731.

The spokesman for this experiment is George Gollin. Here two regenerators were utilized simultaneously to optimally study the interference terms. One of these was totally active (consisting of pure plastic scintillator). This enabled a significant reduction in inelastic processes and this technique will be employed in future efforts.

The analysis is in an advanced state and preliminary results are just now being studied. The analysis here is considerably more subtle than for the rare decay modes.

Experiment E799

In 1989, E799, a high sensitivity search for rare kaon decays, in particular, the $K_L \rightarrow \pi^0 l^+ l^-$ (where l = e, μ) modes, was proposed by the E731 group with Y. Wah and T. Yamanaka as cospokesmen. These K_L decay modes, although quite rare, have both direct and indirect CP violating contributions, such that "ε'/ε" is of order of unity! The interesting direct piece is a short-distance effect which can be reliably calculated. The indirect CP amplitude (the ε piece) and the CP conserving amplitude from a $\pi^0\gamma\gamma$ intermediate state need to be measured or calculated reliably so that the direct CP effect, if observed, could be correctly extracted.

Brookhaven and CERN experiments have determined the M_{ee} spectrum[16] in $K^+ \rightarrow \pi^+ e^+ e^-$ and the $M_{\gamma\gamma}$ spectrum[17] in $K_L \rightarrow \pi^0\gamma\gamma$. It is encouraging that these, together with Chiral Perturbation Theory, support the notion that both the indirect and CP conserving pieces are quite small. Thus the $K_L \rightarrow \pi^0 l^+ l^-$ transition is an exciting opportunity to look into direct CP violation, one that is complimentary to the studies in the 2π decays. Signals at the level of 10^{-11} are expected; it is also true that, at that level, there will be a background[18] from radiative Dalitz decays, $K_L \rightarrow l^+ l^- \gamma\gamma$. Ultimately, these modes will likely be more useful in extracting the parameters of the CKM matrix, particularly the parameter η in the unitarity triangle. This is because of the difficulty on hadronic matrix elements in the 2π decay. These kaon decays in fact are better in this regard than are the decays of the B meson.

Because of the high energy of the kaon beam and thus very high acceptances for multi-body final states, E799 has good sensitivity to many rare decay modes. An eight week data test run (Phase-I) was completed in January 1992. Although the detector is similar to that of E731, there were a few changes and upgrades, the most important of which was the implementation of a new trigger processor. An average trigger rate reduction of about 14 was achieved with the processor.

We list here some of the recent results from the on-going analysis.

1. A (90% confidence) limit[19] on the $K_L \rightarrow \pi^0 e^+ e^-$ decay mode of 4.3×10^{-9}. The result is a small improvement over our earlier one (obtained from E731 data) and over a result[20] from a BNL-Yale experiment. One is still far from the level predicted from the Standard Model. The acceptance as a function of ee invariant mass is considerably flatter than in other searches.
2. A limit[21] on the $K_L \rightarrow \pi^0 \mu^+ \mu^-$ decay of 5.1×10^{-9}. Although again far from the expected level, this represents an improvement of over a factor of 200 compared to the best previous result. This mode is particularly background free.
3. For the $K_L \rightarrow \pi^0 \nu\nu$ mode, an improvement of about a factor of three over the previously published one has been obtained and this will soon be submitted for publication.
4. The $\pi^0 \rightarrow e^+e^-$ transition[22] has been cleanly observed: the branching ratio for this process is observed to be 7.6×10^{-8}, based upon a signal of 9 events over a background of about 1 event. It is worthwhile to mention here that a new technique using "tagged" π^0 from the copious $K_L \rightarrow 3\pi^0$ decay was used to measure this branching ratio which has a contentious history; our result is now consistent with predictions which are dominated by unitarity through the $\gamma\gamma$ channel. This is a most promising technique for rare π^0 decay physics (e.g. $\pi^0 \rightarrow \mu^+e^-$ which is mentioned below). There is also a recent result[23] on this decay mode from Brookhaven. Professor Salam has written[24] on the subject of this decay.
5. The decay $K_L \rightarrow \mu^+ \mu^- \gamma$ has been cleanly observed with a signal of a few hundred events. The branching ratio is $(3.9 \pm 0.3) \times 10^{-7}$. This is essentially a first observation although one good candidate event was observed by an earlier experiment. This mode is very important in the extraction of the short distance contribution to the $K_L \rightarrow \mu^+ \mu^-$ decay. This result should soon be published.

6. A clean signal of 27 decays of the $K_L \rightarrow e^+e^-e^+e^-$ decay has been observed[25]; the previous best experiment[26] saw 6 events. The branching ratio is $(3.6 \pm 0.7) \times 10^{-8}$; while ultimately one wants to study the form factor here, a measurement of the branching ratio is the first step.

7. A limit[27] of 8.1×10^{-9} has been obtained for the $\pi^0 \rightarrow \mu^+e^-$ decay; the mode is background free and the result is a factor of two better than previously published.

8. The first observation of the $K_L \rightarrow \pi^+\pi^-e^+e^-$ decay mode with a signal of about 20 events and no background has been made. There is now some interest[28] in using this decay for the study of CP violation.

The Future

The KTeV collaboration (Elmhurst, Fermilab, Osaka, Rice, Rutgers, UCLA, UCSD, Chicago, Colorado) at Fermilab has as its goal the performance of the same measurements that have been done over the last several years with a fully optimized detector and beam, making major inroads upon all of the problems encountered with the last detector. A much cleaner and more intense neutral beam will be also be built.

The goal is to measure ε'/ε to a precision of 10^{-4} (E832) and study a large variety of rare decay modes of the neutral kaon at the 10^{-11} level (E799-II).

The detector consists in a large pure CsI electromagnetic detector with "digital photomultipliers" (a 8-bit multi-ranging flash ADC inside each PM base); a drift chamber spectrometer; an extensive veto system for rejecting the $K_L \rightarrow 3\pi^0$ background under the $K_L \rightarrow 2\pi^0$ signal; a TRD system for good π/e separation; and an elaborate trigger and data acquisition system to name some of the major components. For E832, the same technique (2 simultaneous beams with a regenerator in one) is employed. For E799, both beams will be vacuum. The new detector hall is scheduled to be completed in November of 1994 and the date for first beam is July, 1995.

The KTeV electromagnetic calorimeter is the heart of the new detector. It is a 1.9 meter square wall of pure CsI scintillating crystals. The calorimeter, shown in Figure 1, is made up of 3100 individual CsI crystals, each 50 cm or 27 radiation lengths long. The crystals are arranged with 2232 2.5 cm square crystals in the center, and 868 5.0 cm square crystals surrounding the smaller crystals. It should provide much better energy resolution than the lead glass of E731.

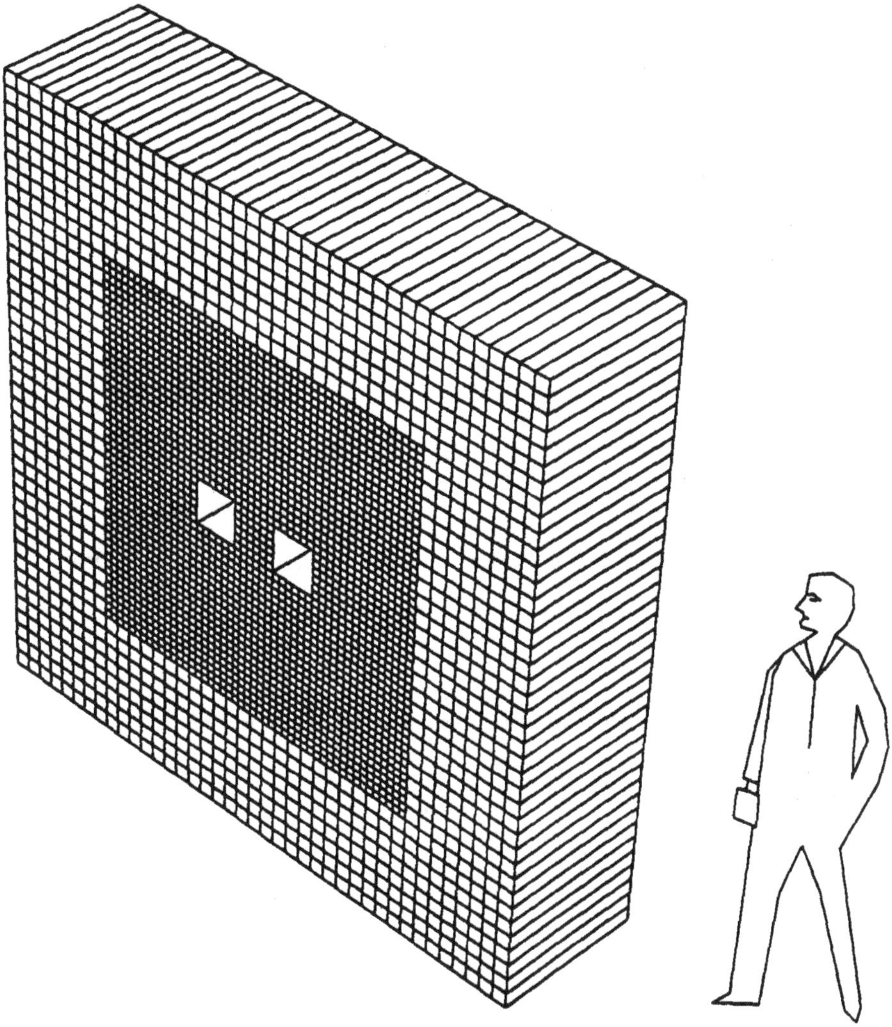

Figure 1 Schematic of the KTeV CsI calorimeter.

Conclusion

With the new effort described at Fermilab, together with the new CERN experiment NA48, the BNL program, the DAPHNE project at Frascati, a possible effort at Novosibirsk, and possibly KAON in Vancouver, it is clear that the physics of kaon decays remains a fruitful subject. We can expect many interesting results in the latter part of the 1990's and beyond.

The author would like to thank the organizers of this wonderful conference commemorating the career of Abdus Salam.

References

[1]J.H. Christenson, J.W. Cronin, V.L. Fitch, and R. Turlay, *Phys. Rev. Lett.* **13** (1964) 138.

[2]A. Salam and J. Strathdee, *Nature* **252** (1974) 569.

[3]L. Wolfenstein, *Phys. Rev. Lett.* **13** (1964) 562.

[4]M. Kobayashi and T. Maskawa, *Prog. Theor. Phys.* **49** (1973) 652.

[5]L. Wolfenstein, *Phys. Rev. Lett.* **51** (1983) 1945.

[6]L.K. Gibbons et al., Measurement of the CP-Violation Parameter Re(ε'/ε), *Phys. Rev. Lett.* **70** (1993) 1203.

[7]G.D. Barr et al., A New Measurement of Direct CP Violation in the Neutral Kaon System, CERN-PPE/93-168.

[8]M. Ciuchini *et al.*, University of Rome "La Sapienza" preprint 911 (Nov. 23, 1992).

[9]L.K. Gibbons et al., New Measurements of the Neutral Kaon Parameters Δm, τ_S, $\Phi_{00}-\phi_{+-}$, and Φ_{+-}, *Phys. Rev. Lett.* **70** (1993) 1199.

[10]T. Nakada, Plenary talk at the Lepton Photon Conference, Cornell, 1993.

[11]S.V. Somalwar et al., Measurement of the Quadratic Slope Parameter in the $K_L \rightarrow 3\pi^0$ Decay Dalitz plot, *Phys. Rev. Lett.* **68** (1992) 2567.

[12]G.E. Graham et al., Search for the Decay $K_L \rightarrow \pi^0 \nu\nu$, *Phys. Lett.* **B295** (1992) 169.

[13]G. Makoff et al., A Study of the Decay $K_L \rightarrow \pi^0 \pi e \nu$, *Phys. Rev. Lett.* **70** (1993) 1591.

[14]E. Ramberg et al., Simultaneous Measurement of the K_S and K_L Decays into $\pi^+\pi^-\gamma$, *Phys. Rev. Lett.* **70** (1993) 2525.

[15]E. Ramberg et al., Measurement of the CP-Violation Parameter $\eta_{+-\gamma}$ in Neutral Kaon Decays, *Phys. Rev. Lett.* **70** (1993) 2529.

[16]C. Alliegro et al., *Phys. Rev. Lett.* **68** (1992) 278.

[17]G. Barr, et al., *Phys. Lett.* **B284** (1992) 440.

[18]H.B. Greenlee, *Phys. Rev.* **D42** (1990) 3724.

[19]D.H. Harris et al., A Limit on the Branching Ratio of $K_L \rightarrow \pi^0 e^+ e^-$, submitted to Physical Review Letters.

[20]K.E. Ohl et al., *Phys. Rev. Lett.* **64** (1990) 2755.

[21]D.H. Harris et al., A Limit on the Branching Ratio of $K_L \rightarrow \pi^0 \mu^+ \mu^-$, submittted to Physical Review Letters.

[22]K.S. McFarland et al., Measurement of the Branching Ratio of $\pi^0 \rightarrow e^+e^-$ from π^0's Produced by $K_L \rightarrow 3\pi^0$ Decays in Flight, *Phys. Rev. Lett.* **71** (1993) 31.

[23]A. Deshpande et al., *Phys. Rev. Lett.* **71** (1993) 27.

---

[24]J. Pati and A. Salam, *Phys. Rev. D* **11** (1975) 1137.

[25]P. Gu et al., A Measurement of the Branching Ratio of $K_L \rightarrow$ eeee, submittted to Physical Review Letters.

[26]M.R. Vagins et al., *Phys. Rev. Lett.* **71** (1993) 35.

[27]P. Krolak et al., A Limit on the Lepton-Family Number Violating Process $\pi^0 \rightarrow \mu^+e^-$, submitted to Physical Review Letters.

[28]LM.Sehgal and M. Wanninger, *Phys. Rev. D* **46** (1992) 1035;
LM. Sehgal and M. Wanninger, *Phys. Rev. D* Erratum (1992).

RESULTS FROM THE FIRST YEAR AT HERA

J. FELTESSE
DAPNIA-SPP, CE-Saclay
91191 *Gif-sur-Yvette Cedex, France*

Abstract

In 1992 the two experiments H1 and ZEUS started to record data on electron proton collisions produced at HERA, the first $e\,p$ collider. Based on an integrated luminosity of about 25 nb^{-1} the very first physics results are presented. The measured total photoproduction cross section rises smoothly with increasing energy. The onset of hard collisions in photoproduction is observed. Hints of gluon components of real photons are seen. First measurements of the proton structure function $F_2(x, Q^2)$ in the new kinematic region $10^{-4} < x < 10^{-2}$ shows a substantial increase as x decreases. Jets measured in deep inelastic scattering are found to be in agreement with models combining first-order QCD matrix elements and higher-order corrections approximated by leading-log parton shower. New limits on leptoquarks, leptogluons and excited leptons are reported.

1. Introduction

On May 31st 1992, the H1 collaboration recorded the first identified deep inelastic event produced at the new electron-proton collider HERA. The squared momentum transfer of the interaction was $Q^2 = 18$ GeV^2, a value in the range of previous lepton-nucleon scattering experiments on fixed targets. But in this event the fraction of the proton momentum carried by the struck quark was $x = 0.0004$, a value hundred times smaller than had ever before been achieved. Eight years after the approval of HERA a new domain of physics was opened by the HERA experiments. H1 [1] and ZEUS [2], two international collaborations of 340 and 450 physicists respectively, have built two multi-purpose detectors which were ready for data taking from the first day of luminosity. Each experiment recorded data for an integrated luminosity of about 25 nb^{-1} during the first year of operation.
In this talk, after a comment on HERA performance in 1992, we present the first physics results on :

-Photoproduction of almost real photons

-Deep Inelastic Scattering (DIS)

-Searches beyond the Standard Model

This lecture has been written in summer 1993 and includes the last results published up to that time by the two HERA experiments.

	Running value in 1992	Design value
electron energy	26.7 GeV	30 GeV
proton energy	820 GeV	820 GeV
electron current per bunch	300 μA	300 μA
proton current per bunch	200 μA	800 μA
circulating electron current	3 mA	60 mA
circulating proton current	2 mA	160 mA
number of bunches	10	210
peak luminosity	$3 \cdot 10^{29}$ $cm^{-2}s^{-1}$	$1.5 \cdot 10^{31}$ cm^{-2} s^{-1}

Table 1: HERA parameters

2. The HERA Storage Ring

The large storage ring HERA (Hadronen-Elektronen-Ring-Anlage) was completed at DESY (Hamburg) in November 1990 and progressively commissioned in 1991, leading to the first electron-proton collisions in October 1991 when the two large experiments were still in a parking position outside the beam lines. HERA consists of two accelerators, one for electrons and one for protons, installed one above the other in a 6.3 km long ring tunnel. The warm magnets and RF cavities of the electron rings provide electron beams of about 27 GeV energy at nominal intensity. The magnets of the proton ring are superconducting and provide proton beams of 820 GeV energy. More details on the various components and the steps of the commissioning phase of the HERA machine can be found in [3, 4]. In a lecture focused on physics results from the first year a few comments on the performance of the machine have some interest. In Table 1 we compare the values achieved during the data taking period in fall 1992 to the design values. The proton current per bunch was a factor four lower than design. This has been traced to instabilities in the injectors and may take some time to cure. The total circulating current of the electron beam was the main limitation to the luminosity in 1992. The usual lifetime of the electron beam was about 10 hours but dropped drastically to a few minutes when the current exceeded 3 mA. This problem was cured at the end of 1992 by replacing 20 cm of the beam pipe in one straight section. It was then possible to accumulate up to 23 mA of electrons just before the 1992 winter shut down.

By increasing the number of bunches one expects an increase in luminosity of an order of magnitude in 1993 without a significant increase of the individual bunch intensities.

As a last remark let us point out that during machine studies a transverse electron polarisation of 58 ± 5 % has been observed. A spin-rotator magnet system, to turn the spin from vertical to longitudinal orientation, will be installed for the experiment HERMES [5], where the electron beam of HERA interacts with an internal gas target.

3. The H1 and ZEUS Detectors

The H1 and ZEUS detectors are large multi-purpose detectors of a size comparable to those at LEP (see Fig.1 and 2). The tracking systems and the calorimeters are more extended in the forward (proton) direction due to the asymmetry of the proton and electron beam energies. The backbone of both the H1 and the ZEUS detectors are their calorimeters. ZEUS uses a compensating uranium-scintillator calorimeter which provides the best possible energy resolution for hadrons ($35\%/\sqrt{E}$ on test beams) but may be sensitive to aging effects. The energy resolution for electrons is expected to be $18\%/\sqrt{E}$. The H1 calorimeter uses plates of lead and stainless steel as absorber and liquid argon for readout. The liquid argon provides a very stable calibration (better than 0.5 % per year) [1] and allows a high transverse and lateral granularity. The energy resolution, measured on test beams, is $12\%/\sqrt{E}$ for electrons and $50\%/\sqrt{E}$ for hadrons. The calorimeter coverage is completed with a small calorimeter with copper absorber and silicon pad readout in the very forward direction ($\theta < 4^0$) and a conventional electromagnetic lead-scintillator sandwich calorimeter in the backward direction ($\theta > 151^0$).

In the HERA experiments the luminosity is determined by measuring the Compton process $ep \rightarrow ep\,\gamma$ at small scattering angle. The electron and the photon are detected in counters situated at 33 m and 100 m respectively from the interaction point.

4. Total Photon-Proton Cross Section

In most of the interactions at HERA the electron interacts with the proton through the exchange of a virtual photon. When the scattered electron is emitted at very low angle and detected in the electron counter the negative mass squared of the virtual photon, Q^2, is close to zero (10^{-7} GeV$^2 < Q^2 < 10^{-2}$ GeV^2), the virtual photon is almost real. Thus it is possible to study collisions of almost real photons with electrons and protons at high energies.

The total photon-proton cross section is predicted to rise with increasing energy from its value of 118 μb at a centre of mass energy of 18 GeV, the highest energy achieved in fixed target experiments. Increase of the total cross sections with increasing centre of mass energy has been predicted and observed in hadron-hadron collisions. Depending on the specific model the photon-proton total cross section is predicted to lie in the range from 145 μb to 800 μb at 200 GeV [6].
H1 and ZEUS have measured the total photon-proton cross section at a centre of mass energy around 200 GeV [7, 8]. The two measurements yield a value of about 160 μb

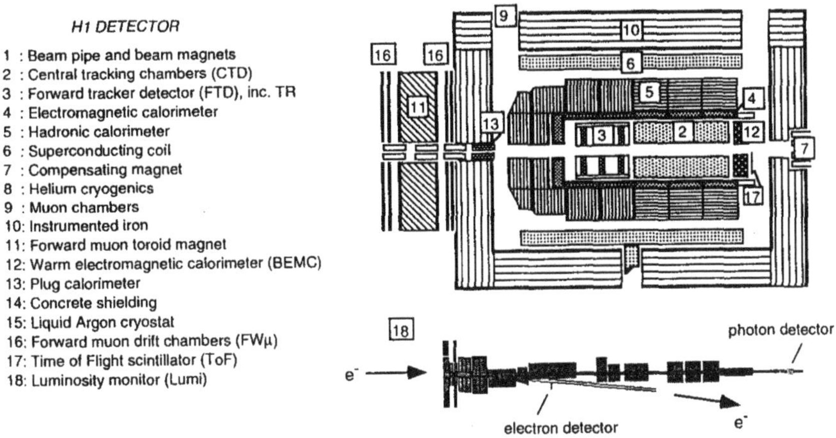

Figure 1: Schematic transversal view of the H1 detector. Note that the luminosity detector, downstream in the electron beam is not to scale.

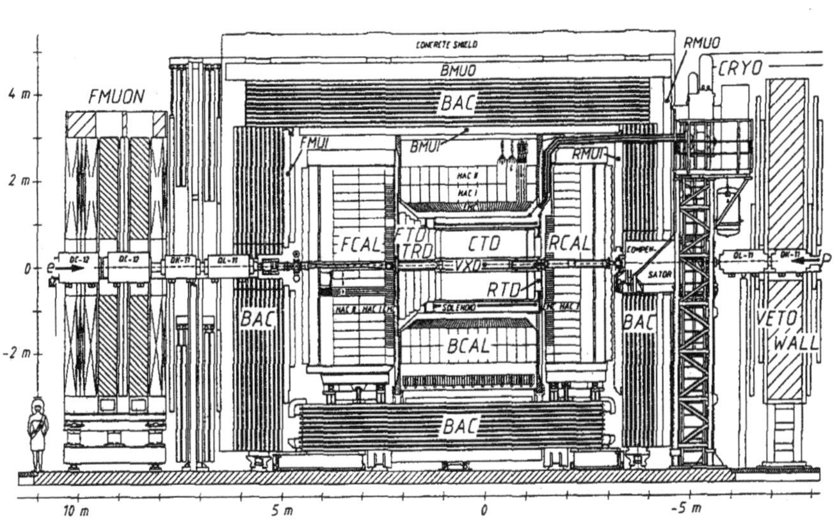

Figure 2: Cross section of the ZEUS detector.

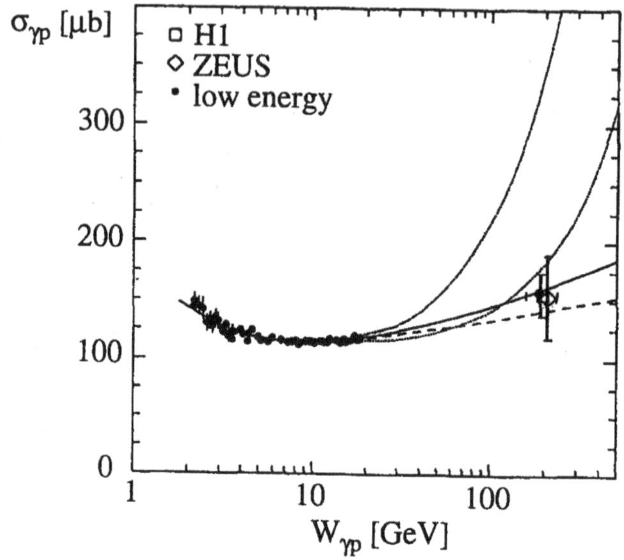

Figure 3: The total γp cross section at low energy together with those measured at HERA by H1 and ZEUS. The solid curve represents a Regge based fit of low energy data [9]. The dashed curve is the prediction of the ALLM parametrization [10] . The dotted lines are obtained using PYTHIA Monte Carlo [14] with the DG parametrization [13] of the photon structure function for $p_T^{min} = 1.4$ GeV/c (upperline) and for $p_T^{min} = 2.0$ GeV/c (lower line).

(Fig.3) in excellent agreement with the parametrizations based on Regge motivated models [9, 10]. The result does not support extreme models which predict a strong rise of the cross section [12].

The systematic errors in this measurement amount to 12% (21%) for H1(ZEUS). The dominant source of errors are a 6% (11%) uncertainty on the luminosity determination and a 10% (15%) uncertainty on the acceptance calculation. There is a good hope to reduce the luminosity error below 5% by a better understanding of the beam phase space. The acceptance of the main detector is process dependent. It is particularly sensitive to the amount of diffractive events which are hardly seen in the detector. Therefore it will be very difficult to reduce the total error on the total cross section below 5%.

5. Jet Production In Photoproduction

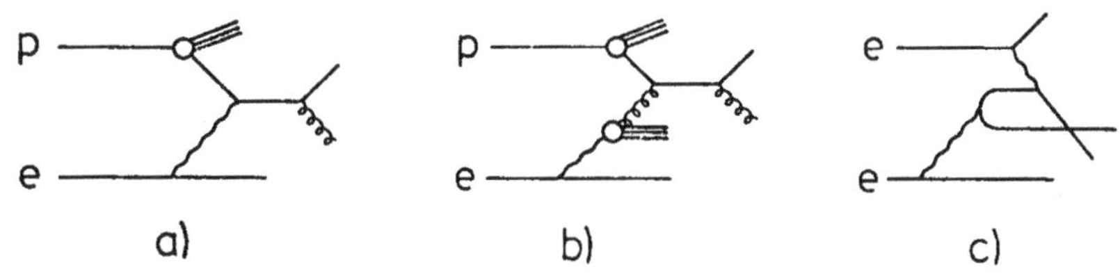

Figure 4: a) Direct photon processes in *ep* scattering. b) Resolved photon process in *ep* scattering. c) Deep inelastic electron-photon process in $e^+ e^-$scattering.

The cross section is dominated by soft scattering processes where the photon interacts as a hadron meson with the proton (Vector Dominance Model). But at large transverse momentum, the photon behaves as a point-like object and interacts directly with a quark or a gluon of the proton, thus producing jets of hadrons through hard scattering processes. These are called direct photon processes (Fig.4a). Like in hadrons, there is also the possibility that a parton in the photon interacts with a parton in the proton. These are called resolved photon processes (Fig.4b). In $e^+ e^-$ colliders, the measured photon-photon processes include mainly deep inelastic electron-photon scattering where an almost real photon probes the quark content of a highly virtual photon (Fig.4c). Recent experimental analyses of jet production in photon-photon collisions [11] show sensitivity to the gluon content of the photon but are limited by theoretical uncertainties in the calculation of the jet cross section ar relatively low E_t.

The parametrizations of the photon structure functions derived from these measurements constrain mainly the quark and antiquark contributions but allow for large difference in the gluon content. At HERA, through the photoproduction of jets, a parton of the proton probes the quark and the gluon content of an almost real photon.

The onset of hard scattering processes can be seen in the p_T spectrum of single charged particles in the central region of the laboratory system (Fig.5). The clear departure from an exponential law at large p_T is a well known signature for hard collisions. An other clear indication of hard scattering processes is the jets structure of the final state. Very clean back to back two-jet events have been observed in the H1 [17] and ZEUS detectors [16] (see for example Fig.6a). Some events show also a two-jet structure plus additional activity in the electron direction associated with the photon remnant (see Fig. 6b).

Already with an accumulated luminosity of about 25 nb^{-1} quantitative studies were possible. The rate of high E_t jets depends on the density of partons issued from the photon. Sensitivity to the photon structure function has been studied [20] in leading-order calculations using three QCD parametrizations which differ

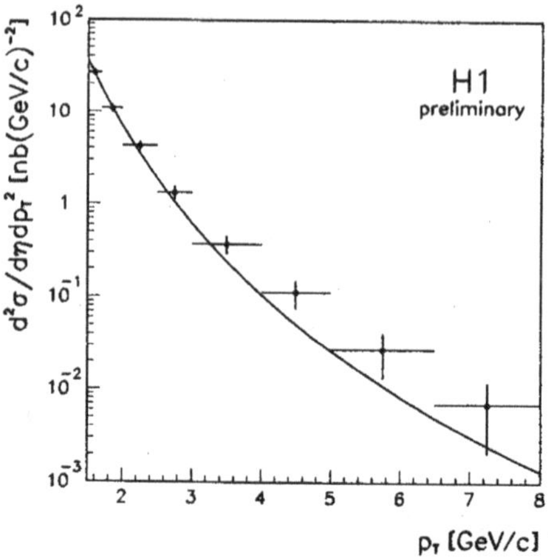

Figure 5: Invariant cross section for single charged particles with $|\eta| < 1$ in the kinematic region of $Q^2 < 0.1$ GeV^2 compared to the shape of the full NLO calculation [15].

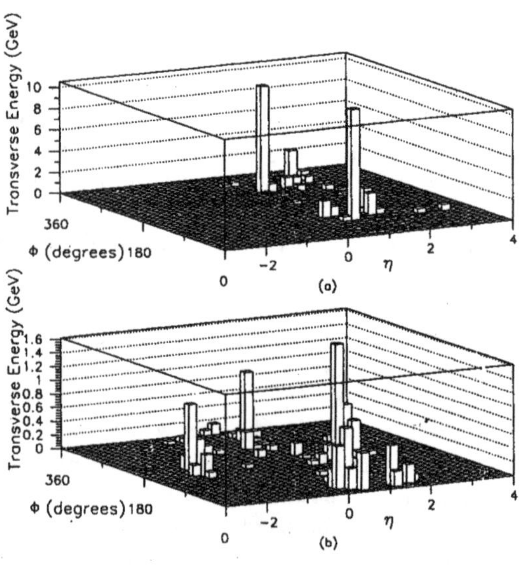

Figure 6: Transverse energy in $\eta - \phi$ space (η stands for pseudo rapidity and ϕ for azimuth angle) in the laboratory frame. Events observed in the ZEUS detector [16] with a) a two-jet structure (b) a two-jet structure plus additional activity in the incident electron direction.

mainly in the gluon densities: GRV-LO [18], LAC2 and LAC3 [19] (Fig.7).The LAC3 prametrization is clearly disfavored. The sensitivity to the gluon content of the photon can be demonstrated by comparing the data with the predicted cross section due to parton processes initiated by quarks, and not by gluons, in the photon (dotted line) using the GRV-LO parametrizations.

In summary, it has been already possible in 1992 to see the onset of hard collisions in photon proton interactions and to have a first look at the momentum distribution of partons, especially of gluons originating from the photon. It will be possible with more luminosity to measure the quark and the gluon content of the photon from the study of jet productions.

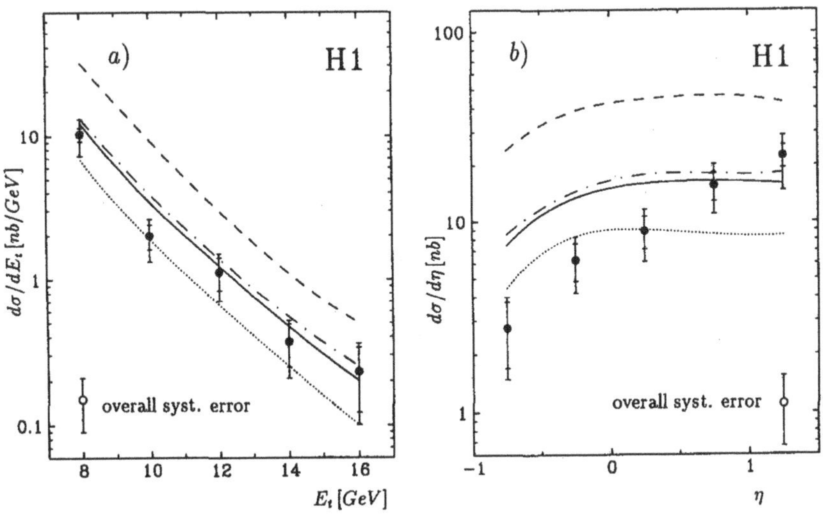

Figure 7:
Inclusive jet E_t spectrum (a) integrated over the pseudorapidity interval $-1.0 < \eta < 1.5$ and inclusive η spectrum (b) for jets with $E_t > 7$ GeV [20]. The inner error bars represent the statistical errors, the outer error bars the statistical and bin-to-bin systematic errors added in quadrature. The overall systematic uncertainty of $\pm 40\%$ is indicated. The curves show leading-order QCD calculations in the framework of the PYTHIA event generator using the photon structure functions LAC-3 (dashed line), LAC-2 (dashed-dotted line), GRV-LO (full line) and GRV-LO, but excluding the gluons originating from the photon (dotted line).

6. Deep Inelastic Scattering

When the mass of the virtual photon exceeds a few GeV^2, the scattering is deeply inelastic. The study of deep inelastic scattering on fixed targets has established the parton model and has long been recognised as one of the best quantitative tests of QCD [21]. At HERA a new regime in deep inelastic scattering will be investigated. The highest Q^2 reached in the data should be a few 10^4 GeV^2 at large x and the lowest x is 10^{-4} at $Q^2 \simeq 5$ GeV^2, thus extending by two orders of magnitude the accessible domain both in Q^2 and x. In this new small x regime the QCD evolution equations must go beyond the usual Leading Log LL(Q^2) approximations and may reach a region of saturation [22] [23]. Small x deep inelastic scattering probes QCD in a region at the frontier between the well known perturbative domain and the almost unknown non-perturbative domain. The squared invariant mass W^2 of the hadronic final state reaches also a few 10^4 GeV^2, close to two orders of magnitude higher than achieved in previous experiments, thus allowing detailed investigations of multi-jet production in deep inelastic scattering.

6.1 $F_2(x, Q^2)$

The cross section at small x is about 50 nb so that, based on only 25 nb^{-1} of data, first measurements of the structure function $F_2(x, Q^2)$ have been possible in this new region. The kinematics of the inclusive deep inelastic scattering process $ep \to eX$ at centre of mass energy $\sqrt{s}$ is determined by two independent Lorentz-invariant variables, conventionally chosen to be two of x, Q^2 and y, the fractional energy transfer to the photon. The kinematic variables are related to the four-momentum vectors l, l' and P of the incoming lepton, the outgoing lepton and the incoming proton respectively by :

$$\begin{aligned} Q^2 &= -q^2 = -(l - l')^2 \\ x &= Q^2/(2Pq) \\ y &= (Pq)/(Pl) \\ s &= 2Pl \end{aligned} \tag{1}$$

At HERA, in contrast with fixed target deep inelastic experiments reconstruction of Q^2 and y follow not only from reconstruction of the scattered electron variables :

$$Q_e^2 = 4E_e E_e' \cos^2(\theta_e/2) \tag{2}$$

$$y_e = 1 - \frac{E_e'}{E_e} \sin^2(\theta_e/2) \tag{3}$$

where E_e' is the energy of the scattered electron, θ_e the polar angle measured relative to the proton beam direction and E_e is the energy of the incident electron, but also from the final state hadron for which, for example :

$$y_h = \frac{\sum_h (E_h - p_{zh})}{2E_e} \tag{4}$$

where E_h is the energy of a hadron and p_{zh} its momentum component along the incident proton direction.

The H1 collaboration has chosen to determine Q^2 from the electron variable (Eq.2) and to use the most accurate determination of y which is the y_e measurement at large y and the y_h measurement at low y. There is significant overlap between both determinations providing a crucial check on systematic errors [24].

The ZEUS collaboration has chosen to use an other combination of electron and hadron variables : the so-called Double-Angle method [25]. x and Q^2 are derived from the angle of the electron scattering angle θ_e and the angle γ_H of a massless object balancing the momentum vector of the electron. This method is only weakly sensitive to the scale errors in the energy measurement of the final state particles.

More details on resolution and systematic errors of the measurement of Q^2 and y are given in the publications of the two collaborations quoted above. Thorough discussions on the relative merits of the various methods can be found in the proceedings of the HERA workshops [26]. In addition to the errors due to the limited resolution of the detectors there are two important sources of systematic errors which affect the measurement of F_2 at HERA : the photoproduction background and the QED corrections.

The main source of background for DIS events is photoproduction at low Q^2 and high y, where the electron scatters through a low angle, remains in most of the cases undetected in the beam pipe but transfers a substantial fraction of its energy to the hadronic final state. A photon, a low energy charged pion or a superposition of both may mimic a scattered electron and thus contaminate the DIS sample. Close to the beam pipe, in a high multiplicity environment, the identification is especially difficult. So far the main tool to reject this background has been to cut on the missing energy in the incident electron direction :

$$E_{miss} = E_e(y_e - y_h) \tag{5}$$

The cut applied on the missing energy is about 10 GeV (11.7 GeV and 9 GeV for H1 and ZEUS respectively). The residual background is negligible at $y \leq 0.4$ but reaches 30% at $y = 0.6$.

To a high accuracy, in the Q^2 range accessed by the first data ($Q^2 < 1000$ GeV^2), the Born cross section for DIS off a proton is given in terms of structure functions as :

$$\frac{d^2\sigma}{dx dQ^2} = \frac{2\pi\alpha^2}{Q^4 x}\left(2\,(1-y) + \frac{y^2}{1+R}\right) F_2(x, Q^2) \tag{6}$$

where R is the cross section ratio of longitudinally and transversely polarised photon, $R = \sigma_L/\sigma_T$. The contribution from R is a small effect in the present region of measurement ($< 8\,\%$) within the present statistics. A potentially much larger effect is due to QED corrections. The measured cross section is not the Born cross section. When the kinematics is determined from the scattered electron variables the ratio between the measured cross section and the Born cross section can be as large as 1.5 [27]. However most of the correction is due to events where a real photon is emitted collinear with the incident electron. These events are largely rejected by the cut on the missing energy. The residual correction is below 10 %. It is also noteworthy that when the cross section is measured in terms of mixed variables or Double-Angle variables the correction is also found to be small even without applying the cut on the missing energy.

The x dependence of F_2 measured by H1 and ZEUS is shown in Fig.8. Both sets of data agree remarkably well. We observe a clear rise of F_2 with decreasing x. Such a rise is not expected from Regge parton models [28]. On the contrary linear evolution equations in perturbative QCD predict a fast growth of the gluon density in the proton in the asymptotic limit $x \to 0$. It is a priori not known at which x value the onset of the asymptotic behaviour should be visible [22]. In any case the fact that F_2 rises at low x could be an indication that the saturation of the parton density in the proton may be detectable at HERA.

For constant values of x, F_2 rises slowly with Q^2 as expected from perturbative QCD. The lack of statistics prevents a sophisticate QCD analysis. Later with more luminosity it will be possible to do a precise measurement of $F_2(x, Q^2)$ from which one could do a measurement of α_s, the strong coupling constant, with a statistical error as low as 4 % [31] and extract the gluon density from the scaling violation with unprecedented accuracy.

6.2 *The Hadronic Final State*

The hadronic final state of deep inelastic events is also a copious source of information for QCD tests. Hadron and jet production are induced by parton emission predicted by QCD. In the naive quark-parton model the transverse momentum of the scattered electron is balanced by a single jet associated with the struck quark, usually called the current jet. Higher-order QCD processes modify this picture. Additional effects of soft gluons and fragmentation are expected together with radiation of hard gluon from the struck quark or production of a quark anti-quark pair by photon-gluon fusion. Experimental signatures of these QCD processes are the widening of the current jet [32, 33] and the production of multi-jet events [34, 35]. Between 10 and 20 % of the DIS events contain 2 + 1 jets (2 high p_t jets and the proton remnant). The production rate is found to rise with Q^2 (Fig.9). Even with very modest statistics there are indications that the observed Q^2 dependence is sensitive to the running of the QCD coupling constant α_s [35].

Jet rates and jet profiles have been thoroughly compared with the various models of the QCD effects. See for example Fig.10. The data are well described by Monte

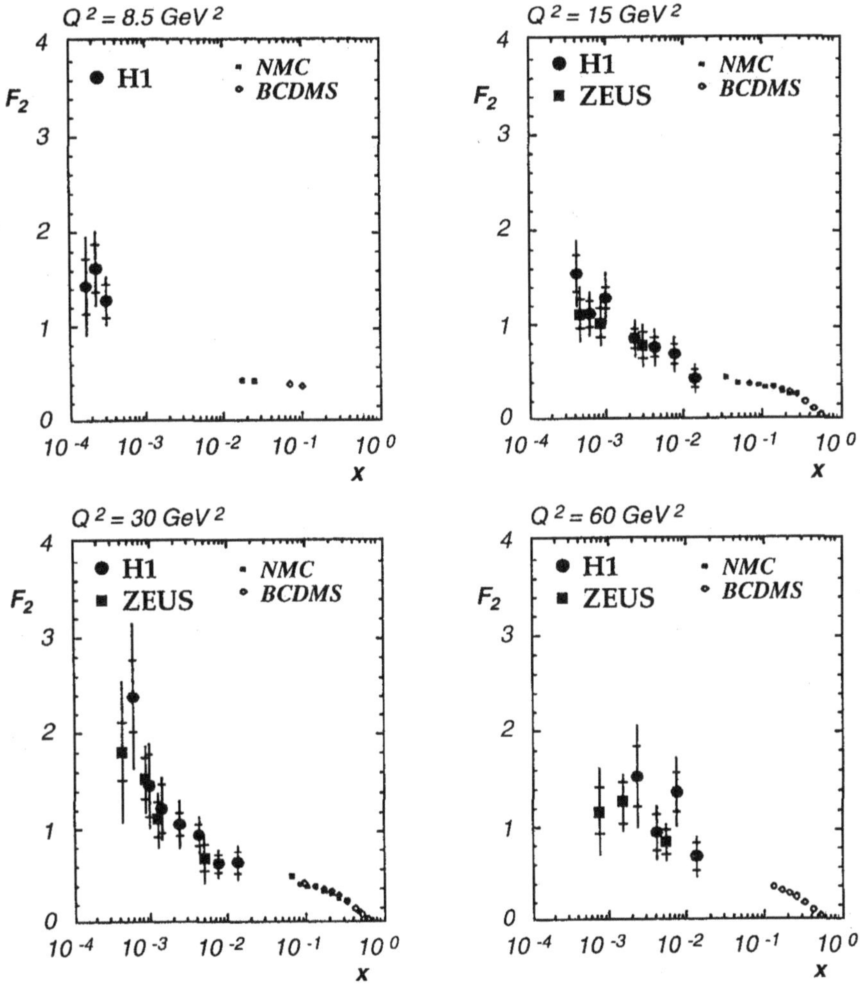

Figure 8: The measured proton structure function F_2 by H1 and ZEUS as a function of x for the four lowest Q^2 bins. The inner error bar is the statistical error and the outer bar shows the systematic error added in quadrature. In addition the H1(ZEUS) points have a normalization uncertainty of 8%(7)%. Data points of the fixed target muon experiments BCDMS [29] and NMC [30] are shown for comparison.

Carlo calculations that include first-order QCD matrix elements plus higher-order QCD corrections approximated by leading-log parton showers, as implemented in the LEPTO Matrix Elements + Parton Shower (MEPS) model [36].

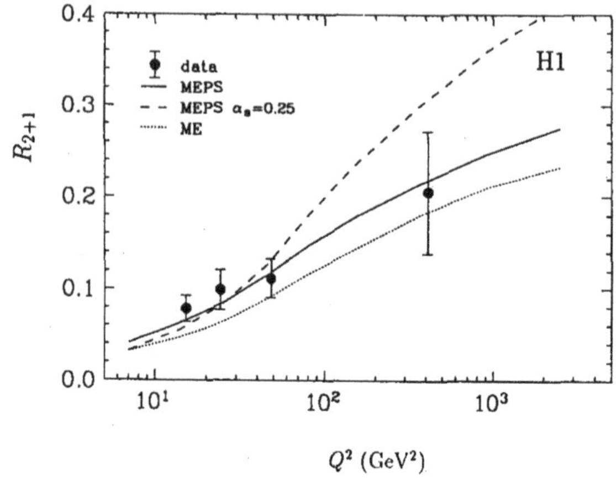

Figure 9: The fraction of (2+1)jet events R_{2+1} as a function of Q^2 measured by H1 [35], together with the expectations of the MEPS model (Matrix Elements + Parton Showers) assuming a running α_s and a constant $\alpha_s = 0.25$ and also with the expectations of the ME model (Matrix Elements but no Parton Showers). The error bars correspond to the statistical errors of data and Monte-Carlo corrections.

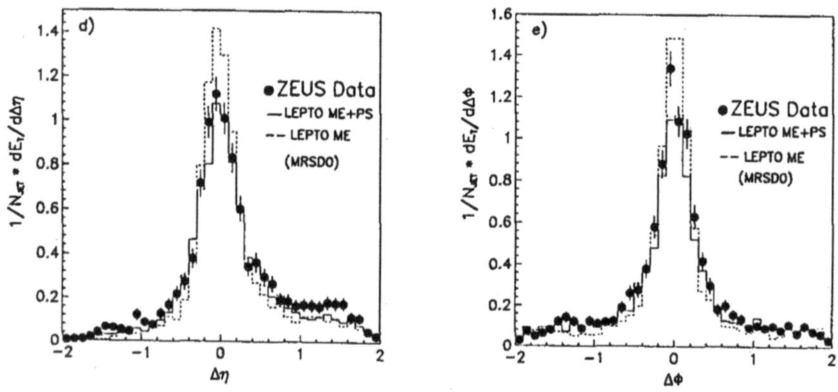

Figure 10: The profile of jets in DIS events together with the expectations of the MEPS and ME models [34].

6.3 *Diffractive Deep Inelastic Scattering Events*

In most of the DIS events, even when the jet issued from the quark hit by the virtual photon is emitted at large angle, there is a significant energy deposit at angles close to the proton direction. This energy flow is due to color transfer between the struck quark and the proton remnant together with initial state radiation and fragmentation of the proton remnant. However it has been observed [37, 24] that there is a sizeable fraction of events in which the hadronic energy deposit closest to the proton beam direction is at large angle. These large rapidity gap events are not described by the usual QCD-inspired models. A 5.4 (6)% fraction of DIS events observed by ZEUS(H1) have no energy clusters > 400 MeV with a pseudorapidity, $\eta = -\ln \tan \theta/2$, above 1.55(1.8), while the fraction of events predicted by Monte Carlo simulations is 0.1% (see Fig.11). Within the limited statistics the rates of large rapidity gap events do not seem to depend significantly on x and Q^2. With more statistics it will be very interesting to study whether these events play a special role in the Q^2 and x dependence of the proton structure function F_2. The large rapidity gap events may also give a clue as to the mysterious nature of the pomeron.

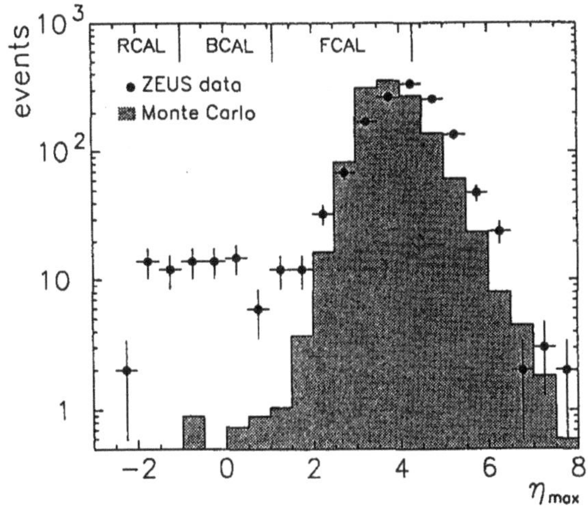

Figure 11: Distribution of η_{max}, the maximum rapidity of a calorimeter cluster for data and Monte Carlo DIS events in the ZEUS detector [37].

7. Beyond the Standard Model

By exploring a new energy domain at 300 GeV in the $e\,p$ centre of mass HERA has an opportunity to detect signals beyond the Standard Model. There are many processes not yet excluded by the present data which could be discovered at HERA. The search for new signals can be divided into two groups : either detecting new particles, for example leptoquarks and excited leptons, or indirect detection of new dynamics, through the appearance of small systematic deviations from the expected Standard Model cross sections. Examples of new dynamics are exchange of new bosons W' and Z' or compositeness of leptons and quarks. High luminosity is necessary for detecting small deviations but the hunt for new particles has started from the very first days of $e\,p$ collisions.

At HERA leptoquarks, with baryon and lepton number conserving couplings, can be can be produced by fusion of the incident electron and a parton from the proton forming an s-channel resonance. The final state consists of a lepton and a jet coming from the decay of the resonance together with the proton fragments in the beam pipe and thus cannot be distinguished event per event from a standard DIS event (Fig.12). A leptoquark of mass M would be detected as a narrow peak in the x distribution at $x_0 = M^2/s$ where x_0 is the fractional momentum of the quark which couples to the leptoquark. Neglecting the DIS-LQ interference, the production cross section of a scalar leptoquark would be :

$$\sigma \quad = \quad \frac{\pi}{4M^2}\lambda^2 q(x_0, M) \tag{7}$$

where $q(x_0, Q^2)$ is the quark density in the proton and λ is a dimensionless coupling constant. The usual reference value for λ is 0.31 which corresponds to an electromagnetic coupling, i.e. $\lambda = \sqrt{4\pi\alpha}$. Based on about 25 nb^{-1} no signal has been found in the neutral and charged currents samples of the two HERA experiments [38, 39]. For couplings equal to that of electromagnetic interactions new lower limits at 95% Confidence Level can be assigned, ranging from $M \geq 145$ GeV to $M \geq 192$ GeV for leptoquarks resulting from the fusion of an electron and a quark and from 98 to 121 GeV for leptoquarks formed with an antiquark. Note that the limits of 98 GeV found on leptoquarks coupling to $e\,\bar{d}$ can be also interpreted in R-parity violating supersymmetry [40] as limits on stop, the scalar top quark.

Leptogluons can also be produced as narrow s-channel resonances at HERA and would look similar to a leptoquark except that the jet would be from a gluon instead of a quark. To lowest order the cross section depends only on the gluon density in the proton and on $(M/\Lambda)^2$, where M is the leptogluon mass and Λ is a scale parameter [41]. H1 excludes the scale region $\Lambda \leq 1.8$ TeV for $M = 100$ GeV and $\Lambda \leq 100$ GeV for $M = 200$ GeV [38].

Excited leptons can be produced in principle and the following decays could be detected : $e^* \to e\,\gamma, e^* \to e\,Z, e^* \to \nu\,W, \nu^* \to \nu\,\gamma, \nu^* \to \nu\,Z, or\ \nu^* \to e\,W.$
No evidence was found by H1 [38] and ZEUS [42] for the production of e^* or ν^* . For

masses above 45 GeV (75 GeV), cross sections above 180 pb (160 pb) were ruled by ZEUS (H1) with 95% Confidence Level. Even higher cross sections were rejected for ν^*.

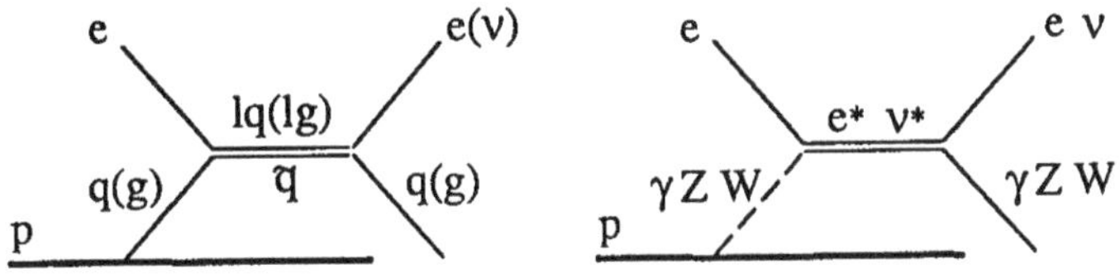

Figure 12: Diagrams describing the formation of new electron-parton (leptoquark, leptogluon, R-parity violating squark) and electron-gauge excited leptons.

8. Conclusion

In 1992, HERA physics has started fruitfully for H1 and ZEUS. It has been a pleasant surprise that based on 25 nb^{-1} such interesting results have been obtained on photoproduction of almost real photons and deep inelastic scattering at very low x together with new limits on the physics beyond the standard model. With more luminosity more processes will be accessible to H1 and ZEUS. Examples are heavy quark production at low Q^2 and neutral or charged current scattering at high Q^2 in the few 10^4 GeV^2 region.

References

[1] H1 Collaboration, T. Ahmed et al., DESY 93-103 .

[2] ZEUS Collaboration, The ZEUS Detector, *Technical Proposal* (1986) and The ZEUS Detector, *Status Report 1989.*

[3] DESY-HERA 81-10 , July 1981;
B.H. Wiik,*Proc.*10th *SLAC Summer Institute*, ed. A. Mosher, 1982, p. 233, and *Proc. XXVI Int. Conf. High Energy Physics*, Dallas, 1992;
G.A. Voss,em Proc. First Euro. Acc. Conf., Rome, 1988, p.7.

[4] G. Wolf, DESY 92-190.

[5] The HERMES Collaboration, DESY-PRC 90-01 , 1990.

[6] G.A. Schuler, *Proc. Workshop on Physics at HERA* (DESY, 1992) p.461;
A. Levy, *Proc. Workshop on Physics at HERA* (DESY, 1992) p.481;
S. Levonian, *Proc. Workshop on Physics at HERA* (DESY, 1992) p.499

[7] ZEUS Collaboration, M. Derrick et al., *Phys. Lett.* **B293** (1992) 465.

[8] H1 Collaboration, T. Ahmed et al., *Phys. Lett.* **B299** (1993) 374 and H1 Collaboration, M. Erdmann et al., DESY 93-077.

[9] A. Donnachie and P.V. Landshoff, *Phys. Lett.* **B296** (1992) 277.

[10] H. Abramowicz et al., *Phys. Lett.* **B269** (1991) 465.

[11] AMY Collaboration, R. Tanaka et al., *Phys. Lett.* **B277** (1992) 215.

[12] M. Drees and F. Halzen, *Phys. Rev. Lett.* **61** (1988) 275 ;
R. Gandhi and J. Sarcevic, *Phys. Rev.* **D44** (1991) R10.

[13] M. Drees and K. Grassie, *Z. Phys.* **C28** (1985) 451.

[14] H.-U. Bengtsson and T. Sjöstrand, Comput. Phys. Commun. **46** (1987) 43 ;
T. Sjöstrand, PYTHIA at HERA, *Proc. Workshop on Physics at HERA* (DESY, 1992) p.1405.

[15] F.M. Borzumati, B.A. Kniehl and G. Kramer, DESY 93-034.

[16] ZEUS Collaboration, M. Derrick et al., *Phys. Lett.* **B297** (1992) 404.

[17] H1 Collaboration, T. Ahmed et al., *Phys. Lett.* **B299** (1993) 374.

[18] M. Glück, E. Reya and A. Vogt, *Phys. Rev.* **D46** (1992) 1973.

[19] H. Abramowicz, K. Charchula and A. Levy, *Phys. Lett.* **B269** (1991) 458.

[20] H1 Collaboration, I. Abt et al., *Phys. Lett.* **B314** (1993) 436.

[21] G. Altarelli, *Ann. Rev. Nucl. Part. Sci.* **39** (1989) ;
M. Virchaux, Invited talk to the Workshop "QCD 20 years later" Aachen, 9-13 June 1992, Saclay Preprint DAPNIA/SPP 92-30.

[22] L.V. Gribov, E.M. Levin and M.G. Ryskin, *Nucl. Phys.* **B188** (1981) 555; *Phys. Rep.* **100** (1983).

[23] A.H. Mueller and J. Qiu, *Nucl. Phys.* **B268** (1986) 427.

[24] H1 Collaboration, I. Abt et al., *Nucl. Phys.* **B407** (1993) 515.

[25] ZEUS Collaboration, M. Derrick et al., *Phys. Lett.* **B316** (1993) 412.

[26] J. Feltesse, in *Proc. HERA Workshop* (DESY, 1987) p.33 ;
S. Bentvelsen et al., *Proc. Workshop on Physics at HERA* (DESY, 1992) p.23 ;
K.C. Hoeger, *Proc. Workshop on Physics at HERA* (DESY, 1992) p.43 ;
G. Bernardi and W. Hildesheim, *Proc. Workshop on Physics at HERA* (DESY, 1992) p.79.

[27] D. Yu. Bardin et al., *Z.Phys.* **C42** (1989) 679 ; M.Bōhm and H. Spiesberger, *Nucl. Phys.* **B294** (1987) 1091.

[28] A. Donnachie, P.V. Landshoff, M/C-th 93/11, DAMTP 93-23.

[29] BCDMS Collaboration, A.C. Benvenuti et al., *Phys. Lett.* **B223** (1989) 489.

[30] NMC Collaboration, P.Amaudruz et al., *Phys. Lett.* **B295** (1992) 159 and Erratum to CERN-PPE/92-124, CERN, April 1993.

[31] J. Blümlein et al., *Proc. HERA Workshop* (DESY, 1987) p.67

[32] H1 Collaboration, T. Ahmed et al., *Phys. Lett.* **B298** (1993) 469.

[33] ZEUS Collaboration, M. Derrick et al., *Z.Phys.* **C59** (1993) 231.

[34] ZEUS Collaboration, M. Derrick et al., *Phys. Lett.* **B306** (1993) 158.

[35] H1 Collaboration, I. Abt et al., DESY 93-137.

[36] M. Bengtsson, G. Ingelman and T. Sjöstrand, *Nucl. Phys.* **B301** (1988) 554.

[37] ZEUS Collaboration, M. Derrick et al., *Phys. Lett.* **B315** (1993) 481.

[38] H1 Collaboration, I. Abt et al., *Nucl. Phys.* **B396** (1993) 3.

[39] ZEUS Collaboration, M. Derrick et al., *Phys. Lett.* **B306** (1993) 173.

[40] T. Kon and T. Kobayashi, *Phys. Lett.* **B270** (1991) 81.

[41] J.Bijnens, *Proc. Workshop on Physics at HERA* (DESY, 1992) p.819

[42] ZEUS Collaboration, M. Derrick et al., *Phys. Lett.* **B316** (1993) 207.

II. Beyond the Standard Model

BEYOND THE STANDARD MODEL

John Ellis

Theoretical Physics Division, CERN
CH - 1211 Geneva 23

1 Introduction

The following are some of the outstanding problems of the Standard Model that lead us to search beyond it. **Unification:** are the different forces that we see all aspects of the same unified interaction? A grand unified theory that combines them may lead to proton decay, magnetic monopoles and neutrino masses. **Flavour:** why are there so many different types of matter particles? Perhaps quarks and leptons are composite? **Mass:** what are the origins of the particle masses? Are they due to an elementary Higgs boson? and is this accompanied by supersymmetric particles? **Theory of Everything:** this should solve all of the problems listed above, as well as the fundamental problems of quantum gravity. The reconciliation of quantum mechanics with general relativity is probably the biggest problem in theoretical physics, and the only candidate we have for a consistent quantum theory of gravity is the superstring. One aspect of the gravity of this problem can be gauged by considering the one-graviton exchange amplitude

$$A \sim G_N E_1 E_2 \tag{1}$$

The growth of this amplitude with energy leads to uncontrollable divergences when one tries to calculate quantum corrections

$$\frac{\delta A}{A} \sim (G_N \Lambda^2)^p \tag{2}$$

My belief is that the reconciliation of quantum mechanics with general relativity is going to require profound new geometrical notions. Superstring theory does this, since it replaces point particles by extended stringy objects.

Abdus Salam has been one of the pioneers working on all these problems, and looking in many directions beyond the Standard Model. For this reason, it is a particular pleasure for me to review these ideas for this volume dedicated to him.

2 Electroweak Symmetry Breaking

It is well known that generating particle masses in the Standard Model requires spontaneous breaking of the $SU(2) \times U(1)$ gauge symmetry:

$$< 0 \mid X_{I,I_3} \mid 0 > \neq 0 \tag{3}$$

The fact that the relative strength of the neutral-current and charged- current reactions

$$\rho = \frac{m_W^2}{m_Z^2 \cos^2 \theta_W} \simeq 1 \tag{4}$$

tells us that most of the X sitting in the vacuum in Eq. (3) must have electroweak isospin $I = \frac{1}{2}$. This is also what is required to give non-zero fermion masses.

$$m_f \propto < 0 \mid X_{\frac{1}{2},\pm\frac{1}{2}} \mid 0 > \neq 0 \tag{5}$$

The question then is, what is this quantity X? In particular, is it elementary or composite?

In the original formulation of the Standard Model, Glashow, Weinberg and Salam [1] chose an elementary Higgs boson. We know now, however, that there are problems with elementary scalar fields when one calculates loop diagrams. In particular, one finds quadratic divergences in the one-loop corrections to scalar and hence vector boson masses

$$\delta m_{H,V}^2 \propto \frac{g^2}{(2\pi)^4} \int \frac{d^4k}{k^2} = O(\frac{\alpha}{\pi})\Lambda^2 \tag{6}$$

The only known way to tame these quadratic divergences is to invoke supersymmetry [2], which provides an effective cut-off Λ related to the sparticle masses $\tilde{m} \lesssim 1\, TeV$.

The alternative is that X is composite, presumably a fermion-antifermion condensate $\bar{F}F$, by analogy with QCD and the BCS theory of superconductivity. There are two popular ideas for this condensate on the market. One is a condensate of top quarks and antiquarks, bound together by a strong Yukawa couplings [3], and the other is a new class of so-called technicolour fermions, bound together by a new set of superstrong gauge interactions [4].

The $\bar{t}t$ condensate mechanism is described by an effective Lagrangian à la Nambu-Jona-Lasinio that respects the symmetries of the Standard Model, and is made out of fermions alone. In its simplest form, it looks like

$$\mathcal{L}_{NJL} = \bar{\psi}_a D\psi^a - \frac{1}{2}G\left[(\bar{\psi}_a\psi^a)^2 - (\bar{\psi}_a\gamma^5\psi^a)^2\right] \tag{7}$$

Solving this Lagrangian non-perturbatively gives the same physics as the Standard Model, with additional constraints that lead to non-trivial new predictions. In particular, the simplest versions of the model predict [4]

$$m_H \simeq (1\,\mathrm{to}\,2)m_t : m_t \simeq (200\,\mathrm{to}\,250)\,GeV \tag{8}$$

However, there are problems with this simplest realization of the $\bar{t}t$ condensate model. The precision electroweak data seem to prefer a smaller value of m_t. This can in principle be accommodated within the model, at the expense of introducing an epicycle such as supersymmetry [5]. However, this makes the point that the predictions of the model are sensitive to modifications of the Nambu-Jona-Lasinio model in Eq. (7).

The alternative that aroused much interest at one time is technicolour [4], which postulates new hyperstrong interactions with a typical energy scale

$$\Lambda_{TC} \simeq 3000\,\Lambda_{QCD} \tag{9}$$

In this framework, the Higgs boson is a composite technipion made out of technifermions $\bar{T}T$. The model needs just one doublet of technifermions to give masses to the W and Z, but needs extensions [6] in order to give fermions masses. This has led theorists to the conventional one-technigeneration model

$$\begin{array}{ll} (\nu, e) & (u, d) \\ (N, E)_{1,\ldots,N_{TC}} & (U, D)_{1,\ldots,N_{TC}} \end{array} \tag{10}$$

It has been known for some time that this extended technicolour model has problems with flavour-changing neutral interactions [7], and possibly with light charged technipions [8]. The former class of problems could perhaps be resolved by postulating some form of "walking" technicolour model. A more recent development is that such an extended technicolour model seems to be difficult to reconcile with precision electroweak data from LEP and elsewhere. Several authors [9] have proposed a general parametrization of the one-loop vacuum polarization contributions to electroweak radiative corrections

$$T = \frac{\epsilon_1}{\alpha},\ S = \frac{4\sin^2\theta_W}{\alpha}\epsilon_3,\ U = -\frac{4\sin^2\theta_W}{\alpha}\epsilon_2 \tag{11}$$

In the Standard Model, the leading behaviours of T and S are

$$T \simeq \frac{3}{16\pi}\frac{1}{\sin^2\theta_W\cos^2\theta_W}\frac{m_t^2}{m_Z^2} - \frac{3}{16\pi\cos^2\theta_W}\ell n\left(\frac{m_H^2}{m_Z^2}\right) + \ldots,\ S = \frac{1}{12\pi}\ell n\left(\frac{m_H^2}{m_Z^2}\right) + \ldots \tag{12}$$

and U is often neglected. The Standard Model bounds on (m_t, m_H) can be rephrased as bounds on the parameters S, T, U. The results [10] of one global fit to the electroweak data are shown in Fig. 1, along with various technicolour predictions. It seems that the data are difficult to reconcile with such a composite model, and so we return to the original elementary Higgs boson of Glashow, Weinberg and Salam [1].

3 Supersymmetry

The strongest phenomenological motivation for supersymmetry [2] has been the hierarchy/ naturalness problem mentioned above, which was the first argument to suggest that supersymmetric particles should weigh $\lesssim 1TeV$. However, LEP has recently provided two other circumstantial phenomenological motivations for supersymmetry. One is that the analysis of precision electroweak data and radiative corrections suggest that $m_H \lesssim 300GeV$ [10], which is highly consistent with the range $40GeV \lesssim m_H \lesssim 130GeV$ expected [11] in the minimal supersymmetric extension of the Standard Model (MSSM) [12]. The other is the fact that the highly precise LEP measurements of the gauge couplings of the Standard Model, when extrapolated to very high energies, are not in good agreement with minimal non-supersymmetric GUTs, but do agree with minimal supersymmetric GUTs [13].

The Lagrangian of the MSSM [12] consists of a supersymmetric part and a part that breaks supersymmetry softly. In the supersymmetric part, there are the same gauge interactions as in the Standard Model, whilst the Yukawa interactions are given by a superpotential of the form

$$W = \lambda_L L E^c H_1 + \lambda_U Q U^c H_2 + \lambda_D Q D^c H_1 + \mu H_1 H_2 \tag{13}$$

The parameters $\lambda_{L,U,D}$ give masses to the charged leptons, charge-2/3 quarks and charge-−1/3 quarks as in the Standard Model, whilst μ is a new mixing parameter that arises because the MSSM contains two Higgs doublets. Many more new parameters may appear in the supersymmetry-breaking part of the Lagrangian

$$\mathcal{L}_{susyx} \ni -\sum_i m_{0_i}^2 \mid \phi^i \mid^2 - \sum_a M_a \tilde{V}_a \tilde{V}_a - \dots \tag{14}$$

The different scalar masses m_{0_i} and the gaugino masses M_a may be renormalized from two basic input parameters $(m_0, m_{1/2})$ at the GUT scale, as shown in Eq. (19) below.

To complete the picture of the MSSM, we need the effective Higgs potential, which depends at the tree level on two parameters, a pseudoscalar Higgs mass m_A and $tan\beta = \frac{v_2}{v_1} :< 0 \mid H_i \mid 0 >\equiv v_i$. The masses and couplings of all the Higgs bosons are specified by these two parameters at the tree level [12]

$$m_h^2 + m_H^2 = m_A^2 + m_Z^2, \; m_{H^\pm}^2 = m_A^2 + m_{W^\pm}^2,$$
$$m_{h,H}^2 = \frac{1}{2}\left[m_A^2 + m_Z^2 \mp \sqrt{(m_A^2 + m_Z^2)^2 - 4m_Z^2 m_A^2 \cos^2 2\beta}\right] \tag{15}$$

However, radiative corrections are very important [11]. The essential parameters of the MSSM are $(\mu, m_0, m_{1/2}, m_A, tan\beta)$, and these are the parameters to be constrained by experiment.

At the tree level, it can be seen from Eq. (15) that one of the supersymmetric Higgs bosons h must weigh less than m_Z, whilst the pseudoscalar Higgs A must be heavier than the h. However, these restrictions are removed at the one-loop level, as seen in Fig. 2. The h may not be the lightest supersymmetric Higgs boson, and may weigh up to 120 or more GeV. At the moment, the analysis of precision electroweak data including radiative corrections calculated in the MSSM does not give significant constraints on the parameter space for supersymmetric Higgs bosons. Searches at LEP 1 exclude supersymmetric Higgs bosons with $m_{h,A} \lesssim 40\,GeV$. LEP 2 will enable the search for supersymmetric Higgs bosons to be extended over a large range of the $(m_A, tan\beta)$ plane, as seen in Fig. 3. Much of the rest of this plane can be explored using the LHC [14], as also seen in Fig. 3. The rare decay $b \to s\gamma$, which has recently been seen [15], is also sensitive to the supersymmetric Higgs sector, and may in the future give significant constraints on its parameters [16], but at present a cancellation between supersymmetric particles and supersymmetric Higgses is still possible [17].

Most supersymmetric theories have one stable particle which should be present in the Universe today, as a cosmological relic from the Big Bang [18]. Its stability is a consequence of the fact that most supersymmetric theories have a multiplicatively-conserved quantum number called R parity, which takes the value $+1$ for all particles and -1 for all sparticles. Its conservation is linked to baryon B and lepton L number conservation, since R parity may be represented by

$$R = (-1)^{3B+L+2S} \tag{16}$$

We see from this expression that R parity may be violated if lepton number is violated either explicitly through a HL coupling, or spontaneously by a v.e.v for a sneutrino field, however these possibilities are severely constrained by upper limits on L violation.

The conservation of R parity has three important implications: (1) sparticles must be produced in pairs, e.g. $e^+e^- \to \tilde{\mu}^+\tilde{\mu}^-$, (2) heavy sparticles decay into lighter sparticles, e.g. $\tilde{q} \to q\tilde{g}$, and (3) the lightest supersymmetric particle is stable, because it has no available decay mode.

The best candidate for supersymmetric dark matter is a neutralino χ, which is some mixture of neutral spin-1/2 gauginos and higgsinos [18]. It is parametrized by $m_{1/2}$, the unmixed gaugino mass, μ, the Higgs mixing parameter, and $tan\beta$, the ratio of Higgs v.e.v.'s. The composition of the lightest supersymmetric particle simplifies in the limits $m_{1/2} \to 0$, in which case it is approximately a photino $\tilde{\gamma}$, and $\mu \to 0$, in which case it is approximately a higgsino $\tilde{H}$, but in the generic case it is a complicated mixture of gauginos and higgsinos. As seen in Fig. 4, the parameters $m_{1/2}$ and μ are significantly constrained by unsuccessful searches at LEP and CDF, implying that

$$m_\chi \gtrsim (10\,\mathrm{to}\,20)\,GeV \tag{17}$$

As also seen in Fig. 4, there are large regions of the $(m_{1/2}, \mu)$ parameter space where the relic density ρ_{LSP} may be close to the critical density of the Universe.

4 Unification of the Fundamental Forces

Grand unification of the strong, weak and electromagnetic forces was pioneered by Jogesh Pati and Abdus Salam [19]. The logarithmic variation of coupling strengths permits their unification at an exponentially high energy

$$m_{GUT} = m_{proton} \exp\left(\frac{O(1)}{\alpha_{em}}\right) \tag{18}$$

where α_{em} is the fine structure constant. The hypothesis of grand unification predicts relations between the gauge couplings measured at low energies, as well as between different particle masses, and may also predict possible new phenomena such as proton decay, neutrino masses and magnetic monopoles. The simplest GUT model is the $SU(5)$ model of Georgi and Glashow [20], but the approach of gauge couplings towards unification is largely independent of the specific GUT model, depending only on the light particle spectrum.

Figure 5 shows that the GUT predictions for $sin^2\theta_W$ are qualitatively very successful. Expanding the vertical scale by a factor of 10, we find a small difference between the predictions of non-supersymmetric and supersymmetric GUTs, with the latter being closer to experiment. Expanding the vertical scale by another factor of 10, we find a possible slight suggestion that the data favour a smaller value of $sin^2\theta_W$, but this discrepancy is not significant. The predictions shown in Fig. 5 were obtained assuming that all the sparticles weigh m_Z, but this is unrealistic. A careful treatment [21] of the sparticle thresholds should take into account the fact that sparticles have different masses, renormalized from the input parameters $(m_0, m_{1/2})$

$$\begin{aligned} &m_{\tilde{q}}^2 \simeq 6m_{\frac{1}{2}}^2 + m_0^2, \qquad m_{\tilde{l}_L}^2 \simeq 0.5m_{\frac{1}{2}}^2 + m_0^2, \qquad m_{\tilde{l}_R}^2 \simeq 0.15m_{\frac{1}{2}}^2 + m_0^2, \\ &m_{\tilde{W}} \simeq \frac{\alpha_2}{\alpha_{GUT}} m_{\frac{1}{2}}, \qquad m_{\tilde{g}} \simeq \frac{\alpha_3}{\alpha_{GUT}} m_{\frac{1}{2}}, etc. \end{aligned} \tag{19}$$

Moreover, massive particles near the GUT threshold could have a significant effect on the unification of the couplings. Figure 6 shows that, even if this effect is neglected, but a one-standard-deviation variation is allowed in the other input parameters, it is not possible [21],[22] at the present time to fix with precision $m_{1/2}$, unless α_3 (m_Z) is very large.

5 Superstring

This is the only candidate we have for a Theory of Everything (TOE). According to this theory, point-like elementary particles become loops of string, providing a quantum theory of gravity which is finite in perturbation theory. It is now known that string models can be formulated directly in four dimensions [23], and there are many consistent four-dimensional heterotic string theories. In the favoured scenaria, the left-moving degrees of freedom carry supersymmetry, while the right-moving degrees of freedom have internal quantum numbers [24]. Supersymmetry plays an essential role in ensuring that there are no tachyons, i.e., there are no states with $m^2 < 0$, and that the vacuum with zero cosmological constant is stable at least at the classical level. The resulting low-energy effective field theory is a supergravity model of the no-scale type [25], i.e., there are no regions of parameter space where the effective potential is negative. Possible gauge groups that have been studied in string models include

$SU(3) \times SU(3) \times SU(3)$ [26], $SU(3) \times SU(2) \times U(1)^n$[27], and $SU(5) \times U(1)^m$[28]. In each case, it is believed that these observable gauge interactions are accompanied by hidden-sector non-Abelian gauge interactions that may be the dynamical origin of supersymmetry breaking and a possible source of metastable dark matter [29].

In the presence of very many apparently consistent string models, the problem is which one has been chosen by Nature. Here the important existential question is: to GUT or not to GUT? By the former option, I mean looking for a large simple gauge group such as $SU(5)$, whereas the alternative is a group of the form $SU(3) \times SU(2) \times U(1)^n$ in which no further unification occurs at the field-theoretical level beyond that in the Glashow-Weinberg-Salam model. We [28] have studied GUT-like string models for the following reasons. They offer the prospect of naturally small neutrino masses and an elegant scenario for baryogenesis, not to mention the possibility of observable baryon decay, whilst non-GUT models do not in general explain naturally the small neutrino masses and may cause baryons to decay too quickly or never.

The essential obstacle to constructing a string GUT is that (almost) all GUTs require adjoint or larger Higgs representations, for example, Georgi-Glashow $SU(5)$ requires an adjoint **24** of Higgses, or even larger representations **75, 50, $\overline{\mathbf{50}}$** in some versions, and the next model $SO(10)$ needs a **45** or larger Higgs representation. The problem [30] is that four-dimensional field theories obtained from string using $k = 1$ Kac-Moody algebras on the world sheet do not contain adjoint or larger Higgs representations.[1] For example, Calabi-Yau compactifications [31] of the original ten-dimensional heterotic string give gauge groups which are subgroups of E_6, that only have **1**, **27** and $\overline{\mathbf{27}}$ representations, not **78**. The solution of this problem is the one known GUT that does not require adjoint or larger Higgs representations, namely flipped supersymmetric $SU(5)$ [28].

The gauge group of this model is $SU(5) \times U(1)$, and the normal quark and lepton matter fields are assigned to the following representations

$$F_i = (\mathbf{10}, \frac{\mathbf{1}}{\mathbf{2}}),\ \bar{f}_i = (\bar{\mathbf{5}}, -\frac{\mathbf{3}}{\mathbf{2}}),\ l_i^c = (\mathbf{1}, \frac{\mathbf{5}}{\mathbf{2}}) : i = 1, ..., N_g \tag{20}$$

The particle assignments within the **10** and $\bar{\mathbf{5}}$ representations are different from those in conventional $SU(5)$:

$$F_1 = \begin{pmatrix} 0 & d^c & -d^c & \vdots & d & u \\ -d^c & 0 & d^c & \vdots & d & u \\ d^c & -d^c & 0 & \vdots & d & u \\ \cdots & \cdots & \cdots & \cdots & \cdots & \cdots \\ d & d & d & \vdots & 0 & \nu^c \\ u & u & u & \vdots & -\nu^c & 0 \end{pmatrix},\ \bar{f}_1 = \begin{pmatrix} u^c \\ u^c \\ u^c \\ \cdots \\ \nu \\ e^- \end{pmatrix} \tag{21}$$

[1]This argument applies also to the $SU(2)$ of the Standard Model. Thus our present knowledge of string theory predicts $I = \frac{1}{2}$ for the Standard Model Higgs and hence predicts successfully the neutral-to-charged current ratio (4).

and similarly for the heavier generations. Note that everywhere that a u-quark appears in conventional $SU(5)$, here there is a d-quark, and vice versa, and that likewise the ν and e are flipped. This is possible because the $U(1)_Y$ of hypercharge in the Standard Model is a linear combination of the external $U(1)$ gauge group factor, as well as the normal hypercharge $U(1)$ in normal $SU(5)$. Therefore, the sums of the charges in GUT representations do not have to add up to zero.

The great attraction of this flipped $SU(5)$ model is that all the symmetry breaking can be done with small Higgs representations. Thus the following representations suffice to break $SU(5) \times U(1) \rightarrow SU(3) \times SU(2)_L \times U(1)_Y$:

$$H = (\mathbf{10}, \frac{1}{2}),\ \bar{H} = (\bar{\mathbf{10}}, -\frac{1}{2}) \tag{22}$$

As you can see from the particle assignments in Eq. (21), the **10** of $SU(5)$ contains a field with the quantum numbers of the ν^c, and if this acquires a v.e.v., the Standard Model gauge group is left invariant. The second stage of symmetry breaking in which $SU(2)_L \times U(1)_Y \rightarrow U(1)_{em}$ is assured by the following Higgs representations

$$h = (\mathbf{5}, -\mathbf{1}),\ \bar{h} = (\bar{\mathbf{5}}, \mathbf{1}) \tag{23}$$

In addition, the model features extra singlet fields ϕ_i.

The flipped $SU(5)$ model has many interesting features, including an elegant and economical mechanism for suppressing proton decay through dimension-five operators, but here I will just discuss two of its important properties, namely a "prediction" for m_t, and the possibilities for neutrino masses and baryogenesis. As we have already discussed, there are many possible four-dimensional string models, which can be regarded as corresponding to different Higgs vacua. One can in general predict Yukawa couplings in any of these models (vacua), and they are normally [32] of the same order of magnitude as the gauge coupling at the string scale close to the Planck mass

$$\lambda = (\text{constant}) \times g \tag{24}$$

For example, in the variant of flipped $SU(5)$ derived [28] from superstring

$$\lambda = \sqrt{2}\, g \tag{25}$$

for all Yukawa couplings λ that are non-vanishing. The top quark could in principle be a mixture between states with $\lambda \neq 0$ and $\lambda = 0$, so that the Yukawa coupling of the physical top quark is given by

$$\lambda_t = \sqrt{2}\, g \cos\theta_t \tag{26}$$

where θ_t is a mixing angle to be determined. It is a general feature of such models that the Yukawa coupling λ_t is renormalized between normalization scales $Q \simeq m_P$ and m_W. There is an interesting feature of the corresponding renormalization group equations, namely that there is an approximate infra-red-stable fixed point [33]. This means that for any finite value of λ_t at the Planck scale,

$$m_t \lesssim 220\, GeV \tag{27}$$

In the particular case of flipped $SU(5)$ where Eq. (25) holds, we find [28]

$$m_t \lesssim 190\, GeV \tag{28}$$

This range of top quark masses compares very well with the range inferred from low-energy experimental data from LEP and other accelerators [10]. It is also consistent with the lower limit on m_t provided by the FNAL Tevatron collider. Indications are that the physical value of m_t may be quite close to the infrared-stable fixed point. The deeper puzzle is not why m_t is so large, but rather why the other quark masses are so much smaller. The answer may well be that the other quark masses originate from higher-order terms in the effective superpotential [34], so that

$$m_f/m_t = O\left(\frac{<0\mid \phi \mid 0>}{m_P}\right)^n : n \geq 1 \tag{29}$$

where $< 0 \mid \phi \mid 0 >$ is smaller than m_P.

It is a generic feature of GUTs that they predict very small neutrino masses, via a seesaw neutrino mass matrix [35]. In the particular case of flipped $SU(5)$, we find [28]

$$(\nu, \nu^c, \phi)_i \begin{pmatrix} 0 & m_u^T & 0 \\ m_u & 0(V) & 0(V) \\ 0 & 0(V) & 0(V) \end{pmatrix}_{ij} \begin{pmatrix} \nu \\ \nu^c \\ \phi \end{pmatrix}_j \tag{30}$$

which yields a light mass eigenstate

$$m_{\nu_i} \simeq m_u^2/M_i \tag{31}$$

where M_i is the effective ν^c mass given by the bottom right-hand 2×2 submatrix in Eq. (30). The scaling law in Eq. (31) enables one to correlate

$$\delta m^2_{\nu_{e,\mu}} \sim 10^{-5}\, eV^2 \tag{32}$$

as suggested by the MSW [36] solution to the solar neutrino problem [37] with

$$m_{\nu_\tau} \sim 10\, eV \tag{33}$$

as required by cosmological models with an admixture of Hot Dark Matter. Moreover, one can use [38] the estimate

$$\sin 2\theta_{e\mu} \sim 10^{-2} \tag{34}$$

used in the MSW explanation of the solar neutrino deficit to predict [38]that

$$\sin 2\theta_{\mu\tau} \sim 10^{-3}\, \text{to}\, 10^{-4} \tag{35}$$

which is within the range accessible to a new round of experiments being prepared at CERN [39]. Another appealing feature of this model is that heavy ν^c decays can generate the cosmological baryon asymmetry [40]. Their decays $\nu^c \to \ell + H$ may generate a small lepton asymmetry in the presence of CP violation, which non-perturbative electroweak effects then recycle [41] to give a baryon asymmetry. This mechanism can easily lead to

$$n_B/n_\gamma \simeq (2\, \text{to}\, 4) \times 10^{-10} \tag{36}$$

as suggested by calculations of homogeneous Big Bang nucleosynthesis. Thus flipped $SU(5)$ may indeed realize the hopes that we had for a string GUT.

6 Indications from Cosmology

Inflation [42] is a very promising solution to many profound cosmological problems. It can explain the observed homogeneity and isotropy of the Universe over distances much larger than the apparent horizon $d = c \times t_0$, where t_0 is the present age of the Universe. It can also explain the near-flatness of the Universe, corresponding to the observation that the present density is close to the critical density:

$$H^2 = \left(\frac{\dot{a}}{a}\right)^2 = \frac{8\pi G_N}{3}\rho - \frac{k}{a^2} : \Omega \equiv \rho/\rho_c \simeq 1 \tag{37}$$

This approximate flatness occurs because the last (curvature) term in Eq. (37) is exponentially suppressed during the inflationary epoch. As a result of this feature, one can also understand how the Universe has got to be so old and large, whereas a generic solution of Einstein's equation only grows to a scale size $a \sim 1/m_P, t \sim 1/m_P$. Moreover, unwanted particles such as monopoles and gravitinos are inflated away beyond the observable horizon. As is well known, the basic idea of inflation is an epoch of exponential expansion driven by an effective cosmological constant

$$H = \sqrt{\frac{\Lambda}{3}} : \Lambda = 8\pi G_N \rho_0 \tag{38}$$

The interest in inflation has been increased recently by the observation using the COBE satellite [43] of fluctuations in the microwave background radiation.

Inflationary cosmology suggests that density and metric perturbations originate in the very early Universe as quantum or thermal fluctuations in the scalar field driving inflation [44]. These quantum fluctuations cause different parts of the Universe to expand at different rates, for example during the roll-down from different initial field values. These quantum fluctuations yield a Gaussian random field of perturbations $\delta\rho/\rho$ which have similar magnitudes at different scale sizes, as long wanted by astrophysicists. The magnitude of these perturbations is related to the value of the field energy driving inflation

$$\frac{\delta\rho}{\rho} \sim G_N\mu^2 : \rho_0 = V \sim \mu^4 \tag{39}$$

The COBE data that indicate $\frac{\delta\rho}{\rho} \sim 10^{-5}$ correspond to a value

$$\mu \simeq 10^{16}\, GeV \tag{40}$$

which is remarkably close to the scale of supersymmetric grand unification. Time will tell whether this is just a coincidence.

The COBE observations cast new light on the problem of dark matter. Inflation suggests that the present density of the Universe is very close to the critical density $\Omega = \rho/\rho_c \simeq 1$. However, observations suggest that $\Omega_{visible} \sim 0.01$. On the other hand, homogeneous cosmological Big Bang nucleosynthesis requires a somewhat larger density of baryons, nevertheless with $\Omega_B \lesssim 0.1$. Observations of galactic rotation curves and the virial theorem suggest that the density of matter in galactic haloes $\Omega_H \sim 0.1$. Mathematically, the halo dark matter could be baryonic, most likely in the form of "Jupiters". Indeed, two recent experiments have found tantalizing indications of possible microlensing phenomena which might be due to such dark matter in the halo of our own galaxy [45]. However, the Big Bang nucleosynthesis estimates indicate that baryons cannot give the critical density, and so there must be non-baryonic dark matter, even if there is not much of it in our galactic halo.

Non-baryonic dark matter particles are conventionally divided into two categories: hot and cold, according to whether the particles were relativistic or not at the time when galaxy-size density perturbations started to grow. Until recently, most models of structure formation in the Universe based on inflationary perturbations favoured a dominance by Cold Dark Matter, as could be provided, for example, by the lightest supersymmetric particle. However, a detailed comparison of COBE and other large-scale data with data from smaller scales indicates that a pure Cold Dark Matter model is no longer favoured. Instead, a mixture of Hot and Cold Dark Matter seems to be preferred [46]

$$\Omega_{Cold} \simeq 0.7,\ \Omega_{Hot} \simeq 0.3,\ \Omega_B \lesssim 0.1 \tag{41}$$

The most natural candidate for the Hot Dark Matter is a τ neutrino weighing about $10\,eV$, as mentioned in the previous section. We can only hope that laboratory experiments will soon find evidence for neutrino masses and/or some type of Cold Dark Matter particle.

7 Summary

We are not yet sure which way to go beyond the Standard Model, but I believe that, as seen in Fig. 7, we have discovered recently various signposts pointing us in one particular direction, that favoured by Abdus Salam over the years. The LEP measurement of $sin^2\theta_W$, the success of simple GUT predictions for m_b/m_τ [47], the suggestions from the solar neutrino deficit that neutrinos might have small but non-zero masses, the astrophysical and cosmological preference for some form of Hot Dark Matter as inferred from COBE data all point us in the direction of GUTs. Moreover, the tentative indications from LEP on the mass of the Higgs boson and the LEP measurements of $sin^2\theta_W$ indicate a role for supersymmetry, as suggested to resolve the naturalness problem. Thus, I believe that we should head in the direction of supersymmetric GUTs, and that beyond them lie some TOE derived from string. Let us hope that experiments soon provide us with more definite proof that this is the right direction to follow beyond the Standard Model.

References

[1] S.L. Glashow, *Nucl. Phys.* **22** (1961) 579;

S. Weinberg, *Phys. Rev. Lett.* **19** (1967) 1264; A. Salam, Proc. 8th Novel Symposium, Stockholm 1968, ed. N. Svartholm (Almqvist and Wiksells, Stockholm, 1968), p. 367.

[2] P. Fayet and S. Ferrara, *Physics Reports* **32** (1977) 251.

[3] Y. Nambu, in Proc. XI Int. Symp. on Elementary Particle Physics, eds. Z. Ajduk, S. Pokorski and A. Trautman (World Scientific, Singapore, 1989);
A. Miranski, M. Tanabashi and K. Yamauraki, *Mod. Phys. Lett.* **A4** (1989) 1043 and *Phys. Lett.* **221** (1989) 177;
W.A. Bardeen, C.T. Hill and M. Lindner, *Phys. Rev.* **D41** (1990) 1647.

[4] E. Farhi and L. Susskind, *Physics Reports* **74C** (1981) 227.

[5] M. Lindner, Heidelberg preprint HD-THEP-93-19 (1993).

[6] S. Dimopoulos and L. Susskind, *Nucl. Phys.* **B155** (1979) 237:
E. Eichten and K.D. Lane, *Phys. Lett.* **90B** (1980) 125.

[7] S. Dimopoulos and J. Ellis, *Nucl. Phys.* **B182** (1981) 505.

[8] P. Binetruy, S. Chadha and P. Sikivie, *Phys. Lett.* **107B** (1981) 425.

[9] M. Peskin and T. Takeuchi, *Phys. Rev. Lett* **65** (1990) 964;
G. Altarelli and R. Barbieri, *Phys. Lett.* **B253** (1991) 161.

[10] J. Ellis, G.L. Fogli and E. Lisi, CERN preprint TH.6967/93 (1993).

[11] Y. Okada, M. Yamaguchi and T. Yanagida, *Progr. Theor. Phys* **85** (1991) 1;
J. Ellis, G. Ridolfi and F. Zwirner, *Phys. Lett.* **B257** (1991) 83;
H.E. Haber and R. Hempfling, *Phys. Rev. Lett.* **66** (1991) 1815.

[12] H.E. Haber and G.L. Kane, *Phys. Rep.* **117** (1985) 75.

[13] P. Langacker, University of Pennsylvania preprint UPR 0238T (1990);
J. Ellis, S. Kelley and D.V. Nanopoulos, *Phys. Lett.***B249** (1990) 441;
U. Amaldi, W. de Boer and H. Furstenau, *Phys. Lett.***B260** (1991) 447.

[14] Z. Kunszt and F. Zwirner, *Nucl. Phys.* **B385** (1992) 3.

[15] R. Ammar et al., *Phys. Rev. Lett.* **71** (1993) 674.

[16] V. Barger, M.S. Berger and R.J.N. Phillips, *Phys. Rev. Lett.* **70** (1993) 1368.

[17] R. Barbieri and G.F. Giudice, *Phys. Lett.***B309** (1993) 86.

[18] H. Goldberg, *Phys. Rev. Lett.* **50** (1983) 1419:
J. Ellis, J.S. Hagelin, D.V. Nanopoulos, K.A. Olive and M. Srednicki, *Nucl. Phys.* **B238** (1984) 453.

[19] J. Pati and A. Salam, *Phys. Rev. Lett.* **31** (1973) 661, and *Phys. Rev.* **D8** (1973) 1240, **D10** (1974) 275.

[20] H. Georgi and S.L. Glashow, *Phys. Rev. Lett.* **32** (1974) 438.

[21] J. Ellis, S. Kelley and D.V. Nanopoulos, *Nucl. Phys.* **B373** (1992) 55.

[22] F. Anselmo, L. Cifarelli, A. Petermann and A. Zichichi, CERN preprints PPE-91-123 (1991), TH.6429/92 (1992), PPE-92-103 (1992).

[23] K.S. Narain, *Phys. Lett.* **169B** (1986) 41;
W. Lerche, D. Lüst and A.N. Schellekens, *Nucl. Phys.* **B287** (1987) 477;
H. Kawai, D.C. Lewellen and S.-H.H. Tye, *Nucl. Phys.* **B288** (1987) 1.

[24] I. Antoniadis, C. Bachas and C. Kounnas, *Nucl. Phys.* **B289** (1987) 87;
A. Antoniadis and C. Bachas, *Nucl. Phys.* **B298** (1988) 586.

[25] For a review, see:
A.B. Lahanas and D.V. Nanopoulos, *Phys. Rep.* **145** (1987) 1.

[26] B.R. Greene, K.H. Kirklin, P.J. Miron and G.G. Ross, *Phys. Lett.* **B180** (1988) 69; *Nucl. Phys.* **B278** (1986) 667 and **B292** (1987) 606; *Phys. Lett.***B192** (1987) 111.

[27] L.E. Ibáñez, J.E. Kim, H.P. Nilles and F. Quevedo, *Phys. Lett.***B191** (1987) 287;
A. Font, L.E. Ibáñez, H.P. Nilles and F. Quevedo, *Phys. Lett.***B210** (1988) 101;
A. Font, L.E. Ibáñez and F. Quevedo, *Phys. Lett.***B217** (1989) 272; *Phys. Lett.***B224** (1989) 79.

[28] I. Antoniadis, J. Ellis, J.S. Hagelin and D.V. Nanopoulos, *Phys. Lett.***B194** (1987) 213 and **B231** (1989) 65.

[29] J. Ellis, J. Lopez and D.V. Nanopoulos, *Phys. Lett.***B247** (1990) 257;
J. Ellis et al., *Nucl. Phys.***B373** (1992) 283.

[30] H. Dreiner et al., *Phys. Lett.***B216** (1989) 283.

[31] P. Candelas, G. Horowitz, A. Strominger and E. Witten, *Nucl. Phys.* **B258** (1985) 46.

[32] E. Witten, *Phys. Lett.***155B** (1985) 151.

[33] C.T. Hill, *Phys. Rev.***D24** (1981) 691;
B. Pendleton and G.G. Ross, *Phys. Lett.***98B** (1981) 291.

[34] J. Lopez, D.V. Nanopoulos and K.J. Yuan, *Nucl. Phys.* **B399** (1993) 654.

[35] T. Yanagida, Proc. Workshop on the Unified Theory and the Baryon Number in the Universe (KEK, Japan, 1979);
R. Slansky, Talk at the Sanibel Symposium, Caltech preprint CALT-68-709 (1979).

[36] L. Wolfenstein, *Phys. Rev.***D17** (1978) 2369;
S.P. Mikheyev and A.Yu. Smirnov, *Nuovo Cimento***9C** (1986) 17.

[37] P. Anselmann et al., *Phys. Lett.***B285** (1982) 390, and references therein.

[38] J. Ellis, J. Lopez and D.V. Nanopoulos, *Phys. Lett.***B292** (1992) 189.

[39] CHORUS Collaboration, N. Armenise et al., CERN-SPSC/90-42 (1990);
NOMAD Collaboration, P. Astier et al., CERN-SPSC/91-21 (1991).

[40] J. Ellis, D.V. Nanopoulos and K.A. Olive, *Phys. Lett.***B308** (1993) 70, and references therein.

[41] M. Fukugita and T. Yanagida, *Phys. Lett.***174B** (1986) 45.

[42] A.A. Starobinsky, *Phys. Lett.***91B** (1980) 99;
A. Guth, *Phys. Rev.***D23** (1981) 349.

[43] G.F. Smoot et al., *Ap. J. Lett.* **396** (1992) L1;
E.L. Wright et al., *Ap. J. Lett.* **396** (1992) L13.

[44] J. Bardeen, P.J. Steinhardt and M.S. Turner, *Phys. Rev.***D28** (1983) 679;
A.H. Guth and S.-Y. Pi, *Phys. Rev. Lett.***49** (1982) 1110;
A.A. Starobinsky, *Phys. Lett.***117B** (1982) 175;
S.W. Hawking, *Phys. Lett.***115B** (1982) 295.

[45] C. Alcock et al., *Nature***365** (1993) 621;
E. Aubourg et al., *Nature***365** (1993) 623.

[46] See, for example:
G. Efstathiou, J.R. Bond and S.D.M. White, *Mon. Not. R. Ast. Soc.***258** (1992) 1P;
M.S. Turner, FNAL Preprint Conf-92/313-A (1992).

[47] M. Chanowitz, J. Ellis and M.K. Gaillard, *Nucl. Phys.***B128** (1977) 506.

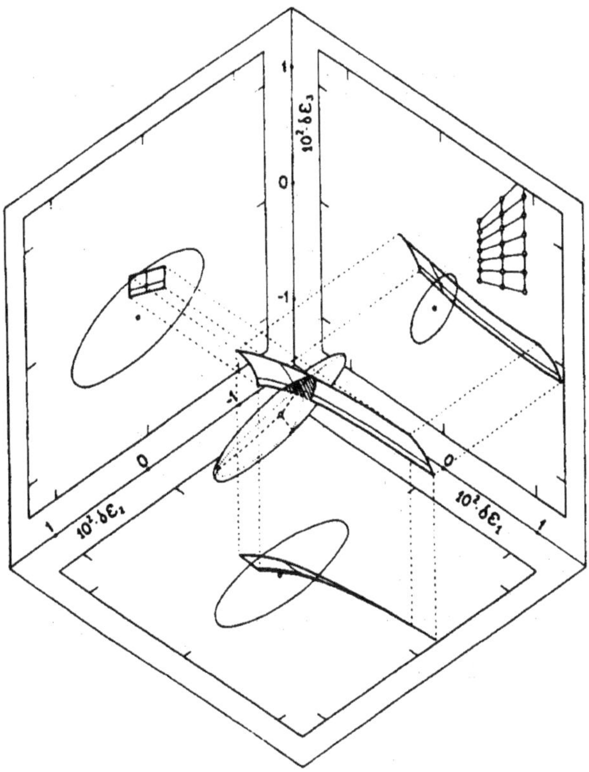

Fig. 1. LEP constraints [10] on $(S,T,U)(\varepsilon_{1,2,3})$ confronted with some technicolour predictions.

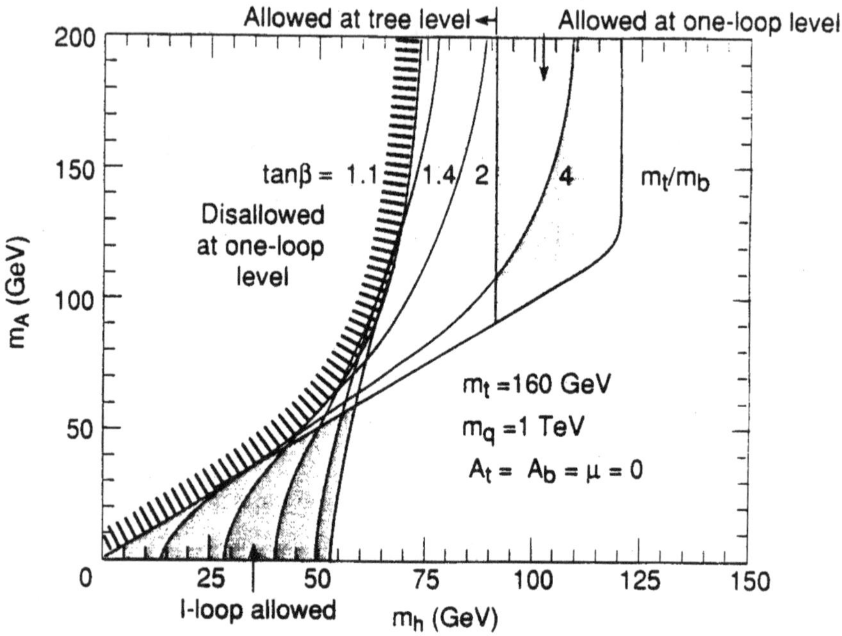

Fig. 2. The (m_h, m_A) parameter plane for Higgs bosons in the MSSM, including radiative corrections [11].

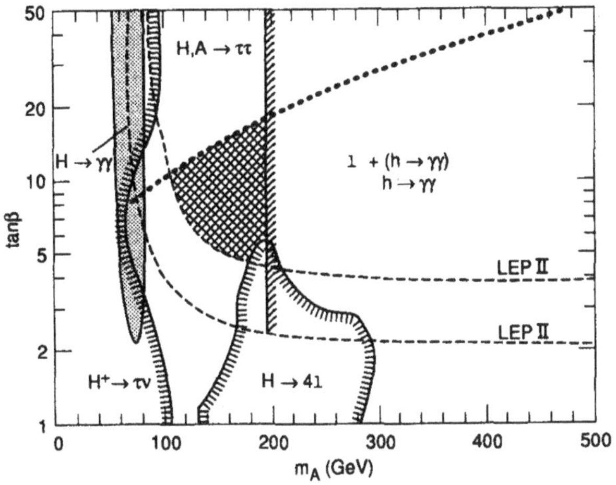

Fig. 3. Regions of the $(m_A, \tan\beta)$ plane accessible to experiments at LEP II and the LHC [14].

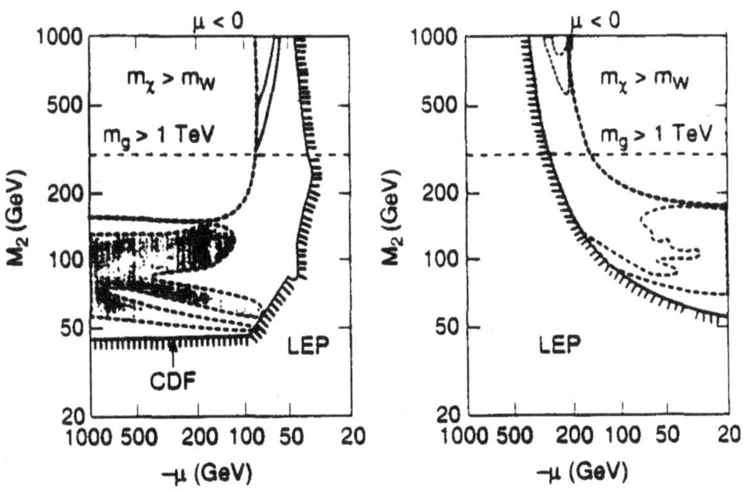

Fig. 4. Regions of the $(\mu, m_{1/2})$ plane excluded by LEP and CDF searches, and those in which an "interesting" relic density $0.1 < \Omega h^2 < 1$ is possible for $m_\chi < m_W$.

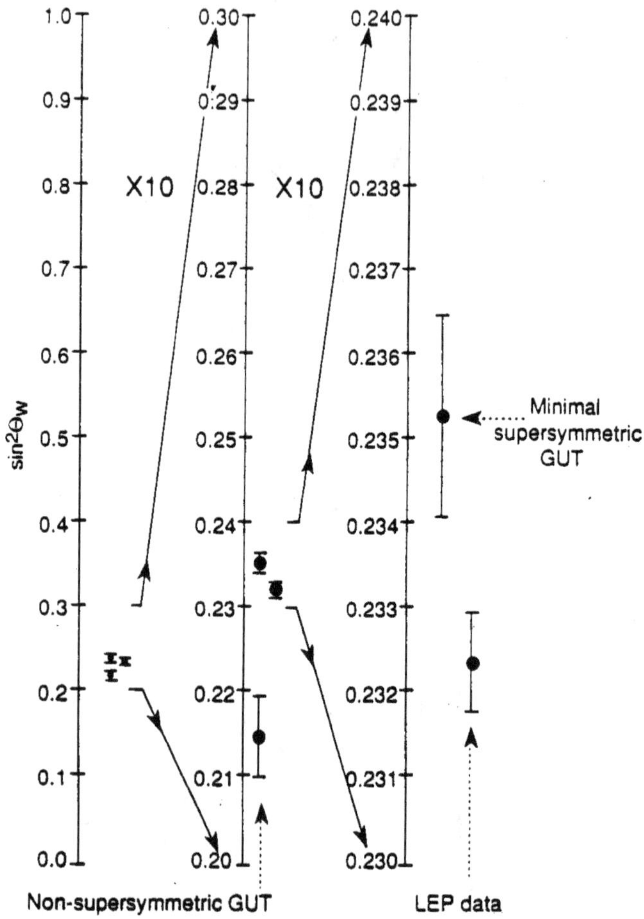

Fig. 5. Different GUT predictions for $\sin^2\theta_W$ confronted with LEP data.

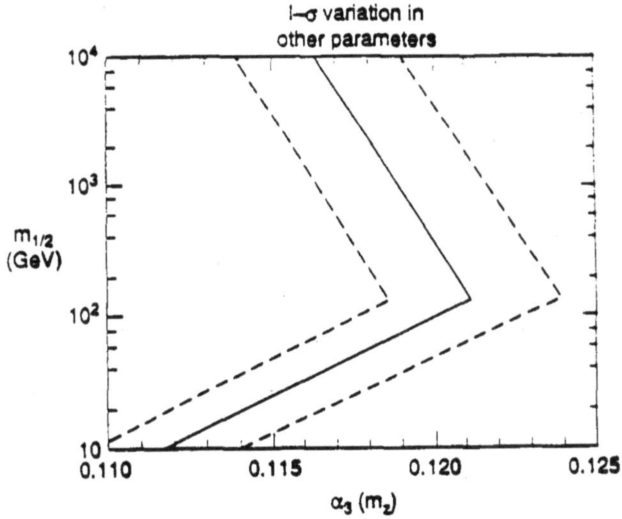

Fig. 6. LEP data do not yet enable us to fix with precision the supersymmetry-breaking scale $m_{1/2}$ [21], [22].

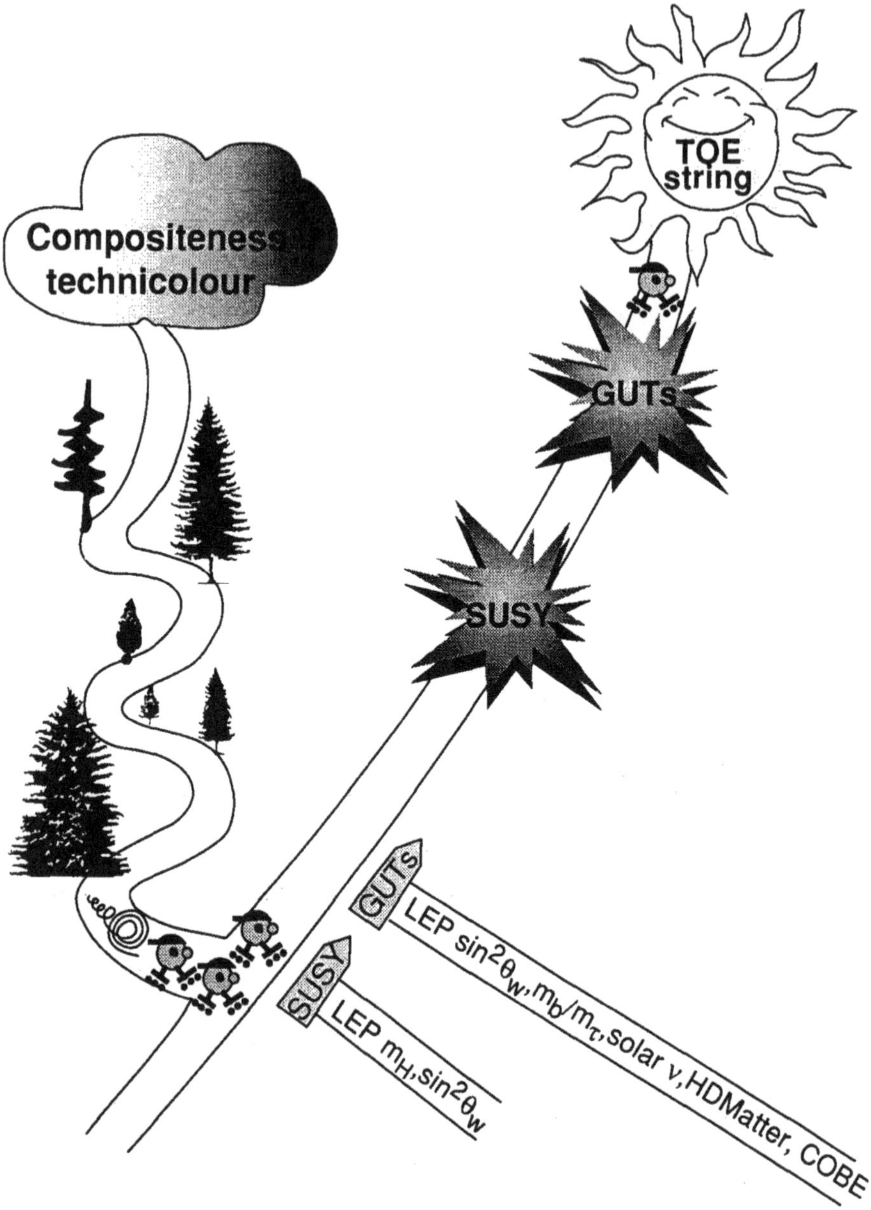

Fig. 7. Signposts pointing the road to take beyond the Standard Model, following Abdus Salam.

Dynamical Electroweak Breaking: Issues and Challenges

R .D. Peccei

Department of Physics, University of California, Los Angeles, CA 90024

Abstract

After briefly remarking on alternatives for breaking the electroweak symmetry, I discuss the implication that recent precision experiments at LEP have for the symmetry breaking sector. The difficulties associated with generating fermion masses when the electroweak symmetry is broken dynamically are exposed and an alternative to the walking technicolor - extended technicolor scenario is suggested, based on models of substructure. Failures and lessons from trying to incorporate families and mixing in these latter schemes are also discussed.

It is a great pleasure and a singular honor for me to be talking at this "Salamfest", particularly on a topic that is so closely connected to one of Salam's central contributions to physics: the theory of the electroweak interactions. The model for the electroweak interactions put forth more than 25 years ago by Glashow, Salam and Weinberg [1], as more and more data was found to be in agreement with its predictions, has made a transition from model, to theory, to paradigm. LEP has provided the last chapter in this saga of success. The precise comparison of LEP data with the predictions of the Glashow, Salam and Weinberg theory provides overwhelming evidence that the electroweak interactions are indeed described by an $SU(2) \times U(1)$ gauge theory spontaneously broken to $U(1)_{em}$, with some custodial global symmetry guaranteeing that $\rho = 1$ [2]. Remarkably, however, even though the $SU(2) \times U(1)$ theory is well tested, the exact nature of the symmetry breaking mechanism is still essentially unknown. My remarks here will try to address this issue and some of its ramifications, particularly concerning dynamical symmetry breakdown.

As in any symmetry breakdown, the breakdown of the $SU(2) \times U(1)$ electroweak gauge symmetry to $U(1)_{em}$ is governed by an order parameter. What is uncertain, at the moment, is precisely what this order parameter is. Two alternative theoretical speculations exist concerning the nature of this order parameter. Either it is thought to be:

i) the vacuum expectation value (vev) of an elementary scalar field , $< \Phi >$,

or it is assumed that

ii) it is related to a dynamical condensate of fermion - antifermion pairs, $< \bar{T}T >$, formed in a, as yet to be discovered, underlying theory.

Both the physics and philosophy behind these two alternatives are quite distinct. In particular, the first option is quite compatible with weak coupling, while the second option necessarily requires strong coupling.

The simplest way to break $SU(2) \times U(1) \rightarrow U(1)_{em}$ is to introduce a complex scalar doublet Φ into the electroweak theory, giving this field an asymmetric potential

$$V(\Phi) = \lambda(\Phi^\dagger\Phi - \frac{v^2}{2})^2 \quad , \tag{1}$$

which leads to the breakdown. The order parameter $< \Phi >$ is then directly related to the scale $v \simeq (\sqrt{2}G_F)^{-1/2} \simeq 250\ GeV$ introduced in the potential and the breakdown occurs irrespective of the strength λ of the potential. In supersymmetric versions of the Glashow, Salam and Weinberg model, where the presence of scalar fields is natural [3], the quartic scalar coefficients in fact are simply related to the $SU(2)$ and $U(1)$ couplings, so that in these cases these couplings are clearly weak [4]. Even without invoking supersymmetry, perturbative control of the theory argues for a weak effective coupling at the scale of the symmetry breakdown, $\lambda(v)$, so that the Landau pole in this coupling constant evolution is beyond the Planck mass, where surely new physics will enter [5]. [1]

Just as scalar vevs are naturally connected with weak coupling, having the $SU(2)\times U(1) \rightarrow U(1)_{em}$ order parameters be related to a fermion-antifermion condensate, $< \bar{T}T >$, necessarily involves strong coupling. This is particularly clear if the interactions which gives rise to the fermion - antifermion condensates are those of a non Abelian gauge theory, as in technicolor [6]. The running coupling constant squared of such theories $\alpha_T(q^2)$, just like that of QCD, is weak at short distances but becomes strong as q^2 decreases. One can define a characteristic scale for these theories as the scale where $\alpha_T\ (q^2)$ becomes unity

$$\alpha_T(\Lambda_T^2) = 1 \quad . \tag{2}$$

Condensate formation occurs precisely because the gauge interactions become strong. Thus the size of the order parameter $< \bar{T}T >$ will be related to the scale where $\alpha_T\ (q^2)$ becomes strong, and one expects

$$< \bar{T}T > \sim \ \Lambda_T^3 \quad . \tag{3}$$

The physics of the symmetry breakdown of the electroweak interactions is most clearly manifested in the amplitudes for longitudinal gauge boson scattering. Thus, it is in $W_L W_L$ scattering where one may eventually see a distinction between the weak coupling and strong coupling alternatives discussed above. At very high energies $E >> M_W$, it is possible to establish a direct connection - through the, so called, equivalence theorem of Cornwall,

[1]In turn, weak coupling, necessitates having a light Higgs scalar in the spectrum, since $M_H^2 = 2\lambda\ (v)\ v^2$.

Levin and Tiktopolous [7] - between the amplitudes of $W_L W_L$ scattering and those of their corresponding Goldstone bosons fields, w. [2] Namely

$$A_{W_L W_L} = A_{ww} + 0(\frac{M_W}{E}) \tag{4}$$

Clearly in the case where the order parameter is the elementary doublet vev $< \Phi >$, because the scalar self coupling λ is weak, the Goldstone boson scattering amplitudes will also be weak and one is led to expect weak interactions for the longitudinal gauge bosons. On the other hand, if the symmetry breakdown is dynamical, as a result of condensate formation, the Goldstone bosons interactions in the underlying theory are strong and thus one expects that there will be strong interactions among the $W_L' s$.

Even though the equivalence theorem is a high energy theorem, so that it cannot be strictly applied near threshold, there should not be much difference at low energy between the strongly coupled and weakly coupled symmetry breaking sectors. As is well known [9], the threshold behaviour of Goldstone boson interactions depends only on the metric of the coset space of the breakdown and not on any other details. Thus, independently of whether the Goldstone bosons w are strongly or weakly coupled, they have the same threshold interactions. Consequently, one expects that also the $W_L' s$ should have the same interactions at low energy, independently of exactly how the $SU(2) \times U(1)$ symmetry is broken.

The distinction between the two symmetry breaking mechanisms discussed, however, should become apparent at CM energies in the $W_L W_L$ subsystem of order $\hat{E}_{cm} \simeq \sqrt{16\pi v^2} \simeq 1.7\ TeV$ [10]. At these energies, in the strongly coupled case one expects resonance formation, but nothing particularly spectacular for the weakly coupled case. To get to these energies in the $W_L W_L$ subsystem one needs the CM energies of the LHC and SSC, because of the energy degradation which occurs as one goes from protons, to quarks and, finally, to $W's$. Many studies have been done to see if signals of strong $W_L W_L$ scattering, such as resonance formation, could be visible at these future machines [11]. The difficulty, of course, is to dig out these signals from the background arising from ordinary two-W production. Although after cuts there are not many events left, in all cases examined signals of strong

[2]These are the fields which are absorbed in the Higgs mechanism [8] and serve to give the $W's$ mass.

$W_L W_L$ scattering are detectable at SSC energies, with these signals being somewhat more marginal for the LHC.

Although the issue of what triggers the $SU(2) \times U(1)$ breaking will be eventually answered when the LHC and SSC become operational, it is interesting to ask whether one can tell anything already now about this question from the high precision electroweak data that has been gathered at LEP. LEP energies are quite "low" to be really directly sensitive to the symmetry breaking sector, but how the symmetry is broken can influence radiative corrections. In the elementary scalar case, as was pointed out first by Veltman [12], the dependence of the radiative corrections on the Higgs scalar mass, which typifies the symmetry breakdown, is given schematically by

$$\text{rad. corr.} \sim \frac{\alpha}{\pi} \ell n\ M_H/M_Z + (\frac{\alpha}{\pi})^2 (\frac{M_H}{M_Z})^2 \ . \tag{5}$$

Thus, at leading order, one feels only **logarithmic** effects. Quadratic dependence on the Higgs mass is only felt at $0(\alpha^2)$ and is negligible if the Higgs scalar is light, as expected in this scenario.

The effects of a possible strong coupling symmetry breaking sector could be felt at LEP through modifications to the gauge propagators - the, so called, **oblique corrections**[13]. Since the typical scale where these effects should begin to be significant is of order $\sqrt{16\pi v^2} >> M_Z$, it is sensible to expand the vacuum polarization tensors for the gauge fields in a power series in q^2, retaining just the first two terms [14]:

$$\Pi_{AB}\ (q^2) = \Pi_{AB}(0) + q^2 \Pi'_{AB}\ (0) \tag{6}$$

Here the pair $\{AB\}$ spans the 4 possible gauge field configurations $\{AB = WW; ZZ; \gamma\gamma; \gamma Z\}$. Electromagnetic gauge invariance implies that $\Pi_{\gamma\gamma}\ (0) = \Pi_{\gamma Z}\ (0) = 0$, so that in toto one has 6 possible constants characterizing the vacuum polarization tensors in this approximation. Since one is assuming that the electroweak theory is $SU(2) \times U(1)$, three combinations of these constants can be fixed in terms of the three independent parameters which typify this theory, the two coupling constants g_1 and g_2 and v. [3] The

[3]More practically, instead of g_1, g_2 and v one uses three other related parameters which are more accurately known experimentally: α, M_Z and G_F, the Fermi constant measured in μ decay.

other three combination of parameters can in principle be determined from experiment and then compared to what is expected by different schemes for symmetry breakdown.

A standard set of parameters, denoted by S, T and U by Peskin and Takeuchi [15] and by other equivalent symbols by other authors, has emerged as being the most convenient to perform this analysis. It turns out that, of the three, only S is moderately sensitive to the symmetry breaking sector, because T and U are dominated by other uncertainties. [4] Furthermore, if the symmetry breaking sector conserves an approximate vectorial $SU(2)$ symmetry - as it surely must, since experimentally the ρ parameter is very near unity - one can write S directly in terms of the vacuum polarization tensors for $SU(2)_V$ and $SU(2)_A$ currents, related to the $SU(2) \times U(1)$ currents, facilitating in this way the comparison with models of dynamical symmetry breaking. One finds [14]

$$S = -4\pi\{\Pi'_{VV}(0) - \Pi'_{AA}(0)\} = \int_0^\infty \frac{ds}{s}[v(s) - a(s)] \ , \tag{7}$$

where the second line makes use of a spectral decomposition for the vector and axial vector current correlation functions.

One can extract what values of S (and T) are allowed by comparing electroweak observables [like the hadronic width of the Z] to the theoretical predictions. To do so one needs some reference value of m_t and M_H to fix the expectations of the standard model, with elementary scalar field breaking [5]. That is, for each experimental observable one has formulas of the type [15]

$$O_{\rm exp} \ = \ 0_{sm} \ + \alpha_0 \ S + \beta_0 \ T + \gamma_0 U \tag{8}$$

with α_0, β_0 and γ_0 computable coefficients [e.g. for Γ_Z, using as reference point $M_H = 1\ TeV, m_t = 150\ GeV$, one has $(\Gamma_Z)_{\rm exp} \ = \ 2.487 \pm 0.010\ GeV$; $(\Gamma_Z)_{sm} \ = \ 2.484\ GeV; \alpha_0 \ = \ -9.58 \times 10^{-3}\ GeV$; $\beta_0 = +2.615 \times 10^{-2}\ GeV$; $\gamma_0 = 0$]. A global fit of all precision electroweak data, performed

[4] T depends quadratically on the still unknown top quark mass and so its measured value serves mostly to constrain m_t. U, on the other hand, depends sensitively on the value of the W mass and the present experimental uncertainty in this value precludes using U as a sensitive probe of the symmetry breaking sector.

[5] The dependence of the results on a standard model reference point can actually be avoided [16]. This dependence is, at any rate, very mild, as can be seen from Fig. 1.

by Peskin and Takeuchi [15] is shown in Fig. 1. As can be seen, the combined analysis of all data favors negative values of S with

$$S \simeq -0\,(1)\ . \tag{9}$$

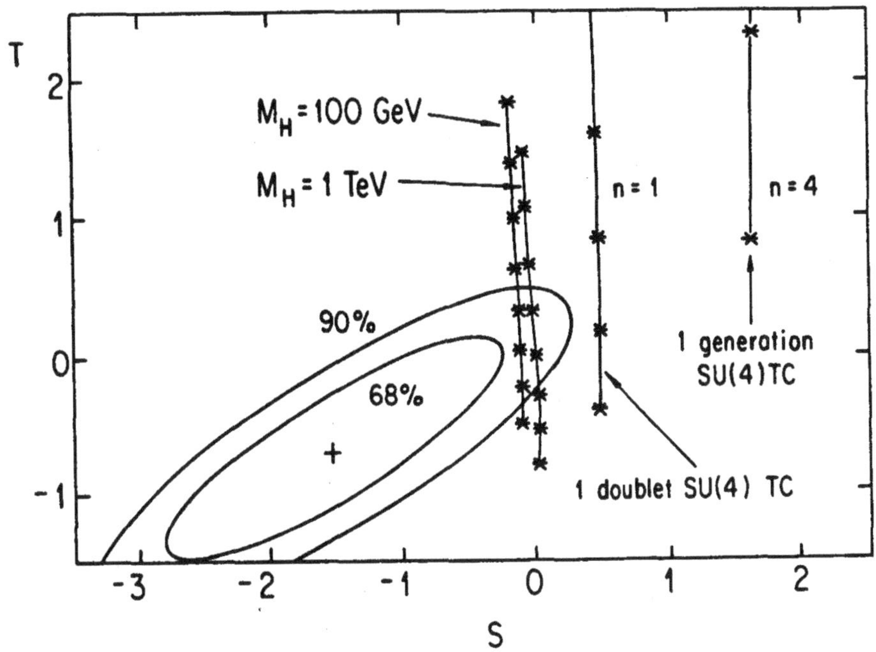

Figure 1: 68% C.L. and 90% C.L. curves in the S and T plane determined by the global fit to all data carried out by Peskin and Takeuchi [15]. Very similar results have also been obtained by [17] [18]. In the figure the predictions of the standard model and of an $SU(4)$ technicolor theory with $n = 1$ and $n = 4$ are shown, as a function of m_t. The crosses in the figure denote m_t values which, starting from the bottom, increase by 20 GeV from $m_t = 90\ GeV$.

This value can be compared to the expectations of various dynamical symmetry breaking models. The simplest of these models is technicolor [6]. Here one imagines that the condensates which break $SU(2) \times U(1)_{em}$ are those of an underlying $SU(N)_T$ theory, whose technifermions come in n

replications of left-handed doublets $\begin{pmatrix} U_i \\ D_i \end{pmatrix}_L, i = 1, 2, \cdots, n$ and right-handed singlets U_{iR}, D_{iR} - much as ordinary quarks do. The condensates

$$< \bar{U}_i U_j > = < \bar{D}_i D_j > = \delta_{ij} \Lambda_T^3 \tag{10}$$

then break $SU(2) \times U(1) \to U(1)_{em}$ precisely as n replicas of scalar doublets Φ_i would. Because this theory is just a "scaled up" version of QCD, with the dynamical scale $\Lambda_T \sim v \sim 250$ GeV, one can estimate the parameter S from QCD. By dominating the vector and axial vector spectral functions in QCD by the ρ and A_1 poles one deduces that [15]

$$S_{QCD} = \frac{6\pi f_\pi^2}{m_\rho^2} \tag{11}$$

and hence for the $SU(N)_T$ theory one has

$$S \simeq n\frac{N}{3} S_{QCD} \simeq n\frac{N}{3}\left(\frac{6\pi f_\pi^2}{m_\rho^2}\right) \simeq 0.08\, nN \tag{12}$$

The vertical lines in Fig. 1 represent the expectations of such a simple technicolor model with $N = 4$ and $n = 1$ or 4 where, however, the QCD spectral integral was computed out without resorting to ρ and A_1 dominance [15] [6]. As can be seen from this figure, the theoretically "more realistic" $n = 4$ technicolor model is about 3σ away from the experimentally allowed region.

Although this result is not very encouraging for dynamical symmetry breaking theories, I do not believe it represents a death knell for these theories. First of all, one can argue, rather convincingly, that in more sophisticated theories with dynamical symmetry breaking of $SU(2) \times U(1)$ - the so called, walking technicolor theories [19], where $\alpha_T(q^2)$ runs slowly - that the integral over the spectral functions in S should converge much more slowly, so that S should be below the estimate given above [20]. Furthermore, if one worries about how fermions get mass in theories where $SU(2) \times U(1)$ is broken dynamically, it becomes quite clear that a naive scaling up of QCD will just not do! I turn to discuss this important point further.

[6]These results agree quite well, however, with those obtained by assuming simple ρ and A_1 dominance.

Simple technicolor models do not address the issue of fermion masses unless one introduces some communication between ordinary fermions, f - the quarks and leptons - and the fermions which condense, T. This is done ordinarily by introducing a second underlying theory - a, so called, extended technicolor theory (ETC) [21] - to connect the f and T fermions together. If these fermions sit in the same ETC representation and the ETC theory is a spontaneously broken gauge theory, broken at a large scale Λ_{ETC}, one generates in this way an effective four-fermion interaction

$$\mathcal{L}_{eff}^{ETC} = \frac{1}{\Lambda_{ETC}^2} (\bar{T}_L T_R)(\bar{f}_L f_R) \quad . \tag{13}$$

This term, when the T fermions condense $< \bar{T}_L T_R > \sim \Lambda_T^3$, gives the fermions f a mass

$$m_f \sim \frac{< \bar{T}_R T_L >}{\Lambda_{ETC}^2} \sim \frac{\Lambda_T^3}{\Lambda_{ETC}^2} \tag{14}$$

Given that the technicolor scale $\Lambda_T \sim v$, the above formula has a generic difficulty. If one wants to generate with it the top quark mass, because top is so heavy, the scale Λ_{ETC} cannot be too large. However, if Λ_{ETC} is only in the $(1 - 10)$ TeV range, then one cannot avoid large flavor changing neutral current interactions (FCNC), which are not observed experimentally [22]. If Λ_{ETC} is large, to avoid the FCNC problem, then one can never generate in this way large enough fermion masses.

Although the above conundrum is not the only problem of ETC theories, this problem can be ameliorated by walking technicolor models (WTC) [19], which have more realistic dynamics. Furthermore, WTC theories can also ameliorate some of the other endemic problems of ETC, like having too light pseudo Goldstone bosons. So, if one believes that the symmetry breakdown of $SU(2) \times U(1)$ to $U(1)_{em}$ is dynamical, it is much more sensible to focus on WTC models. Let me briefly indicate what is the underlying idea in these models. The identification of Λ_T, the scale of the condensate, with v, the scale related to the W mass, is borrowed from QCD. In QCD, indeed, both the scale which measures the size of the quark condensate $< \bar{u}u >$ - which triggers chiral symmetry breakdown - and that which measures the pion decay constant f_π - associated with the coupling of the broken chiral currents to the Goldstone pion fields - are the same. That is

$$< \bar{u}u > \sim f_\pi^3 \sim \Lambda_{QCD}^3 \tag{15}$$

In WTC models, it turns out that the techniquark condensates $< \bar{T}T >$ have a scale which is much bigger than v^3. Hence, one can obtain rather large masses for sizable ETC scales Λ_{ETC}, avoiding the FCNC $\leftrightarrow m_t$ conundrum.

Both $< \bar{T}T >$ and v, the technipion decay constant, are related to the techniquark self energy $\Sigma(p)$, but probe different momentum scales in this function. Graphically, the relation of $< \bar{T}T >$ and v^2 to $\Sigma(p)$ can be easily inferred from Fig. 2. Retaining only the leading momentum behaviour - neglecting logarithmic terms - one has [23]

$$< \bar{T}T > \sim \int \frac{d^4p}{p^2}\Sigma(p) \sim \int dp^2 \Sigma(p) \ ,$$

while

$$v^2 \sim \int \frac{d^4p}{(p^2)^2}\Sigma^2(p) \sim \int \frac{dp^2}{p^2}\Sigma^2(p) \ . \tag{16}$$

For an asymptotically free theory the chirality breaking self energy $\Sigma(p)$ falls at large momentum as [24]

$$\Sigma(p) \sim \frac{\Sigma(0)^3}{p^2} \ , \tag{17}$$

where again I have neglected logarithmic factors. Here $\Sigma(0)$ serves as the order parameter for the breakdown of the global chiral symmetries of the theory. Thus, one expects that it be of order of the dynamical scale of the technicolor theory: $\Sigma(0) \sim \Lambda_T$.

Because the integral over $\Sigma(p)$ which enters for v^2 is (essentially) convergent, one sees that v^2 probes the region of the techniquark self energy near $p^2 = 0$,

$$v \sim \Sigma(0) \sim \Lambda_T \ . \tag{18}$$

On the other hand, the condensate $< \bar{T}T >$ feels much more the large momentum structure of $\Sigma(p)$. In particular, the region of p^2 important for fermion mass generation is $p^2 \sim \Lambda_{ETC}^2$ and one has

$$m_f \sim \frac{< \bar{T}T >}{\Lambda_{ETC}^2} \sim \frac{1}{\Lambda_{ETC}^2}\int^{\Lambda_{ETC}^2} dp^2\Sigma(p^2) \simeq \Sigma(\Lambda_{ETC}) \ . \tag{19}$$

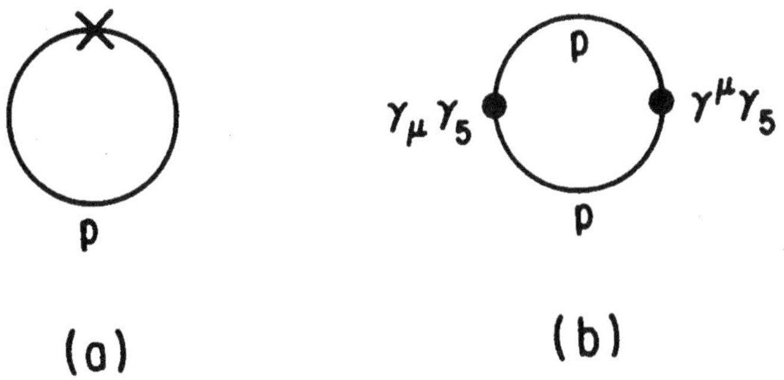

Figure 2: Graphs relating $< \bar{T}T >$ (a) and v^2 (b) to the techniquark self energy $\Sigma(p)$.

The value of $\Sigma(\Lambda_{ETC})$ is not known a priory, without a dynamical calculation. If one assumes that at this scale the technifermion self energy already has achieved its asymptotic value $\Sigma(p) \sim \Sigma(0)^3/p^2$, then indeed one reproduces for m_f the naive estimate of before:

$$m_f \sim \Sigma(\Lambda_{ETC}) \sim \frac{\Sigma(0)^3}{\Lambda_{ETC}^2} \sim \frac{\Lambda_T^3}{\Lambda_{ETC}^2} \quad . \tag{20}$$

The assumption one makes in walking technicolor models is that the dynamics is such that at $\Lambda_{ETC}, \Sigma(\Lambda_{ETC})$ is still much above its ultimate asymptotic value. Thus one can have large masses for fermions even for large values of Λ_{ETC}, solving the FCNC $\leftrightarrow m_t$ conundrum.

To obtain a behaviour of the type

$$\Sigma(\Lambda_{ETC}) >> \frac{\Sigma(0)^3}{\Lambda_{ETC}^2} \tag{21}$$

it is necessary that all physical quantities evolve slowly with momentum - hence the moniker walking for these theories. This can be achieved if the underlying theory is very nearly not asymptotically free or, perhaps, a fixed point theory [19]. Furthermore, if at Λ_{ETC} the coupling constant of the walking technicolor theory is still rather strong, so that $\Sigma(p)$ does not take yet its perturbative asymptotic value, it is clear that one cannot simply decouple the ETC and the WTC dynamics from each other. The presence of the ETC interactions can therefore influence the self energy functions for different technifermions differently, which may provide a physical rationale for some of the observed hierarchies in the quark and lepton mass matrices [25].

Rather than pursue further ETC/WTC theories here, I would like to indicate an alternative possibility in which also the dynamics of fermion mass generation and that of $SU(2) \times U(1)$ breaking are naturally interlocked. This possibility is realized in composite technicolor models, where both the light fermions we see (the quarks and leptons) and the fermions which condense (the technifermions) are bound states of some fundamental preon theory. In these kinds of theories, the preon dynamics produces effective 4-fermion interactions involving technifermions and light fermions which are analogous to the, perhaps more familiar, ETC interactions. The formation of technifermion condensates, combined with the presence of these interactions, then allows mass to be generated for the light fermions.

In what follows I want to describe a toy attempt in this direction, developed in collaboration with S. Khlebnikov [26]. Our model produces one generation of quarks, say t and b, as preon bound states [7]. These states are almost point-like, since their size is much smaller than their Compton's wavelength

$$< r_q > << \frac{1}{m_q} \; . \tag{22}$$

However, one obtains a top-bottom mass hierarchy precisely because top is more extended than bottom,

$$\frac{m_t}{m_b} \sim \frac{< r_t >^2}{< r_b >^2} \; . \tag{23}$$

The root cause for quark mass generation in the model is the formation of certain $SU(2) \times U(1)$ breaking condensates of other fermionic preon bound states, which act precisely as techniquarks. Thus, in the model, the W and Z get mass dynamically by the same condensates which also generate quark masses. As it will become clear below, the model is rather uneconomical.

[7]Leptons are not included in the model, but there are no hypercharge anomalies since these are cancelled by the presence of the techniquark bound states in the spectrum.

However, it is an interesting toy laboratory in which to study a number of dynamical issues. Furthermore, the difficulties one encounters in trying to incorporate families in extensions of this model are also quite illustrative.

The model is based on a chiral gauged preon model. One can argue that certain of the global chiral symmetries in the model are preserved in the binding. As a result of these global chiral symmetries, there are a number of massless bound states B in the spectrum. These states, however, acquire mass when one gauges some subset of the preons in a vector-like manner, as a result of condensate formation. After this further gauging, the original massless bound states B split into three different set of states

$$B = \{q, T, M\} \quad , \tag{24}$$

with q being the quarks (t, b), T being techniquarks and M being megaquarks. The metacolor gauge interaction at the preon level leads to the formation of $<\bar{M}M>$ condensates which, in turn, serve to produce effective ETC interactions between the quarks and techniquarks. The technicolor gauge interaction then, through the formation of $<\bar{T}T>$ condensates, leads to the appearance of the W and Z masses and of mass terms for the quarks - these latter masses originating as a result of the effective ETC interactions.

In more detail, the preon dynamics of the model we considered is based on replicas of an $SU(6)$ preon theory with 10 Weyl preons in the fundamental representation, $F_a(a = 1, \cdots, 10)$, and one preon in the symmetric conjugate representation, $\bar{S}$. Such a theory is chiral, but has no gauge anomalies. It has a nominal chiral global symmetry

$$G = SU(10) \times U(1)_Q \tag{25}$$

where $U(1)_Q$ is the anomaly free combination of F and $\bar{S}$ fermion number. However, one can argue dynamically [27] that a smaller symmetry than G is preserved in the binding, namely

$$H = SU(6) \times SU(4) \times U(1)'_Q \quad . \tag{26}$$

Furthermore, one can identify the set of bound states B which are massless by matching the H anomalies at the preon level with those at the bound state level [28]. These states are readily seen to comprise two different kinds of states

$$B = \begin{cases} B_1 \sim & (15;\, 1,\, \text{-}5/3) \\ B_2 \sim & (6;\, 4;\, \text{-}5/6) \end{cases} \tag{27}$$

where, in the above, I have given the transformation of these states under H.

The bound states B feel effective interactions among each other, as a result of their common underlying preonic structure [8]. There are H invariant dimension 6 interactions, scaling as Λ_c^{-2}, with Λ_c being the dynamical scale of the preon theory,

$$\mathcal{L}_9 = \frac{1}{\Lambda_c^2}(\bar{B}\gamma^\mu\lambda B)(\bar{B}\gamma_\mu\lambda B) \tag{28}$$

and dimension 9 interactions, scaling as Λ_c^{-5}

$$\mathcal{L}_9 = \frac{1}{\Lambda_c^5}\; B_1 B_1 \bar{B}_2 \bar{B}_2 \bar{B}_2 \bar{B}_2 \quad . \tag{29}$$

These latter interactions are not so important in the one generation model under study, but they play an important role in multigeneration models, since they violate individual B_1 and B_2 number.

The actual model studied in [26] is based on 3 replicas of these $SU(6)$ preon models, with the individual models constructed so that among their B states one generates, respectively, a t_R state, a b_R state and the $\begin{pmatrix} t \\ b \end{pmatrix}_L$ doublet of states. The first two $SU(6)$ models are identical in content with their 10 (right-handed) Weyl preons feeling different gauge interactions. One gauges an $SU(3)_c \times SU(3)_T \times SU(4)_M$ group of color, technicolor and metacolor interactions at the preon level, with the 10 F_a preons transforming under this group as

$$F_a = \{(\bar{3},1,1) \oplus (1,\bar{3},1) \oplus (1,1,4)\} \quad . \tag{30}$$

The final $SU(6)$ preon theory has a doubled set of preons - 20 F_a and 2 $\bar{S}$ - so as to be able to introduce at the preon level the electroweak $SU(2) \times U(1)$ interactions. These preons are (left-handed) Weyl states and are organized

[8] The states B have the following schematic preonic structure: $B_{ab} \sim F_a^T \sigma_2 \sigma_\mu F_b \sigma^\mu \bar{S}$.

in doublets of $SU(2)$, except for the preons which feel the metacolor interactions, which are $SU(2)$ singlet states. Thus, in the last preon theory, the preons transform as [9]

$$F_a = \{(\bar{3},1,1,2) \oplus (1,\bar{3},1,2) \oplus 2(1,1,4,1)\}; \; \bar{S} = (1,1,1,2) \quad . \tag{31}$$

The B_1 and B_2 massless bound states of each of these 3 $SU(6)$ preon theories can be classified immediately in terms of $SU(3)_c \times SU(3)_T \times SU(4)_M$. Since the antisymmetric combination of two $\bar{3}'s$ is a 3, one has

$$B_1 \sim \{3,1,1) \oplus (1,3,1) \oplus (\bar{3},\bar{3},1)\}$$

and

$$B_2 \sim \{(\bar{3},1,4) \oplus (1,\bar{3},4)\} \tag{32}$$

Thus the B_1 states contains both quarks and techniquarks, while the B_2 states are the only massless bound states which feel metacolor.

Gauging the metacolor group will cause the formation of condensates of the B_2 states, produced by the right-handed $SU(6)$ preon theories, with those produced by the left-handed $SU(6)$ preon theory. These condensates tie the left and right theories together, but preserve $SU(2) \times U(1)$. Specifically, one has two such condensates forming, [10].

$$< \bar{B}_2^{t_R} \, B_2^{1L} > \sim \Lambda_4^3 \quad ; \quad < \bar{B}_2^{b_R} B_2^{2L} > \sim \Lambda_4^3 \tag{33}$$

These metacolor condensates, combined with the effective residual interactions (28) of the preon theory, give rise to a form of ETC interactions. How these ETC-like interactions arise is sketched schematically in Fig. 3. Because the individual $SU(6)$ preon theories have their own intrinsical dynamical scales, the resulting ETC interactions need not be the same for top and bottom. Indeed, as a result of the metacolor condensate formation one obtains

$$\begin{aligned} \mathcal{L}_{\text{eff ETC}} &= \frac{\Lambda_4^2}{\Lambda_L^2 \Lambda_{t_R}^2} \, (\bar{B}_1^L \gamma_\mu \lambda \, B_1^L)(\bar{B}_1^{t_R} \gamma^\mu \lambda B_1^{t_R}) \\ &+ \frac{\Lambda_4^2}{\Lambda_L^2 \, \Lambda_{b_R}^2} \, (\bar{B}_1^L \gamma_\mu \lambda B_1^L)(\bar{B}_1^{b_R} \gamma^\mu \lambda B_1^{b_R}) \end{aligned} \tag{34}$$

[9]All preons also have appropriate $U(1)$ quantum number[26].

[10]The notation employed here is to append a superscript t_R, b_R or L to indicate to which preon theory the bound states belong. Note that there are 2 B_2^L bound states formed, denoted by B_2^{1L} and B_2^{2L}, respectively.

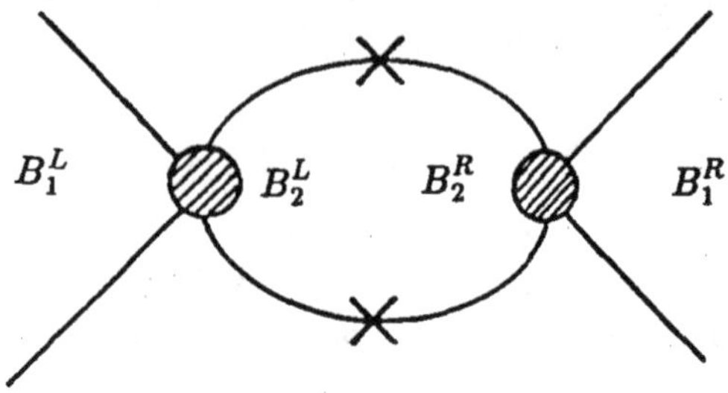

Figure 3: Generation of effective ETC interactions, through metacolor condensation

The B_1 states, recall, contain both quarks and techniquarks. Thus the above effective ETC interactions, once one turns on the technicolor interactions, will generate masses for the quarks. Although the technicolor condensates are expected to preserve a vectorial $SU(2)$ symmetry [29], the resulting top and bottom quark masses will be different since the effective ETC interactions for these states have different strength. If the scale of the technicolor condensate is Λ_T, the one obtains

$$m_t \sim \left(\frac{\Lambda_T^3 \Lambda_4^2}{\Lambda_L^2}\right) \frac{1}{\Lambda_{t_R}^2} \quad ; \quad m_b \sim \left(\frac{\Lambda_T^3 \Lambda_4^2}{\Lambda_L^2}\right) \frac{1}{\Lambda_{b_R}^2} \tag{35}$$

The masses of the top and bottom quarks will be small compared to the various preonic dynamical scales Λ_L, Λ_{t_R} and Λ_{b_R} - the compositeness scales - provided that these scales are large compared to the metacolor scale Λ_4 and the technicolor scale Λ_T. If this is so, these states will appear for all purposes as effectively elementary when probed with energies of $0(100\ GeV)$. The particular hierarchy between m_t and m_b in the model is due to the difference in the dynamical scales of the two (right-handed) preon theories, Λ_{t_R} and

Λ_{b_R}. These scales typify the physical size of top and bottom

$$< r_t > \sim \frac{1}{\Lambda_{t_R}} \;\; ; \;\; < r_b > \sim \frac{1}{\Lambda_{b_R}} \; . \tag{36}$$

Thus Eq. (35) contains the interrelation between mass and size alluded to earlier in Eq. (23).

It is easy to imagine a mechanical extension of this model, so as to introduce families of quarks [26]. To get N_f generations, all one has to do is to make N_f copies of the $SU(6)^3$ preon model discussed above. In this extended model one can reproduce the hierarchy of the observed quark mass spectrum by assuming appropriate dynamical scales for all the various right-handed preon theories.. That is,

$$m_f \sim \frac{1}{\Lambda^2_{f_R}} \sim < r_f >^2 \; . \tag{37}$$

However, this naive extension of the toy one family model has many problems. These are both of a theoretical and a phenomenological nature. On the theoretical side, because this $\left(SU(6)\right)^{3N_f}$ preon theory has so many preons, at the preon level, all the vectorial interactions - color, technicolor and metacolor - are **not** asymptotically free. Thus, it is no longer clear whether one has control of the dynamics or that the assumed techniquark and metaquark condensates really occur. On the phenomenological side, perhaps even worse disasters occur. The extended theory [26] for $N_f = 3$ has a natural $[U(1)_V]^3$ family symmetry which, if it is not broken somehow, prevents generating any quark mixing matrix at all. Thus, one has a quark mass hierarchy, but the Cabibbo, Kobayashi Maskawa matrix $V(_{CKM})$ is identically equal to unity:

$$V_{CKM} = 1 \tag{38}$$

It is, perhaps, useful to explain what prevents quark mixing in the model, as this is quite generic of models where generations are obtained by just blindly copying what happens in one family. For each generation a vectorial $U(1)$ is preserved in the binding, under which all preons that carry color have charge $+1$, and all preons which carry technicolor carry charge -1. Naively, one would presume that this $[U(1)_V]^3$ symmetry is also preserved by all condensates. However, because of the chiral nature of the effective interactions experienced by the B_1 and B_2 bound states and because metacolor itself is probably strong at the compositeness scale, [11] it is not clear whether the Vafa - Witten theorem applies [29]. Thus, it is probable that the metacolor condensates do not respect the $[U(1)_V]^3$ family symmetry.

To get family mixing it is necessary that the metacolor condensates be not family diagonal

$$< \bar{B}_{2i}^{u_R} B_{2j}^{1L} > \sim \Lambda_4^3 \, \Sigma_{ij}^1 \; ; \; < \bar{B}_{2i}^{d_R} \, B_{2j}^{2L} > \sim \; \Lambda_4^3 \Sigma_{ij}^2 \tag{39}$$

with

$$\Sigma_{ij}^1 \neq \delta_{ij} \; ; \; \Sigma_{ij}^2 \neq \delta_{ij} \tag{40}$$

Unfortunately, this assumption is not enough. Although these condensates appear to break $[U(1)_V]^3$ to a vectorial family symmetry $U(1)_V$, the dynamics of the theory is such that there remains a discrete family symmetry, which ultimately prevents quark mixing. To understand this point, consider the analog to Eq. (34) which ensues after the formation of the nondiagonal metacolor condensates given above. Although these interactions now connect B_1 bound states of different families, e.g.

$$\mathcal{L}_{\text{eff ETC}} = \frac{\Lambda_4^2}{\Lambda_{Li}^2 \, \Lambda_{u_Rj}^2} \; (\bar{B}_{1i}^L \gamma^\mu \lambda B_{1i}^L)(\bar{B}_{ij}^{u_R} \gamma_\mu \, \lambda B_{1j}^{u_R}) \tag{41}$$

they still preserve the individual B_{1j} - numbers since they always involve $\bar{B}_{1j} \; B_{1j}$ combinations.

To transmit the breaking of family number that occurs through the metacolor condensates, one needs to make use of the dimension 9 interactions of Eq. (29) which involve only B_1 fermions and B_2 antifermions (or vice versa). Using these interactions, and the metacolor condensates (39), then indeed one obtains effective interactions which break explicitly $[U(1)_V]^3$. Schematically, these interactions take the form

$$\mathcal{L}_{B_1 \text{ break}} = \frac{\Lambda_4^{16}}{\Lambda_{Li}^{14} \, \Lambda_{u_Rj}^5, \; \Lambda_{d_Rk}^5} \; B_{1i}^L \; B_{1i}^L \; B_{1i}^L B_{1i}^L \bar{B}_{1j}^{u_R} \bar{B}_{ij}^{u_R} \; \bar{B}_{1k}^{d_R} \; \bar{B}_{1k}^{d_R} \tag{42}$$

Unfortunately, although the above interaction breaks the $[U(1)_V]^3$ family symmetry, because of the structure of the theory there remains a $(Z_2)^3$ discrete family symmetry! This is a pity because otherwise, given the high

[11]Recall that metacolor is not asymptotically free in multigenerational models at the compositeness scale.

powers of Λ involved, the B_1 breaking Lagrangian would really act as a small dynamical perturbation to generate quark mixing.

Protective family symmetries or their remnants, like the $(Z_2)^3$ symmetry encountered above, are a generic feature of these simple replica models. These symmetries prevent quark mixing to occur, although one can generate arbitrary diagonal mass matrices. This problem may be solved in models where family structure occurs more naturally. Given the intriguing interrelation between the mass of the quarks and their extent, one can well imagine that in a more realistic model the largest of the diagonal elements in the mass matrices of quarks would involve a $c-t$ transition. In turn this leads one to speculate that, if there are FCNC in more realistic theories, these will be most strongly felt in the $c-t$ channel (and possibly in the $b-s$ channel) [26].

Attempts to generate fermion masses in models of dynamical symmetry breakdown of $SU(2)\times U(1)$ are salutary exercises. They serve to remind one that if one believes that the electroweak theory is broken dynamically, then resolving the issue of fermion masses is absolutely crucial. Without a a satisfactory solution to the fermion mass problem, it is difficult to give credence that $SU(2)\times U(1)$ is broken by some condensate of a yet to be discovered underlying theory. Continuing failures to satisfactorily resolve the fermion mass problem give further impetus to the idea that the electroweak theory of Glashow, Salam and Weinberg is broken down simply by a scalar vev. Before surrendering by default to this idea, however, it seems worthwhile to continue exploring some of these composite technicolor models, particularly since the dynamics of combined chiral and vectorial gauge theories is very rich and could still hide some interesting surprises. In doing so, one would be following the example of Salam, who was not afraid many times in his career to follow unfashionable paths, which eventually led to profound later insights. Whether the same will occur in this instance remains to be seen.

This work was supported in part by the Department of Energy under grant No. DE-FG03-91ER40662 TASK C

References

[1] S. L. Glashow, Nucl. Phys. 22 (1961) 579; A. Salam in **Elementary**

Particle Theory ed. N. Svartholm (Almquist and Wiksell, Stockholm 1968); S. Weinberg, Phys. Rev. Lett. 19 (1967) 1264

[2] See, for example, L. Rolandi, Proceedings of the XXVI International Conference on High Energy Physics, Dallas, Texas, August 1992, ed. J. R. Sanford, AIP Conference Proceeding No. 272 (AIP, New York, 1993)

[3] K. Wilson, as quoted in L. Susskind, Phys. Rev. D20 (1979) 2619; E. Gildener and S. Weinberg, Phys. Rev. D13 (1976) 3333; G 't Hooft in **Recent Developments in Gauge Theories**, Cargese Lectures 1979, eds. G 't Hooft et al (Plenum Press, New York, 1980).

[4] For a recent discussion, see for example, F. Zwirner, in **Texas/Pascos '92: Relativistic Astrophysics and Particle Cosmology**, eds C. W. Akerlof and M. A. Srednicki; Ann. New York Acad. Sci. 688 (1993) 55

[5] L. Maiani, G. Parisi and R. Petronzio, Nucl. Phys. B136 (1978) 115

[6] S. Weinberg, Phys. Rev. D13 (1976) 974; L. Susskind, Phys. Rev. D20 (1979) 2619

[7] J. M. Cornwall, D. M. Levin and G. Tiktopolous, Phys. Rev. D10 (1974) 1145; D11 (1975) 972 (E)

[8] P. W. Higgs, Phys. Lett 12 (1964) 132; Phys. Rev. Lett. 13 (1964) 508; F. Englert and R. Brout, Phys. Rev. Lett. 13 (1964) 321; G. S. Guralnik, C. Hagen and T. W. B. Kibble, Phys. Rev. Lett. 13 (1964) 585

[9] S. Coleman, J. Wess, and B. Zumino, Phys. Rev. 177 (1969) 2239; C. Callan, S. Coleman, J. Wess and B. Zumino, Phys. Rev. 177 (1969) 2247

[10] B. W. Lee, C. Quigg and H. B. Thacker, Phys. Rev. D16 (1977) 1519

[11] For recent discussions, see for example, the contributions of M. S. Chanowitz and J. Bagger in the Proceedings of the XXVI International

Conference on High Energy Physics, Dallas, Texas, August 1992, ed. J. R. Sanford, AIP Conference Proceedings No. 272 (AIP, New York, 1993)

[12] M. Veltman, Nucl. Phys. B123 (1977) 89; Acta Physica Polonica B12 (1981) 432

[13] B. W. Lynn, M. E. Peskin and R. G. Stuart in **Physics at LEP**, Vol. 1, eds J. Ellis and R. D. Peccei, CERN 86-02 (1986); D. Kennedy and B. W. Lynn, Nucl. Phys. B232 (1989) 1; M. Consoli, W. Hollik and F. Jegerlehner in **Z Physics at LEP** Vol. 1, eds G. Altarelli, R. Kleiss and C. Verzegnassi; CERN 89-08 (1989)

[14] M. Peskin and T. Takeuchi, Phys. Rev. Lett. 65 (1990) 964; D. Kennedy and P. Langacker, Phys. Rev. Lett. 65 (1990) 2967; G. Altarelli and R. Barbieri, Phys. Lett. B253 (1991) 161

[15] M. Peskin and T. Takeuchi, Phys. Rev. D46 (1992) 381

[16] G. Altarelli, R. Barbieri and F. Caravaglios, CERN preprint CERN TH 6770 93

[17] D. Kennedy and P. Langacker, Phys. Rev. D44 (1991) 1591

[18] G. Altarelli, R. Barbieri and S. Jadach, Nucl. Phys. B369 (1992) 3

[19] B. Holdom, Phys. Rev. D24 (1981) 1441; Y. Yamawaki, M. Bando and K. Matumoto, Phys. Rev. Lett. 56 (1986) 1335; T. Akiba and T. Yanagida, Phys. Lett. 169B (1986) 432; T. Appelquist, D. Karabali and L. Wijewardhana, Phys. Rev. Lett. 57 (1986) 982

[20] T. Appelquist and G. Triantaphyllou, Phys. Lett B278 (1992) 345; S. D. Hsu and R. Sundrum, Nucl. Phys. B. 391 (1993) 127

[21] S. Dimopoulos and. L. Susskind, Nucl. Phys. B155 (1979) 237; E. Eichten and K. Lane, Phys. Lett. 90B (1980) 237

[22] S. Dimopoulos and J. Ellis, Nucl. Phys. B182 (1981) 505

[23] For a discussion, see for example, B. Holdom in Proceedings of the 1991 Nagoya Spring School on Dynamical Symmetry Breaking, ed. K. Yamawaki (World Scientific, Singapore 1992)

[24] K. D. Lane, Phys. Rev. D10 (1974) 2605; H. D. Politzer, Nucl. Phys. B117 (1976) 397

[25] For a discussion of this point and a semirealistic model, see for example, B. Holdom, Phys. Lett. B246 (1990) 169

[26] S. Khlebnikov and R. D. Peccei, Phys. Rev. D48 (1993) 361

[27] I. Bars and S. Yankielowicz, Phys. Lett. 101B (1981) 159

[28] G. 't Hooft, Ref. [3]

[29] C. Vafa and E. Witten, Nucl. Phys. B234 (1984) 173

TESTING EXTENSIONS OF THE STANDARD MODEL

PAUL H. FRAMPTON
Institute of Field Physics
Department of Physics and Astronomy
University of North Carolina
Chapel Hill, NC 27599-3255, USA

ABSTRACT

After some preliminary discussion of the standard model of particle theory, two extensions are described. Firstly, the aspon model of CP violation predicts vector-like quarks and the aspon gauge boson at SSC energies. Second, the 3-3-1 model predicts exotic quarks, dileptonic gauge bosons and a Z′ all at SSC energies. Finally, some general discussion is given of the role for model building until the SSC is in commission.

It is an honor to contribute to this Festschrift for Abdus Salam who is one of the great men of particle theory. I do still recall appreciatively his writing me an encouraging letter about my research some twenty-five years ago at about the same time he was proposing his standard electroweak theory which is amongst the most exciting physics to be created in my lifetime.

The standard model (SM) of particle theory [1] is well established as correct at present accelerator energies. Because no data exist which definitively contradict the SM, or which guide the theory forward, we may say that

i) Any extension of the SM is a long-shot
ii) Any model disagreeing with SM at present accelerator energies is wrong

In the SM are twelve gauge bosons corresponding to the generators of SU(3) x SU(2) x U(1). A theme of the present article is:

Will additional massive gauge bosons be discovered at the SSC?

The fermion menu of the SM is well-known. There are 15 helicity states per family: 12 quarks and 3 leptons as follows:

Family 1: $(u^\alpha\, d^\alpha)_L\ \bar{u}_{\alpha L}\ \bar{d}_{\alpha L}\quad (\alpha = 1,2,3)$

$(\nu_e\, e^-)_L\ e^+_L$

Family 2: $(c^\alpha\, s^\alpha)_L\ \bar{c}_{\alpha L}\ \bar{s}_{\alpha L}$

$(\nu_\mu\, \mu^-)_L\ \mu^+_L$

Family 3: $(t^\alpha\, b^\alpha)_L\ \bar{t}_{\alpha L}\ \bar{b}_{\alpha L}$

$(\nu_\tau\, \tau^-)_L\ \tau^+_L$

At the time of writing, the top quark has not been established although rumors emanate from the CDF and D0 groups at Fermilab. A second question is therefore:

Will additional fermions (beyond the top quark) be discovered at the SSC?

The final sector of the SM is the Higgs. In the minimal version with one complex doublet $(\phi^+\ \phi^0)$ of Higgs scalars there is one physical Higgs (H). Its mass is in the range 60 GeV < M_H < 1 TeV. H is the least understood particle in the SM. The energy and luminosity of the SSC are such that H is detectable.

The most depressing scenario is where the SSC discovers only an elementary H conforming to the standard model. Do we then build an even larger collider than the SSC?

The most popular extension of the SM is surely low-energy supersymmetry. A reason for not accepting fully low-energy supersymmetry is: why have the partners not already shown up? It is a fine idea for gravity at the Planck scale. Concerning the oft-cited gauge-hierarchy problem: (i) It is a theorist-made problem - nothing is known of physics above a TeV (ii) The Higgs H may be composite $(\bar{f}f)$ (iii) Supersymmetry does not really solve the gauge hierarchy problem even if it exists since the small ratio of the weak scale to the GUT scale must still be put in by hand. Thus, MSSM (Minimal Supersymmetric Standard Model) is just one of many testable and motivated "long-shot" extensions of the standard model.

The most obvious shortcoming of the standard model is its large number of parameters. There are 9 fermion masses, 4 quark flavor mixing parameters, and 4 electroweak parameters (α_1, α_2, M_W, M_H) giving already 17. In QCD there are two parameters α_3 and $\bar{\theta}$ making a total of 19, surely an unacceptably large number for a "final" theory. If neutrinos have non-zero Majorana masses this number is increased from 19 to 28 by three each of masses, mixing angles and CP violation phases. Behind the 19 parameters is a pre-parameter N_f, the number of families. N_f determines the number of mass parameters ($3N_f$) and of flavor mixing parameters $(N_f - 1)^2$.

In the first model described below the motivation is from the CP violating parameters δ and $\bar{\theta}$. In the second model the motivation is from the pre-parameter $N_f = 3$.

Testability of an extension of the SM requires adequate cross-section for production and detection of new particles at the SSC. The full pp cross-section of about 100mb gives 10^{15} collisions/SSC y at the design luminosity of 10^{33} cm^{-2} s^{-1}. Thus a cross-section below 1 fb. means less than one event per month and is too small. The models described here have cross-sections of 1 pb or even 1 nb corresponding to between one event per hour and one event per second at the SSC.

I. The Aspon Model of CP Violation

A crucial issue unanswered by the SM is the strong CP problem. In the aspon model, the solution is linked to a mechanism of CP violation additional to the CKM phase of the miminal SM. It also leads to new particles which are predicted to be below 600 GeV in mass and detectable at the SSC.

The most popular solution of the strong CP problem is the axion model [2-5]. The principal objection to the axion model is that it involves assuming a global symmetry where all fields are transformed according to

$$\phi_q (\underline{x}, t) \rightarrow e^{i\alpha q} \phi_q (\underline{x}, t) \tag{1}$$

Although mathemathically sensible, such a transformation is bizarre *physically* since it implies we are making the transformation for all spacetime points e.g. here - now and on Jupiter in 1492. Global symmetries do not make attractive physics; there are less intuitive objections based on wormholes and black holes. More physical is a local or gauge symmetry

$$\phi_q (\underline{x}, t) \rightarrow \exp(i\alpha(\underline{x},t)q)\, \phi_q(\underline{x}, t) \tag{2}$$

The axion model symmetry cannot be so gauged because it is necessarily color anomalous in order to solve the strong CP problem.

If we rule out the axion, the most promising direction seems to be to *assume* CP symmetry is spontaneously broken. This means CP is assumed to be a symmetry of the lagrangian, unlike in the SM where CP is explicitly broken. This distinguishes CP from P which is explicitly broken. Why CP should be so different from P might be related to the fact that CP is more weakly violated.

The assumption of CP symmetry of the lagrangian means that for the parameter $\bar{\theta}$

$$\bar{\theta} = \theta - \arg(\det M) \tag{3}$$

in which

θ = parameter multiplying CP violating term in QCD lagrangian

M = quark mass matrix

$\bar{\theta}$ = physical parameter

There is a lot of theory underlying Eq(3) which can be found in the standard review articles [6]. From the upper limit on the neutron electric dipole moment we know that $\bar{\theta} < 2 \times 10^{-10}$ which is an unacceptable fine tuning. The assumption of CP symmetry means that $\theta = 0$. In the aspon model, arg (det M) = 0 at tree level but becomes non-zero at one loop.

Now we are ready to specify the particle content of the minimal aspon model [7]. The standard model gauge group is extended to:

$$SU(3) \times SU(2) \times U(1) \times U(1)_{new} \tag{4}$$

New fermions are needed: an SU(2) doublet of vector-like quarks

$$\begin{pmatrix} U \\ D \end{pmatrix}_L \quad \begin{pmatrix} U \\ D \end{pmatrix}_R$$
$$Q_{new} = +1 \quad Q_{new} = -1 \tag{5}$$

Finally, two complex singlet Higgs scalars

$$\chi^{\alpha} \qquad \alpha = 1,2$$
$$Q_{new} = +1 \tag{6}$$

All particles of the SM have $Q_{new} = 0$. Two scalars are needed in (6) because otherwise the phase of a complex VEV can be rotated away by a $U(1)_{new}$ gauge transformation.

With these $U(1)_{new}$ assignments the up-quark mass matrix is now

$$\begin{pmatrix} m_{ij} & F_i \\ O & M \end{pmatrix} \tag{7}$$

where m_{ij} $(1 \leq i, j \leq 3)$ and M are real and where $F_i = \sum_\alpha h_i^{(\alpha)} \langle\chi^\alpha\rangle$ are complex. Here $h_i^{(\alpha)}$ are the Yukawas in $h_i^{(\alpha)} \bar{q}_L^i Q_R \chi^\alpha$. The form of (7) means that arg det $(M_u) = 0$. Similarly, arg det $(M_D) = 0$ and hence from (3) $\bar{\theta} = 0$ at tree level. The VEVs of the χ^α which break spontaneously CP symmetry are

$$\langle\chi^\alpha\rangle = \rho_\alpha e^{i\phi_\alpha} \quad \phi_1 - \phi_2 \neq 0 \tag{8}$$

It is useful to define $x_i = |F_i/M|$ as the relevant expansion parameter in the diagonalization of the quark mass matrix. There are two distinct paths. If x_i is of order one, the CKM phase can explain CP violation. Now $x_i \sim h_i <\chi> M^{-1}$ and for $\bar{\theta}$ sufficiently small at one loop level $h_i \leq 10^{-3}$. This means that $M \leq 10^{-3} |<\chi>|$ which is a mini-hierachy between the vector-quark mass and the CP breaking scale.

To avoid this mini-hierachy, the aspon model assumes $|<\chi>| = M$ and hence $x_i \leq 10^{-3}$. In this case an additional mechanism of CP violation is necessary. Such is provided by exchange of the $U(1)_{new}$ gauge boson -the aspon- as in Figure 1.

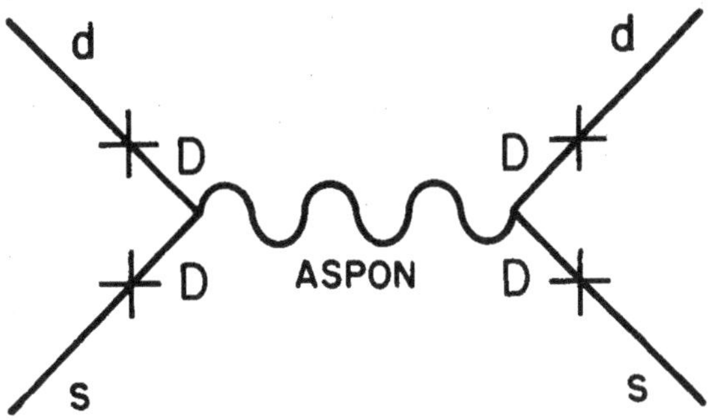

Figure 1

A detailed estimation of CP violation parameters in the K° and B° systems gives

$$|<\chi>| < 2\text{TeV} \tag{9}$$

If $g_{new} \le e$ then

$$M(A), M(Q) \le 600 \text{ GeV} \tag{10}$$

and hence accessible to the SSC.

The cross-section for $pp \to Q\bar{Q}$ is shown in Figure 2

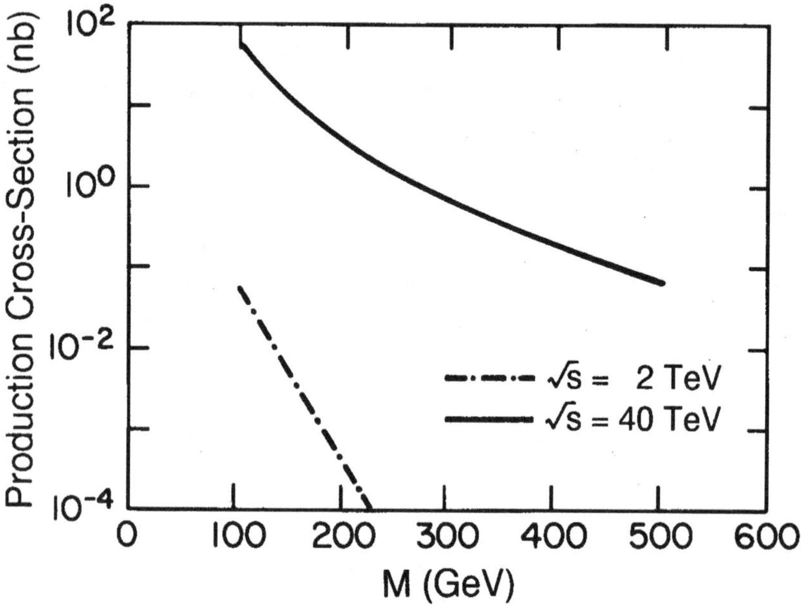

Figure 2

The solid line in Figure 2 is for SSC ($\sqrt{s} = 40$ GeV) while the dotted line is for Fermilab ($\sqrt{s} = 2$ TeV).

For aspon production pp $\rightarrow Q\bar{Q}A$ the relevant Feynman diagrams are in Figure 3

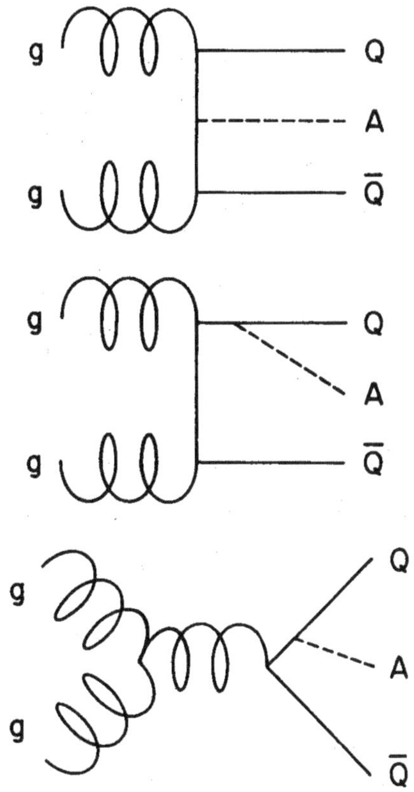

Figure 3

The resulting cross-sections are shown in Figure 4

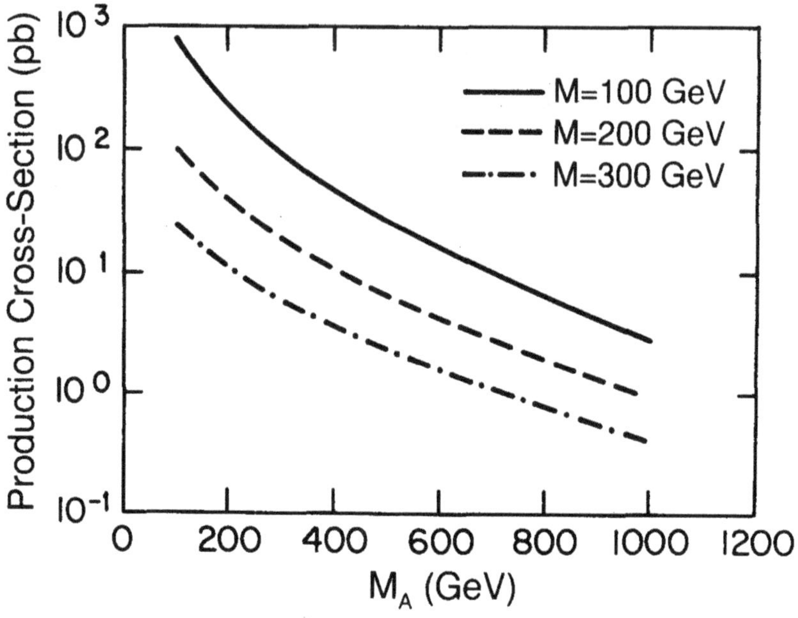

Figure 4

The signatures for decay of Q e.g. pp $\rightarrow U\overline{U}$ are that U decays democratically $U \rightarrow d^iW \rightarrow d^i\, l\nu$ so one needs two isolated charged leptons and two or more jets plus missing energy due to neutrinos. To distinguish from $t\bar{t}$, b-jet identification could be helpful although (F. Paige, private communication) is not crucial.

For the aspon, its decay width depends strongly on the mass M_A. If $M_A > 2M, \Gamma_A \sim 1$ GeV; if $2M > M_A > M, \Gamma_A \sim 1$ MeV; and if $M_A < M, \Gamma_A \sim 1$ keV. The aspon can be sought in the same experiment as pp $\rightarrow \bar{t}t\,H$. Two sharp distinctions between A and H are that $\Gamma_A << \Gamma_H$ and $A \rightarrow 2\gamma$ is forbidden by Yang's theorem.

Q-onia have a smaller production cross-section than open flavor Q and would thus not be the discovery mode but would be interesting to investigate.

The 3-3-1 Model

As a second extension of the SM, I consider one motivated by the issue of the 3 families. The 3-3-1 model can be combined straightforwardly with the aspon mechanism described above for CP violation, but to introduce the 3-3-1 model the following are motivations:

i) Consistency of gauge field theory - unitarity and renormalizability - require anomaly cancellation. This requirement alone is able to fix all the electric charges for one quark-lepton family and hence to explain charge quantization including the electric neutrality of the H atom.

ii) Point i) does not explain the replication of families but is sufficiently impressive to suggest that $N_f = 3$ families is explicable by anomaly cancellation in an extension of the SM. This requires that each family have a non-vanishing anomaly and that the three families are not all treated similarly.

iii) A striking feature of the SM mass spectrum is the top quark mass suggesting that the third family be treated differently and the anomaly cancellation be in the pattern $+1+1-2$ for the three families.

iv) To find the (-2) in (iii) we see that the quark charges give

$$\frac{Q(u)}{Q(d)} = -2 \tag{11}$$

Anomaly cancellation in the 3-3-1 model depends on (11).

v) The electroweak gauge group will be extended from SU(2) x U(1) to SU(3) x U(1) giving 5 extra gauge bosons. One is a neutral Z_2. The other four are a doublet of dileptons (Y^{--} Y^{-}) with lepton number $L = +2$ and their antiparticles (Y^{++} Y^{+}) with $L = -2$. Such dileptons appeared previously in stable-proton GUTs.

With these motivations we are ready to define the 3-3-1 model:

The SM gauge group is extended to [9,10]

$$SU(3) \times SU(3)_L \times U(1)_X \tag{12}$$

The leptons are assigned democratically to X = 0 antitriplets

$$\begin{pmatrix} e^- \\ \nu_e \\ e^+ \end{pmatrix}_L \quad \begin{pmatrix} \mu^- \\ \nu_\mu \\ \mu^+ \end{pmatrix}_L \quad \begin{pmatrix} \tau^- \\ \nu_\tau \\ \tau^+ \end{pmatrix}_L \tag{13}$$

For the first two families the left-handed quark fields are in X = – 1/3 triplets

$$\begin{pmatrix} u^\alpha \\ d^\alpha \\ D^\alpha \end{pmatrix}_L \quad \begin{pmatrix} c^\alpha \\ s^\alpha \\ S^\alpha \end{pmatrix}_L \tag{14}$$

with right-handed singlets. In the third family, the left-handed quarks are in an X = + 2/3 antitriplet

$$\begin{pmatrix} b^\alpha \\ t^\alpha \\ T^\alpha \end{pmatrix} \tag{15}$$

Anomaly cancellation is complete for this arrangement. For X^3, $(3_L)^2X$, and $(3_L)^3$ each family has non-zero anomaly and the three families cancel with anomalies proportional to (+ 1 + 1 – 2) as anticipated.

The Higgs sector comprises a triplet Φ^α (X = + 1) with $\langle\Phi^\alpha\rangle = U\,\delta^{\alpha 3}$ breaking to the SM and giving mass to D,S,T and the gauge bosons Y and Z_2. The electroweak breaking requires two triplets $\langle\phi^\alpha(X = 0)\rangle = V\,\delta^{\alpha 2}$ giving masses to d,s,t and (antisymmetrically in family space) to the charged leptons and $\langle\phi'^\alpha\,(X = -1)\rangle = V'\,\delta^{\alpha 3}$ giving mass to u,c,b. Finally, there is a sextet $H^{\alpha\beta}$ (X = 0) with $\langle H^{13}\rangle \neq 0$ to give a symmetric contribution to the mass matrix of the charged leptons.

The new scale U has both a lower bound and (somewhat unexpectedly) an upper bound. The lower bound on $M(Z_2)$ arises from the minimal Higgs structure, already described, which requires $M(Z_2)$ greater than about 4M(Y) which, in turn, is bounded below M(Y) > 230 GeV from data on polarized muon decay.

The upper bound arises because embedding of SU(2) x U(1) in SU(3) x U(1) requires that the electroweak mixing angle satisfy $\sin^2\theta < 1/4$. As the limit $\sin^2\theta = 1/4$ is approached the $U(1)_X$ becomes strongly coupled. To avoid this one finds

$$900 \text{ GeV} < M(Z_2) < 1500 \text{ GeV} \quad (16)$$

$$230 \text{ GeV} < M(Y) < 370 \text{ GeV} \quad (17)$$

are the permitted ranges. When the $U(1)_X$ is extrapolated to higher energy a Landau pole occurs less than one order of magnitude above $M(Z_2)$. This strongly suggests that further gauge bosons are needed above $M(Z_2)$ to render this coupling more asymptotically free.

Discussion

I have described two motivated and testable examples of extending the standard model. Others include:

- heavy leptons. E.g. vector-like leptons $\begin{pmatrix} N \\ E \end{pmatrix}_L$ $\begin{pmatrix} N \\ E \end{pmatrix}_R$. If $M_E < M_N$, the charged variety may be long-lived and distinguishable from a muon with good nanosecond timing at SSC

- axigluons. Chiral color theory. The axigluon mass is known to exceed 300 GeV. It reveals itself as a jacobian peak at half its mass in the single-jet inclusive distribution

These extensions of the standard model are, however, less motivated as are the axion, left-right electroweak models and many others in the literature.

Low-energy supersymmetry, technicolor and the two models herein are *motivated* and *testable* SM extensions. These four models are the only ones known which have *both* these two qualities.

One problem *in* the SM itself is the nature of the Higgs boson - elementary or composite $\bar{f}f$ where f = unknown fermion? The most economic idea is that f = top quark as proposed by Nambu [11] and Bardeen et al. [12]

In extensions of the SM, predictions are less confident as the spin decreases, as follows:

Spin One. Massive gauge bosons e.g. aspon, dilepton, Z′, axigluon, ... These are geometric, being associated with the gauge group, and the most confident prediction.

Spin 1/2. Additional fermions are usually present. These are algebraic, being associated with anomaly cancellation which gives some arbitrariness. [The W,Z discoveries (geometric) were more fundamental to the SM than the t (algebraic)].

Spin 0. Extended Higgs sector present, with elementarity uncertain. It is unknown whether *any* spin zero elementary particle exists, so this is the least certain sector.

The years until commissioning of the SSC seem an ideal time to try to invent extensions of the SM which are motivated by features which are only accommodated rather than explained in the SM and which are testable at SSC. While each extension is necessarily a long-shot, and only one can be correct, the more possibilities we anticipate the more likely the correct one will be in hand by the time the SSC experiments are producing their data.

This work was supported in part by the U.S. Department of Energy under Grant No. DE-FG05-85ER-40219.

References

1. A. Salam, in Proceedings of the VIII Nobel Symposium, edited by N. Svartholm (Almqvist and Wiksells, Stockholm, 1968) p. 367.
   S.L. Glashow, *Nucl. Phys.* **22**, 579 (1961)
   S. Weinberg, *Phys. Rev. Lett.* **19**, 1264 (1967)
2. S. Weinberg, *Phys. Rev. Lett.* **40**, 223 (1978)
3. F. Wilczek, *Phys. Rev. Lett.* **40**, 279 (1978)
4. J.E. Kim, *Phys. Rev. Lett.* **43**, 103 (1979)
5. M. Dine, W. Fischler and M. Srednicki, *Phys. Lett.* **B104**, 199 (1981)
6. J.E. Kim, *Phys. Reports* **150**, 1 (1987)
7. P.H. Frampton and T.W. Kephart, *Phys. Rev. Lett.* **66**, 1666 (1991)
8. P.H. Frampton and D. Ng, *Phys. Rev.* **D43**, 3034 (1991)
9. P.H. Frampton, *Phys. Rev. Lett.***69**, 2889 (1992)

10. F. Pisano and V. Pleitez, *Phys. Rev.* **D46**, 410 (1992)
11. Y. Nambu, Talk at the Kasimirz Conference (1988)
V.A. Miransky, M. Tanabashi and K. Yamawaki, *Mod. Phys. Lett.* **A4**, 1043 (1989)
12. W.A. Bardeen, C.T. Hill and M. Lindner, *Phy. Rev.* **D41**, 1647 (1990)

The Abdus Salam Dream

A. ZICHICHI

CERN, 1211 Geneva 23, Switzerland

ABSTRACT

The frontiers in Physics are dominated by the contributions of Abdus Salam in terms of new basic ideas. These contributions have their most fascinating synthesis in the following question: does the Superworld exist? The only possible answer to this question is to review the most basic results of Super-Grand unification. We focus the attention on two specific models and their predictions. These two models represent an example of a direct comparison between a traditional unified-theory and a string-inspired approach to the solution of the many open problems of the Standard Model. Viable models must satisfy *all* available experimental constraints and be as simple as theoretically possible. The two well defined supergravity models, $SU(5)$ and $SU(5) \times U(1)$, can be described in terms of only a few parameters (five and three respectively) instead of the more than twenty needed in the MSSM model, *i.e.*, the Minimal Supersymmetric extension of the Standard Model. A case of special interest is the strict no-scale $SU(5) \times U(1)$ supergravity where all predictions depend on only one parameter (plus the top-quark mass). A general consequence of these analyses is that the Superworld can be at the verge of discovery, lurking around the corner at present and near future facilities. This is in fact the projection of Salam's dream in the real axis.

CONTENTS

1 Introduction

One of the most important and original contributions to the advancement of physics is the invention by Abdus Salam of Superspace. The unification of all forces of Nature being his great dream, it is for me a great pleasure to contribute to his 70th anniversary celebration by reviewing the present status of this new frontier of our Science.

Physics has been and continues to be a source of new ideas, new concepts, new problems. Our understanding of the foundations of Space, Time, Matter, Forces, Charges, has drastically changed. The Einstein four-dimensional Space-Time has developed into a Superspace where the number and the properties of the original dimensions is now 10-bosonic plus 32-fermionic. These frontiers of Physics are dominated by the Salam dream: the existence of the Superworld.

Purpose of this lecture is to review the present status of this dream. A dream with many appealing features: the boson-fermion equivalence, the unification of all gauge couplings and masses, consequently of all forces, including gravity, and the ultimate goal of discovering the Theory of Everything (TOE). A large number of papers has been published on this subject and many unjustified claims have been mixed with real achievements. In order to distinguish between these two classes clarity is badly needed especially when experimental consequences are involved. For example, to predict the energy level where the Superworld should show up is one of the most exciting problems of modern physics. The effort needed to implement a project, in terms of people and financial resources, is so vast that experimentalists themselves need to judge the value of a "prediction". It has often been said that the energy level needed to discover the first evidence of the Superworld is inaccessible with the present facilities: Fermi-Lab, LEP(I,II), HERA, Gran Sasso, SKK. The primary purpose of this lecture is to show that the Salam dream can be around the corner at present and near future facilities.

2 The lesson from LEP: two Supergravity Models

The LEP e^+e^- collider at CERN has been in operation since 1989 with a center-of-mass energy around the Z-boson mass ($M_Z \approx 91\,GeV$). During this time a vast amount of data has been collected on many different decay modes of the Z. Analyses of the totality of the data in the context of the Standard Model of electroweak interactions show no deviations from expectations, which have been tested at the level of one-loop and for some quantities two-loop corrections [1]. In particular, the weak mixing angle ($\sin^2\theta_W(M_Z)$) and the QCD running coupling ($\alpha_3(M_Z)$) have been determined with unprecedented accuracy. This picture-perfect agreement between theory and experiment has had several consequences. One of these is the immediate rejection of large classes of models which predicted new physics below the electroweak scale.

The lesson from LEP is therefore how to reconcile the successes of the Standard Model with the compelling reasons for the existence of new physics beyond it.

An obvious shortcoming of the Standard Model is the ad-hoc nature of the many parameters involved: the quark and lepton masses, the quark mixing angles, and the CP violating phase. However, even if explanations for these parameters could be found within physics not so far from the electroweak scale, a problem would show up when extrapolating the theory to much larger energies. This so-called *gauge hierarchy problem* is manifest in theories with elementary scalar fields, such as the Higgs boson in the Standard Model. The reason is that radiative corrections to the Higgs-boson mass become extremely large if the theory also contains very massive particles, as is the case in unified theories, or any theory which attempts to incorporate gravity in a common framework. This is a "problem" for the theory, as opposed to a terminal disease, in that there is a grossly undesirable solution which involves fine-tuning parameters to large numbers of decimal places, order-by-order in perturbation theory.

Particle physics seems to always find solutions to its problems by invoking the existence of larger symmetries. The proposed solutions to the above two problems of the Standard Model are no different: unified models and supersymmetry, respectively. The former postulate that the Standard Model is actually embedded in a larger group structure which includes $SU(3)_C \times SU(2)_L \times U(1)_Y$ but which is only fully manifest at very large energies, where new degrees of freedom are excited and the symmetry group is effectively enlarged to *e.g.*, $SU(5), SU(5)\times U(1), SO(10)$. In the larger theory relations among the parameters arise, which help explain some of the regularities observed in the Standard Model parameters, such as the quantization of the electric charge and some relations among the quark and lepton masses. In fact, the existence of three light generations of quarks and leptons had already been predicted in this context in studies of the m_b/m_τ ratio [2]. An immediate consequence of the larger symmetry is that the gauge couplings of the Standard Model are "unified", *i.e.*, take the same value at the unification scale (E_{GUT}), even though they are measured to be different numbers at the weak scale. By-products of great experimental interest are the new interactions among the quarks and leptons which may be induced by the heavy degrees of freedom. Among these, the most model-independent one leads to the prediction of proton decay, which must occur at a rather slow rate ($\tau_p \gtrsim 10^{32}\,y$) to avoid conflict with experimental limits.

Supersymmetry solves the gauge hierarchy problem by predicting the existence of partners for the ordinary particles which differ by half-a-unit of spin, *e.g.*, electron($\frac{1}{2}$)-selectron(0), photon(1)-photino($\frac{1}{2}$). The radiative corrections to the Higgs-boson mass are now tamed by the fact that the partners of the heavy particles have an opposite effect on the correction. However, since supersymmetry cannot be an unbroken symmetry (otherwise the sparticles would have the same mass as their Standard Model partners and no such sparticles have been observed) the cancellation is not perfect, but up to the typical mass splitting of particles and their superpartners. If this splitting can be predicted to be no larger than $\sim 1\,TeV$, then the gauge hierarchy problem will remain solved. In fact, it is this further requirement which demands that we enlarge the symmetry even more, by making supersymmetry a local symmetry called *supergravity*. In this class of theories the mass splittings can be explicitly computed in terms of very few parameters. This ability is of great experimental interest, since a large class of (*e.g.*, collider and rare) low-energy processes can be calculated in terms of few parameters, providing close knit correlations among the several experimental predictions, which may otherwise appear arbitrary.

We will consider two such supergravity models: the minimal $SU(5)$ [3] and $SU(5)\times U(1)$ [4]. Conceptually, the $SU(5)$ model is a typical (and the simplest) grand unified supersymmetric model which predicts the existence of proton decay at a rate which should be on the verge of being observed. In fact, its non-supersymmetric ancestor, the Georgi-Glashow $SU(5)$ model [5], was ruled out experimentally on the basis of its incorrect (too short) prediction for the proton lifetime. Theoretically, this model also suffers from the gauge hierarchy problem, which is present in the absence of supersymmetry. In the supersymmetric $SU(5)$ model, the unification of gauge couplings occurs at a scale $E_{GUT} \sim 10^{16}\,GeV$, and the proton lifetime problem of non-supersymmetric $SU(5)$ is naturally solved. As we now well know, unification of the gauge couplings also does not work in non-supersymmetric $SU(5)$, while it works very well in its supersymmetric counterpart. The goodness of the prediction for the m_b/m_τ relation alluded to above is also maintained with the addition of supersymmetry [2].

The $SU(5)\times U(1)$ supergravity model is best motivated in the context of superstrings, where it is found that $SU(5)\times U(1)$ (in contrast with $SU(5)$) is relatively easy to obtain. In this model the unification of couplings should occur at the string scale ($E_{GUT} \sim 10^{18}\,GeV$). This fact appears to require a non-minimal spectrum of particles at intermediate scales, in contrast with the $SU(5)$ model where the "big desert" picture is assumed to hold. Furthermore, the $SU(5)\times U(1)$ supergravity model allows a further reduction of the unknown parameters of the theory, making its predictions very sharp. Moreover, these two models allow to compare possible implementations of supersymmetry: one ($SU(5)$) as example of traditional unified theories, the other ($SU(5)\times U(1)$) as example of string-inspired theories.

3 Convergence of gauge couplings: geometry versus physics

The three gauge couplings of the Standard Model, as measured at the Z-boson mass scale, can be expressed in terms of: α_e, $\sin^2\theta_W$, and α_3. Experimentally we know that (see *e.g.*, [6]),

$$\alpha_e^{-1}(M_Z) = 127.9 \pm 0.1, \tag{1}$$

$$\alpha_3(M_Z) = 0.120 \pm 0.010, \tag{2}$$

$$\sin^2\theta_W(M_Z) = 0.2324 \pm 0.0006. \tag{3}$$

The $U(1)_Y$ and $SU(2)_L$ gauge couplings are related to these by $\alpha_1 = \frac{5}{3}(\alpha_e/\cos^2\theta_W)$ and $\alpha_2 = (\alpha_e/\sin^2\theta_W)$. In numbers:

$$\text{At} \quad Q = M_Z \qquad \left\{ \begin{array}{lcl} \alpha_1^{-1} & = & 58.91 \pm 0.16 \\ \alpha_2^{-1} & = & 29.72 \pm 0.080 \\ \alpha_3^{-1} & = & 8.33 \pm 0.69 \end{array} \right. \tag{4}$$

The three gauge couplings evolve with increasing values of the scale Q in a logarithmic fashion, and may become equal at some higher scale, signaling the possible presence of a larger gauge group. However, this need not be the case: the three gauge couplings may meet and then depart again. Conceptually, the presence of a unified group is essential in the discussion of unification of couplings. In this case, the newly excited degrees of freedom will be such that all three couplings will evolve together for scales $Q > E_{GUT}$, and one can then speak of a unified coupling. (For a recent review and extensive references to the current literature see *e.g.*, Ref. [7].)

The running of the gauge couplings is prescribed by a set of first-order non-linear differential equations: the renormalization group equations (RGEs) for the gauge couplings. In general, there is one such equation for each dynamical variable in the theory (*i.e.*, for each gauge coupling, Yukawa coupling, and sparticle mass). These equations give the rate of change of each dynamical variable as the scale Q is varied. For the case of a gauge coupling, the rate of change is proportional to (some power of) the gauge coupling itself, and the coefficient of proportionality is called the *beta function.* The beta functions encode the spectrum of the theory, and how the various gauge couplings influence the running of each other (a higher-order effect). Assuming that all supersymmetric particles have a common mass M_{SUSY}, the RGEs (to two-loop order) are:

$$\frac{d\alpha_i^{-1}}{dt} = -\frac{b_i}{2\pi} - \sum_{j=1}^{3} \frac{b_{ij}\alpha_j}{8\pi^2}. \tag{5}$$

where $t = \ln(Q/E_{GUT})$, with Q the running scale and E_{GUT} the unification mass. The one-loop (b_i) and two-loop (b_{ij}) beta functions are given by

$$b_i = \left(\tfrac{33}{5}, 1, -3\right), \tag{6}$$

$$b_{ij} = \begin{pmatrix} \frac{199}{25} & \frac{27}{5} & \frac{88}{5} \\ \frac{9}{5} & 25 & 24 \\ \frac{11}{5} & 9 & 14 \end{pmatrix}. \tag{7}$$

These equations are valid from $Q = M_{SUSY}$ up to $Q = E_{GUT}$. For $M_Z < Q < M_{SUSY}$ an analogous set of equations holds, but with beta functions which reflect the non-supersymmetric nature of the theory (*i.e.*, with all the sparticles decoupled),

$$b_i' = \left(\tfrac{41}{10}, -\tfrac{19}{6}, -7\right), \tag{8}$$

$$b_{ij}' = \begin{pmatrix} \frac{199}{50} & \frac{27}{10} & \frac{44}{5} \\ \frac{9}{10} & \frac{35}{6} & 12 \\ \frac{11}{10} & \frac{9}{2} & -26 \end{pmatrix}. \tag{9}$$

The non-supersymmetric equations are supplemented with the initial conditions given in Eq. (4).

If the above is all the physics which is incorporated in the study of the convergence of the gauge couplings, then it is easy to see that the couplings will always meet at some scale E_{GUT}, provided that M_{SUSY} is tuned appropriately [8, 9, 10, 11]. This is a simple consequence of euclidean geometry, as can be seen from Eq. (5). Neglecting the higher-order terms, we see that as a function of t, α_i^{-1} are just straight lines. In fact, the slope of these lines changes at $Q = M_{SUSY}$, where the beta functions change. The convergence of three straight lines with a change in slope is then guaranteed by euclidean geometry, as long as the point where the slope changes is tuned appropriately. (This fact was pointed out by A. Peterman and myself in 1979 [12].) What is non-trivial about the convergence of the couplings is that with the initial conditions given in Eq. (4), the change in slope needs to be $M_{SUSY} \sim 1\,TeV$ [10].

This prediction [10] for the likely scale of the supersymmetric spectrum (*i.e.*, $M_{SUSY} \sim 1\,TeV$ [10]) is in fact *incorrect* [13]. The reason is simple: the physics at the unification scale, which is used to predict the value of M_{SUSY}, has been ignored completely. In fact, such a geometrical picture of convergence of the gauge couplings is physically inconsistent, since for scales $Q > E_{GUT}$ the gauge couplings will depart again. One must consider a unified theory to be assured that the couplings will remain unified, as shown in Fig. 1. This entails the study of a new kind of effect, namely the influence on the running of the gauge couplings of the degrees of freedom which are excited near the unification scale (*i.e.*, the *heavy threshold effects*) [8, 14]. In fact, the whole concept of a single unification point needs to be abandoned. The upshot of all this is that the theoretical uncertainties on the values of the parameters describing the heavy GUT particles are such that the above prediction for M_{SUSY} [10] is washed out completely [13]. Furthermore, the insertion of a realistic spectrum of sparticles at low energies (as opposed to an unrealistic common M_{SUSY} mass) blurs the issue some more [8, 13]. *Thus, it is perfectly possible to obtain acceptable unification, with supersymmetric particle masses as low as experimentally allowed.* The most complete analysis of a unified theory is shown in Fig. 2. Note the unification of the gauge couplings which continues above E_{GUT}. Notice also that "light" and "heavy" thresholds have been duly accounted for, plus other detailed effects like the evolution of gaugino masses (EGM). This effect has in fact been calculated at two loops [15].

A related point is that LEP data do not uniquely demonstrate that the gauge couplings must unify at a scale $E_{GUT} \sim 10^{16}\,GeV$ [14]. This is probably the simplest conclusion one could draw. However, this conclusion is easily altered by for example considering models with particles at intermediate scales, *i.e.*, by populating the "big desert". In fact, one such simple modification allows the gauge couplings to converge at the string scale $E_{GUT} \sim 10^{18}\,GeV$ instead [4].

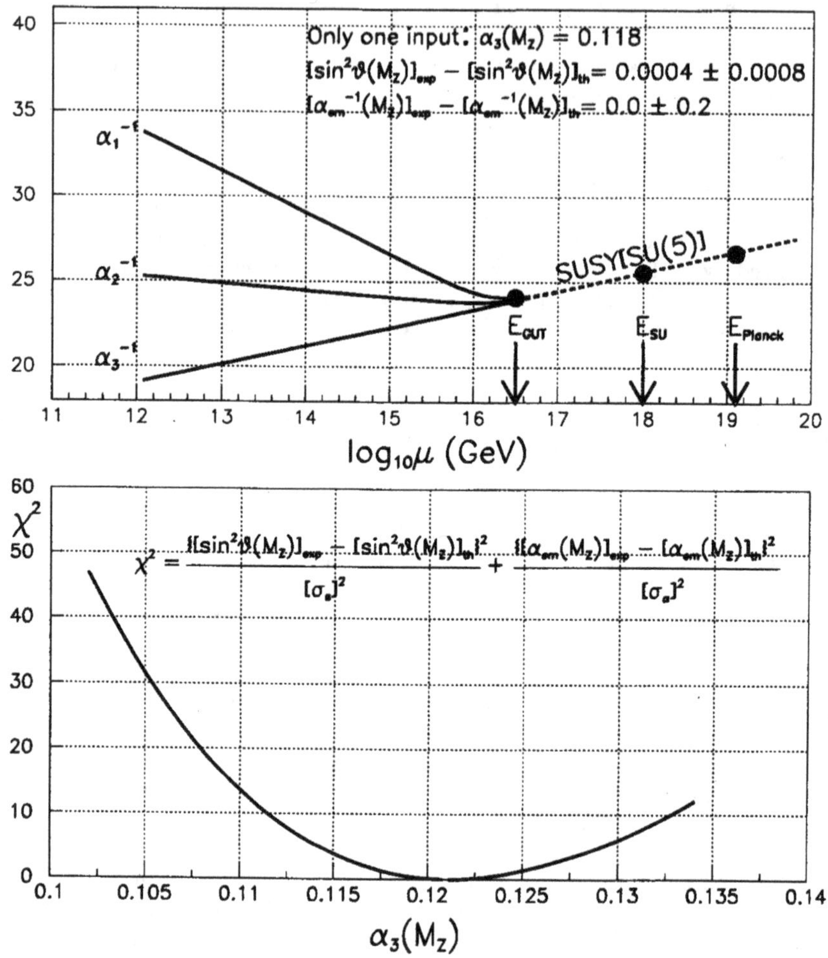

Figure 1: The convergence of the gauge couplings $(\alpha_1, \alpha_2, \alpha_3)$ at E_{GUT} is followed by the unification in a unique α_{GUT} above E_{GUT}. The RGEs include the heavy and light thresholds plus the evolution of gaugino masses. These results are obtained using as input the world-average value of $\alpha_3(M_Z)$ and comparing the predictions for $\sin^2\theta(M_Z)$ and $\alpha_{em}^{-1}(M_Z)$ with the experimental results. The χ^2 constructed using these two physically measured quantities allows to get the best E_{GUT}, $\alpha_{GUT}(E_{GUT})$, and $\alpha_3(M_Z)$ (shown). The RGEs can go down to M_Z without any need of introducing a change of slope at $E \approx 10^3$ GeV as would be required if the various effects mentioned above are neglected.

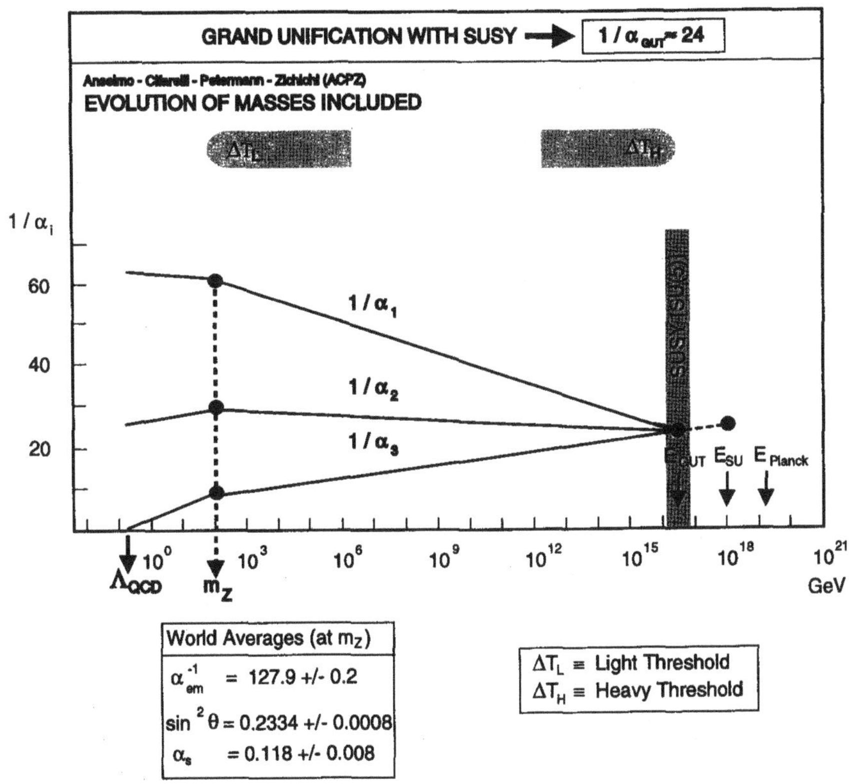

Figure 2: This is the best proof that the convergence of the gauge couplings can be obtained with M_{SUSY} at an energy level as low as M_Z. Notice that the effects of "light" and "heavy" thresholds have been accounted for, as well as the Evolution of Gaugino Masses [14,15]. This figure is Fig. 2 of ref. [16]. E_{SU} is the string unification scale.

4 Constraints from unification

The convergence of the gauge couplings implies that given α_e and α_3, one is able to compute the values of $\sin^2\theta_W$, the unification scale E_{GUT}, and the unified coupling α_U. In lowest-order approximation (*i.e.*, neglecting all GUT thresholds, two-loop effects, and taking $M_{SUSY} = M_Z$) one obtains

$$\ln\frac{E_{GUT}}{M_Z} = \frac{\pi}{10}\left(\frac{1}{\alpha_e} - \frac{8}{3\alpha_3}\right), \tag{10}$$

$$\frac{\alpha_e}{\alpha_U} = \frac{3}{20}\left(1 + \frac{4\alpha_e}{\alpha_3}\right), \tag{11}$$

$$\sin^2\theta_W = 0.2 + \frac{7\alpha_e}{15\alpha_3} . \tag{12}$$

These equations provide a rough approximation to the actual values obtained when all effects are included. Nonetheless, they embody the most important dependences on the input parameters. In Fig. 3 (from Ref. [11]) we show the relation between E_{GUT} and α_3 for various values of $\sin^2\theta_W$. One can observe that:

$$\alpha_3 \uparrow \quad \Rightarrow \quad E_{GUT} \uparrow \quad \textit{for fixed } \sin^2\theta_W \tag{13}$$

$$\sin^2\theta_W \uparrow \quad \Rightarrow \quad E_{GUT} \uparrow \quad \textit{for fixed } \alpha_3 \tag{14}$$

$$\alpha_3 \uparrow \quad \Rightarrow \quad \sin^2\theta_W \downarrow \quad \textit{for fixed } E_{GUT} \tag{15}$$

These are the most important systematic correlations, which are not really affected by the neglected effects. These correlations are evident in Eqs. (10–12) and in Fig. 3. In this figure we also show the lower bound on E_{GUT} which follows from the proton decay constraint. Clearly a lower bound on $\alpha_3(M_Z)$ results, which allows the world-average value. Another interesting result is the anticorrelation between E_{GUT} and M_{SUSY}. This is shown in Fig. 4, where for fixed $\sin^2\theta_W(M_Z)$ we see that increasing $\alpha_3(M_Z)$ increases E_{GUT} (as already noted in Eq. (13)) and decreases M_{SUSY}. Taking for granted this approach (*i.e.*, all supersymmetric particle masses degenerate at M_{SUSY}) for comparison with the large amount of papers published following this logic, in Fig. 5 we see the narrow band left open once the experimental limits on τ_p and M_{SUSY} are imposed. Figure 5 is a guide to understand the qualitative interconnection between the basic experimentally measured quantities, $\alpha_3(M_Z)$, $\sin^2\theta_W(M_Z)$, τ_p, M_{SUSY} and the theoretically desirable E_{GUT}. The experimental lower bounds on the proton lifetime $(\tau_p)_{exp}$ and on M_{SUSY} produce two opposite bounds (lower and upper, respectively) on the unification energy scale E_{GUT}. Note that, in order to make definite predictions on the lightest detectable supersymmetric particle, a detailed supergravity model is needed. The study of the correlations between the basic quantities, as exemplified in this figure, is interesting but should not be mistaken as example of prediction for the superworld. In particular, the introduction of the quantity M_{SUSY} is really misleading.

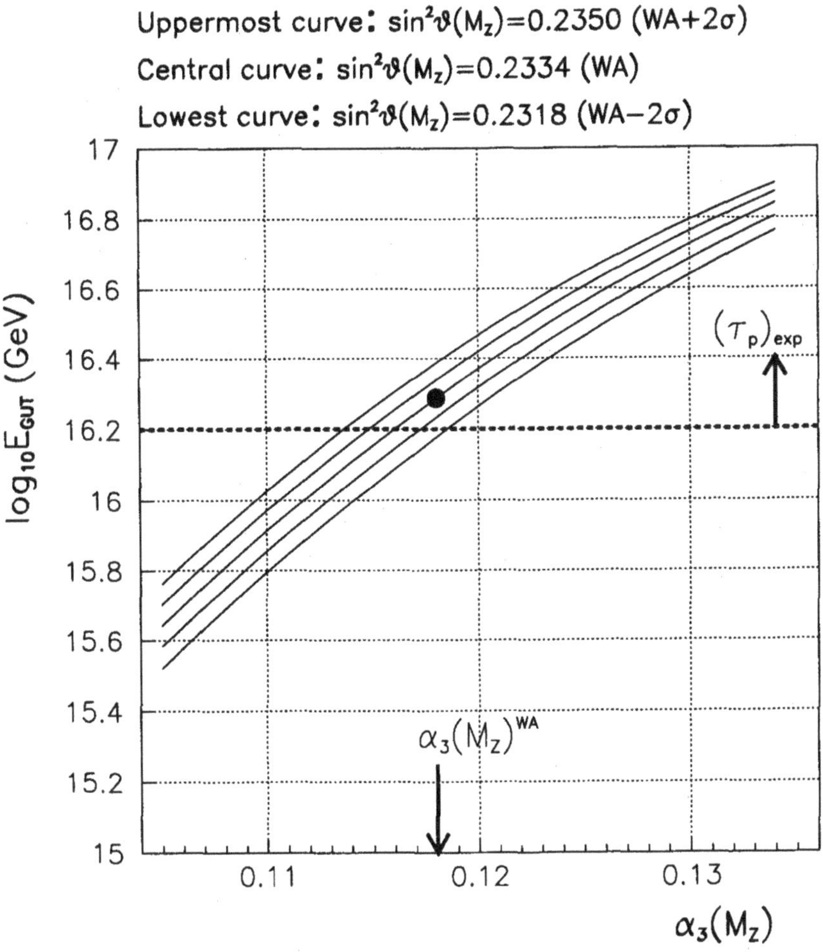

Figure 3: The unification scale E_{GUT} versus $\alpha_3(M_Z)$ for various values of $\sin^2\theta_W(M_Z)$ within $\pm 2\sigma$ of the world-average value. Also indicated is the lower bound on E_{GUT} from the lower limit on the proton lifetime.

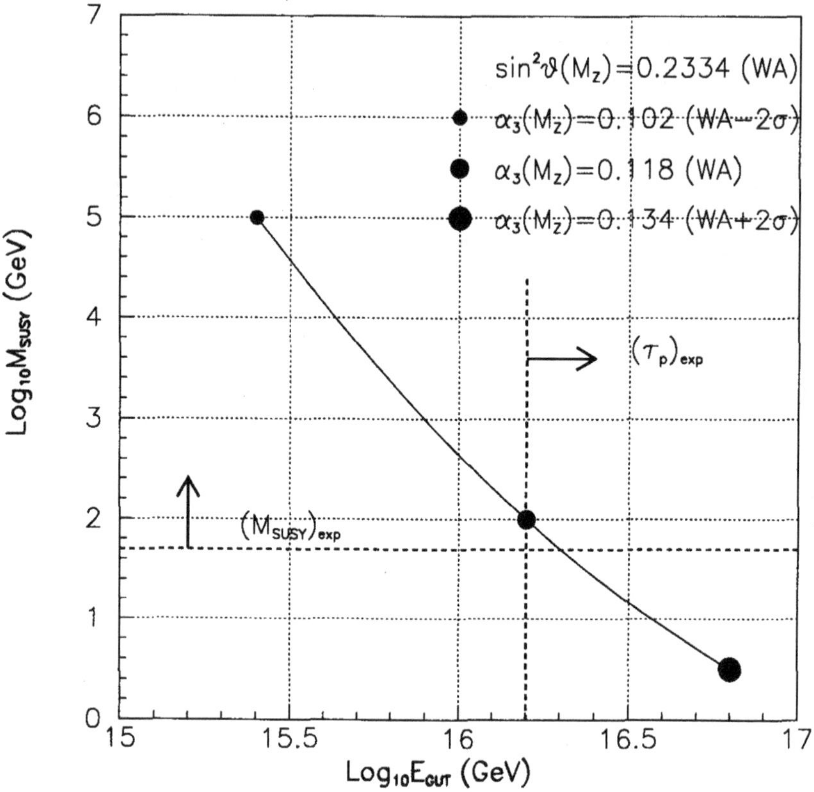

Figure 4: The unification scale E_{GUT} versus M_{SUSY} for different values of $\alpha_3(M_Z)$ and fixed $\sin^2\theta_W(M_Z)$. Note the anticorrelation between M_{SUSY} and E_{GUT}. The experimental lower bound on M_{SUSY} is shown. The lower bound on E_{GUT} from Fig. 1 is also indicated.

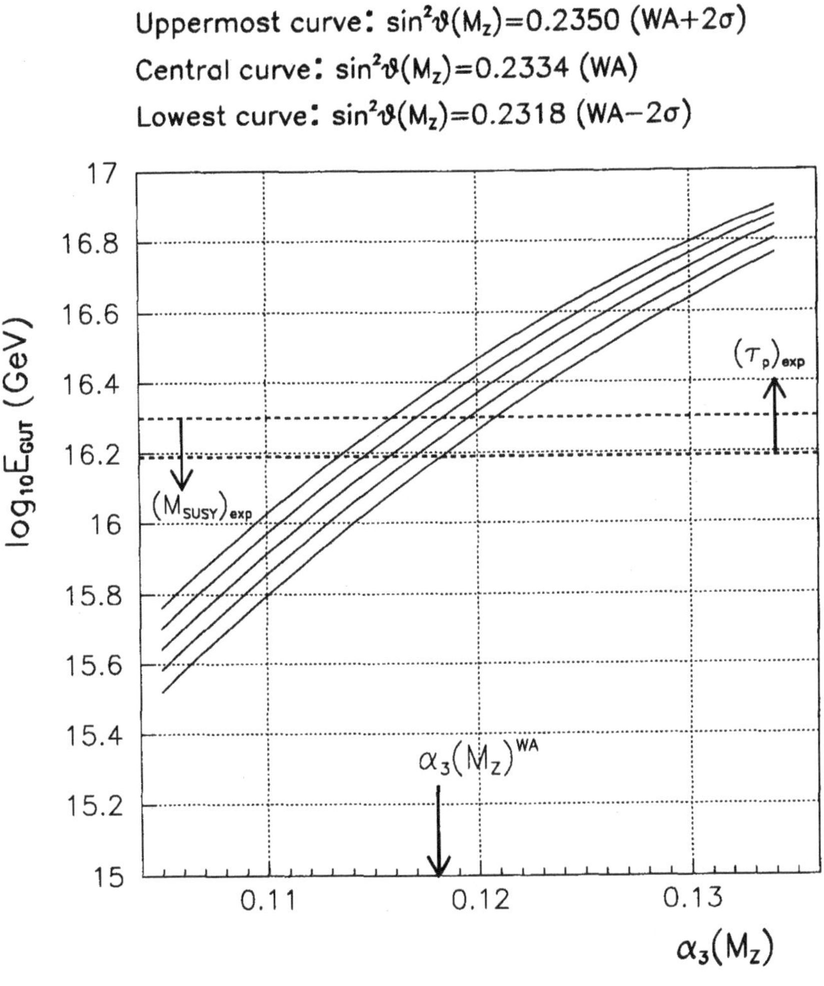

Figure 5: The correlation between all measured quantities, $\alpha_3(M_Z)$, $\sin^2\theta_W(M_Z)$, τ_p, the limits on the lightest detectable supersymmetric particle (here represented by M_{SUSY}) and the unification energy scale E_{GUT}.

5 The origin of M_{SUSY} and the need for local supersymmetry

The calculations which we have described so far, attempted to determine the value of M_{SUSY}, or more properly the supersymmetric particle spectrum, by fitting the spectrum to obtain the "best possible" unification picture. This program did not succeed because of the large inherent uncertainties in the physics at the GUT scale. Nevertheless, for a given GUT model, it should be possible to compute the GUT threshold effects and attempt the "best fit" procedure to deduce the corresponding light supersymmetric spectrum. This picture is not very satisfying since one would like to know why the supersymmetric spectrum should be the way the fit would require it to be. In other words, the real question is: what determines the values of the sparticle masses? And why should these be below $\sim 1\,TeV$, so that the gauge hierarchy problem is not re-introduced?

As mentioned in Sec. 2, considering a theory with supergravity (instead of global supersymmetry) provides the means to compute the masses of the sparticles [3]. This framework assumes that supersymmetry breaking occurs in a "hidden sector" of the theory, where "gravitational particles" (those introduced when the supersymmetry was made local) may grow vacuum expectation values (vevs) which break supersymmetry spontaneously in the hidden sector. These vevs are best understood as induced dynamically by the condensation of the supersymmetric partners of the hidden sector particles when the gauge group which describes them becomes strongly interacting at some large scale. The splitting of the particles and their partners would then be generated, and would be the order of the condensation scale ($\sim 10^{12-16}\,GeV$). However, such huge mass splittings will not be immediately transmitted to the "observable" (the normal) sector of the theory, since the two sectors only communicate through gravitational interactions. The dampening in the transmission mechanism is such that the splittings in the observable sector are usually much more suppressed than those in the hidden sector, and suitable choices of hidden sectors may yield realistic low-energy supersymmetric spectra. This picture of hidden and observable sectors becomes completely natural in the context of superstrings, where models typically contain both sectors and one can study explicitly the predicted spectrum of supersymmetric particles at low energies.

In a large number of models, the supersymmetric particle masses at the unification scale are also "unified". This situation is called *universal soft-supersymmetry-breaking*, and the masses of all scalar partners (*e.g.*, squarks and sleptons) take the common value of m_0, the gaugino (the partners of the gauge bosons) masses are given by $m_{1/2}$, and there is a third parameter (A) which basically parametrizes the mixing of stop-squark mass eigenstates at low energies. The breaking of the electroweak symmetry is obtained dynamically in the context of these models, through the so-called *radiative electroweak symmetry breaking mechanism*, which involves the top-quark mass in a fundamental way [17]. After all these well motivated theoretical ingredients have been incorporated, the models depend on only five parameters: $m_{1/2}, m_0, A$, the top-quark mass (m_t), and the ratio of the two Higgs vacuum expectations values ($\tan\beta$) [18].

A rather interesting framework occurs in the so-called *no-scale* scenario [19, 20, 17], where all the scales in the theory are obtained from just one basic scale (*i.e.*, the unification scale or the Planck scale) through radiative corrections. These models have the unparalleled virtue of a vanishing cosmological constant at the tree-level *even after supersymmetry breaking*, and in their unified versions predict that the universal scalar masses and trilinear couplings vanish (*i.e.*, $m_0 = A = 0$) and the universal gaugino mass ($m_{1/2}$) is the only seed of supersymmetry breaking. Moreover, this unique mass can be determined in principle by minimizing the vacuum energy at the electroweak scale. The generic result is $m_{1/2} \sim M_Z$ [19, 20], in agreement with theoretical prejudices (*i.e.*, "naturalness"). Furthermore, no-scale supergravity is obtained in the infrared limit of superstrings [21].

More generally, in generic supergravity models the five-dimensional parameter space is constrained by phenomenological requirements, such as sparticle and Higgs-boson masses not in conflict with present experimental lower bounds, a sufficiently long proton lifetime, a sufficiently old Universe (a cosmological constraint on the amount of dark matter in the Universe today), various indirect constraints from well measured rare processes, etc. In addition, further theoretical constraints can be imposed which give m_0 and A as functions of $m_{1/2}$, and thus reduce the dimension of the parameter space down to just three. In what follows we will focus on some specific supergravity models which are so constrained that precise experimental predictions can be made.

6 Details on the chosen supergravity models: SU(5) and SU(5)xU(1)

The two supergravity models that we have chosen belong to the class of models we just described. However, these models are even more predictive than the run-of-the-mill supergravity model because further well motivated theoretical assumptions are made.

6.1 The minimal SU(5) supergravity model

In this model the gauge group is $SU(5)$, which completely contains the Standard Model as a subgroup. This implies that all three Standard Model gauge couplings should evolve from low energies and become equal at (and above) the unification scale E_{GUT} (up to heavy threshold corrections). This scale has some dependence on the uncertainties on the various parameters; one usually obtains $E_{GUT} \sim 10^{16}\,GeV$. The Standard Model particles and their superpartners are assigned to the $\overline{5}$ and 10 representations:

$$\bar{5} = \{d^c, L = \begin{pmatrix} e \\ \nu_e \end{pmatrix}\}, \qquad 10 = \{Q = \begin{pmatrix} u \\ d \end{pmatrix}, u^c, e^c\}. \tag{16}$$

The heavy GUT particles which are excited near the unification scale include the 24 representation of gauge bosons (and gauginos) which contains the twelve Standard Model gauge bosons (8 gluons, $W^\pm, Z, \gamma$) plus twelve heavy (charged and colored) gauge bosons which mediate the $SU(5)$ gauge interactions. There is also a 24 representation of Higgs bosons (and their superpartners), and the vacuum expectation value of the neutral component of this set effects the $SU(5) \to SU(3)_C \times SU(2)_L \times U(1)_Y$ gauge symmetry breaking. The minimal $SU(5)$ spectrum also includes a pair of pentaplet representations (5,$\bar{5}$) which contain the two Higgs doublets of the low-energy supersymmetric theory, but also a pair of colored Higgs triplets.

Perhaps the most decisive property of this model is the prediction for the proton lifetime. Proton decay can be mediated in two ways: by exchange of heavy gauge or Higgs bosons (dimension-six operators), or by the exchange of heavy Higgsinos (dimension-five operators) [22, 23]. The dimension-six operators predict a proton lifetime proportional to M_U^4 or M_H^4 respectively, and the expected values of these mass scales make this contribution rather small. On the other hand, the dimension-five operators entail a proton lifetime proportional to $M_{\tilde{H}}^2$, which requires $M_{\tilde{H}} > E_{GUT}$ and some strong constraints on the supersymmetric spectrum (if it is light enough to be observable). In fact, the three-dimensional soft-supersymmetry-breaking parameter space $(m_{1/2}, m_0, A)$ is constrained in such a way that the squarks and sleptons should be heavy, while the neutralinos and charginos should be light [24].

Another important constraint on the parameter space of the minimal $SU(5)$ supergravity model is provided by cosmological considerations. The models we consider include a discrete symmetry called *R-parity* which has value +1 for the ordinary particles, and −1 for the supersymmetric particles. This implies that supersymmetric particles must always be created or destroyed in pairs, *i.e.*, at each vertex there are always none, two, or four supersymmetric particles. In particular, the *lightest supersymmetric particle* (LSP) must be stable in these models. Astrophysical considerations require this particle to be neutral and colorless [25]. The only two candidates are the lightest neutralino (a linear combination of the weakly interacting neutral sparticles: photino, zino, two Higgsinos) and the sneutrino. It turns out that it is the neutralino which is the lightest one. Since the neutralinos are stable, they must have pair-annihilated away quite efficiently in the early Universe, otherwise their present relic abundance would be too large, and the Hubble parameter would correspond to a Universe younger than the oldest known stars.

The cosmological constraint can be usually satisfied as long as the supersymmetric particles are not too heavy. However, the proton decay constraint requires heavy squarks and sleptons and thus the constraint is relevant. In fact, the only particles which mediate pair-annihilation efficiently are the Z-boson and the lightest Higgs boson, which are light. In practice, the neutralino mass must be near half the masses of these particles so that resonant s-channel annihilation occurs [26]. The resulting effect on the parameter space of the model is quite severe [27].

6.2 The SU(5)xU(1) supergravity model

The $SU(5) \times U(1)$ ("flipped $SU(5)$") gauge group [28] differs from $SU(5)$ in several ways. The Standard Model particles are also assigned to the $\overline{5}$ and 10 representations, but in a "flipped" way (*i.e.*, $u^c \leftrightarrow d^c$, $e^c \leftrightarrow \nu^c$) relative to the minimal $SU(5)$ case,

$$\bar{5} = \{u^c, L = \binom{e}{\nu_e}\}, \qquad 10 = \{Q = \binom{u}{d}, d^c, \nu^c\}, \qquad 1 = e^c. \tag{17}$$

Note that a right-handed neutrino ν^c has appeared naturally. This leads to an automatic *see-saw* mechanism to generate small neutrino masses in this model [29]. The heavy GUT spectrum of fields includes the 24 representation of gauge boson plus the $U(1)$ gauge boson of $SU(5) \times U(1)$. Unlike $SU(5)$, the symmetry breaking Higgs fields are contained in the 10,$\overline{10}$ representations (which have neutral components because of the "flipped" assignment). This property is central to the appeal of $SU(5) \times U(1)$ as a paradigm of a string model [30], since larger Higgs representations (like the 24) are not easily obtainable in string model building. There is also a pair of Higgs pentaplets 5,$\overline{5}$ which contain the two Higgs doublets of the low-energy theory, and the heavy Higgs triplets.

Proton decay is a concern that is easily dismissed in the typical $SU(5) \times U(1)$ models: the dimension-six proton decay operators are small as usual, while the perilous dimension-five operators are strongly suppressed. The latter is a direct consequence of the $SU(5) \times U(1)$ symmetry as it applies to the solution of the *doublet-triplet splitting problem*. The Higgs triplets are not only heavy but also cannot mediate the dangerous proton decay diagram [29, 31].

The model we have studied is most simply understood in the context of string model building when the gauge couplings of the Standard Model unify at the string scale $E_{GUT} \sim 10^{18}\,GeV$ [32]. This entails the addition of another 10, $\overline{10}$ representations to the heavy GUT spectrum, otherwise the gauge couplings would unify as in the minimal $SU(5)$ model. (Specific string models with this property have also been constructed [33].)

For the soft-supersymmetry-breaking parameters we have used two string-inspired scenarios, which allow one to determine two of the parameters as functions of the third one. We have considered,

(i) the *no-scale* model: $m_0 = A = 0$

(ii) the *dilaton* model: $m_0 = \frac{1}{\sqrt{3}} m_{1/2}$, $A = -m_{1/2}$ [34].

Note that the dimension of the parameter space is now reduced to just three: $m_t, \tan\beta$, and $m_{1/2}$, where $m_{1/2} \propto m_{\tilde{g}}$. This implies that the sparticle masses are (up to $\tan\beta$-dependent effects) proportional to the gluino mass. The corresponding approximate proportionality coefficients are shown in Table 1. For both these supersymmetry breaking scenarios there is a yet more restrictive possibility which we call the *strict no-scale* and the *special dilaton* scenarios. A further requirement allows one to deduce the value of $\tan\beta$ as a function of m_t and $m_{1/2}$, yielding highly predictive two-parameter models.

With these choices for the supersymmetry breaking parameters, one determines that the cosmological constraint does not restrict the models in any way [32, 35]. In fact, these two models offer a natural explanation for the required amount of dark matter in the Universe.

Table 1: The approximate proportionality coefficients to the gluino mass, for the various sparticle masses in the two supersymmetry breaking scenarios considered for $SU(5) \times U(1)$ supergravity.

	no-scale	dilaton
$\tilde{e}_R, \tilde{\mu}_R$	0.18	0.33
$\tilde{\nu}$	0.18 − 0.30	0.33 − 0.41
$2\chi^0_1, \chi^0_2, \chi^\pm_1$	0.28	0.28
$\tilde{e}_L, \tilde{\mu}_L$	0.30	0.41
$\tilde{q}$	0.97	1.01
$\tilde{g}$	1.00	1.00

7 The projection of Salam's dream on the real axis

As remarked above, in a generic supergravity model one expects a large degree of correlation among the predictions for, *e.g.*, the sparticle and Higgs masses and the rates for the various experimental processes of interest. These generic models are further constrained by the non-observation of any sparticle or Higgs boson. Interestingly enough, such five-parameter models are still not very predictive in that many possible trends of correlations are possible. In contrast, in the two specific supergravity models described in Sec. 6, the correlations become sharp and of much more experimental interest.

Without imposing any further constraints on the models, besides the present collider bounds on sparticle and Higgs masses and the proton decay and cosmological constraints as applicable, we have calculated rates for the following processes at the indicated facilities:

$$\begin{array}{ll}
Tevatron\ [36] & p\bar{p} \to \chi_1^\pm \chi_2^0,\quad \chi_1^\pm \to \chi_1^0 l^\pm \nu_l,\quad \chi_2^0 \to \chi_1^0 l^+ l^-,\quad l = e,\mu \quad \text{"}trileptons\text{"} \\
LEPI\ [37] & e^+e^- \to Zh \to f\bar{f}h \\
LEP\ II\ [38] & \left\{\begin{array}{l}
e^+e^- \to \chi_1^+\chi_1^-,\quad \chi_1^+ \to \chi_1^0 q\bar{q}',\quad \chi_1^- \to \chi_1^0 l^- \nu_l,\ l = e,\mu \quad \text{"}mixed\text{"} \\
e^+e^- \to \chi_1^+\chi_1^-,\quad \chi_1^+ \to \chi_1^0 l^+ \bar{\nu}_l,\quad \chi_1^- \to \chi_1^0 l^- \nu_l,\ l = e,\mu \quad \text{"}dilepton\text{"} \\
e^+e^- \to \tilde{l}_R^+ \tilde{l}_R^-,\quad \tilde{l}_R^+ \to \chi_1^0 l^+,\quad \tilde{l}_R^- \to \chi_1^0 l^-,\quad l = e,\mu,\tau \quad \text{"}dilepton\text{"} \\
e^+e^- \to Z^* h \to f\bar{f}h,\quad h \to \chi_1^0\chi_1^0, b\bar{b}, c\bar{c}, \tau^+\tau^-, gg
\end{array}\right. \\
HERA\ [39] & \left\{\begin{array}{l}
e^-p \to \tilde{e}_R^- \chi_1^0 + p,\quad \tilde{e}_R^- \to \chi_1^0 e^- \quad \text{"}elastic\ selectron - neutralino\text{"} \\
e^-p \to \tilde{\nu}_e \chi_1^- + p,\quad \chi_1^- \to \chi_1^0 e^- \bar{\nu}_e \quad \text{"}elastic\ sneutrino - chargino\text{"} \\
e^-p \to \tilde{e}_R^- \chi_1^0 + X,\quad \tilde{e}_R^- \to \chi_1^0 e^- \quad \text{"}DI\ selectron - neutralino\text{"} \\
e^-p \to \tilde{\nu}_e \chi_1^- + X,\quad \chi_1^- \to \chi_1^0 e^- \bar{\nu}_e \quad \text{"}DI\ sneutrino - chargino\text{"}
\end{array}\right.
\end{array}$$

All the above processes are kinematically accessible in the $SU(5) \times U(1)$ models. In the minimal $SU(5)$ model the sleptons are heavy and neither the slepton pair production at LEPII nor any of the indicated processes at HERA are allowed.

We now give a sample of the actual results obtained for the most important processes listed above. These are shown in Figures 6,7,8,9.

More recently it has been realized that indirect experimental constraints on the $SU(5) \times U(1)$ models exist and can be quite significant. These constraints come generally from three sources: (i) the experimentally allowed range for $B(b \to s\gamma)$ as recently determined by the CLEO Collaboration [42]; (ii) the long-standing value for the anomalous magnetic moment of the muon [43]; and (iii) the precision LEP measurements of the electroweak parameters in the form of the allowed range for the ϵ_1 parameter [44]. (Only the last constraint is relevant in the minimal $SU(5)$ supergravity model because of its relatively heavier spectrum.)

The first two constraints exclude regions of the parameter space of the $SU(5) \times U(1)$ models which span all allowed values of the chargino mass, when viewing the parameter space in the $(m_{\chi_1^\pm}, \tan\beta)$ plane for fixed m_t. On the other hand, the ϵ_1 constraint basically implies an upper bound on m_t: $m_t \lesssim 165\,GeV$, unless the chargino is very light ($m_{\chi_1^\pm} \lesssim 70\,GeV$) in which case the upper bound on m_t can be relaxed up to $m_t \lesssim 180\,GeV$. As an example of the effect of the constraints, in Fig. 10 we show the parameter space for $m_t = 130\,GeV$ (when the ϵ_1 constraint is not restrictive) for the no-scale model [45]. Clearly the region of parameter space accessible to LEPII searches ($m_{\chi_1^\pm} \lesssim 100\,GeV$) has become quite constrained.

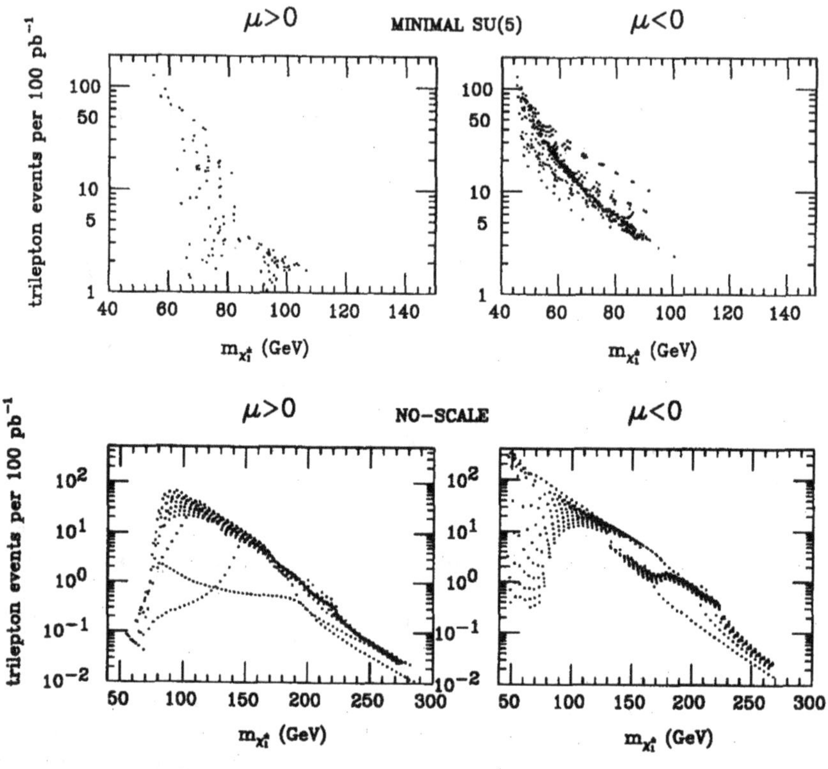

Figure 6: The number of trilepton events at the Tevatron per $100\,pb^{-1}$ in the minimal $SU(5)$ model and the no-scale $SU(5)\times U(1)$ model (for $m_t = 130\,GeV$). Note that with $200\,pb^{-1}$ and 60% detection efficiency it should be possible to probe basically all of the parameter space of the minimal $SU(5)$ model, and probe chargino masses as high as $175\,GeV$ in the no-scale model. Upper bounds on the trilepton cross section ($\sigma \cdot B < (0.6-1)$ pb) have been recently announced by the CDF [40] and D0 [41] Collaborations.

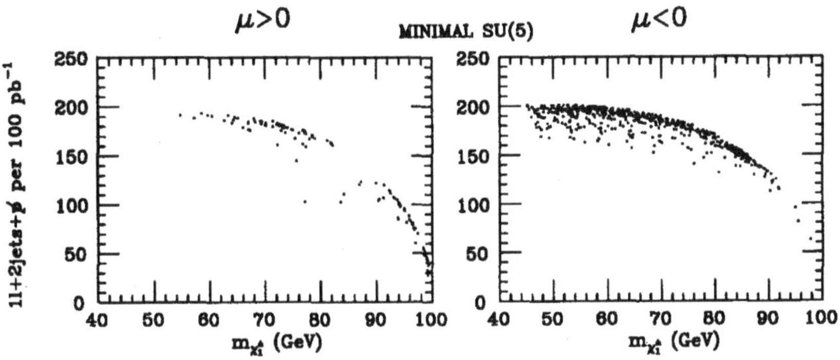

Figure 7: The number of "mixed" events (1-lepton+2jets+$\not{p}$) events per $\mathcal{L} = 100\,pb^{-1}$ at LEPII versus the chargino mass in the minimal $SU(5)$ model.

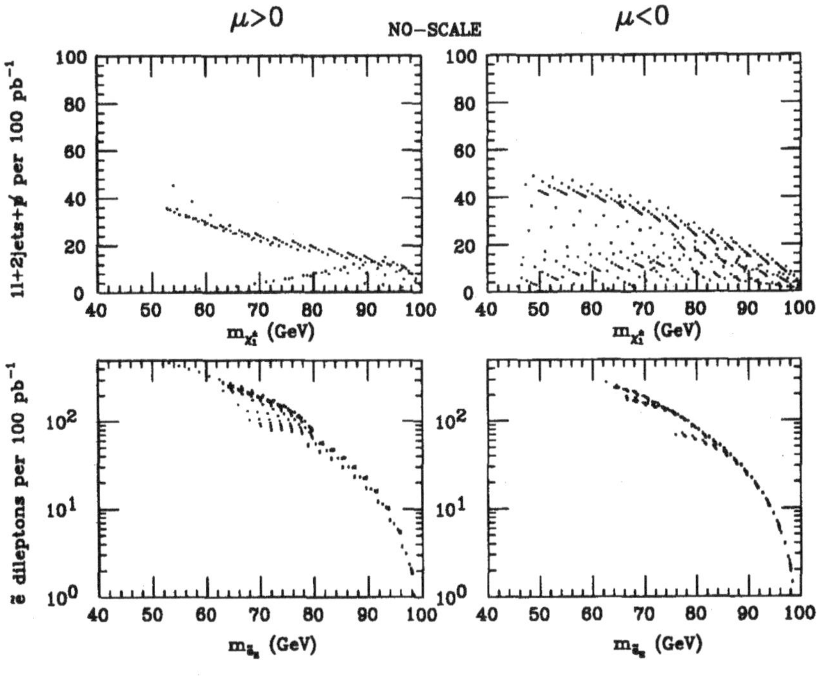

Figure 8: The number of "mixed" events (1-lepton+2jets+$\not{p}$) events per $\mathcal{L} = 100\,pb^{-1}$ at LEPII versus the chargino mass in the no-scale model (top row). Also shown (bottom row) are the number of di-electron events per $\mathcal{L} = 100\,pb^{-1}$ from selectron pair production versus the lightest selectron mass.

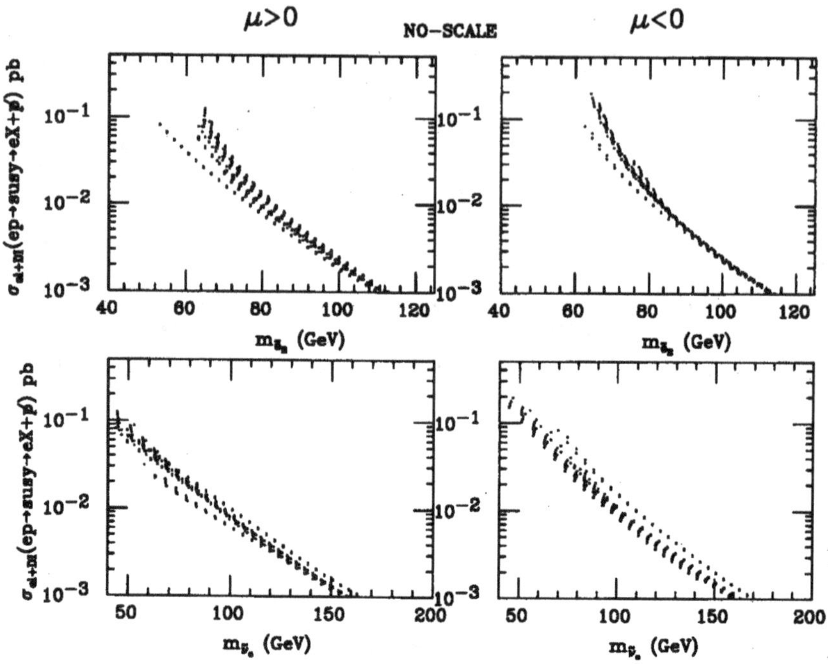

Figure 9: The elastic plus deep-inelastic total supersymmetric cross section at HERA ($ep \rightarrow susy \rightarrow eX + \not{p}$) in the no-scale model versus the lightest selectron mass ($m_{\tilde{e}_R}$) and the sneutrino mass ($m_{\tilde{\nu}}$). The short- and long-term limits of sensitivity are expected to be $10^{-2}\,pb$ and $10^{-3}\,pb$ respectively.

8 Conclusions

The dream of Abdus Salam can well be "around the corner" and at the verge of discovery at present and near future facilities, such as the Tevatron, HERA, and LEP (I and II). In order to have a correct and realistic view on this dream all experimental constraints on supersymmetric unified models need to be consistently imposed. Moreover, theoretical models should be as simple as the best physics motivations allow.

Acknowledgements

The results reported in this review paper would not have been possible without the dedicated work of all my collaborators: F. Anselmo, L. Cifarelli, J.L. Lopez, D.V. Nanopoulos, A. Peterman, H. Pois, and Xu Wang. To all of them I would like to express my sincere appreciation and gratitude.

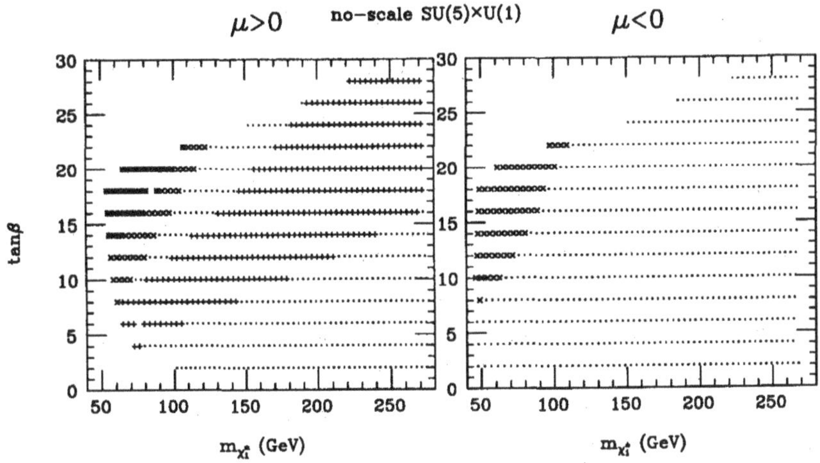

Figure 10: The parameter space of the no-scale $SU(5) \times U(1)$ supergravity model for $m_t = 130\, GeV$. Points denoted by periods satisfy all presently known experimental constraints, whereas those denoted by pluses violate the limits on $B(b \to s\gamma)$ and those denoted by crosses violate the limits on $(g-2)_\mu^{susy}$.

References

[1] For a recent review see *e.g.*, G. Altarelli, CERN-TH.6867/93.

[2] A. J. Buras, J. Ellis, M. K. Gaillard, D. V. Nanopoulos, Nucl. Phys. B **135** (1978) 66; D. V. Nanopoulos and D. A. Ross, Nucl. Phys. B **157** (1979) 273, Phys. Lett. B **108** (1982) 351, and Phys. Lett. B **118** (1982) 99.

[3] For reviews see *e.g.*, R. Arnowitt and P. Nath, *Applied N=1 Supergravity* (World Scientific, Singapore 1983); H. P. Nilles, Phys. Rep. **110** (1984) 1.

[4] For a recent review see J. L. Lopez, D. V. Nanopoulos, and A. Zichichi, CERN-TH.6926/93 and Texas A & M University preprint CTP-TAMU-33/93.

[5] H. Georgi and S. Glashow, Phys. Rev. Lett. **32** (1974) 438.

[6] See *e.g.*, P. Langacker and N. Polonsky, Phys. Rev. D **47** (1993) 4028.

[7] J. L. Lopez, Erice 93 Subnuclear Physics School Lecture, Texas A & M University preprint CTP-TAMU-42/93.

[8] J. Ellis, S. Kelley, and D. V. Nanopoulos, Phys. Lett. B **249** (1990) 441; Phys. Lett. B **260** (1991) 131; Nucl. Phys. B **373** (1992) 55.; Phys. Lett. B **287** (1992) 95.

[9] P. Langacker and M.-X. Luo, Phys. Rev. D **44** (1991) 817.

[10] U. Amaldi, W. de Boer, and H. Fürstenau, Phys. Lett. B **260** (1991) 447.

[11] F. Anselmo, L. Cifarelli, A. Peterman, and A. Zichichi, Nuovo Cim. **104A** (1991) 1817 and **105A** (1992) 1025.

[12] A. Zichichi, EPS Conference, Geneva 1979, Proceedings; and Nuovo Cimento Rivista **2** (1979) 1. This is the first instance in which it was pointed out that supersymmetry may play an important role in the convergence of the gauge couplings.

[13] F. Anselmo, L. Cifarelli, A. Peterman, and A. Zichichi, Nuovo Cim. **105A** (1992) 581; F. Anselmo, L. Cifarelli, and A. Zichichi, Nuovo Cim. **105A** (1992) 1357.

[14] F. Anselmo, L. Cifarelli, and A. Zichichi, Nuovo Cim. **105A** (1992) 1335.

[15] F. Anselmo, L. Cifarelli, A. Peterman, and A. Zichichi, Nuovo Cim. **105A** (1992) 1201.

[16] A. Zichichi, "Understanding where the supersymmetry threshold should be", CERN-PPE/92-149, 7 September 1992.

[17] For a review see A. B. Lahanas and D. V. Nanopoulos, Phys. Rep. **145** (1987) 1.

[18] See *e.g.*, S. Kelley, J. L. Lopez, D. V. Nanopoulos, H. Pois, and K. Yuan, Nucl. Phys. B **398** (1993) 3.

[19] J. Ellis, A. Lahanas, D. V. Nanopoulos, and K. Tamvakis, Phys. Lett. B **134** (1984) 429.

[20] J. Ellis, C. Kounnas, and D. V. Nanopoulos, Nucl. Phys. B **241** (1984) 406, Nucl. Phys. B **247** (1984) 373.

[21] E. Witten, Phys. Lett. B **155** (1985) 151.

[22] S. Weinberg, Phys. Rev. D **26** (1982) 287; N. Sakai and T. Yanagida, Nucl. Phys. B **197** (1982) 533.

[23] J. Ellis, D. V. Nanopoulos, and S. Rudaz, Nucl. Phys. B **202** (1982) 43; B. Campbell, J. Ellis, and D. V. Nanopoulos, Phys. Lett. B **141** (1984) 229; K. Enqvist, A. Masiero, and D. V. Nanopoulos, Phys. Lett. B **156** (1985) 209; P. Nath, A. Chamseddine, and R. Arnowitt, Phys. Rev. D **32** (1985) 2348; P. Nath and R. Arnowitt, Phys. Rev. D **38** (1988) 1479.

[24] R. Arnowitt and P. Nath, Phys. Rev. Lett. **69** (1992) 725; P. Nath and R. Arnowitt, Phys. Lett. B **287** (1992) 89.

[25] J. Ellis, J. S. Hagelin, D. V. Nanopoulos, K. A. Olive and M. Srednicki, Nucl. Phys. B **238** (1984) 453.

[26] J. L. Lopez, D. V. Nanopoulos, and A. Zichichi, Phys. Lett. B **291** (1992) 255.

[27] J. L. Lopez, D. V. Nanopoulos, and H. Pois, Phys. Rev. D **47** (1993) 2468; R. Arnowitt and P. Nath, Phys. Lett. B **299** (1993) 58 and **307** (1993) 403(E); P. Nath and R. Arnowitt, Phys. Rev. Lett. **70** (1993) 3696; J. L. Lopez, D. V. Nanopoulos, and K. Yuan, Phys. Rev. D **48** (1993) 2766.

[28] S. Barr, Phys. Lett. B **112** (1982) 219, Phys. Rev. D **40** (1989) 2457; J. Derendinger, J. Kim, and D. V. Nanopoulos, Phys. Lett. B **139** (1984) 170.

[29] I. Antoniadis, J. Ellis, J. Hagelin, and D. V. Nanopoulos, Phys. Lett. B **194** (1987) 231.

[30] I. Antoniadis, J. Ellis, J. Hagelin, and D. V. Nanopoulos, Phys. Lett. B **231** (1989) 65.

[31] J. Ellis, J. Hagelin, S. Kelley, and D. V. Nanopoulos, Nucl. Phys. B **311** (1988/89) 1.

[32] J. L. Lopez, D. V. Nanopoulos, and A. Zichichi, CERN-TH.6667/92, Texas A & M University preprint CTP-TAMU-68/92 (to appear in Phys. Rev. D).

[33] J. L. Lopez, D. V. Nanopoulos, and K. Yuan, Nucl. Phys. B **399** (1993) 654.

[34] V. Kaplunovsky and J. Louis, Phys. Lett. B **306** (1993) 269.

[35] J. L. Lopez, D. V. Nanopoulos, and A. Zichichi, CERN-TH.6903/93, Texas A & M University preprint CTP-TAMU-31/93.

[36] J. L. Lopez, D. V. Nanopoulos, X. Wang, and A. Zichichi, Phys. Rev. D **48** (1993) 2062.

[37] J. L. Lopez, D. V. Nanopoulos, H. Pois, X. Wang, and A. Zichichi, Phys. Lett. B **306** (1993) 73.

[38] J. L. Lopez, D. V. Nanopoulos, H. Pois, X. Wang, and A. Zichichi, Phys. Rev. D **48** (1993) 4062.

[39] J. L. Lopez, D. V. Nanopoulos, X. Wang, and A. Zichichi, Phys. Rev. D **48** (1993) 4029.

[40] Talk given by Y. Kato (CDF Collaboration) at the 9th Topical Workshop on Proton-Antiproton Collider Physics, Tsukuba, Japan,October 1993.

[41] Talk given by J. T. White (D0 Collaboration) at the 9th Topical Workshop on Proton-Antiproton Collider Physics, Tsukuba, Japan,October 1993.

[42] J. L. Lopez, D. V. Nanopoulos, and G. Park, Phys. Rev. D **48** (1993) R974; J. L. Lopez, D. V. Nanopoulos, G. Park, and A. Zichichi, Texas A & M University preprint CTP-TAMU-40/93 (to appear in Phys. Rev. D).

[43] J. L. Lopez, D. V. Nanopoulos, and X. Wang, Texas A & M University preprint CTP-TAMU-44/93 (to appear in Phys. Rev. D).

[44] J. L. Lopez, D. V. Nanopoulos, G. Park, H. Pois, and K. Yuan, Phys. Rev. D **48** (1993) 3297.

[45] J. L. Lopez, D. V. Nanopoulos, G. Park, X. Wang, and A. Zichichi, in preparation.

III. Astro-Particle Physics and Cosmology

SOLAR NEUTRINO EXPERIMENTS: NEW PHYSICS?

JOHN N. BAHCALL
Institute for Advanced Study, Princeton, New Jersey 08540, USA

ABSTRACT

Physics beyond the simplest version of the standard electroweak model is required to reconcile the results of the chlorine and the Kamiokande solar neutrino experiments. None of the 1000 solar models in a full Monte Carlo simulation is consistent with the results of the chlorine or the Kamiokande experiments. Even if the solar models are forced artificially to have a ^{8}B neutrino flux in agreement with the Kamiokande experiment, none of the fudged models agrees with the chlorine observations. This comparison shows that consistency of the chlorine and Kamiokande experiments requires some physical process that changes the *shape* of the ^{8}B neutrino energy spectrum. The GALLEX and SAGE experiments, which currently have large statistical uncertainties, differ from the predictions of the standard solar model by 2σ and 3σ, respectively. The possibility that the neutrino experiments are incorrect is briefly discussed.

1. New Physics is Required if Solar Neutrino Experiments are Correct.

Four solar neutrino experiments[1-4] yield results different from the combined predictions of the standard solar model and the standard electroweak model with zero neutrino masses. The question physicists most often ask each other about these results is: Do these experiments—if correct—require new physics beyond the standard electroweak model? For the purpose of this discussion, I shall mean by the standard electroweak model the simplest version in which neutrinos are massless and neutrino flavors are conserved.

Hans Bethe and I have argued[5] that new physics is required, that some process beyond the enormously successful electroweak model is required to change the shape of the ^{8}B neutrino spectrum. I will review this argument below and then make some concluding remarks on whether the solution to the solar neutrino problem is more likely to be new physics or wrong experiments. The discussion will proceed by assuming that the standard electroweak model is correct, i.e., nothing happens to solar neutrinos while they are traveling to the earth from the point at which they are created in the sun. We shall see that this assumption leads us to a contradiction if the chlorine and Kamiokande solar neutrino experiments are correct.

The basis for our investigation is a collection of 1000 precise solar models[6] in which each input parameter (the principal nuclear reaction rates, the solar composition, the solar age, and the radiative opacity) for each model was drawn randomly from a normal distribution with the mean and standard deviation appropriate to that variable. The uncertainties in the neutrino cross sections for chlorine and for gallium were included by assuming a normal distribution for each of the absorption cross sections with its estimated mean and error.

This Monte Carlo study automatically takes account of the nonlinear relations among the different neutrino fluxes that are imposed by the coupled partial differential equations of stellar structure and by the boundary conditions of matching the observed solar luminosity, heavy element to hydrogen ratio, and effective temperature at the present solar age. This is the only Monte Carlo study to date that uses large numbers of standard solar models that satisfy the equations of stellar evolution. Just as detailed Monte Carlo calculations are necessary in order to understand the relative and absolute sensitivities of complicated laboratory experiments, a full Monte Carlo calculation is required to determine the interrelations and absolute values of the different solar neutrino fluxes. The sun is as complicated as a laboratory accelerator or a laboratory detector, for which we know by painful experience that detailed simulations are necessary. For example, the fact that the ^{8}B flux may be crudely described as $\phi(^8\mathrm{B}) \propto T_{\mathrm{central}}^{18}$ and $\phi(^7\mathrm{Be}) \propto T_{\mathrm{central}}^{8}$ does not specify whether the two fluxes increase and decrease together or whether their changes are out of phase with each other. The actual variations of the calculated neutrino fluxes are determined by the coupled partial differential equations of stellar evolution and the boundary conditions, especially the constraint that the model luminosity at the present epoch be equal to the observed solar luminosity.

Figures 1a–c show the number of solar models with different predicted event rates for the chlorine solar neutrino experiment, the Kamiokande (neutrino-electron scattering) experiment, and the two gallium experiments (GALLEX and SAGE). For the chlorine experiment, which is sensitive to neutrinos above 0.8 Mev, the solar model with the best input parameters predicts[6] an event rate of about 8 SNU. None of the 1000 calculated solar models yields a capture rate below 5.8 SNU, while the observed rate is[1]

$$< \phi\sigma >_{\mathrm{Cl\ exp}} = \ (2.2 \pm 0.2)\ \mathrm{SNU}, \quad 1\sigma \text{ error.} \tag{1}$$

The discrepancy that is apparent in Figure 1a was for two decades the entire "solar neutrino problem."

Figure 1a implies that something is wrong with either the standard solar model, with the chlorine experiment, or with the standard electroweak description of the neutrino. I assume in this part of the discussion that the solar neutrino experiments are correct with their indicated uncertainties. Therefore, we must try to decide here between the validity of the standard solar model and the standard electroweak theory. At first glance, the choice is clear: the standard electroweak model has been checked in exquisite detail by laboratory experiments; the standard solar model has tested extensively only in the outer regions in which no neutrinos are produced. "Surely," says our intuition, "the standard electroweak model must be correct and the standard solar model must be wrong." Not so, I think.

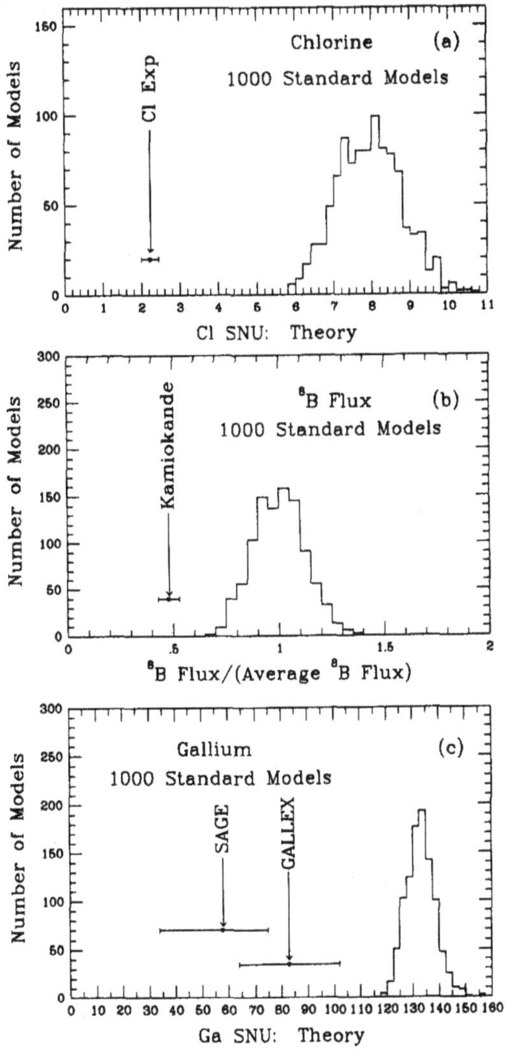

FIG. 1. 1000 Solar Models versus Experiments. The number of precisely-calculated solar models that predict different solar neutrino event rates are shown for the chlorine (Figure 1a), Kamiokande (Figure 1b), and gallium (Figure 1c) experiments. The solar models from which the fluxes were derived satisfy the equations of stellar evolution including the boundary conditions that the model luminosity, chemical composition, and effective temperature at the current solar age be equal to the observed values[6]. Each input parameter in each solar model was drawn independently from a normal distribution having the mean and the standard deviation appropriate to that parameter. The experimental error bars include both statistical errors (1σ) and systematic uncertainties, combined linearly.

Figure 1b shows the number of solar models with different ^{8}B neutrino fluxes. For convenience, we have divided each ^{8}B flux by the average ^{8}B flux so that the distribution is peaked near 1.0. The rate measured for neutrinos with energies above 7.5 MeV by the Kamiokande II and III experiments is[2]

$$< \phi(^8B) > = [0.48 \pm 0.05(1\sigma) \pm 0.06(\mathrm{syst})] < \phi(^8B) >_{\mathrm{Average}} \tag{2}$$

for recoil electrons with energies greater than 7.5 MeV. Here $< \phi(^8B) >_{\mathrm{Average}}$ is the best-estimate theoretical prediction. None of the 1000 standard solar models lie below 0.65 $< \phi(^8B) >_{\mathrm{Average}}$. If one takes account of the Kamiokande measurement uncertainty($\pm$0.08) in the Monte Carlo simulation, one still finds that none of the solar models are consistent with the observed event rate. This result provides independent support for the existence of a solar neutrino problem.

Figure 1c shows the number of solar models with different predicted event rates for gallium detectors and the recent measurements by the SAGE[3,7] ($58^{+17}_{-24} \pm 14$(syst) SNU) and the GALLEX ($83 \pm 19(1\sigma) \pm 8$(syst) SNU) collaborations[4]. With the current large statistical errors, the results differ from the best-estimate theoretical value[6] of 132 SNU by approximately 2 σ (GALLEX) and 3.5 σ (SAGE). The gallium results provide modest support for the existence of a solar neutrino problem, but by themselves do not constitute a strong conflict with standard theory.

Can the discrepancies between observation and calculation that are summarized in Figure 1 be resolved by changing some aspect of the solar model? This is difficult to do. A comparison of Figures 1a–1b shows that the discrepancy with theory is energy dependent. The larger discrepancy occurs for the chlorine experiment, which is sensitive to lower neutrino energies than is the Kamiokande experiment.

The key result to be used here is a detailed investigation—assuming no new physics beyond the standard electroweak model—which shows[8] that, to an accuracy of one part in 10^5, the energy spectrum of the ^{8}B solar neutrinos must have the same shape as the spectrum determined from laboratory nuclear physics experiments. The invariance of the energy spectrum allows us to compute the rate of neutrino capture in the chlorine experiment—independent of any considerations of solar models—provided only that we know from the Kamiokande experiment the flux of the higher energy ($>$ 7.5 MeV) ^{8}B neutrinos. In this process, we ignore the expected contributions to the chlorine experiment, which has a threshold of only 0.8 MeV (an order of magnitude less than the Kamiokande experiment), from ^{7}Be, CNO, and pep neutrinos.

Using the empirical result obtained for the Kamiokande experiment[2], one finds that the predicted rate *from ^8B neutrinos alone* in the chlorine experiment is 6.20 SNU (from the standard model)$\times$0.48 (from the Kamiokande measurement), or

$$< \phi\sigma >_{\mathrm{Cl;\ Kamiokande\ only}} = [3.0 \pm 0.3(1\sigma) \pm 0.4(\mathrm{syst})\ ]\ \mathrm{SNU}. \tag{3}$$

This minimum rate exceeds by more than 2σ the observed chlorine rate. The contributions of the other neutrino fluxes are more robustly calculated than the ^{8}B flux which yields the minimum rate given above. Therefore, these other fluxes make more acute the disagreement between calculated and observed rates. However, one cannot apply to these other fluxes the same argument—based upon the invariance of the shape of the energy spectrum—given above since they are below threshold for the Kamiokande experiment.

Figure 2 is the central result upon which Hans and I based our conclusions. This figure provides a quantitative expression of the difficulty in reconciling the Kamiokande and chlorine experiments by changing solar physics. We constructed Figure 2 using the same 1000 solar models as were used in constructing Figure 1, but for Figure 2 we artificially replaced the ^{8}B flux for each standard model by a value drawn randomly for that model from a normal distribution with the mean and the standard deviation measured by Kamiokande (see Eq. (2)). This *ad hoc* replacement is motivated by the fact that the ^{7}Be (p,γ) ^{8}B cross section is the least accurately measured of all the relevant nuclear fusion cross sections and by the remark that the ^{8}B neutrino flux is more sensitive to solar interior conditions than any of the other neutrino fluxes. The peak of the resulting distribution is moved to 4.7 SNU (from 8 SNU) and the full width of the peak is decreased by about a factor of three. The peak is displaced because the measured (i.e., Kamiokande) value of the ^{8}B flux is smaller than the calculated value. The width of the distribution is decreased because the error in the Kamiokande measurement is less than the estimated theoretical uncertainty ($\approx$ 12.5%) and because ^{8}B neutrinos constitute a smaller fraction of each displaced rate than of the corresponding standard rate.

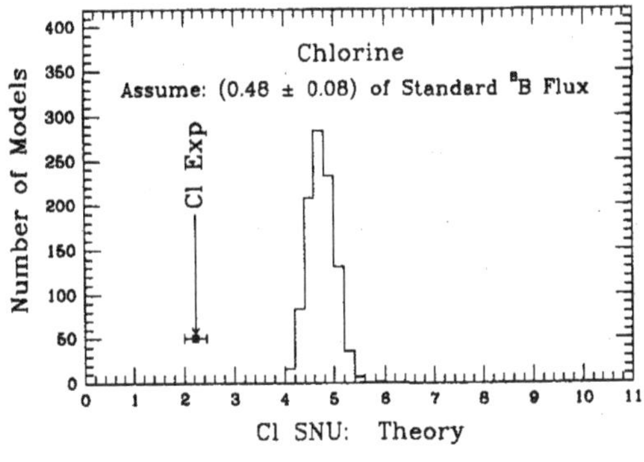

FIG. 2. 1000 Artificially Modified Fluxes. The ^{8}B neutrino fluxes computed for the 1000 accurate solar models were replaced in the figure shown by values drawn randomly for each model from a normal distribution with the mean and the standard deviation measured by the Kamiokande experiment[2].

Figure 2 was constructed by assuming that something is seriously wrong with the standard solar model, something that is sufficient to cause the ^{8}B flux to be reduced to the value measured in the Kamiokande experiment. Nevertheless, there is no overlap between the distribution of fudged standard model rates and the measured chlorine rate. None of the 1000 fudged models lie within 3σ (chlorine measurement errors) of the experimental result.

The results presented in Figures 1–2 suggest that new physics is required beyond the standard electroweak theory if the existing solar neutrino experiments are correct within their quoted uncertainties. Even if one abuses the solar models by artificially imposing consistency with the Kamiokande experiment, the resulting predictions of all 1000 of the "fudged" solar models are inconsistent with the result of the chlorine experiment(see Figure 2).

The difference between the minimum rate of 3.0 SNU for the chlorine experiment obtained by making use of the invariance of the ^{8}B neutrino spectrum and the larger values shown in Figure 2 depends upon our understanding of the solar interior. In the future, it will be possible to use solar neutrinos to test electroweak theory independent of solar models by measuring the energy spectrum of the ^{8}B neutrinos with the Super-Kamiokande[9], the SNO[10], and the ICARUS experiments[11] and by measuring the ratio of charged to neutral currents with the SNO experiment[10].

2. Are the Chlorine and Kamiokande Experiments Correct?

New solar neutrino experiments presently under development will decide if the current solar neutrino experiments are correct. The next generation of experiments, SNO, Super-Kamiokande, BOREXINO, and ICARUS, will settle this question. However, being a physicist at a festive occasion for physicists, I cannot resist summarizing my thoughts on the matter.

First of all, the strong reasons for caution. Neutrino experiments are fiendishly difficult. There have been a number of reported results of experiments with laboratory neutrinos that have—upon further investigation—subsequently required major revisions or outright rejection. Also, as a matter of historical record, we almost never learn new physics by trying to resolve astronomical puzzles. New physics has in the past come from the rigorously-defined conditions of laboratory experiments or from new theoretical formalisms, not from the mushier arguments of astrophysics.

Nevertheless, I think it is likely that the solution of the solar neutrino problem will be new physics rather than either revisions of the standard solar model or corrections to current solar neutrino experiments. Here are my reasons.

The predictions of the standard solar model are robust. The absolute values of the event rates for the chlorine experiment that I and my colleagues have calculated for the past 25 years are within the currently-quoted (modest) theoretical errors. This fact is illustrated in Figure 1.2 of the book *Neutrino Astrophysics*. Recent precise measurements of thousands of eigenfrequencies of the solar pressure waves are in essential agreement with predictions of the standard solar model[6]. Unfortunately, the fact that accurate measurements of the pressure modes are largely confined—at present—to the outer half of the solar mass limits the significance, for our purposes, of confirmatory helioseismological measurements.

The chlorine and Kamiokande experiments disagree with each other independent of solar models. Assuming the invariance of the shape of the ^{8}B neutrino energy spectrum that is promised by the standard electroweak model[8], the calculated rate for the ^{8}B neutrinos that are observed in the Kamiokande experiment exceeds the observed rate in the chlorine experiment (see Eq. 3). This discrepancy is made much more severe when the contributions of the other, more robustly calculated neutrino fluxes are included in the predictions for the chlorine experiment (see Fig. 2).

The chlorine and Kamiokande experiments have been carefully examined by many non-affiliated researchers and have stood the test of time. They have also been subject to a variety of internal tests of self-consistency[1,2].

Finally, there is a simple and beautiful extension of the standard electroweak model that violates the conventional assumptions of zero neutrino masses and flavor conservation and does so in a way that solves the solar neutrino problem. The well-known MSW mechanism[12] can account for all the known solar neutrino experiments with a neutrino mass difference (of order 0.003 eV) that is consistent with current theoretical ideas regarding physics that might lie beyond the standard model[13]. Many other explanations of the solar neutrino problem that involve new, but apparently less "natural", physics have also been advanced.

I am willing to bet that the next generation of solar neutrino experiments will establish current suspicions that new physics is required to solve the solar neutrino problem. But, the wonderful thing about our subject is that in a few years we will not have to depend upon plausibility arguments; the experiments under development will determine the solution of the solar neutrino problem.

3. Acknowledgements

This work was supported in part by the NSF via grant PHY 92-45317 at I.A.S.

4. References

1. R. Davis Jr., in *Proc. of Seventh Workshop on Grand Unification, ICOBAHN'86*, edited by J. Arafune (World Scientific, Singapore, 1987), p. 237; R. Davis Jr., K. Lande, C. K. Lee, P. Wildenhain, A. Weinberger, T. Daily, B. Cleveland, and J. Ullman, in *Proceedings of the 21st International Cosmic Ray Conference*, Adelaide, Australia, 1990, in press; J. K. Rowley, B. T. Cleveland, and R. Davis Jr., in *Solar Neutrinos and Neutrino Astronomy*, edited by M. L. Cherry, W. A. Fowler, and K. Lande (American Institute of Physics, New York, 1985), Conf. Proceeding No. 126, p. 1.
2. K. S. Hirata *et al.*, Phys. Rev. Lett. **63** (1989) 16; **65** (1990) 1297.

3. A. I. Abazov *et al.*, Phys. Rev. Lett. **67** (1991) 332.
4. P. Anselmann *et al.*, Phys. Lett. **B285** (1992) 376.
5. J. N. Bahcall and H. A. Bethe, Phys. Rev. Lett. **65** (1990) 2233; J. N. Bahcall and H. A. Bethe, Phys. Rev. **D47** (1993) 1298.
6. J. N. Bahcall and R. K. Ulrich, Rev. Mod. Phys. **60** (1988) 297; J. N. Bahcall, *Neutrino Astrophysics* (Cambridge University Press, Cambridge, England, 1989).
7. V. N. Gavrin, et al., *XXVI International Conference on High Energy Physics*, Dallas, Texas (1992).
8. J.N. Bahcall, *Phys. Rev.* **D44** (1991) 1644.
9. Y. Totsuka, in *Proc. of the International Symposium on Underground Physics Experiments*, edited by K. Nakamura (ICRR, Univ. of Tokyo, 1990), p. 129.
10. G. Aardsma et al., Phys. Lett. **B194** (1987) 321.
11. J. N. Bahcall, M. Baldo-Ceollin, D. Cline, and C. Rubbia, Phys. Lett. **B178** (1986) 324.
12. S. P. Mikheyev and A. Yu. Smirnov, Sov. J. Nucl. Phys. **42** (1986) 913; L. Wolfenstein, Phys. Rev. **D17** (1979) 2369.
13. M.Gell-Mann, P. Ramond, and R. Slansky, in *Supergravity: Proc. of Supergravity Workshop* Stony Brook, New York, ed. P. Van Nieuwenhuizen and D. Z. Freedman (Amsterdam: North-Holland, 1979) p. 315; T. Yanagida, Prog. Theor. Phys. **B315** (1978) 66.

NEUTRINOS IN ASTROPHYSICS AND COSMOLOGY

DAVID N. SCHRAMM
The University of Chicago
5640 S. Ellis Avenue
Chicago, IL 60637
and
NASA/Fermilab Astrophysics Center
Fermi National Accelerator Laboratory
Box 500, Batavia, IL 60510-0500

ABSTRACT

Neutrinos have provided the major consummated connection between particle physics and astrophysics and cosmology. This talk emphasizes the three principle areas of activity of neutrinos at this interface, namely: (1) solar neutrinos, (2) supernova neutrinos, and (3) cosmological neutrinos. While there are other theoretical venues, such as cosmic ray neutrinos, the above three all have real experimental results as well as theoretical frameworks. This talk, dedicated to Abdus Salaam, briefly summarizes the current state of these active areas and the important role that experiment as well as electro-weak theory has played in the current situation.

Introduction

For this conference dedicated to Abdus Salaam, I would like to focus my attention on the role of neutrinos in astrophysics and cosmology. In particular, Abdus Salaam's work on electroweak theory has naturally found fruitful applications in astrophysical and cosmological environments. This talk will focus its attention on those three topics where experiments (as well as theory) have consummated the marriage of particle physics with astrophysics and cosmology. These three topics are:

(1) solar neutrinos;
(2) supernova neutrinos;
(3) cosmological neutrinos.

The rest of this talk will briefly describe the current situation of each. Obviously, there are other areas where neutrinos have an impact on astrophysics (e.g. cosmic rays and high energy point sources), but I choose to focus on the above topics since they have significant experimental results as well as theoretical motivations.

Solar Neturinos

The deficit of the solar neutrino flux with respect to the prediction of the standard solar models, observed by the current solar neutrino experiments, is one of the major issues of modern physics[1,2] Currently, the Homestake experiment,[3] the Kamiokande II and III experiments,[4] the SAGE experiment,[5] and the GALLEX experiment,[6] all observed neutrino fluxes lower than the predictions of the standard solar model of Bahcall and Ulrich (hereafter BUSSM)[1] or the standard solar model of Turck-Chièze *et al.* (hereafter TSSM).[7] The current solar neutrino experimental status and the theoretical predictions are shown in Table 1 and Table 2. The following discussion draws heavily on a recent paper by Shi and Schramm.[8]

TABLE 1
Current Solar Neutrino Experimental Status

	Homestake	Kamiokande	Gallium exp.
Detector material	C_2Cl_4	H_2O	$GaCl_3$-HCl (GALLEX) Ga(SAGE)
Fiducial mass	610 ton	680 ton	30 ton Ga (GALLEX) 57 ton (30t before 1991) (SAGE)
Reaction	$^{37}Cl(\nu_e, e^-)^{37}Ar$ ν_e only	ν-e scattering $\nu_e, \nu_\mu, \nu_\tau, \bar{\nu}$	$^{71}Ga(\nu_e, e^-)^{71}Ge$ ν_e only
Threshold	0.814 MeV[a]	7.5 MeV[b]	0.233 MeV[a]
Type	Radiochemical Time Integral Energy Integral	Direct Counting Real Time Spectroscopic	Radiochemical Time Integral Energy Integral
BUSSM prediction (theoretical range)	7.9±2.6 SNU*	1.0±0.37[†]	132^{+20}_{-17} SNU
TSSM prediction (1σ error)	6.4±1.3 SNU	0.75±0.20[†]	123±7 SNU
Event rate	$\sim$ 0.5/day	$\sim$ 0.3/day	$\sim$0.3/day
Observed flux	2.28±0.23 SNU	0.46±0.05 ± 0.06(K II)[†] 0.55±0.07 ± 0.06(K III)[†]	83±19 ± 8 SNU (GALLEX) 58^{+17}_{-24} ± 14 SNU (SAGE)

[a] The threshold energy of the incoming neutrinos.
[b] The threshold energy of the recoil electrons.
[†] Normalized by the prediction of the BUSSM.
*1 SNU= 10^{-36} event/target atom/second.

TABLE 2
Solar Neutrino Fluxes Predicted by BUSSM and TSSM

Neutrino Soucrces	BUSSM [TSSM[†]] Prediction		
	^{37}Cl exp. (SNU)	^{71}Ga exp. (SNU)	Kamiokande (×BUSSM)
pp	0 [0]	70.8 [71.1]	0 [0]
pep	0.2 [0.2]	3.0 [2.8]	0 [0]
^{7}Be	1.1 [1.0]	34.4 [30.6]	0 [0]
^{8}B	6.1 [4.6]	14.0 [9.3]	1.0 [0.75]
^{13}N	0.1 [0.1]	3.8 [3.9]	0 [0]
^{15}O	0.3 [0.4]	6.0 [6.5]	0 [0]
Total	7.9 [6.4]	132 [123]	1.0 [0.75]

[†]Based on 1993 model of Turck-Chièze *et al.* [7].

There are two basic approaches used to explain the deficit. One is to use new neutrino physics, and the other is to modify the solar model. In the first approach, the MSW matter mixing is the most natural and robust scenario of new physics,[9] although vacuum neutrino oscillations are also allowed but with a narrower range in the parameter space.[10] Under the MSW matter mixing scheme, solar neutrinos (ν_e) mix with other neutrino species (ν_μ or ν_τ, or new hypothetical sterile neutrinos). On their way out of the Sun, ν_e's encounter a resonance inside the Sun and flip their flavor. To reproduce the fluxes observed by the current solar neutrino experiments, the neutrino mixing parameters are restricted to the diagonal and vertical regions of a triangle in the parameter space ($\Delta m^2 = m_2^2 - m_1^2$ vs. $sin^2 2\theta_{ex}/cos2\theta_{ex}$, where m_1 and m_2 are the neutrino mass eigenvalues and θ_{ex} is the mixing angle between ν_e and ν_x), assuming standard solar models.[11,12]

For the second approach, the simplest technique is to modify the astrophysical input parameters sufficiently to fit the experimental results.[2] The current standard solar models have large uncertainties in certain input parameters. Different choices of input parameters could yield different neutrino fluxes, especially for ^{8}B neutrinos. This is manifested in the comparison of the predictions of the BUSSM and the TSSM (shown in Table 2). The differences in the neutrino fluxes in these two models are mainly caused by the different rates adopted for the reaction ^{7}Be(p,γ)^{8}B and the reaction ^{3}He(^{3}He,2p)^{4}He.[1,7,13] When the same input values are used, the two models (and in fact, the standard solar models of Sienkiewicz *et al.*,[14] Filippone and Schramm,[15] and of the Princeton-Yale collaboration[13]) all agree with each other within 0.1 SNU for the ^{37}Cl experiment.[13] Similar agreements are also found for other experiments. Therefore, simply because of the uncertainties in the input parameters instead of the underlying physics of the solar models or the calculation methods exploited, different predictions for the solar neutrino experiments can be drawn. However such input parameter variations cannot cause the less temperature sensitive ^{7}Be and especially *pp* neutrino fluxes to be significantly decreased relative to the ^{8}B neutrino flux. Thus, for example, if it is eventually shown that the ^{7}Be neutrino flux is depleted more than the ^{8}B neutrino flux, then new neutrino physics is required. At present, the comparison of the ^{37}Cl experiment (see ^{8}B neutrinos and ^{7}Be neutrinos) to the Kamiokande experiments (see ^{8}B neutrinos only) and GALLEX as well as SAGE (see *pp*, ^{7}Be and ^{8}B neutrinos) might, at face value, imply a greater ^{7}Be neutrino reduction, and thus new neutrino physics. But one should be cautious here.

Besides the theoretical uncertainties in the solar models, questions have occassionally been cast on the solar neutrino experiments themselves.[16] Because of the weakness of the interaction between neutrinos and matter, it is very difficult to measure the incoming solar neutrino fluxes. Current experiments utilize two general methods to measure neutrino flux:[2] the radiochemical method and the neutrino-electron scattering method. So far, even with detector fiducial masses of 30–600 tons, the average counting rates of the current radiochemical solar neutrino experiments are of the order 0.1/day. To extract the real events from such small signals and such large volumes of material is an extremely challenging task. Careful calibrations of the experiments are essential. However, among the five current solar neutrino experiments, only the Kamiokande ν-e scattering experiment has been truly calibrated.[17] Therefore, the low rates with respect to the standard solar models of the current radiochemical experiments have to be taken as preliminary. However, such a statement should in no way detract from the impressive experimental techniques used in each of the experiments and the open and imformed manner in which the experiments have attempted to check each potential source of error. In this latter regard, the 20-year pioneering experiment of Davis is particularly notable.

Given all these factors that could cause ambiguity in the solar neutrino problem, it is important to attempt to explore the following question: How necessary is new neutrino physics?

It has been accurately argued by Bahcall and Bethe[18] that the disagreement between the results of the ^{37}Cl experiment (29% $\pm$ 3% the BUSSM prediction) and the Kamiokande II (46% $\pm$ 8% the BUSSM prediction) implies neutrino oscillations. Obviously, assuming the full validity of both experiments, if nothing happens after the solar neutrinos are produced, the ^{8}B neutrino contribution to the ^{37}Cl experiment is at most 2.3$\pm$0.2 SNU, corresponding to 38%$\pm$4% the BUSSM prediction. It marginally agrees with the Kamiokande II result (the problem is even more severe for the Kamiokande III result). But if ν_e oscillates into ν_μ or ν_τ, the remaining 62% or so now changed into ν_μ or ν_τ will interact with the electrons via neutral currents, with $\sigma(\nu_{\mu\ or\ \tau}\text{-e})/\sigma(\nu_e\text{-e}) = 1/6 \sim 1/7$.[2] So the total rate inferred from the Homestake experiment for the Kamiokande II would be $38\% + 62\% \times 1/7 \approx 47\%$ the BUSSM prediction, which agrees with the Kamiokande II measurement extremely well.

Because the ^{37}Cl experiment observes both the ^{8}B neutrinos and the ^{7}Be neutrinos, and the Kamiokande II experiment detects only the ^{8}B neutrinos, the above agreement implies that nearly no ^{7}Be neutrinos are observed in the ^{37}Cl experiment. If the ^{7}Be neutrinos are entirely depleted and the ^{8}B neutrinos are depleted by about a half to two-thirds, as implied by the Kamiokande and Homestake experiments, the gallium experimental yield would be $\sim$ 90 SNU, provided there is no *pp* neutrino reduction, which would be in good agreement with the GALLEX and SAGE results. In fact, without resorting to any new physics, if the chlorine experiment yield is smaller than or equal to the yield of the Kamiokande experiments with respect to the prediction, the ^{7}Be neutrinos have been entirely missing.

No known astrophysical solution can explain a higher ^{7}Be neutrino reduction than ^{8}B neutrino reduction relative to the standard solar model prediction. A lower central temperature in the Sun apparently doesn't help because the ^{8}B neutrinos are more temperature sensitive than the ^{7}Be neutrinos as mentioned before. In terms of nuclear physics, the ^{7}Be neutrinos and the ^{8}B neutrinos are produced by two different channels of the ^{7}Be nuclear reaction (^{7}Be(e,ν_e)^{7}Li for ^{7}Be neutrinos; ^{7}Be$(p,\gamma)^8$B and subsequent ^{8}B β-decay for ^{8}B neutrinos).[2] If one supresses the ^{7}Be neutrinos by supressing the overall ^{7}Be production, the ^{8}B neutrinos will be reduced by the same factor. If one suppresses the branching ratio that produces ^{7}Be neutrinos, ^{8}B neutrino flux will increase, opposite to the experimental implication. Therefore, the current experimental results, at their face value, suggest new physics, if not in particle physics, then in the more fundamental nuclear physics. Thus, once Bahcall and Bethe assumed the validity of both the Homestake and Kamiokande experiments, they were led to new physics.

Since the two most controversial experiments,[16] the Homestake and SAGE, happen to yield the largest reductions with respect to the standard solar models, it is interesting to see how the situation of astrophysical solution vs. new neutrino physics could change if, say, unanticipated systematic errors did exist in either or both of these experiments. For example, in the chlorine experiment, if only runs after the pump failures are considered, which yield an average of 2.95±0.34 SNU, the above ^{7}Be neutrino vs. ^{8}B neutrino argument for new physics may not be strong when the experimental and theoretical uncertainties are fully taken into account. For SAGE, some may argue that it might be safer to use only their latest runs, which yield $85^{+22}_{-32} \pm 20$ SNU.

In order to show the sensitivity of the MSW solution to the experimental results relative to a modified solar model in being able to solve the solar neutrino deficit, we treat the solar core temperature T_c, the neutrino mixing parameters Δm^2 and $sin^2 2\theta/cos 2\theta$ (we assume a two-family mixing between ν_e and non-sterile ν_x) as three free parameters and follow Bludman et al.[19] to do a χ^2 test to see how well the experimental data can be fit by adjusting these parameters. The MSW mixing is represented by parameters Δm^2 and $sin^2 2\theta/cos 2\theta$. The astrophysical solutions are represented by lower T_c's with respect to T_c(BUSSM), where the neutrino fluxes are assumed to follow the T_c dependence as calculated by Bahcall and Ulrich in reference 1. Generally, the core temperature can be as low as $\sim 1.54 \times 10^7$ K ($\sim 0.98 T_c$(BUSSM)) in the uncertainty range of standard solar models.[2] Additional T_c reductions may be achieved by non-standard solar models. But below $0.98 T_c$(BUSSM), the temperature dependence of different neutrino sources may vary. Therefore, for illustrative purposes, Shi and Schramm developed a "tuned" model to show behavior when T_c goes beyond the standard solar model range. In it they have replaced the ^{7}Be$(p,\gamma)^8$B rate in BUSSM with the rate calculated by Johnson[20] from Filippone's[21] experiment. So when T_c=T_c(BUSSM), the standard solar model yields 6.9 SNU instead of 7.9 SNU. Furthermore, they reduced the metalicity z to 0.015 from 0.0196 and thereby obtained 4.5 SNU for ^{37}Cl (see Table 3).

TABLE 3
Major Different Inputs and the Resultant T_c and ^{37}Cl Rate for Different Standard Solar Models

Models	$S_0(^7\mathrm{Be}+p)$	$S_0(^3\mathrm{He}+^3\mathrm{He})$	Z	T_c	^{37}Cl rate
BUSSM	24.3 eV·b	5.15 MeV·b	0.0196	1.564×10^9 K	7.9 SNU
TSSM	21 eV·b	5.57 MeV·b	0.0197	1.551×10^9 K	6.4 SNU
Bahcall *et al* 1992 (with helium diffusion)	22.4 eV·b	5.57 MeV·b	0.0196	1.567×10^9 K	8.0 SNU
"Tuned" Shi & Schramm 1993	20.2 eV·b	5.57 MeV·b	0.0150	1.545×10^9 K	4.5 SNU

The MSW solution with astrophysical uncertainties, for both the two-family and three-family mixing, has been discussed by Shi, Schramm and Bahcall.[12] Fig. 1 shows the two family MSW solution and the three family MSW solution for 1000 solar models[1] whose input parameters are chosen within the theoretical uncertainty range via a Monte Carlo method.

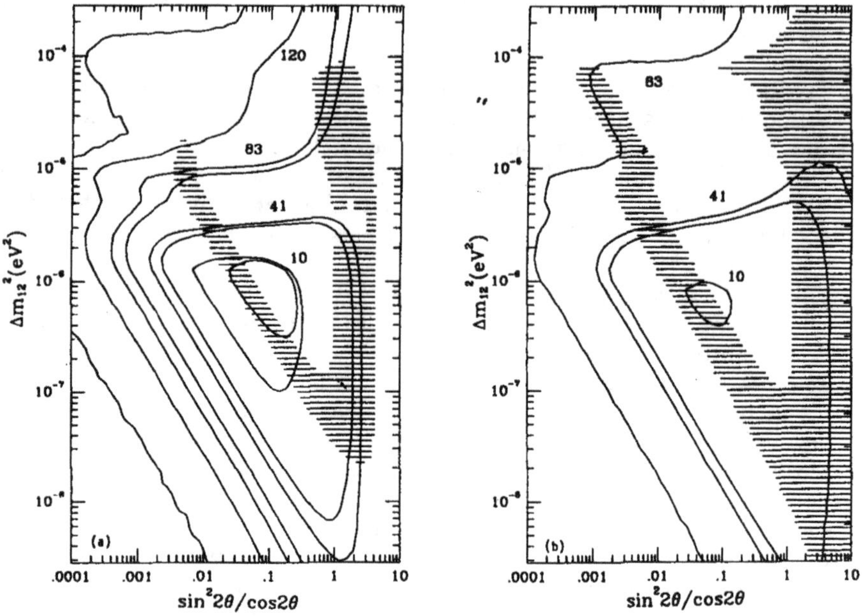

Figure 1. Assuming two-family MSW mixing, the shaded region is the MSW solution allowed by both the ^{37}Cl experiment and the Kamiokande II experiment at 95% C. L., using Monte Carlo simulations of the theoretical uncertainties in 1000 solar models. The iso-SNU bands for the ^{71}Ga experiment are also shown in solid lines (GALLEX gives 83±21 (1σ) SNU). (a) two-family mixing; (b) three-family mixing with ϕ=0.5 rad (ϕ is the mixing angle between ν_e and ν_τ).

In the three-family case, we assume a mass hierachy between three generations and a large mixing angle between ν_e and ν_τ ϕ =0.5 rad. Small ν_e-ν_τ mixing three-family MSW solutions are intermediate cases between the two shown. The conclusion is that the three-family MSW solution doesn't alter the basic picture of the two-family MSW solution. Therefore we consider only the two-family MSW solution in our test. By calculating the χ^2 of the two-family MSW solution and low T_c astrophysical solutions for the current solar neutrino experiments, we can find out how well they fit the experimental data. In the test, we have included both the experimental errors and the theoretical errors coming from the $S(0)$ factor of the reaction ^{7}Be$(p,\gamma)^8$B. The theoretical uncertainties of Z and opacities have been represented by different choices of T_c. We obtain χ^2 for the Homestake experiment, the Kamiokande II, the Kamiokande III, GALLEX and SAGE. In order to show how the Homestake experiment and SAGE can affect our result, we have done the test for two cases: (1) the experimental result as shown in Table 1; (2) same as case (1) except a higher Homestake rate 2.95±0.34 SNU (which is the average rate over the period after the pump failures); and (3) same as (1) except a SAGE rate of $85^{+22}_{-32} \pm 20$ SNU (which is the average rate of their recent less controversial 6 runs).

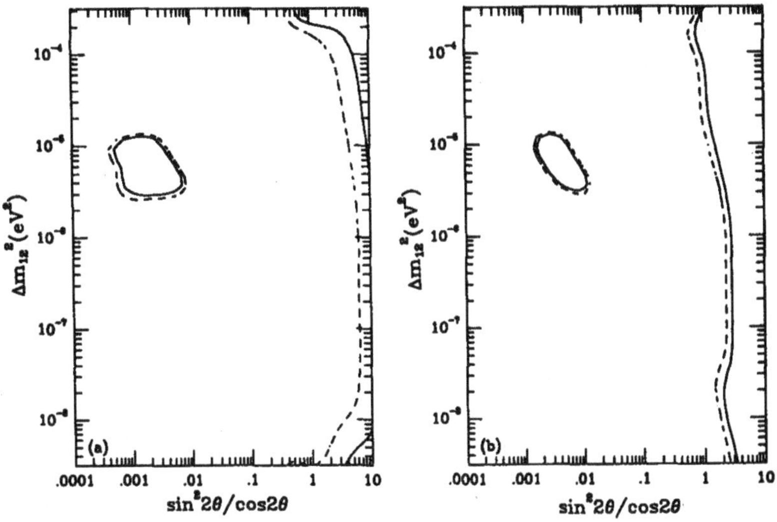

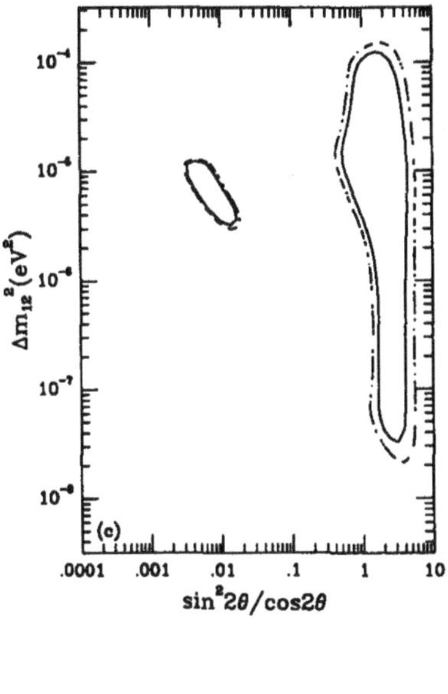

Fig. 2 Assuming two-family MSW mixing and case (1), the χ^2 contours for the five running solar neutrino experiments. The solid line: 90% C. L.; The short-dashed line: 95% C. L.. (a) T_c/T_c(BUSSM)=0.98; (b) T_c/T_c(BUSSM)=1.00; (c) T_c/T_c(BUSSM)=1.02;

In case (1), the MSW solution with varying core temperatures has a minimal χ^2 of 0.89, when T_c/T_c(BUSSM)=1.04. Figure 2 shows the χ^2 contours with C. L.'s of 90% and 95% respectively in the MSW mixing parameter space for different T_c/T_c(BUSSM)'s. Two MSW solutions exist: the diagonal solution and the vertical solution, which agrees with the usual MSW analysis. When assuming the BUSSM, the diagonal solution is more favored than the vertical one. But as T_c increases, the likelihood of the vertical region also increases. For comparison, the likelihood of the astrophysical solutions without neutrino mixing has a maximum of 0.04 when T_c/T_c(BUSSM)=0.95.

In case (2), the χ^2 of the cooler Sun solutions decreases. The maximal likelihood of the cooler Sun solutions is 0.08 at T_c/T_c(BUSSM)$\approx$ 0.96, corresponding to a T_c of 1.50×10^7 K. Our modified standard solar model yields a likelihood of 0.04.

In case (3), the likelihood of cooler Sun solutions reaches a maximum of 0.23 when T_c/T_c(BUSSM)$\approx$ 0.95. The likelihood of our modified standard solar model is 0.03.

If one uses the higher yields for both the Homestake experiment and SAGE as in cases (2) and (3), the cooler Sun solutions can only be ruled out at 57% C. L. Similarly, our modified solar model solution is only ruled at 80% C. L.

Given only five experimental data, it is of course not very meaningful to fit them with three parameters. Obviously more experiments need to be done to rule out decisively any of the alternative solutions. But if the current solar neutrino experimental results are taken at face value as in Table 1 (i.e., "as is"), we find that the astrophysical solutions may be disfavored, especially for standard solar models. But if either the chlorine result or the gallium result is higher than listed in Table 1, the cooler Sun solutions may fit the data fairly well. Therefore the astrophysical solution to the solar neutrino problem may, given the current experimental situation, remain plausible and viable.

In order to resolve the different solutions, in addition to the calibrations needed to be done for current experiments, information from next generation experiments seems to be necessary. The additional information should include: (1) the spectral shape of the incoming solar neutrinos; (2) the statistically significant data on the neutral current component in the scattering events; (3) the diurnal and seasonal variations of signals; and (4) limits on possible antineutrino components in the solar neutrino flux.

The spectral shape of the solar neutrino flux provides one possible way to distinguish the new neutrino physics solution from the astrophysical solution. Different solutions generate different neutrino spectra at the detectors. The astrophysical solution will reduce the ^{8}B neutrino spectrum uniformly with respect to the BUSSM. Other neutrino fluxes remain more or less the same. In the MSW solution, the horizonal region ($\Delta m^2 \sim 10^{-4}\mathrm{eV}^2$) which has been ruled out so far would show almost complete depletion at high energies and nearly no depletion at low energies; among the two regions that are allowed by the current experiments, the diagonal region ($2\times10^{-6}\mathrm{eV}^2 < \Delta m^2 < 10^{-4}\mathrm{eV}^2$, $\Delta m^2 sin^2 2\theta \sim 10^{-7}\mathrm{eV}^2$) has a larger depletion rate at low energies than at high energies, while the vertical solution (10^{-8}eV^2 < Δm^2 < 10^{-4}eV^2, $sin^2 2\theta \gtrsim 0.5$) yields a uniform depletion over the whole energy range. The long wavelength vacuum mixing solution ($\Delta m^2 \sim 10^{-10}$eV^2, $sin^2 2\theta > 0.7$), also shows a distortion in spectrum with respect to the BUSSM, with the distortion depending on the detail of the mixing parameters.[10] The spin-flavor precession neutrino magnetic moment solution ($\mu_\nu \sim 10^{-11}\mu_B$) is similar to the MSW solution,[22,23] that has three regions exhibiting different spectrums. The neutrino decay solution, in which the solar neutrinos decay on their way to the earth, will have a lower depletion rate at higher energies because the life-time of neutrinos is proportional to their energies.[24]

So far, only the Kamiokande experiments have spectral resolutions of ~22% at 10 MeV.[17] But because of its low statistics, the spectrum obtained by the Kamiokande II experiment rules out only a small region of the horizontal branch of the MSW solution which can also be ruled out by the combination of the ^{37}Cl experiment and the Kamiokande II experiment. The spectrum shows no appreciable distortion with respect to the BUSSM and is consistent with the diagonal and the vertical MSW solutions, as well as the astrophysical solution, and all other solutions that have a uniform depletion over the ^{8}B neutrino spectrum above 8 MeV.

The neutral current interaction is indiscriminate to different flavors. Therefore, its component of the scattering events includes contributions from ν_e, ν_μ and ν_τ. The ν_e contribution, however, can be calculated from charged current events, which involve only ν_e. Any neutral current events in excess of the ν_e contribution clearly imply the presence of neutrinos of other flavors, into which the solar neutrinos oscillate. So far only the Kamiokande experiments can see neutral current events, but they cannot distinguish them from the charged current events.

If solar neutrinos oscillate into sterile neutrinos, the neutral current test cannot resolve neutrino oscillations from astrophysical solutions. Spectral distortions and time variations are then tests for this resolution. But since a large portion of the mixing parameter space of the neutrino oscillation solutions (including the MSW mixing, the vacuum mixing and the spin-flavor precession) yields no appreciable spectral distortion and time variation, the currently available tests cannot distinguish them from astrophysical solutions. However, the constraint on extra neutrino species from the Big Bang Nucleosynthesis prohibits large angle MSW mixing between ν_e and sterile neutrinos.[25,26]

Solar neutrino experiments of the next generation could provide the additional information to resolve the solar neutrino problem. Right now four experiments have been approved for construction or are proposed for construction: the Super Kamiokande, SNO (Sudbury Neutrino Observatory), Borexino, and the Iodine experiment. They are all direct counting real-time experiments except the Iodine experiment, which is radiochemical. A summary of the four experiments is in Table 4.[28] By 1996, when, at the very least, SNO and Super Kamiokande are operating, we may finally know the nature of the solar neutrino problem.

TABLE 4
Future Solar Neutrino Experiments

	SNO	Super Kamiokande	Borexino	^{127}I exp.
Detector material	D_2O	H_2O	Liquid Scintillator	NaI
Fiducial mass	1 kton	50 kton	100 ton	1 kton
Reaction	ν_e+d→p+p+e$^-$ ν-e scattering ν+d→ ν+p+n $\nu_e, \nu_\mu, \nu_\tau, \bar{\nu}$	ν-e scattering $\nu_e, \nu_\mu, \nu_\tau, \bar{\nu}$	ν-e scattering $\nu_e, \nu_\mu, \nu_\tau, \bar{\nu}$	$\nu_e + ^{127}I \rightarrow e^- + ^{127}Xe$ ν_e only
Threshold	~5 MeV	~5 MeV	0.25 MeV	0.789 MeV
Type	Direct Counting Real Time Spectroscopic Directional	Direct Counting Real Time Spectroscopic Directional	Direct Counting Real Time Spectroscopic Non-directional	Radiochemical Time Integral Energy Integral Non-directional
Energy resolution				-
(10 MeV)	14%	14%	4%	-
(1 MeV)	-	-	7.5%	-
Event rate estimated by BUSSM	~ 30/day	~ 40/day	~ 50/day	~ 10/day

Supernova Neutrinos

From their discovery, neutral currents have played a prime role in gravitational collapse calculations. For over 20 years, it has been known that the gravitational collapse events, thought to be associated with Type II supernovae and neutron star or black hole formation, are copious producers of neutrinos. In fact, the major form of energy transport in these objects comes from neutrino interactions. It has long been predicted that the neutrino fluxes produced by these events would be high enough that if an event occurred within the Galaxy, it could be detected. The following discussion of the neutrinos borrows heavily from Schramm (1987) and Schramm (1991)[28].

The key role of neutral currents is to emit all species of neutrinos and anti-neutrinos even though the core temperature is well below the muon rest mass. This point was noted by Schramm and Arnett[29] shortly after the discovery of neutral currents. However, at that time the main emphasis on neutral currents was their possible role in ejecting the outer part of the star, thereby creating a supernova (see ref. 30) via coherent neutrino scattering. Unfortunately, this role for neutral currents did not come to fruition. Some mistakenly think that neutral currents led to neutrino trapping in the core. However, Mazurek[31] first found neutrino trapping using only charged current interactions. Neutral currents merely enhanced the effect. To put the neutral current role in perspective, let us review in general the role of neutrinos in gravitational collapse.

It has been well established in the models of Arnett[32] and Weaver et al.[33] that massive stars with $M \gtrsim 8M_\odot$ evolve to an onion-skin configuration with a dense central iron core of about the Chandrasekhar mass surrounded by burning layers of silicon, oxygen, neon, carbon, helium, and hydrogen. Collapse inevitably occurs when no further nuclear energy can be generated in the core. The collapsing iron core mass is always about $1.4 \pm 0.2 M_\odot$. (Some authors emphasize the variation about $1.4 M_\odot$; however, for general understanding the key is a close convergence to the Chandrasekhar mass, not the variance.)

Bethe and Browne[34] and Baron et al.[35] have argued that, provided the equation of state of matter above nuclear density is very soft, stars in the mass range $10 \leq M \lesssim 16 M_\odot$ with cores slightly below the $1.4 M_\odot$ Chandrasekhar mass may explode due to the prompt exit of the shock wave formed after the core bounces upon reaching supra-nuclear density. For stars with $16 \lesssim M \leq 80 M_\odot$, the shock wave stalls on its exit from the core and becomes an accretion shock. Wilson et al.[u9] have shown that such stars will eventually ($\sim$ 1 second later) eject their envelope as a result of neutrino heating in the region above the neutrinosphere and below the shock. (The delayed ejection can also occur in the lower mass collapses if the initial bounce does not produce an explosion.) The success of such delayed ejection seems to depend critically on the calculational details. Wilson[37] has shown that small variations in the numerical treatment of the neutrino transport can make or break the possibility of ejection. Obviously the above scenarios are sensitive to the stiffness of the core equation of state which is still poorly known at and above nuclear mass densities. As was first emphasized by Arnett and Schramm,[38] the ejecta have a composition which fits well with the observed 'cosmic' abundances for the bulk of the heavy elements. Prior to SN 1987A, the emphasis was on the ejecta and the ejection mechanism, not on the neutrino emission, and yet, it is the emission where neutral currents are the most significant.

Regardless of the details of collapse, bounce, and explosion, it is clear that to form a neutron star the binding energy, $\epsilon_B \approx 2 \times 10^{53}$ ergs must be released. The total light and kinetic energy of a supernova outburst is about 10^{51} ergs. Thus, the difference must come out in some invisible form, either neutrinos or gravitational waves. It has been shown[39] that gravitational radiation can at most carry out 1% of the binding energy for reasonable collapses because neutrino radiation damps out the non-sphericity of the collapse (see Kazanas and Schramm[40,41]). Thus, the bulk ($\gtrsim$ 99%) of the binding energy comes off in the form of neutrinos.

It is also well established[30] that for densities greater than about 2×10^{11} g/cm^3, the core is no longer transparent to neutrinos. Thus, as Mazurek[31] first established, the inner core has its neutrinos degenerate and in equilibrium with the matter. Because of the trapping of neutrinos, the neutrino emission time and thus the collapse time scale is governed by neutrino diffusion times the order of seconds rather than hydrodynamic free fall times which would be the order of milleseconds.

For electron neutrinos, the 'neutrinosphere' has a temperature such that the average neutrino energy is around 10 MeV. This was established once it was realized that the collapsing iron core mass is $\sim 1.4 M_\odot$, due to the role of the Chandrasekhar mass in the pre-supernova evolution fixing the scale. Since the μ and τ neutrinos and their antiparticles interact only at these temperatures via neutral rather than charged current weak interaction, their neutrinosphere is deeper within the core. Therefore, their spectra are hotter than that of the electron neutrinos. The electron antineutrino opacity will initially be dominated by charged current scattering off protons, but as the protons disappear, it will shift to neutral current domination. Thus, the effective temperature for $\bar{\nu}_e$'s changes from that for ν_e's to that for ν_μ and ν_τ's.

The average emitted neutrino energy is actually quite well determined[28,29,42] for the peak of the neutrino distribution and is very insensitive to model parameters. The peak emission occurs at the highest temperature for which neutrinos can still free stream out of the star. The Fermi-Dirac (F-D) temperature of this peak varies as the $\sim 1/5$ power of the model dependent parameters, thus yielding a well determined value (see ref. 1) regardless of the input.

For ν_e this yields

$$T_{\nu_e} \simeq 3.5\text{MeV}$$

or an average energy

$$\langle E_{\nu_e} \rangle \approx 10\text{MeV}.$$

This is in good agreement with detailed numerical results. As mentioned above, for $\bar{\nu}_e$'s the average energy increases with time. The time averaged value is about 15MeV.

While the general scenario for collapse events is well established, the detailed mechanism for the ejection of the outer envelope in a supernova as the core collapses to form a dense remnant continues to be hotly debated. Therefore, most theorists working on collapse prior to SN 1987A have focused on these details in an attempt to solve the mass-ejection problem. As a result, most of the pre-1987 papers in the literature are concerned with the role played by neutrinos internal to the stellar core, rather than the nature of the fluxes which might be observed by a neutrino detector on earth. In particular, while it has been known since the early 1970's that the average energy of the emitted neutrinos was about 10 MeV, with neutrino luminosities of a few 10^{52} ergs/sec, the detailed nature of the emitted spectra was only recently explored in detail by Mayle, Wilson, and Schramm[43,44]. Their calculation emphasized the high-energy neutrinos which are easier to detect. The diffusion approximation used in most collapse calculations does not treat the high-energy tail of the spectrum accurately.

In addition to the basic energetic arguments, there is the neutronization argument (see ref. 1, and references therein). The collapsing core has $\sim 10^{57}$ protons that are converted to neutrons via

$$p + e^- \rightarrow n + \nu_e$$

to form a neutron star. (This process is also called deleptonization by some authors.) Each ν_e, so emitted from the core, carries away on the average 10 MeV, thus around 1.3×10^{52} ergs are emitted by neutronization ν_e's. this is $\lesssim 10\%$ of the binding energy. The remainder of the neutrinos come from pair processes such as

$$e^+ + e^- \rightarrow \nu_i + \bar{\nu}_i$$

where $i = e$, μ, or τ, with ν_μ and ν_τ production occurring via neutral currents, and ν_e via both charged and neutral currents. (See review by Freedman et al.[3].) It is curious to note that the neutral current role focused upon in the 1970's was the coherent scattering of heavy nuclei ($\sigma \propto A^2$) and its possible role in the ejection. These early papers (c.f. reference 29) also recognized the neutral current emission of all species but did not emphasize it due to the preoccupation with ejection mechanisms and presumed unlikelihood of ever expecting to see a neutrino burst. The coherent scattering ended up not playing a significant role as the value of $\sin^2\theta_W$ dropped to its present value and the value of the eventual flux times σ was insufficient to do the job (see reference 30).

Some fraction ($\lesssim 50\%$) of neutronization occurs as the initial shock hits the neutrinosphere (the remainder occurs on a neutrino diffusion timescale). The pair ν's come from the 'thermally' radiating core on a diffusion timescale. The timescale for an initial neutronization ν_e burst will be much less ($\lesssim 10^{-2}$ sec) than the diffusion time ($\sim$ seconds) that governs the emission of the bulk of the flux. Some so-called 'advection/convection' models increase the initial ν_e burst by convecting high-T, degenerate core material out. These models have higher-energy ν_e's with larger fluxes and suppress the $\bar{\nu}_e$ fluxes.

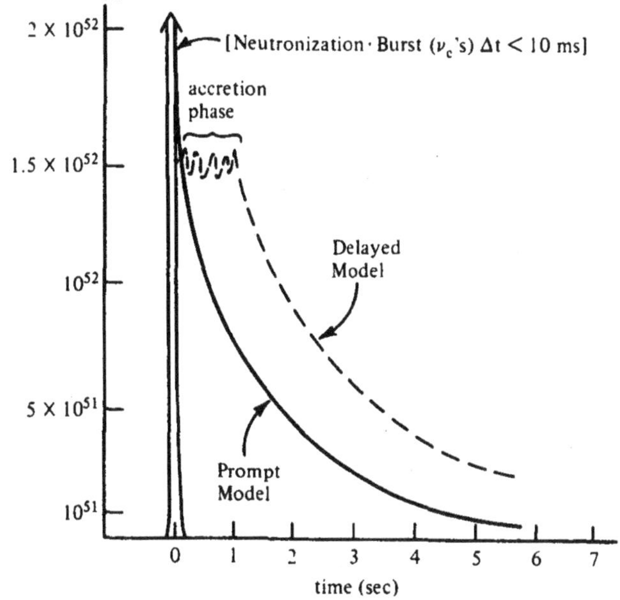

Figure 3. A schematic showing the neutrino emission time profile for prompt and delayed models.

As to the time scale for the bulk neutrino emission, in the "detailed" explosion models, more than half of the thermal neutrino emission comes out in the first one or two seconds for the delayed ejection models due to the high neutrino luminosity during the accretion phase. For prompt models, the exponentially falling Kelvin-Helmholz cooling starts immediately. Thus, the bulk of the emission also occurs early but for prompt models the neutrino luminosity does not stay uniformly high for as long as it does in accretion models. In both models, once the ejection has occurred, the remainder of the neutrinos come out over many seconds as the hot, newborn, neutron star cools down via Kelvin-Helmholz neutrino cooling to become a standard 'cold' neutron star. Figure 3 is a schematic summarizing this. Burrows and Lattimer[45] carried out detailed cooling calculations prior to SN 1987A. Most other authors cut off their calculations after the bulk of the neutrino emission occurred and mass ejection was established. Detailed models for the bulk of the neutrino emission (c.f. Mayle et al.[43]) seem to find that the pair processes yield an approximate equipartition of energy in the different species. The ν_μ and ν_τ's have a higher energy per ν. Thus, their flux is down to preserve this equipartition.

Despite the explosive mechanism, for stars in the mass range $10 \lesssim M \lesssim 16\mathrm{M}_\odot$ the most distinctive structure in the neutrino signal is the initial neutronization burst. However, in the delayed explosions seen by Wilson et al.,[36] for stars with $M \geq 16M_\odot$, besides the burst, the neutrino luminosity shows an oscillatory behavior superimposed on an almost constant neutrino luminosity during the post-bounce pre-ejection accretion phase. The oscillations in luminosity are related to oscillations in the mass accretion rate onto the proto-neutron star. The physical nature of the instability that is responsible for the oscillations in luminosity and mass-accretion rate is described in Wilson[37] and in more detail in Mayle.[47] After the envelope is ejected, the luminosity will smoothly decrease as the remaining binding energy is emitted during the Kelvin-Helmholz cooling. Models without the accretion phase go directly from the neutronization burst to Kelvin-Helmholz cooling. Those models thus have the $\bar{\nu}_e$ emission fall off with a single characteristic cooling time. However, models with an accretion phase have a high average emission rate for a second or so after the neutronization burst before the mass ejection and onset at the cooling phase with its dropping emission.

It is important to remember that the mean neutrino energy and total emitted energy depend only on the initial iron-core mass and are otherwise independent of the explosive mechanism. Because the opacity is less for the ν_μ and ν_τ's, they are emitted from deeper in the core where temperature is higher. Thus, they have a higher average energy. The calculations of Mayle et al.[43,44] and also Bruenn[47] and Burrows[48] find $E_{\nu_\mu} \simeq E_{\nu_\tau} \approx 2E_{\nu_e}$. As mentioned above, in the Mayle et al. work, the easier-to-observe $\bar{\nu}_e$'s start out with energy comparable to ν_e's and gradually shift over to the $\nu_\mu - \nu_\tau$ energy as their emission continues from progressively deeper in the core.

Before SN 1987A, it was obvious that a supernova, if detected by its neutrinos, would constrain neutrino properties. In particular, if the neutrinos got here, we'd have a lifetime limit. If the time pulse were not too spread out, that would mean a mass limit on those neutrino types that were clearly identified. Also, from the number of $\bar{\nu}_e$ counts, one could constrain N_ν since if N_ν was large, the fraction of thermally produced $\bar{\nu}_e$'s would go down. In addition, neutrino mixing could be constrained by detecting different types and comparing with Mikheyev–Smirnov[49] matter mixing, as parameterized to solve the solar neutrino problem, $\nu_e \to \nu_\mu$ (or ν_τ), and ν_μ (or ν_τ) $\to \nu_e$, but nothing happens in the antineutrino sector. Such mixing would eliminate seeing the initial ν_e burst, but give higher energies to the later, thermal ν_e since they'd be mixed ν_μ's (see Walker and Schramm[50]). Of course, non-solar Mikheyev–Smirnov can be used if antineutrino mixing is seen. All of these effects have been examined with the data from SN 1987A. Future collapse will have much better statistics and stronger statements might be made.

On February 23, 1987, definitive neutrino detections were reported at Kamioka[51] and IMB[52] detectors. Thus, unquestionably, extra solar system neutrino astronomy has been born!

For the sparse data of Kamioka and IMB, 4 observables may be useful:

1. The total number of events which can be used to estimate the emitted luminosity.
2. The average energy of the events which when deconvoluted with detector efficiencies can be used to estimate a F-D temperature assuming a constant emission temperature.
3. The duration of the burst which gives hints about diffusion timescales, softness or stiffness of equations of state and possibly gives hints regarding prompt versus delayed models.
4. Angular distributions which give hints about electron scattering events versus proton absorption which in turn may tell something about high energy suprathermal distribution tails or other emission physics.

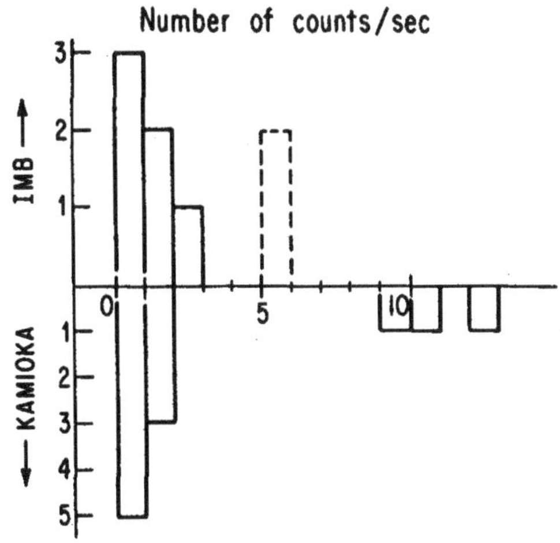

Figure 4. The ν counting rates for IMB/Kamioka.

Figure 4 is a plot showing the energy and timing of the Kamioka and IMB events.

(Kamioka's event no. 6 is ignored as being below their criteria for a definitive event.) Note that almost all the counts concentrate in the first few seconds, as one expects in collapse models. A reasonable tail, as predicted by theory (see Figure 3), yields low but finite rates after ten seconds. Such rates following the bulk early emission from an accretion phase have little difficulty in producing apparent gaps in counts due to the problems of small number statistics (c.f., Bahcall[53] et al. or Mayle and Wilson[54]). Note also that the IMB late counts (dashed lines) nicely fill in the 6 second gap in the Kamioka data.

To convert a mean neutrino energy to an effective temperature requires assuming that the emitted ν spectrum was well described by F-D statistics. Mayle et al. argue that this is a reasonable assumption, but, as already mentioned, they did find that their models had a higher tail at high energies than a simple, single-temperature model would yield. Thus, one might expect the IMB temperature to be slightly higher than the Kamioka temperature due to its weighting on the high-energy events. Schramm[28] and others have shown how one can deconvolute an F-D distribution with thresholds and efficiencies to derive an effective temperature for the emitted neutrinos from the measured energies of the events assuming a single temperature model.

The inferred emitted energy, ϵ_{ν_e} in $\bar{\nu}_e$'s is obtained (c.f. reference 28) by integrating the $\bar{\nu}_e + p \rightarrow n + e^+$ cross section over the F-D distribution with appropriate efficiencies and comparing it to the observed number of counts. The total emitted energy, ϵ_T, can be related to $\epsilon_{\bar{\nu}_e}$ by:

$$\epsilon_T \approx \frac{2N_\nu \epsilon_{\bar{\nu}_e}}{(1 - fn)}.$$

It is found that a concordant region has $T_{\bar{\nu}} \sim 4.5$ MeV and $\epsilon_T \sim 3 \times 10^{53}$ ergs[28]. Similar conclusions were reached by Sato and Suzuki[55] and Bahcall et al.[53] using a different treatment. Once T and ϵ_T are determined, one can use the luminosity–temperature relationship to solve for the radius, R, of the neutrinosphere and obtain, in our case, a few tens of kilometers in reasonable agreement with the standard models.

The total time spread of the IMB/Kamioka events (see Figure 4) shows that ν-emission (or at least detection) lasted for $\sim$ 10 sec. The duration of neutrino emission varies in different collapse models due to the equation of state and the total mass of the collapsing core (is it slightly greater or less than $1.4M_\odot$?) and the dynamics (prompt vs. accretion). Longer timescales favor soft equations of state and higher core masses (1.4 - 1.6 vs. 1.2 - $1.3M_\odot$) and thus favor accretion versus prompt mechanisms. However, until we have a collapse in our Galaxy with a more detailed time evolution of the ν-signal, it will be hard to make detailed statements on the collapse mechanism.

The detection of neutrinos from a reasonably well understood source 170,000 light years away makes for useful constraints on neutrinos and other weakly interacting particles that might have been emitted. Unfortunately, the statistics of so few counts have limited the accuracy of some of these arguments so that some of the more interesting ones such as neutrino mass or number of flavors do no better than current terrestrial experiments. However, a future collapse in our Galaxy with 1000's of counts will be able to make significant improvements and may be our only way of getting limits on m_{ν_μ} and m_{ν_τ} that are of cosmologically significance.[56]

Supernova SN 1987a has confirmed that our basic understanding of gravitational collapse and neutrino emission is on firm ground. If it were not for neutral currents (and Abdus Salaam), this event would have been impossible to interpret since it would not have been able to be easily related to the binding energy of the forming neutron star.

Cosmological Neutrinos

Neutrinos have had two major roles in cosmology:

(1) big bang nucleosynthesis (BBN) and the number of flavors, and

(2) neutrinos as dark matter.

Let us first review the role in BBN which leads us naturally to the need for non-baryonic dark matter and hence the third topic.

TABLE 5
Light Element Abundances

Element (isotope)	Predicted Primordial Abundance	When Observed
^{1}H	~76% by mass	1960's
^{4}He	~24% by mass	1960's
^{2}H/H	2×10^{-5}	1970's
^{7}Li/H	10^{-10}	1980's
Neutrinos	$N_\nu = 3$	1990

A major pillar of the Big Bang (along with the Hubble expansion and the microwave background) is the light element abundance measurements and, most recently, the measurement at LEP of the number of families of neutrinos. In particular, the Big Bang predicts[57] that when the universe was at a temperature of $\sim$ 10 billion degrees and was about one second old, it should have started nuclear processes that would eventually yield certain well-specified abundances for the light isotopes (see Table 5). The abundances of these light elements have all been accurately determined to be in impressive agreement with the Big Bang predictions to the accuracy of the measurements. Of special interest is that even the one part in 10 billion for lithium works. Furthermore, the Big Bang predicted that the abundances would fit well only if there were no more than three families of neutrinos, and that was exactly what was observed at LEP. Thus, as is frequently emphasized, the Big Bang has made a variety of detailed predictions based on the nature of the universe at $\sim$ 1 sec, and the predictions have been confirmed by observation and experiment. The BBN neutrino arguments are relevant for SUSY models since they prohibit additional light ($\lesssim 1$ MeV) weakly interacting particles.

A recent detection of ^{6}Li in a metal-poor star[58] has confirmed that the Spite Lithium Plateau is probably not the result of stellar depletion but is a true measurement for the primordial value.[59] This gives us even greater confidence in the basic BBN arguments and we can use the Li constraint as well as the D and ^{3}He constraint on the density. The light elements with abundances ranging from $\sim 76\%$ for hydrogen to one part in 10^{10} for lithium all fit with the cosmological predictions, with the one adjustable parameter being the baryon density,

$$\Omega_B \sim 0.05 \pm 0.03.$$

Thus, if one prefers a universe with $\Omega_{TOTAL} = 1$, one must also demand that the bulk of the matter in the universe (greater than 90%) be something other than baryons. A favorite non-baryonic candidate is the LSP.

Recent attempts to find alternatives to this conclusion by introducing an inhomogeneous baryon distribution at the nucleosynthesis epoch have ended up reaching essentially the same constraint on Ω_B as in the standard homogeneous model.[60]

The baryon density, Ω_B, deduced from the abundances of light elements is within the range of dynamical estimates of the mean mass density in and around the bright parts of galaxies, including the dark massive halos associated with galaxies[61] (see Figure 5). The IRAS/GA point (IRAS is the Infrared Astronomy Satellite and GA stands for the Great Attractor) shows that we are beginning to have some experimental/observational evidence that Ω_{TOTAL} exceeds Ω_B and that some non-baryonic dark matter is truly needed. (This evidence comes from the large scale velocity flows implied by the IRAS survey and the GA.)

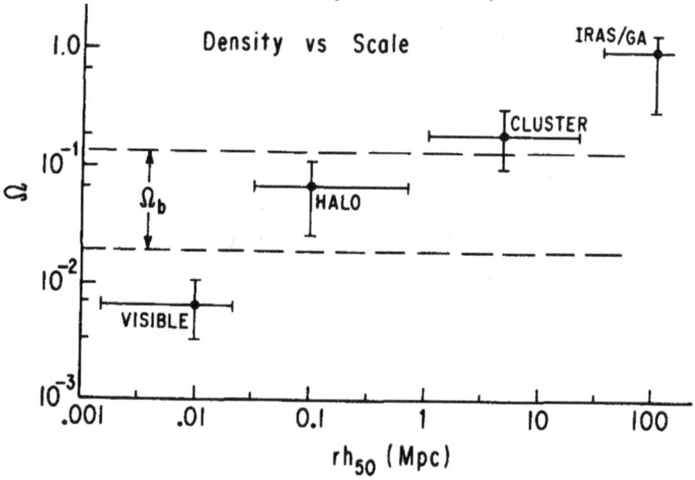

Figure 5. The density paramater Ω versus the scale on which it is determined. Note that the baryon density is consistent with galactic halo densities but exceeds visible matter densities. Note also that only on the largest scales is there any hint that Ω exceeds Ω_B.

An interesting additional bit of information on the distribution and relative proportions of matter in the universe has come from the ROSAT x-ray satellite. In particular, Mushotzky[62] has shown that rich clusters of galaxies have a mass fraction, $\sim 0.3h_{50}^{-3/2}$, of hot x-ray gas. Since $\Omega_{cluster} \sim 0.2$, this yields $\Omega_{baryon} \sim 0.06$ for these clusters in reasonable agreement with BBN (but only for $h_{50} = H_0/50 \sim 1$). However, such a high ratio of baryons to non-baryons in rich clusters and yet an overall low cosmic average would imply that even on the scale of clusters, baryons are preferentially being selected. Such selection is not present in most models for cluster formation and may turn out to be an important discriminator of structure formation models.

Let us now turn to the possible role of neutrinos as the non-baryonic dark matter.

To make the observed objects in the universe such as galaxies, clusters, stars, planets and people requires something beyond just some combination of baryons and hot neutrinos and/or cold exotic dark matter. There also had to be some "seeds" to get the matter, both baryonic and exotic, to begin clumping in some fashion. Having additional exotic matter does help accelerate the clumping process to form observable lumps, but all models require some sort of "seed" or density fluctuation to get from the smooth, early universe to the lumpy universe in which we live today. All seeds that are capable of producing the observed objects, even the most carefully contrived ones, inevitably induce fluctuations in the microwave background on different angular scales at the minimal level of $\sim 10^{-6}$ and in most cases at the level of $\sim 10^{-5}$. (This is sometimes noted with the analogy that "you can't make an omelet without cracking eggs." i.e., you can't make galaxies without disturbing the microwave background.)

A "model" consists of some assumption about the seeds and some assumptions about Ω_B plus Ω_{HDM} plus Ω_{CDM} and, for some people, assumptions about Λ. (HDM and CDM refer to hot and cold non-baryonic dark matter, hot being rapidly moving at galaxy formation and cold being slow at that epoch.)

The general cause of the microwave fluctuations in apparent temperature from density fluctuations is merely gravity. If the density is higher than the average, then the gravitational field is higher and hence the background radiation climbing out of that higher potential well will be redshifted. (Note that regions of high density which eventually form structure would have produced temperature fluctuations to the low side, whereas low density fluctuations would yield "hot spots.") Different structure formation models have different distributions of the gravitational seeds or density fluctuations and hence predict different patterns and amplitudes for the temperature variations.

As is well-known, the COBE sattelite found

$$\frac{\Delta T}{T} \sim 10^{-5}$$

for angular scales from 7° to 180°.

Of course, no scientific observation is believable until it is checked. Fortunately, the large scale COBE anisotropy has now been dramatically checked by a balloon experiment designed and built by a collaboration led by Stephan Meyer[63] of MIT (and Chicago). Their balloon mission also measured the anisotropy at large angular scales, and they found an anisotropy that both correlated with the directions of the COBE data and also had about the same size fluctuations. This dramatic confirmation means that we now seem to know that there really were primordial fluctuations on large angular scales. This has been further confirmed by the 2nd year of COBE data which shows essentially the same result as the first year, but with better statistics.[64]

With the large angular scales now relatively fixed (to experimental accuracy), attention has now turned to the smaller angular scales that directly correspond to the scales that eventually formed galaxies, clusters, etc.

Three groups recently announced somewhat ambiguous results about the 1-to-2 degree scale anisotropies. One of the groups,[65] led by Lubin of UC Santa Cruz, reported results from their previous year's South Pole observations. The second group, known as the MAX collaboration,[66] reported on their balloon-borne measurements. The third group is the Center for Astronomical Research at Antarctica which carried out two anisotropy experiments, one at small scales ($\lesssim$ 0.5 deg.), $\frac{\Delta T}{T} < 10^{-5}$, and the other at 2.5 deg. $\frac{\Delta T}{T} \sim 3 \times 10^{-5}$. While the anisotropies observed are at levels not too different from that predicted by Figure 3, the relative levels reported were not uniform in all directions, hinting either to non-gaussianity or to source contamindation, but the error bars are also still quite large when fit to COBE's large angle data. Hopefully, with more South Pole and balloon data, the important 1-to-2 degree scales will be unambiguously mapped out in the near future. All microwave data on these scales can be directly compared with how galaxies cluster on the equivalent scales.

Several major surveys of the 3-dimensional positions of galaxies have mapped out some interesting nearby structure. For example, the Center for Astrophysics[67] survey has found a "Great Wall" of galaxies stretching for about 150 Mpc, and the APM survey[68] has also found evidence for very large structures of galaxies.

One general feature of all of these galaxy surveys is that they seem to find more really big structures than might have been estimated if galaxies were randomly distributed and if the mass of the systems really traced the emitted light. However, many have argued that it is possible that light is not a good tracer of the underlying mass distribution. Such a shift between light and mass distributions (biasing) makes it difficult to compare unambiguously galaxy distributions with theories. Furthermore, different types of galaxies (or clusters) may have different biasing factors and, at present, the statistics on large separations are poor since the numbers of objects studied is limiting. This latter point should be alleviated with the Sloan Digital Sky Survey (a Chicago-Princeton-Fermilab-Johns Hopkins-Japanese collaboration) when the 3-D positions for about a million galaxies will be determined (as contrasted with current surveys that see at most about ten thousand).

One way of comparing the microwave anisotropy data with the galaxy data is to look at the relative power implied on different separation scales. Because of the horizon cut-off, it is expected with flat gaussian density seeds that the power will peak on scales of about 1 or 2 degrees (see Figure 6) which today would correspond to distance scales of a few hundred Mpc. The dotted line in Figure 6 shows this behavior. The absolute amplitude of the initial fluctuations is very uncertain so the dotted curve is free to be moved up or down. For Figure 6, it has been fixed by the additional requirement that it fit the large scale data of COBE. The curves for the different galaxy surveys may be moved up or down, but not left or right via biasing and/or selection effects. Note that if the dotted curve is made to fit the small scale end of the galaxy data, then the large scale end doesn't fit. Note also that if we use the COBE fit at large scales and fit the IRAS or APM data at a couple of hundred Mpc, then the small scale IRAS and APM data falls below the dotted curve. Thus, no simple constant biasing can enable the dotted curve to work. Furthermore, there are hints from the radio galaxy[69] and cluster data[70] that there may be even more power near 100 to 200 Mpc, although the uncertainties are large and biasing may move the points coherently up and down.

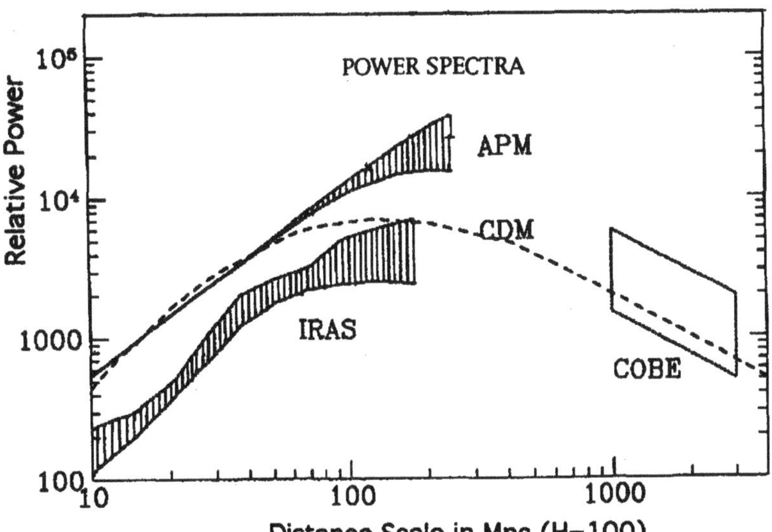

Figure 6. The relative power of producing structures on different scales. The dotted curve is for a traditional CDM model with gaussian density seeds. The galaxy data from different samples can be moved up and down via biasing and/or selection effects but the shapes are fixed. This figure was prepared by Jean Quashnock of the University of Chicago.

This conflict with the simple "CDM model" of an $n = 1$, gaussian density seed spectrum with CDM and constant biasing has led to several alternatives, all designed to fit the COBE data on large scales and the shape of the galaxy data on smaller scales. The possible models are shown in Table 6. As seen from the table, the surviving models each involve some assumptions beyond or instead of the traditional CDM constant biased model. For example, one can retain the gaussian density fluctuation picture by either adding a variable scale-dependent biasing (that goes negative at small scales!), or by adding some HDM to smooth out the small scale fluctuations and thereby decrease the small scale power, or by tilting the initial spectrum enough to fit the small scales but not violate the COBE error box. Another alternative is to go to topological defects. These can be either cosmic strings if the defects form in the early universe or almost any defect if the defects form after the microwave radiation decouples, but in this latter case, the COBE spectrum still requires some gaussian primordial fluctuations. The advantage of the defects plus HDM is that they tend to give a larger bump in the power spectrum near the epoch of galaxy formation which leads to an easier fit to the rich cluster and radio galaxy data (although this is the most uncertain part of the structure data and many models choose to ignore it at present).

TABLE 6
Models After COBE with Critical Total Density and ~ 5% Baryons

Model Name	Non-Baryonic Dark Matter	Seed Type	Biasing
Modified CDM	CDM	flat-gaussian density	variable, going negative at small scales
Mixed (MDM)	CDM plus HDM	flat-gaussian density	negligible, except for cluster data
Tilted	CDM	slightly tilted spectrum increasing towards large scale	negligible, except for cluster data
Cosmic Strings	HDM	cosmic strings	negligible, may even fit cluster data
Late-Time Phase Transition	HDM	topological defects plus background flat-gaussian density	negligible

Note: models with cosmological constant or pure baryons are excluded here for reasons mentioned previously in text. Also, it is not obvious how many of the above models will fit the cluster x-ray gas analysis of Mushotzky.[62]

Obviously, HDM appears as a viable solution in three of the entries in Table 6. Thus, neutrinos are still a possible candidate for non-baryonic dark matter.

This exciting and rapidly moving field may even come to some resolution of the problem of structure formation in the universe in the not-too-distant future. This resoution will come by an interdisciplinary effort of particle physicsts, nuclear physicists, radio, IR, optical, UV and x-ray astronomers and theorists. It is exciting to be doing cosmology in this "Golden Age" where progress really occurs.

Acknowledgements

I would like to thank Xiangdong Shi for his significant contributions to the solar neutrino section of this paper. This work was supported in part by the NSF, DOE and NASA at the University of Chicago and by the DOE and by NASA through grant NAGW 2381 at Fermilab.

References

1. J.N. Bahcall and R.K. Ulrich, *Rev. Mod. Phys.* **60** (1988) 297.
2. J.N. Bahcall, *Neutrino Astrophysics* Cambridge University Press, 1989).
3. J.K. Rowley, B.T. Cleveland, and R. Davis, Jr., in *Neutrino '88*, ed. J. Schneps et al. (World Scientific, Singapore, 1989), p. 518;
   R. Davis in *Frontiers of Neutrino Astrophysics*, eds. Y. Susuzi and K. Nakamura (Universal Academy Press, Tokyo, 1993).
4. K.S. Hirata et al., *Phys. Rev. Lett.* **65** (1990) 1297;
   Y. Suzuki, in *Frontiers of Neutrino Astrophysics*, eds. Y. Susuzi and K. Nakamura (Universal Academy Press, Tokyo, 1993).
5. A. I. Abazov, *et al.*, *Phys. Rev. Lett.* **67** (1991) 3332;
   V. N. Govrin, talk at the 26th Intl. Conf. on High Energy Physics, Dallas, 1992.
6. P. Anselmann et al., *Phys. Lett. B* **285** (1992) 376.
7. S. Turck-Chièze, S. Cahn, M. Cassé, and C. Doom, *Ap. J.* **335** (1988) 415;
   S. Turck-Chièze and I. Lopes, *Ap. J.* **408** (1993) 347;
   S. Turck-Chièze et al., *Phys. Rep.* **230** (1993) 59. In the 1993 model of Turck-Chièze *et al.*, a correction was made relative to the original 1988 model. The prediction of ^{37}Cl capture rate increases from the original 5.8 SNU to 6.4 SNU, with the increase mainly in ^{8}B neutrinos.
8. X. Shi and D. N. Schramm, *Particle World*, in press (1993).
9. L. Wolfenstein, *Phys. Rev.* D **17** (1978) 2369;
   S. P. Mikheyev and A. Yu. Smirnov, *Sov. J. Nucl. Phys.* **42** (1985) 913;
   T. K. Kuo and J. Panteleone, *Rev. Mod. Phys.* **61** (1989) 937.
10. V. Barger, R. J. N. Phillips, and K. Whisnant, *Phys. Rev.* D **43** (1991) 1110;
    P. I. Krastev and S. T. Petcov, *Phys. Lett. B* **245** (1992) 85.
11. S. A. Bludman, D. C. Kennedy, and P. G. Langacker, *Nucl. Phys.* B **374** (1992)

373.
12. X. Shi, D. N. Schramm and J. N. Bahcall, *Phys. Rev. Lett.*, **69** (1992) 717.
13. J. N. Bahcall and M. H. Pinsonneault, *Rev. Mod. Phys.* **64** (1992) 885.
14. Sienkiewicz, R., J. N. Bahcall, and B. Paczyński, *Ap. J.*, **348** (1990) 641.
15. B. W. Filippone and D. N. Schramm, *Ap. J.*, **253** (1982) 393.
16. D. R. O. Morrison, *Particle World*, **3** (1992) 30;
D. R. O. Morrison, CERN-PPE/92-109.
17. K. S. Hirata *et al.*, *Phys. Rev.* D **44** (1991) 2241.
18. J. N. Bahcall and H. A. Bethe, *Phys. Rev. Lett.* **65** (1990) 2233.
19. S. A. Bludman, N. Hata, D.C. Kennedy, and P.G. Langacker, *Phys. Rev. D* **47** (1993) 47.
20. C. W. Johnson *et al.*, *Ap. J.*, **392** (1992) 31420.
21. B. W. Filippone *et al.*, *Phys. Rev.* C **28** (1983) 2222.
22. C.-S. Lim and W. J. Marciano, *Phys. Rev. D* **37** (1988) 1368;
23. X. Shi, D. N. Schramm, R. Rosner and D. S. Dearborn, *Comments on Nuclear and Particle Physics* **21** (1993) 151.
24. A. Acker, S. Pakvasa and J. Pantaleone, *Phys. Rev.* D **43** (1991) 1754.
25. K. Enqvist *et al.*, NORDITA-91/71 P.
26. X. Shi, D. N. Schramm and B. Fields, *Phys. Rev. D*, in press (15 Sept. 1993).
27. For a review of current and future solar neutrino experiments, see Ch. 10–14 of ref. 2.
28. D. Schramm, *Comments Nuc. Part. Phys.* **A17** (1987) 239.
D. Schramm in *Neutrino Physics*, ed. K. Winter (Cambridge University Press, 1991) p. 629.
29. D. Schramm and W.D. Arnett, *Ap.J.* **198**, 629 (1974).
30. D.Z. Freedman, D.N. Schramm, and D. Tubbs, *Ann. Rev. Nucl. Sci.* **27** (1977) 167; and references therein.
31. T. Mazurek, *Ap. J. Lett.* **207** (1976) L87.
32. W.D. Arnett, in *Explosive Nucleosynthesis* (University of Texas Press, Austin, 1973).
33. T. Weaver, S. Woosley, and G. Fuller, in *Numerical Astrophysics*, eds. J. Centrella, J. LeBlanc and R. Bowers (Jones and Bartlett, 1984) p. 374.
34. H. Bethe and G. Browne, *Scientific American* **252** (1985) 60.
35. E. Baron, et al., *Phys. Rev. Lett.* **55** (1985) 126.
36. J. Wilson, et al., *Ann. N.Y. Acad. Sci.* **262** (1975) 54.
37. J. Wilson, talk at Aspen Winter Meeting on SN 1987A (1988).
38. W.D. Arnett and D.N. Schramm, *Ap. J.Lett.* **184** (1973) L47;
T.A. Weaver and S. Woosley, in *Nuclear Astrophysics: Proceedings of the Caltech Centennial Year Nuclear Astrophysics Symposium in Honor of William A. Fowler's 80th Birthday*, eds. D.N. Schramm and S.E. Woosley, *Phys. Rep.* **227** (1993) 65.
39. S. Shapiro, in *Gravitational Radiation*, ed. L. Smarr (Oxford University Press, Oxford, 1978).
40. D. Kazanas and D.N. Schramm, *Nature* **262** (1976) 671.

41. D. Kazanas and D.N. Schramm, *Ap. J.* **214** (1977) 819.
42. D.N. Schramm, in *Proc. 2nd Dumand Symposium* (1976) p. 87.
43. R. Mayle, J. Wilson, and D.N. Schramm, *Ap. J.* **318** (1987) 288.
44. D.N. Schramm, J. Wilson, and R. Mayle, in *Proc. 1st International Conference on Underground Physics*, in *Il Nuovo Cimento* **9C** (1986) 443.
45. A. Burrows and J. Lattimer, "The Birth of Neutron Stars," preprint (1985).
46. R. Mayle, Ph.D. Thesis, University of California, Berkeley (available as Lawrence Livermore preprint UCRL 53713).
47. S. Bruenn, Florida Atlantic U. preprint (1988).
48. A. Burrows, U. of Arizona preprint (1988).
49. S.P. Mikheyev and A.Yu. Smirnov, *Il Nuovo Cimento* **9C** (1986) 17;
L. Wolfenstein, *Phys. Rev.* **D17** (1978) 2369.
50. T. Walker and D.N. Schramm, *Phys. Lett.* **195B** (1987) 331;
E. Symbalisty, J. Wilson, and D. Schramm, *Ap. J. Lett.* **291** (1985) L11.
51. A. Burrows, University of Arizona preprint (1987);
Hirata et al. *Phys. Rev. Lett.* **58** (1987) 1490.
52. W.D. Arnett, University of Chicago preprint (1987);
Bionta et al. *Phys. Rev. Lett.* **58** (1987) 1494.
53. J. Bahcall, T. Piran, W. Press, and D. Spergel, *Nature* **327** (1987) 692.
54. R. Mayle and J. Wilson, Livermore preprint (1987).
55. K. Sato and H. Suzuki, Tokyo preprint UTAP-52 (1987).
56. L. Krauss, P. Romanelli, D. Schramm, and R. Lehrer, *Nucl. Phys. B* **380** (1992) 507.
57. T. Walker, G. Steigman, D. N. Schramm, K. Olive, and H.-S. Kang, *Ap.J.* **376** (1991) 51.
58. V. Smith, D. Lambert, and R. Nissen, R., *Ap.J.* **408** (1993) 262.
59. K. Olive and D. N. Schramm, *Nature* **360** (1992) 434.
60. H. Kurki-Suonio, R. Matzner, K. Olive, and D. N. Schramm, *Ap.J.* **353** (1990) 406.
61. J. R. Gott, J. Gunn, D. N. Schramm, and B. Tinsley, *Ap.J.* **194** (1974) 543.
62. R. Mushotzky, in *Texas/PASCOS December 92: Relativistic Astrophysics and Particle Cosmology*, eds. C. Akerlof and M. Srednicki, *Annals of the N.Y. Academy of Sciences* **688** (1993) 184.
63. S. Meyer, in *Texas/PASCOS December 92: Relativistic Astrophysics and Particle Cosmology*, eds. C. Akerlof and M. Srednicki, *Annals of the N.Y. Academy of Sciences* **688** (1993) 795, 801.
64. E. Wright, in *Texas/PASCOS December 92: Relativistic Astrophysics and Particle Cosmology*, eds. C. Akerlof and M. Srednicki, *Annals of the N.Y. Academy of Sciences* **688** (1993) 836.
65. P. Lubin et al., in *Texas/PASCOS December 92: Relativistic Astrophysics and Particle Cosmology*, eds. C. Akerlof and M. Srednicki, *Annals of the N.Y. Academy of Sciences* **688** (1993) 804.
66. MAX collaboration, in *Texas/PASCOS December 92: Relativistic Astrophysics and Particle Cosmology*, eds. C. Akerlof and M. Srednicki, *Annals of the N.Y.*

Academy of Sciences **688** (1993) 809.
67. M. S. Vogeley, C. Park, M. Geller, and J. Huchra, *Ap.J. Lett.* **391** (1992) L5.
68. J. A. Peacock, G. Efstathiou et al., *MNRAS* **253** (1991) 1.
69. J. A. Peacock and D. Nicholson, *MNRAS* **253** (1991) 307.
70. J. Bahcall and R. Soniera, *ApJ* **270** (1983) 20.

THE PRESENT STATUS OF THE DECAYING NEUTRINO THEORY

D.W. SCIAMA

International Centre for Theoretical Physics, Trieste, Italy
and
International School for Advanced Studies, Trieste, Italy

ABSTRACT

The present status of the decaying neutrino theory is reviewed. Three recent developments are highlighted:

(a) the proposal that the dark matter in rich clusters of galaxies is mainly baryonic

(b) the implications of the hypothesis that decay photons are the solution of the C^o/CO ratio problem

(c) the strong supporting evidence from recent observations with the Hubble Space Telescope.

1. Introduction

It is a personal pleasure as well as an honour to be speaking at this Salamfest. Nearly 50 years ago I was studying mathematics and physics at Cambridge. Just ahead of me was a rising young star in these subjects - Abdus Salam. Of course we did not realise then just how high he would rise.

Abdus's work is characterised by a combination of strong formal theory and close links with experiment. In particular his Standard Model is amazingly successful experimentally. I thought that I would talk about various astronomical and cosmological observations which point to the need to go beyond the Standard Mode. These observations have led me to propose a particular version of the decaying neutrino theory, whose present status I review here.

2. The Decaying Neutrino Theory

If neutrinos have non-zero rest mass one would expect a more massive neutrino type ν_1 to decay into a photon and a less massive neutrino type ν_2:

$$\nu_1 \to \gamma + \nu_2 \ .$$

Conservation of energy and momentum in the decay tells us that the energy E_γ of the photon in the rest frame of ν_1 is given by

$$E_\gamma = \frac{1}{2}\, m_1 \left(1 - \frac{m_2^2}{m_1^2}\right) \ .$$

It is very likely that $\frac{m_2^2}{m_1^2} \ll 1$. In that case we would have the simple result

$$E_\gamma = \frac{1}{2}\, m_1 \ .$$

Cowsik [1] pointed out that even if the lifetime for this decay is much longer than the age of the universe, the photon flux from the cosmological distribution of neutrinos pair created in the hot big bang might be appreciable, simply because of the great length of the line of sight through the universe. This point has led to a considerable literature which I will not go into here. As I write this article I am completing a book which gives a detailed discussion of this question and is entitled Modern Cosmology and the Dark Matter problem.

The point of departure for this article was my suggestion [2] that decay photons from dark matter neutrinos in our galaxy might be responsible for the otherwise puzzling widespread ionization of hydrogen in the interstellar medium. The key point here is that the ionization potential of hydrogen is 13.6 eV, so we require

$$E_\gamma > 13.6 \text{ eV}$$

and therefore

$$m_1 > 27.2 \text{ eV}$$

This gives us the first of the many coincidences in this subject because (a) the Tremaine–Gunn phase space constraint for neutrinos to be bound in the galaxy leads to essentially the same lower limit (b) the cosmological density ρ_{ν_1} of ν_1 neutrinos is such that

$$m_1 = 93\Omega_{\nu_1} h^2 \text{ eV}$$

where as usual $\Omega_{\nu_1} = \rho_{\nu_1}/\rho_{crit}$ and h is the Hubble parameter given by

$$H_0 = 100h \text{ km./sec/Mpc}, \qquad \left(\frac{1}{2} < h < 1\right) \ .$$

If $\Omega_{\nu_1} \sim 1$ and $h \sim \frac{1}{2}$ (which would be compatible with the observed age of the universe) we would have

$$m_1 \sim 23 \text{ eV} .$$

Thus if h is only slightly greater than $\frac{1}{2}$ we would again have

$$m_1 > 27.2 \text{ eV} ,$$

and the decay photon would be able to ionize hydrogen.

Encouraged by these coincidences I evaluated the decay lifetime τ needed to account for the observed electron density in the interstellar medium ($n_e \sim 0.03 \text{ cm}^{-3}$) and obtained

$$\tau \sim 2 \times 10^{23} \text{secs.}$$

This choice of lifetime leads to another coincidence. Let us evaluate the intergalactic ionizing flux F of decay photons from the cosmological distribution of neutrinos of density n_{ν_1}. A naive estimate would be

$$F \sim \frac{n_{\nu_1}}{\tau} \frac{c}{H_0} .$$

With $n_{\nu_1} \sim 100 \text{ cm}^{-3}, \tau \sim 2 \times 10^{23}$ secs and $c/H_0 \sim 2 \times 10^{28}$ cm we would obtain

$$F \sim 10^7 \text{ photons cm}^{-2}\text{sec}^{-1} .$$

At first sight this result is a disaster because there is an observational upper limit on F, namely

$$F <\sim 10^6 \text{ cm}^{-2} \text{ sec}^{-1} .$$

We can try to solve this problem by using the fact that the decay photons will be red shifted as they propagate, and once their energy is reduced below 13.6 eV they will no longer contribute to the ionizing flux. This effect is simple to include. If one writes

$$E_\gamma = (13.6 + \epsilon) \text{ eV} ,$$

one has

$$F = \frac{n_{\nu_1}}{\tau} \frac{c}{H_0} \frac{\epsilon}{13.6} .$$

We can therefore solve our problem if

$$\epsilon <\sim 1 ,$$

so that

$$E_\gamma <\sim 14.6 \text{ eV} .$$

Hence

$$E_\gamma = 14.1 \pm 0.5 \text{ eV} ,$$

and

$$m_1 = 28.2 \pm 1 \text{ eV} .$$

This gives us a new coincidence since this value of m_1 leads to $\Omega_\nu \sim 1$ for a low value of h. Also the resulting value of F would account for the otherwise puzzling ionization of the intergalactic medium and Lyman α clouds.

We can do even better than this if we use recent observational data to argue that the decay photons must also be able to ionize nitrogen [3]. This requires that

$$E_\gamma > 14.53 \text{ eV} .$$

Remarkably, this is just compatible with our upper limit and would lead to

$$E_\gamma = 14.565 \pm 0.035 \text{ eV} ,$$

and

$$m_1 = 29.13 \pm 0.07 \text{ eV} .$$

Thus the mass of the decaying neutrino ($?\nu_\tau$) would be pinned down to better than 1%. Also if we assume that $\Omega = 1$ when the baryon density is included then we obtain [4]

$$H_0 = 56 \pm 0.5 \text{ km/sec/Mpc}$$

so that the Hubble constant would be pinned down to 1%.

With this background we are now ready to consider three recent developments in these ideas.

3. The Nature of the Dark Matter in Clusters of Galaxies

This decaying neutrino theory is widely regarded as having been disproved by the failure of Davidsen *et al.* [5] to observe a decay line from dark matter neutrinos in the rich cluster of galaxies A665. However, since then Hughes and Tanaka [6] have published detailed observations of the x–ray emission from the hot gas in this cluster and have derived model distributions for the density of the galaxies, the gas and the dark matter. From this, one can show [7,8] that the dark matter is more concentrated to the centre of the cluster than are either the galaxies or the gas. A similar concentration is found in other clusters both from the x-ray data and the gravitational lens data. I find it rather pleasing that general relativity can be used to help demonstrate this excess concentration of the dark matter.

It seems unlikely that neutrinos would be more concentrated than baryons in clusters, since the neutrinos are non-dissipative. It is more likely that most of the dark matter in this case is baryonic. One possibility is that a cooling flow in the hot gas ends up as faint low mass stars [9]. The absence of an observable decay line would then be explained.

By contrast the dark matter in our galaxy has a more extended distribution than the visible matter. This would fit in well with the hypothesis that this dark matter is in the form of neutrinos, but does not demand it. In any case I believe that these considerations make clear that the decaying neutrino theory is still viable.

4. Ultra Violet Radiation in Dense Molecular Clouds

It has been known for a long time that the abundance ratio of atomic carbon to carbon monoxide in the interiors of dense molecular clouds in the galaxy is typically 10^5 times greater than would be expected from simple equilibrium models. This is referred to as the C^o/CO ratio problem. In these models the flux of ultraviolet radiation in the interiors of the clouds is very low because of their great opacity, and so the dissociation rate of CO is also very low. One class of proposed solutions to this problem involves mechanisms for increasing the ultra violet flux in the interiors. These proposals have recently been reviewed by Sorrell [10]. They involve complicated mechanisms which are difficult to assess.

By contrast, decaying neutrinos situated in the clouds would provide a straightforward mechanism, as was pointed out by Tarafdar [11]. This proposal has recently been updated by Sciama [12]. A crucial role in Tarafdar's analysis is played by the opacity of the clouds to the decay photons, since this controls the resulting flux of these photons. The clouds typically have particle densities in the range $10^3 - 10^5$ cm^{-3}, and the vast majority of these particles is in the form of H_2 molecules. The flux of decay photons would only be sufficient to solve the C^0/CO ratio problem if these photons are unable to dissociate the H_2 molecules, so that the opacity of the clouds would be relatively small. Here enters a new coincidence. The photodissociation continuum of H_2 has a threshold at 14.68 eV. Hence we require that

$$E_\gamma < 14.68 \text{ eV} .$$

This constraint is essentially the same as our previous one, which followed from considering the intergalactic ionizing flux ($E_\gamma <\sim 14.6$ eV). The new constraint is more stringent in the sense that the threshold involved has been accurately measured, whereas the previous constraint involved more uncertain quantities (the decay lifetime and the observational limit on the intergalactic flux). If our nitrogen and C^o/CO assumptions are both correct we would then have the essentially exact constraints

$$14.53 < E_\gamma < 14.68 \text{ eV} ,$$

and (if $m_2 \ll m_1$)

$$29.06 < m_1 < 29.36 \text{ eV} .$$

The consequences for the value of the Hubble constant would then be essentially the same as before.

5. Recent HST Observations of the Electron Density in the Interstellar Medium

Strong support for the neutrino decay theory can be derived [13] from the analysis by Spitzer and Fitzpatrick [14] of their observations of a galactic halo star HD93521 using the Hubble Space Telescope. They observed the ultraviolet absorption spectrum of several regions along the line of sight to this star. One of the absorbing species was singly ionized carbon in an excited state, CII*. The excitation is probably due to electron collisions and so leads to the value of the electron density n_e in each absorbing region.

In their analysis Spitzer and Fitzpatrick also used the 21cm emission observations of Danley *et al.* [15] which showed that the atomic hydrogen column density makes each region opaque to Lyman continuum photons. They concluded that the ionized and neutral gas are intimately mixed up together, in constrast to the standard view that these two components of the interstellar medium are located in separate regions (the warm ionized medium or WIM and the warm neutral medium or WNM).

If this is correct we face the same type of problem as we did in the previous section with the molecular clouds. In this case we are dealing with an excess of hydrogen–ionizing photons in an opaque region rather than an excess of CO dissociating photons. Again decaying neutrinos come to the rescue. In fact it was related opacity problems which led to the decaying neutrino theory in the first place, but we have here the first clear observational evidence that the free electrons (which were previously known to exist from $H\alpha$ and pulsar dispersion data) are located in opaque regions.

Further support for the theory comes from the values of n_e derived by Spitzer and Fitzpatrick. They found for the slowly moving absorbing regions (which would be unaffected by shock waves) that n_e is the same in each region to their observing precision of 10%. This is just what would be expected for opaque regions in the decaying neutrino theory, and it was predicted in the original paper [2]. The reason is simple. Since every decay photon produces an ionization locally, to be followed by a recombination, one has the ionization equilibrium equation

$$\frac{n_\nu}{\tau} = \alpha n_e^2 ,$$

where α is the appropriate recombination coefficient. In limited regions of the galaxy n_ν will be nearly constant, and n_e will depend on the gas temperature T (via $\alpha(t)$) only as $T^{0.37}$. Moreover in warm regions, such as those observed by Spitzer and Fitzpatrick, the temperature is fairly constant. Hence n_e should be nearly constant, as is observed.

Moreover, if one assumes that carbon is undepleted in these regions (for which there is some evidence [13]), one finds for the four slowly moving regions $n_e = 0.055 \text{ cm}^{-3}$. This is just the same as the values found for three other regions from an analysis of the pulsar dispersion measures of three pulsars with accurately known (parallactic) distances [16,17]. One would not expect such constancy for other ionization mechanisms. Future HST measurements should provide a stringent test of these ideas.

6. Conclusions

I conclude from this discussion that the evidence in favour of the decaying neutrino theory is rather strong. One should therefore take seriously its other implications. These include the early reionization of the universe and the resulting suppression of anisotropies in the cosmic microwave background on small angular scales [18], and attempts to derive the required decay lifetime from elementary particle physics [19,20].

7. Acknowledgements

I am grateful to the Italian Ministry of Universities and Scientific and Technological Research for their financial support.

8. References

1. R. Cowsik, *Phys. Rev. Lett.* **39** (1977) 784.
2. D.W. Sciama, *Ap. J.* **364** (1990a) 549.
3. D.W. Sciama, *Int. J. Mod. Phys.* **D.1** (1992) 161.
4. D.W. Sciama, *Phys. Rev. Lett.* **65** (1990b) 2839.
5. A.F. Davidsen et al., it Nature **351** (1991) 128.
6. J.P. Hughes and K. Tanaka, *Ap. J.* **398** (1992) 62.
7. D.W. Sciama, M. Persic and P. Salucci, *Nature* **358** (1992) 718.
8. D.W. Sciama, M. Persic and P. Salucci, *PASP* **105** (1993) 102.
9. P.A. Thomas and A.C.Fabian, *Mon. Not. R.A.S.* **246** (1990) 156.
10. W. Sorrell, *Comments Astrophys.* **16** (1992) 123.
11. S.P. Tarafdar, *MNRAS* **252** (1991) 55P.
12. D.W. Sciama, to be published (1993a).
13. D.W. Sciama, *Ap. J.* **409** (1993b) L25.
14. L. Spitzer and E.L. Fitzpatrick, *Ap. J.* **409** (1993) 299.
15. L. Danley et al., *Ap. J. Suppl.* **81** (1992) 125.
16. R.J. Reynolds, *Ap. J.* **348** (1990) 153.
17. D.W. Sciama, *Nature* **346** (1990c) 40.
18. D. Scott, M.J. Rees and D.W. Sciama, *Astron. and Astrophys.* **250** (1991) 295.
19. E. Roulet and D. Tommasini, *Phys. Lett.* **256B** (1991) 218.
20. F. Gabbiani, A. Masiero and D.W. Sciama, *Phys. Lett.* **259B** (1991) 323.

IV. Strings and Quantum Gravity

QUANTUM BACKGROUND INDEPENDENCE IN STRING THEORY

EDWARD WITTEN*

School of Natural Sciences
Institute for Advanced Study
Olden Lane
Princeton, NJ 08540

ABSTRACT

Not only in physical string theories, but also in some highly simplified situations, background independence has been difficult to understand. It is argued that the "holomorphic anomaly" of Bershadsky, Cecotti, Ooguri, and Vafa gives a fundamental explanation of some of the problems. Moreover, their anomaly equation can be interpreted in terms of a rather peculiar quantum version of background independence: in systems afflicted by the anomaly, background independence does not hold order by order in perturbation theory, but the exact partition function as a function of the coupling constants has a background independent interpretation as a state in an auxiliary quantum Hilbert space. The significance of this auxiliary space is otherwise unknown.

* Research supported in part by NSF Grant PHY92-45317.

1. Background Independence And The Holomorphic Anomaly

Finding the right framework for an intrinsic, background independent formulation of string theory is one of the main problems in the subject, and so far has remained out of reach. Moreover, some highly simplified special cases or analogs of the problem, which look like they might be studied for practice, have also resisted understanding.

An important example is the problem of understanding the mirror map in the theory of mirror symmetry. In $(2,2)$ compactification on a Calabi-Yau threefold X, one encounters two sets of renormalizable Yukawa couplings, involving modes coming from $H^{1,1}(X)$ and $H^{2,1}(X)$. These are closely related to two twisted topological field theories that can be constructed for a given Calabi-Yau target space X – the A model and the B model.

Mirror symmetry is a relation between two Calabi-Yau manifolds X and Y, in which the Yukawa couplings involving $H^{1,1}(X)$ are identified with those that involve $H^{2,1}(Y)$, and vice-versa. Equivalently, mirror symmetry exchanges the A model of X with the B model of Y, and vice-versa.

The moduli space of sigma models with a Calabi-Yau target space is locally a product of two factors. One factor, the moduli space $\mathcal{M}_A$ of the A model, is (an open set in) $H^{1,1}(X, \mathbb{C}/2\pi i\mathbb{Z})$. The other factor, the moduli space $\mathcal{M}_B$ of the B model, is the moduli space of complex structures on X. Mirror symmetry therefore implies a natural map between $\mathcal{M}_A(X)$ and $\mathcal{M}_B(Y)$. This seems bizarre because – being related to the linear space $H^{1,1}(X, \mathbb{C})$ – $\mathcal{M}_A$ has a natural "flat" (or really, affine linear) structure, while $\mathcal{M}_B$, the moduli space of complex structures, has no such natural structure.

Candelas et. al. [1] tried to overcome this problem by using the "special coordinates" on $\mathcal{M}_B$. Special coordinates are background dependent. They are defined as follows. Pick a complex structure J_0 on X representing a base-point in $\mathcal{M}_B$. Let Ω_0 be a three-form on X holomorphic with respect to the complex structure J_0. Let J be a variable complex structure on X, and let Ω_J be a holomorphic three-form in the complex structure J, normalized so that if one decomposes Ω_J with respect to the Hodge structure defined by J_0, then the $(3,0)$ part of Ω_J is cohomologous to Ω_0. Then [2] the map from $\mathcal{M}_B$ to the $(2,1)$ part of the cohomology class of Ω_J is locally an isomorphism from $\mathcal{M}_B$ to the linear space $H^{2,1}(X_0, \mathbb{C})$ (here X_0 is X with complex structure J_0).† This gives the desired "flat" structure on $\mathcal{M}_B$, the "special coordinates" being the components of the $(2,1)$ part of Ω_J.

† Upon contraction with Ω_0, $H^{2,1}(X_0, \mathbb{C})$ can be identified with $H^1(X_0, T^{1,0}X_0)$, which is the tangent space to $\mathcal{M}_B$ at its chosen base-point X_0. This is the natural linear space with which one might try to identify $\mathcal{M}_B$ near X_0.

The trouble with this is of course the dependence on the base-point J_0. According to Morrison [3], the mirror of the natural flat structure on $\mathcal{M}_A(X)$ is the flat structure determined on $\mathcal{M}_B(Y)$ by a base-point at infinity. In this context, "infinity" refers to a particular type of degeneration of the complex structure of X; neither existence nor uniqueness of such a degeneration is apparent. (Lack of uniqueness can lead to a multiple mirror phenomenon and topology change [4].) So we come to our first problem:

(1) What is the analog in the A model of choosing a base-point in the B model? Why does the obvious flat structure on the parameter space of the A model correspond to the B model with a base-point at infinity?

Apart from using Calabi-Yau threefolds for compactification of physical string theories, one can use the twisted A or B topological field theories to construct topological string theories which one might think would be a highly simplified laboratory for studying background independence in string theory. Indeed, for open string versions of either the A or B model, there is no problem [5] in identifying the background independent space-time physics, which moreover is local in some important cases. But the natural attempt at extracting an effective space-time field theory for closed topological strings (see §5 of [5]) gives a result that is non-local and is background dependent in the case of the B model. So, even if we put aside the non-locality, we have our second problem of background independence:

(2) What is the origin of the background dependence in the space-time field theory of the closed string B model?

A special case of these problems is so simple that it seems worthy of pointing out separately. The usual analysis of the B model appears to show that in genus zero, in expanding around an arbitrary base-point J_0, the partition function vanishes together with its first two derivatives (while the third derivative gives the Yukawa couplings). So our third question is obviously:

(3) Since background independence appears to require that the choice of base-point J_0 should play no special role, does not vanishing of the genus zero partition function at J_0 require that function to vanish identically?

Another cluster of questions that may be similar involves soluble string theories in $D \leq 2$. At least in the case of $D = 2$, where one has a two dimensional space-time, a graviton-dilaton system with a black hole solution, a tachyon scattering matrix, etc., one would hope to find an intrinsic description of the space-time geometry of these theories. Yet this has proved surprisingly elusive. A possible relation of these problems to those discussed above is suggested by the fact that the $D \leq 2$ string theories can be interpreted via twisted $N = 2$ models, as first shown for $D < 2$ by K. Li [6] and argued more recently for $D = 2$ [7–9].

So our final problem – which we will actually not discuss in this paper – is the following:

(4) Either describe the background independent space-time physics of $D = 2$ (or maybe $D \leq 2$) string theory or describe the obstruction to doing so.

1.1. Role Of The Holomorphic Anomaly

Recently, Bershadsky, Cecotti, Ooguri, and Vafa [10], following earlier work [11,12], described a "holomorphic anomaly" in topological field theories obtained by twisting $N = 2$ models. Their anomaly can be understood as a violation of naive background independence and I will presently argue that it explains most of the puzzles cited above.

Consider an $N = 2$ supersymmetric theory that contains chiral or twisted interactions, in addition to other interactions that we will bury in a Lagrangian L_0. There is no essential loss in considering the chiral case. If W_a are the chiral primary fields, then the Lagrangian

$$L = L_0 - \sum_a t^a \int d^2x\, d^2\theta\, W_a - \sum_a \overline{t}^a \int d^2x\, d^2\overline{\theta}\, \overline{W}_a, \tag{1.1}$$

with complex parameters t^a, describes a family of $N = 2$ theories. One can also consider twisted topological field theories with two of the supersymmetries, the ones that generate shifts in $\overline{\theta}$, identified with BRST operators $Q_\pm$. Then the W_a are the physical observables, and $\int d^2x\, d^2\overline{\theta}\, \overline{W}_a$ is formally irrelevant being $\{Q_+, [Q_-, \overline{W}_a]\}$. After twisting (1.1), it is natural to add the physical observables as perturbations, considering a more general Lagrangian

$$L = L_0 - \sum_a (t^a + u^a) \int d^2x\, d^2\theta\, W_a - \sum_a \overline{t}^a \int d^2x d^2\overline{\theta}\ \overline{W}_a. \tag{1.2}$$

Formally, the BRST machinery appears to show that the topological observables of the topological field theory (1.2) are independent of $\overline{t}^a$, and so are only functions of $t^a + u^a$. This is where the holomorphic anomaly comes in; Bershadsky et. al. show that the topological observables really have a dependence on $\overline{t}^a$ determined by their holomorphic anomaly. I want to interpret this as a failure of background independence. The idea is that the choice of $\overline{t}^a$ determines the original "physical" Lagrangian (1.1), which was then twisted and perturbed by topological observables with coefficients u^a. The dependence on $\overline{t}^a$ means that the particular "physical" theory that one started with is not forgotten. It defines a base-point in the space of theories.

Let us change the notation, renaming $t+u$ as t and $\bar{t}$ as t'. Henceforth $\bar{t}$ denotes the complex conjugate of t; t' is an independent complex variable. Now (1.2) takes a more symmetric-looking form,

$$L = L_0 - \sum_a t^a \int d^2x\, d^2\theta\, W_a - \sum_a t'^a \int d^2x d^2\bar{\theta}\, \overline{W}_a. \tag{1.3}$$

The symmetry between t and t' is broken by the choice of twisting. The t'^a determine a base-point in the space of topological couplings, the point at which $\bar{t}^a = t'^a$ and at which the theory can actually be interpreted as a twisting of a physical theory. The BRST argument appears to show that in the twisted model, the topological observables depend only on the t^a and not on the t'^a, but the anomaly obstructs this.

Now we can rather easily dispose of the first three problems identified above:

(1) The symmetry in structure between A and B models is restored because just as the B model naturally depends on a base-point in the space of complex structures, the A model also depends on a choice of base-point, namely a choice of a point $t'^a \in H^{1,1}(X, \mathbb{C}/2\pi i\mathbb{Z})$. There is therefore a more general family of A-like models than hitherto realized. Bershadsky et. al. show in detail that the traditional A model (in which correlation functions are given by standard instanton sums) corresponds to the case in which $t'^a \to \infty$, that is the case in which the A model is taken to have a base-point at infinity. The mirror of the traditional A model is therefore naturally a B model with base-point at infinity. A more general B model with another base-point would be mirror to an A model with a finite base-point t'^a.

(2) Because of the holomorphic anomaly, background independence of the field theory of the B model is not expected.

(3) The paradox involving the genus zero free energy F_0 is also eliminated once one abandons naive background independence. One is merely left with the statement that if the model is constructed with a base-point t'^a, then F_0 vanishes together with its first two derivatives near $t^a = \bar{t}'^a$.

The fourth problem on our above list – the space-time physics of soluble string theories – should perhaps also be reexamined in light of the holomorphic anomaly, and its possible cousins.

1.2. Salvaging Something From Background Independence

Though the interpretation of the holomorphic anomaly as an obstruction to background independence eliminates some thorny puzzles, it is not satisfactory to simply leave matters at this. Is there some more sophisticated sense in which background independence does hold?

In thinking about this question, it is natural to examine the all orders generalization of the holomorphic anomaly equation, which (in the final equation of their paper) Bershadsky et. al. write in the following form. Let F_g be the genus g free energy. Then

$$\overline{\partial}_{i'}F_g = \overline{C}_{i'j'k'}e^{2K}G^{jj'}G^{kk'}\left(D_jD_kF_{g-1}+\frac{1}{2}\sum_r D_jF_r\cdot D_kF_{g-r}\right). \qquad (1.4)$$

This equation can be written as a linear equation for $Z=\exp\left(\frac{1}{2}\sum_{g=0}^{\infty}\lambda^{2g-2}F_g\right)$, namely

$$\left(\overline{\partial}_{i'}-\lambda^2\overline{C}_{i'j'k'}e^{2K}G^{jj'}G^{kk'}D_jD_k\right)Z=0. \qquad (1.5)$$

This linear equation is called a master equation by Bershadsky et. al.; it is similar in structure to the heat equation obeyed by theta functions. (I am here using the notation of Bershadsky et. al., but later, we will make some changes in notation.)

It would be nice to interpret (1.5) as a statement of some sophisticated version of background independence. In thinking about this question, a natural analogy arises with Chern-Simons gauge theory in $2+1$ dimensions. In that theory, an initial value surface is a Riemann surface Σ. In the Hamiltonian formulation of the theory, one constructs a Hilbert space $\mathcal{H}$ upon quantization on Σ. $\mathcal{H}$ should be obtained by quantizing a certain classical phase space $\mathcal{W}$ (a moduli space of flat connections on Σ). Because the underlying Chern-Simons Lagrangian does not depend on a choice of metric, one would like to construct $\mathcal{H}$ in a natural, background independent way. In practice, however, quantization of $\mathcal{W}$ requires a choice of polarization, and there is no natural or background independent choice of polarization.

The best that one can do is to pick a complex structure J on Σ, whereupon $\mathcal{W}$ gets a complex structure. Then a Hilbert space $\mathcal{H}_J$ is constructed as a suitable space of holomorphic functions (really, sections of a line bundle) over $\mathcal{W}$. We denote such a function as $\psi(a^i;t'^a)$ where a^i are complex coordinates on $\mathcal{W}$ and t'^a are coordinates parametrizing the choice of J.

Now background independence does not hold in a naive sense; ψ cannot be independent of t'^a (given that it is to be holomorphic on $\mathcal{W}$ in a complex structure that depends on t'^a). But there is a more sophisticated sense in which background independence can be formulated [13,14]. The $\mathcal{H}_J$'s can be identified with each other (projectively) using a (projectively) flat connection over the space of space of J's. This connection ∇ is such that a covariantly constant wave function should have the following property: as J changes, ψ should change by a Bogoliubov transformation, representing the effects of a change in the representation used for the canonical commutation relations.

Using parallel transport by ∇ to identify the various $\mathcal{H}_J$'s, one can speak of "the" quantum Hilbert space, of which the $\mathcal{H}_J$'s are realizations determined by a J-dependent choice of representation of the canonical commutators. Background independence of $\psi(a^i; t'^a)$ should be interpreted to mean that the quantum state represented by ψ is independent of the t'^a, or equivalently that ψ is invariant under parallel transport by ∇. Concretely, this can be written as an equation

$$0 = \left(\frac{\partial}{\partial t'^a} - \frac{1}{4} \left(\frac{\partial J}{\partial t'^a} \omega^{-1} \right)^{ij} \frac{D}{Da^i} \frac{D}{Da^j} \right) \psi \tag{1.6}$$

that is analogous to the heat equation for theta functions. We will discuss this equation further in §2.

(1.5) and (1.6) have an obvious similarity. Our goal in the rest of this paper will be to make this similarity more precise, introducing an auxiliary quantum system with quantum Hilbert space $\mathcal{H}$ and interpreting (1.5) as the statement that the vector in $\mathcal{H}$ determined by the partition function Z is independent of a choice of polarization. For X a Calabi-Yau threefold, the phase space of the auxiliary system will be simply $\mathcal{W} = H^3(X, \mathbb{R})$. $\mathcal{W}$ has a natural symplectic structure given by the intersection pairing

$$\omega(\alpha, \beta) = \int_X \alpha \wedge \beta. \tag{1.7}$$

$\mathcal{W}$ has no natural complex structure, but every choice of complex structure on X determines a complex structure on $\mathcal{W}$ via the Hodge decomposition. Then it turns out (at least up to a c-number factor) that (1.5) can be interpreted to mean that the quantum state determined by the partition function Z is independent of the base-point.

Though this interpretation of the holomorphic anomaly is elegant, its rationale remains obscure. What really is the origin of the phase space $\mathcal{W}$, what is the significance of the Hilbert space $\mathcal{H}$, and why should it be possible to interpret the partition function as a J-independent vector in $\mathcal{H}$? In the case of Chern-Simons theory, these questions are answered by appealing to the underlying three dimensional Chern-Simons action and field theory, but in the present case, it is not clear where an answer would come from.

2. Quantum Background Independence

2.1. Quantization

Before plunging into our specific problem, I will first recall some generalities about quantization of linear spaces. (See the introduction to [14] for more detail.) We consider a linear space $\mathcal{W} \cong \mathbb{R}^{2n}$ with a constant symplectic structure, say $\omega = \frac{1}{2}\omega_{ij}dx^i dx^j$, with ω_{ij} a constant invertible matrix and the x^i linear coordinates on $\mathbb{R}^{2n}$. The symbol ω^{-1} will denote the matrix inverse to ω, obeying $\omega_{ij}\omega^{-1jk} = \delta_i{}^k$. To give the problem of quantization its most natural (but perhaps not entirely familiar) formulation [15–18], one begins by introducing a "prequantum line bundle"; this is a unitary line bundle $\mathcal{L}$ with a connection whose curvature is $-i\omega$. Up to isomorphism, there is only one such choice of $\mathcal{L}$. One can take $\mathcal{L}$ to be the trivial unitary line bundle, with a connection given by the covariant derivatives

$$\frac{D}{Dx^i} = \frac{\partial}{\partial x^i} + \frac{i}{2}\omega_{ij}x^j. \tag{2.1}$$

Then one can introduce the "prequantum Hilbert space" $\mathcal{H}_0$ which consists of $\mathbf{L}^2$ sections of $\mathcal{L}$.

A vector in $\mathcal{H}_0$ is represented by a function with a rather general dependence on all $2n$ coordinates x^i. The quantum Hilbert space is instead to be a comparatively "small" subspace of $\mathcal{H}_0$ consisting of functions that depend freely on only n of the coordinates. There is no natural way to choose which n coordinates are allowed; such a choice is called a choice of polarization.

We will consider a polarization defined by a choice of a complex structure J on $\mathcal{W}$ with the following properties:

(a) J is translation invariant, so it is defined by a constant matrix $J^i{}_j$ with $J^2 = -1$.

(b) The two-form ω is of type $(1,1)$ in the complex structure J. In components this means that $J^{i'}{}_i J^{j'}{}_j \omega_{i'j'} = \omega_{ij}$ or equivalently (using $J^2 = -1$)

$$J^{i'}{}_i\omega_{i'j} = -J^{j'}{}_j\omega_{ij'}. \tag{2.2}$$

Since the curvature of the prequantum line bundle $\mathcal{L}$ is proportional to ω, having ω be of type $(1,1)$ means that the $(0,2)$ part of the curvature vanishes, so that the connection on $\mathcal{L}$ endows it with a complex structure.

(c) J is positive in the sense that the metric g defined by $g(v,w) = \omega(v,Jw)$ is positive.

Given such a J, we define the quantum Hilbert space $\mathcal{H}_J$ to consist of vectors in $\mathcal{H}_0$ that are represented by functions that are holomorphic in the complex structure J.

$\mathcal{H}_J$ is a quantization of the symplectic manifold $\mathcal{W}$; we want to exhibit a flat connection over the parameter space of the J's that will enable us to identify the $\mathcal{H}_J$'s. Construction of such a connection enables one to speak of "the" quantum Hilbert space $\mathcal{H}$ which has realizations depending on the choice of J. Actually, the connection will only be projectively flat, so this will only work up to a scalar multiple.

To write down the connection, some notation is useful. First of all, one has projection operators

$$\frac{1}{2}(1 \mp iJ) \tag{2.3}$$

that project onto the $(1,0)$ and $(0,1)$ parts of a vector. It is convenient (as in [14]) to introduce a special notation for vectors that have been so projected. For any vector v^i, we write

$$\begin{aligned} v^{\underline{i}} &= \frac{1}{2}(1 - iJ)^i{}_j v^j \\ v^{\overline{i}} &= \frac{1}{2}(1 + iJ)^i{}_j v^j . \end{aligned} \tag{2.4}$$

Similarly for a one-form w_i, we write

$$\begin{aligned} w_{\underline{j}} &= \frac{1}{2}(1 - iJ)^i{}_j w_i \\ w_{\overline{j}} &= \frac{1}{2}(1 + iJ)^i{}_j w_i . \end{aligned} \tag{2.5}$$

For example, J itself has non-zero components $J^{\underline{i}}{}_{\underline{j}} = i\delta^{\underline{i}}{}_{\underline{j}}$, $J^{\overline{i}}{}_{\overline{j}} = -i\delta^{\overline{i}}{}_{\overline{j}}$ (which means that projections of $J^i{}_j$ and $\delta^i{}_j$ are proportional).

Let $\mathcal{M}$ be the space of J's obeying conditions (a), (b), and (c) above; it is a copy of the Siegel upper half plane. $\mathcal{M}$ has a natural complex structure, defined as follows. The condition $J^2 = -1$ implies that for δJ a first order variation of J, one must have $J \cdot \delta J + \delta J \cdot J = 0$. This means that the non-zero projections of δJ are $\delta J^{\underline{i}}{}_{\overline{j}}$ and $\delta J^{\overline{i}}{}_{\underline{j}}$. We give $\mathcal{M}$ a complex structure by declaring $\delta J^{\underline{i}}{}_{\overline{j}}$ to be of type $(1,0)$ and $\delta J^{\overline{i}}{}_{\underline{j}}$ to be of type $(0,1)$.

Over $\mathcal{M}$ we now introduce two Hilbert space bundles. One of them, say $\mathcal{H}^0$, is the trivial bundle $\mathcal{M} \times \mathcal{H}_0$ whose fiber is the fixed Hilbert space $\mathcal{H}_0$. (Recall that the definition of $\mathcal{H}_0$ was independent of J.) The second is the bundle, say $\mathcal{H}^Q$, whose fiber over a point $J \in \mathcal{M}$ is the Hilbert space $\mathcal{H}_J$. $\mathcal{H}^Q$ is a sub-bundle of $\mathcal{H}^0$; a section of $\mathcal{H}^0$ is an arbitrary function $\psi(x^i; J)$, while a section of $\mathcal{H}^Q$ is a $\psi(x^i; J)$ which for each given J is, as a function of the x^i, holomorphic in the complex structure defined by J:

$$\frac{D}{Dx^{\bar{i}}}\psi = 0. \tag{2.6}$$

(This equation has a dependence on J coming from the projection operators used in defining $x^{\bar{i}}$.)

A connection on $\mathcal{H}^0$ restricts to a connection on $\mathcal{H}^Q$ if and only if its commutator with $D_{\bar{i}}$ is a linear combination of the $D_{\bar{j}}$. For instance, since $\mathcal{H}^0$ is defined as the product bundle $\mathcal{M} \times \mathcal{H}_0$, there is a trivial connection δ on this bundle:

$$\delta = \sum_{i,j} \mathrm{d}J^i{}_j \frac{\partial}{\partial J^i{}_j}. \tag{2.7}$$

Thus, $\psi(x^i; J)$ is annihilated by δ if and only if it is independent of J. We can expand δ in $(1,0)$ and $(0,1)$ pieces, $\delta = \delta^{(1,0)} + \delta^{(0,1)}$, with

$$\begin{aligned} \delta^{(1,0)} &= \sum_{i,j} \mathrm{d}J^i{}_{\bar{j}} \frac{\partial}{\partial J^i{}_{\bar{j}}} \\ \delta^{(0,1)} &= \sum_{i,j} \mathrm{d}J^{\bar{i}}{}_j \frac{\partial}{\partial J^{\bar{i}}{}_j}. \end{aligned} \tag{2.8}$$

The commutation relation

$$[\delta, D_{\bar{i}}] = \left[\delta, \frac{1}{2}\left(D_i + iJ^k{}_i D_k\right)\right] = \frac{i}{2}\mathrm{d}J^k{}_i D_k \tag{2.9}$$

shows that $[\delta, D_{\bar{i}}]$ is not a linear combination of the $D_{\bar{j}}$ and hence that δ does not descend to a connection on $\mathcal{H}^Q$.

Rather, we must add an extra term that reflects the effects of the Bogoliubov transformation on the quantum state. The appropriate connection is actually

$$\begin{aligned}\nabla^{(1,0)} &= \delta^{(1,0)} - \frac{1}{4}\left(\mathrm{d}J\omega^{-1}\right)^{\underline{ij}} \frac{D}{Dx^{\underline{i}}}\frac{D}{Dx^{\underline{j}}}. \\ \nabla^{(0,1)} &= \delta^{(0,1)}.\end{aligned} \tag{2.10}$$

Indeed, the commutator $[\nabla^{(0,1)}, D_{\bar{i}}]$ is a linear combination of the $D_{\bar{j}}$, as one can see by using (2.9). The commutator $[\nabla^{(1,0)}, D_{\bar{i}}]$ vanishes. To see this, one uses, in addition to (2.9), the defining relation

$$[D_{\bar{i}}, D_{\underline{j}}] = -i\omega_{\bar{i}\underline{j}} \tag{2.11}$$

of the prequantum line bundle $\mathcal{L}$, and the relation

$$(\mathrm{d}J\omega^{-1})^{ij} = (\mathrm{d}J\omega^{-1})^{ji}, \tag{2.12}$$

which follows from differentiating (2.2).

So ∇ descends to a connection on $\mathcal{H}^Q$. Now let us compute the curvature of ∇. The $(0,2)$ part of the curvature vanishes trivially, since $\nabla^{(0,1)} = \delta^{(0,1)}$. The $(2,0)$ part of the curvature can be seen to vanish using $[D_{\underline{i}}, D_{\underline{j}}] = 0$ and also (2.9). Let us work out in detail the $(1,1)$ part of the curvature. This is

$$\left[\nabla^{(0,1)}, \nabla^{(1,0)}\right] = \left[\delta^{(0,1)}, -\frac{1}{4}\left(\mathrm{d}J\omega^{-1}\right)^{\underline{ij}} D_{\underline{i}}D_{\underline{j}}\right]. \tag{2.13}$$

The only J dependence that $\delta^{(0,1)}$ can act on is in the projection operators implicit in the definition of the indices $\underline{i}, \underline{j}$. So we make those projection operators explicit:

$$\left(\mathrm{d}J\omega^{-1}\right)^{\underline{ij}} D_{\underline{i}}D_{\underline{j}} = \left(\mathrm{d}J\omega^{-1}\right)^{ij}\frac{1}{2}(\delta^{i'}{}_i - iJ^{i'}{}_i)\frac{1}{2}(\delta^{j'}{}_j - iJ^{j'}{}_j)D_{i'}D_{j'}. \tag{2.14}$$

Inserting this in (2.13), we get

$$\left[\nabla^{(0,1)}, \nabla^{(1,0)}\right] = \frac{i}{8}\left(\mathrm{d}J\omega^{-1}\right)^{\underline{ij}} \delta^{(0,1)} J^{i'}{}_i (D_{i'}D_j + D_j D_{i'}). \tag{2.15}$$

To proceed further, we restrict ∇ and its curvature form to $\mathcal{H}^Q$. According to (2.8), the only non-zero components of $\delta^{(0,1)} J^{i'}{}_j$ are $\mathrm{d}J^{\bar{i}'}{}_{\underline{j}}$. The right hand side of (2.15) can therefore be simplified using the fact that $D_{\overline{i}'}$ annihilates sections of $\mathcal{H}^Q$ and using $[D_{\overline{i}'}, D_j] = -i\omega_{\overline{i}'j}$. One gets finally

$$\left[\nabla^{(0,1)}, \nabla^{(1,0)}\right] = -\frac{1}{8}\mathrm{d}J^i{}_{\overline{k}}\mathrm{d}J^{\overline{k}}{}_i. \tag{2.16}$$

Thus, the curvature is not zero, even when restricted to $\mathcal{H}^Q$. But it is a c-number, that is, it depends only on J and not on the variables x^i that are being quantized. It is possible to eliminate this central curvature by adding to ∇ a one-form that depends on J only or – to formulate this more invariantly – by tensoring $\mathcal{H}^Q$ by the pullback of a line bundle on $\mathcal{M}$ endowed with a connection whose curvature is minus that of ∇.

The fact that the curvature of ∇ is a c-number means that parallel transport by ∇ is unique up to a scalar factor (which moreover is of modulus 1 since the curvature is real or more fundamentally since ∇ is unitary). So up to this factor one can identify the various $\mathcal{H}_J$'s, and regard them as different realizations of "the" quantum Hilbert space $\mathcal{H}$.

2.2. Application To Calabi-Yau Manifolds

The symplectic manifold that we want to quantize is the linear space $\mathcal{W} = H^3(X, \mathbb{R})$, X being a Calabi-Yau threefold. On $\mathcal{W}$ there is a natural symplectic form,

$$\omega(\alpha, \beta) = \int_X \alpha \wedge \beta. \tag{2.17}$$

Every complex structure on X determines a complex structure on $\mathcal{W}$, which can be used to quantize $\mathcal{W}$. So we get a family of quantizations of $\mathcal{W}$, parametrized by the Teichmuller space $\mathcal{T}$ of complex structures on X up to isotopy. We will see that the natural connection on the family of quantum Hilbert spaces over $\mathcal{T}$ is the anomaly equation of Bershadsky et. al.

First, we recall some facts about variation of Hodge structures on X. (A convenient reference is [19].) The usual complex structure on $\mathcal{T}$ is the one in which the $(1,0)$ part of a variation δJ of the complex structure of X is $\delta J^i{}_{\overline{j}}$ – or differently put, in which the holomorphic tangent space to $\mathcal{T}$ is the $\overline{\partial}$ cohomology group $H^1(X, T^{1,0}X)$. Let Ω be a holomorphic three-form on X that varies holomorphically in t. A basis of the complexification of $\mathcal{W}$ is given by

$$\begin{aligned} V_0 &= \Omega \\ V_a &= \frac{\partial \Omega}{\partial t^a} \\ \overline{V}_a &= \frac{\partial \overline{\Omega}}{\partial \overline{t}^a} \\ \overline{V}_0 &= \overline{\Omega}. \end{aligned} \tag{2.18}$$

In practice, we will be working near some base-point $t \in \mathcal{T}$. One can normalize Ω so that, at t, the V_a are of type $(2,1)$ (and hence the $\overline{V}_b$ of type $(1,2)$). If this is done, then ω becomes block diagonal, the non-zero matrix elements being $\omega(V_0, \overline{V}_0)$ and $\omega(V_a, \overline{V}_b)$. The latter are related to the natural metric $g_{\underline{a}\overline{b}}$ on $H^{2,1}(X)$:

$$\omega(V_a, \overline{V}_b) = \int_X \frac{\partial \Omega}{\partial t^a} \wedge \frac{\partial \overline{\Omega}}{\partial \overline{t}^b} = i g_{\underline{a}\overline{b}}. \tag{2.19}$$

We will use on $\mathcal{T}$ a similar notation to that which we used on $\mathcal{M}$ – the $(1,0)$ and $(0,1)$ projections of a vector v^a will be written as $v^{\underline{a}}$, $v^{\overline{a}}$.

The metric $g_{\underline{a}\overline{b}}$ in (2.19) is closely related to the Zamolodchikov metric $G_{\underline{a}\overline{b}}$ on the space $H^1(X, T^{1,0}X)$ of physical states. In fact, there is a natural map from $H^1(X, T^{1,0}X)$ to $H^{2,1}(X)$ by contracting with a holomorphic three-form; G and g are related by this map, so $g_{\underline{a}\overline{b}} = G_{\underline{a}\overline{b}} e^{-K}$, where e^{-K} is the natural metric on the space of holomorphic three-forms.

The Yukawa couplings are

$$C_{\underline{a}\underline{b}\underline{c}} = -\int_X \Omega \wedge \frac{\partial^3 \Omega}{\partial t^a \partial t^b \partial t^c} = \int_X \frac{\partial \Omega}{\partial t^a} \wedge \frac{\partial^2 \Omega}{\partial t^b \partial t^c} \tag{2.20}$$

(and other projections of C vanish). The second derivative $\partial^2 \Omega / \partial t^a \partial t^b$ is a linear combination of forms of type $(3,0)$, $(2,1)$, and $(1,2)$. A non-vanishing contribution on the right hand side of (2.20) comes only from the $(1,2)$ part, which is necessarily a linear combination of the $(1,2)$ forms $\partial \overline{\Omega} / \partial \overline{t}^b$. Comparing coefficients, we find

$$\frac{\partial^2 \Omega}{\partial t^a \partial t^b} = -i C_{\underline{a}\underline{b}\underline{c}} g^{\underline{c}\overline{c}} \frac{\partial \overline{\Omega}}{\partial \overline{t}^c} \quad \text{mod } \mathrm{d}(\ldots) \oplus H^{(2,1)} \oplus H^{(3,0)}. \tag{2.21}$$

We still need a few more formulas. First of all, since Ω is of type $(3,0)$ and J

acts on an index of type $(1,0)$ or $(0,1)$ as multiplication by i or $-i$, one has

$$J^{i'}{}_i J^{j'}{}_j J^{k'}{}_k \Omega_{i'j'k'} = -i\Omega_{ijk}. \tag{2.22}$$

Differentiating this and using the fact that $\partial_a\Omega$ is of type $(2,1)$, we get

$$\frac{\partial J^{i'}{}_i}{\partial t^a}\Omega_{i'jk} + \text{cyclic permutations of } ijk = 2i\frac{\partial\Omega_{ijk}}{\partial t^a}. \tag{2.23}$$

Similarly, using the fact that that $\partial_{\overline{b}}\Omega$ is of type $(2,1)$, we have

$$J^{i'}{}_i J^{j'}{}_j J^{k'}{}_k \partial_{\overline{b}}\Omega_{i'j'k'} = i\partial_{\overline{b}}\Omega_{ijk}. \tag{2.24}$$

Differentiating this, we get

$$\frac{\partial J^{i'}{}_i}{\partial t^a}\frac{\partial\Omega_{i'jk}}{\partial t^b} + \text{cyclic permutations of } ijk = 2i\frac{\partial^2\Omega_{ijk}}{\partial t^a\partial t^b} \quad \text{mod } H^{2,1}\oplus H^{3,0}. \tag{2.25}$$

Combining this with (2.21), we have

$$\frac{\partial J^{i'}{}_i}{\partial t^a}\frac{\partial\Omega_{i'jk}}{\partial t^b} + \text{cyclic permutations} = 2C_{\underline{a}\underline{b}\underline{c}}g^{\underline{c}\overline{c}}\frac{\partial\overline{\Omega}}{\partial\overline{t}^c} \text{mod } d(\ldots)\oplus H^{2,1}\oplus H^{3,0}. \tag{2.26}$$

The complex conjugate of this formula asserts that

$$\frac{\partial J^{i'}{}_i}{\partial\overline{t}^a}\frac{\partial\overline{\Omega}_{i'jk}}{\partial\overline{t}^b} + \text{cyclic permutations} = 2\overline{C}_{\overline{a}\overline{b}\overline{c}}g^{c\overline{c}}\frac{\partial\Omega}{\partial t^c} \text{mod } d(\ldots)\oplus H^{1,2}\oplus H^{0,3}. \tag{2.27}$$

2.3. Complex Structure Of $\mathcal{W}$

We now must study the complex structure of $\mathcal{W}$. An element Θ of $\mathcal{W} = H^3(X,\mathbb{R})$ has an expansion

$$\Theta = \lambda^{-1}\Omega + \sum_a u^a\frac{\partial\Omega}{\partial t^a} + \sum_a \overline{u}^a\frac{\partial\overline{\Omega}}{\partial\overline{t}^a} + \overline{\lambda}^{-1}\overline{\Omega} \tag{2.28}$$

in the basis (2.18), with complex coefficients λ, u^a. The object λ appearing here will turn out to be the string coupling constant.

We now have to make explicit the complex structure of $\mathcal{W}$. $\mathcal{W}$ has no natural complex structure, but every choice of complex structure J on X enables one to pick a complex structure, which I will call $\mathcal{J}$, on $\mathcal{W} = H^3(X, \mathbb{R})$. We choose $\mathcal{J}$ to be the complex structure on $H^3(X, \mathbb{R})$ in which $H^{3,0}$ and $H^{2,1}$ are considered to be of type $(1,0)$. Hence, $\mathcal{J}$ should be represented by an operator on differential forms that multiplies $(3,0)$ and $(2,1)$ forms by i, and multiplies $(0,3)$ and $(1,2)$ forms by $-i$. Such an operator is

$$(\mathcal{J}\Theta)_{ijk} = \frac{1}{2} J^{i'}{}_i J^{j'}{}_j J^{k'}{}_k \Theta_{i'j'k'} + \frac{1}{2}\left(J^{i'}{}_i \Theta_{i'jk} + \text{cyclic permutations of } ijk\right). \tag{2.29}$$

We now need to compute the $\bar{t}$ dependence of $\mathcal{J}$. As in §1, we will use the notation t' for $\bar{t}$. The computation of t' dependence is straightforward:

$$\left(\frac{\partial \mathcal{J}}{\partial t'^a}\Theta\right)_{ijk} = \frac{1}{2}\left(\frac{\partial J^{i'}{}_i}{\partial t'^a} J^{j'}{}_j J^{k'}{}_k \Theta_{i'j'k'} + \frac{\partial J^{i'}{}_i}{\partial t'^a}\Theta_{i'jk}\right) + \text{cyclic permutations of } ijk. \tag{2.30}$$

In particular, if Θ is of type $(1,2)$, this implies

$$\left(\frac{\partial \mathcal{J}}{\partial t'^a}\Theta\right)_{ijk} = \left(\frac{\partial J^{i'}{}_i}{\partial t'^a}\Theta_{i'jk} + \text{cyclic permutations}\right) \mod H^{1,2} \oplus H^{0,3}. \tag{2.31}$$

Taking $\Theta = \partial\overline{\Omega}/\partial t'^b$, and using (2.27), we learn that

$$\frac{\partial \mathcal{J}}{\partial t'^a}\left(\frac{\partial \overline{\Omega}}{\partial t'^b}\right) = 2\overline{C}_{\bar{a}\bar{b}\bar{c}} g^{c\bar{c}} \frac{\partial \Omega}{\partial t^c} \mod \mathrm{d}(\ldots) \oplus H^{1,2} \oplus H^{0,3}. \tag{2.32}$$

2.4. Final Evaluation

To make completely explicit the connection (2.10) for quantization of $\mathcal{W}$, with a family of polarizations parametrized by $\mathcal{T}$, we need to evaluate $(\mathrm{d}\mathcal{J}\omega^{-1})^{\underline{i}\underline{j}}$. The only non-zero matrix element (given that $\underline{i}$ and $\underline{j}$ are to be indices of type $(1,0)$, corresponding to $(3,0)$ or $(2,1)$ forms) comes from (2.32) together with

$$\omega^{-1\bar{b}\underline{a}} = -ig^{\bar{b}\underline{a}}, \tag{2.33}$$

which is equivalent to (2.19). Combining these pieces, we get

$$(\mathrm{d}\mathcal{J}\omega^{-1})^{\underline{i}\underline{j}} D_{\underline{i}} D_{\underline{j}} = -2i \sum_a \mathrm{d}t'^a \overline{C}_{\bar{a}\bar{b}\bar{c}} g^{b\bar{b}} g^{c\bar{c}} \frac{D}{Du^b}\frac{D}{Du^c}. \tag{2.34}$$

The condition that the quantum state represented by a vector $\Psi(\lambda, u; t')$ is independent of t' can be read off from (2.10) and is

$$\left(\frac{\partial}{\partial t'^a} + \frac{i}{2}\overline{C}_{\bar{a}\bar{b}\bar{c}} g^{b\bar{b}} g^{c\bar{c}} \frac{D}{Du^b}\frac{D}{Du^c}\right)\Psi = 0$$
$$\frac{\partial}{\partial \overline{t}'^a}\Psi = 0. \tag{2.35}$$

The main point of this paper is that the first equation in (2.35) practically coincides with the holomorphic anomaly equation (1.5) of Bershadsky et. al.* A factor of $2i$ presumably results from a difference in conventions (for instance, as will be clear momentarily, it can be absorbed in the definition of the string coupling constant). Let us analyze the remaining discrepancies.

First of all, Bershadsky et. al. consider the partition function as a function of the string coupling constant and also the complex structure of X. In quantizing $\mathcal{W}$, our wave functions depend on the variables λ and u introduced in (2.28). λ determines a holomorphic three-form, and in the B model this means that λ should be associated with the string coupling constant. On the other hand, u is a $(2,1)$ cohomology class. To Bershadsky et. al., the natural variables would be the string coupling constant λ and an element t of $H^1(X, T^{1,0}X)$.† In a familiar fashion, one can map $H^1(X, T^{1,0}X)$ to $H^{2,1}(X)$ by contracting with a holomorphic three-form, which in this case should be naturally the one determined by λ. This means that the natural relation between u and t is $t = \lambda u$. In terms of t, the first equation in (2.35) would therefore be

$$\left(\frac{\partial}{\partial t'^a} + \frac{i}{2}\lambda^2\overline{C}_{\bar{a}\bar{b}\bar{c}} g^{b\bar{b}} g^{c\bar{c}} \frac{D}{Dt^b}\frac{D}{Dt^c}\right)\Psi = 0. \tag{2.36}$$

Now we see the natural appearance of the string coupling constant, as in (1.5).

Another important point is that the connection (2.10) for quantization of an affine space is only projectively flat. Therefore, in attempting to study the t' dependence of a wave function Ψ using (2.35), there would be an undetermined factor of modulus unity – a c-number factor in the sense that it depends only on t' and not on t. Of course, this comes from the fact that Ψ is really a section of the prequantum line bundle rather than a function. Perhaps a trivialization of this line bundle is implicit in [10].

⋆ The second equation in (2.35) is also true in their formalism. They consider t and $\bar{t}$ as independent complex variables and consider functions that are holomorphic in $t, \bar{t}$. Given that their $t, \bar{t}$ correspond to our u, t', holomorphicity in $\bar{t}$ is the second equation in (2.35) and holomorphicity in t is (2.6).

† Recall from §1 that once a base-point is picked, the Teichmuller space of X has a natural local isomorphism with the linear space $H^1(X, T^{1,0}X)$ via special coordinates.

In the definition of the partition function $Z = \exp(\frac{1}{2}\sum_{g=0}^{\infty}\lambda^{2g-2}F_g)$, the genus one term F_1 has the following characteristics: (i) the power of λ multiplying it vanishes; (ii) because of zero mode contributions (analogous to those in Ray-Singer analytic torsion) it is not most naturally interpreted as a "function" but as a section of a certain line bundle. A better understanding of F_1 might clarify the role of the prequantum line bundle.

Despite unresolved questions, the resemblance of the holomorphic anomaly equation to the equation (2.36) of quantum background independence is so close that it is hard to believe that it is a coincidence. This relation seems likely to repay further study.

2.5. A Speculation

For instance, it is amusing to speculate along the following lines. Perhaps the phase space $\mathcal{W}$ that has appeared in this discussion should be interpreted as part of the usual classical phase space of a system, and the construction of the auxiliary Hilbert space $\mathcal{H}$ should be regarded as part of the process of quantization. Then what is unusual is that – as opposed to the usual situation in which the choice of a vector in $\mathcal{H}$ is a choice of initial conditions – the physical system here determines a distinguished vector in $\mathcal{H}$, namely the partition function.

Before identifying this as a cosmologist's dream, in which the initial conditions of the universe are uniquely determined by fundamental theory, we should pause to note that in physical applications of string theory, background independence is realized "normally" and one probably does not want to abandon that since the realization of background independence in general relativity is "normal." "Quantum" background independence as we have encountered it in this paper apparently depends on having a non-trivial cohomology of the $b_0, \bar{b}_0$ operators (which leads to considerations of $t - \bar{t}$ fusion and the holomorphic anomaly); this is absent in the usual critical string theories. Though one probably would not want "quantum" background independence for transverse gravitons, perhaps there is a modification of the usual string theories in which the BRST cohomology is such that "quantum" background independence holds for the conformal factor in the space-time metric. Then – blindly imitating what we have found above – quantum background independence might dictate that the dependence of the wave function on the conformal factor should be given by the partition function. It has been argued [20] that under such conditions the cosmological constant would vanish, since under some hypotheses the partition function of the universe diverges at zero cosmological constant.

In any event, though "ordinary" background independence appears to suffice (apart from such exotic speculations) for the usual physical applications of string theory, familiarity with "quantum" background independence may be useful in trying to go off-shell. I am reminded of the BV formalism of quantization, which enters on-shell only in exotic string theories in $D \leq 2$ [21,22], but seems to be very valuable in formulating critical string theories off-shell, even at the classical level [23].

Acknowledgements: I would like to thank P. Aspinwall, M. Bershadsky, H. Ooguri, D. Morrison, and C. Vafa for discussions and helpful suggestions.

REFERENCES

1. P. Candelas, X.C. de la Ossa, P. S. Green, and L. Parkes, "A Pair Of Calabi-Yau Manifolds As An Exactly Soluble Superconformal Theory," Nucl. Phys. **B359** (1991) 21.
2. R. Bryant and P. Griffiths, in *Arithmetic and Geometry*, vol. II. p. 77 (M. Artin and J. Tate, eds.), Birkhauser, Boston (1983).
3. D. Morrison, "Mirror Symmetry And Rational Curves On Quintic Threefolds: A Guide For Mathematicians," J. Am. Math. Soc. **6** (1993) 223.
4. P. Aspinwall, B. Greene, and P. Morrison, "Multiple Mirror Manifolds And Topology Change In String Theory," IASSNS-HEP-93/4.
5. E. Witten, "Chern-Simons Gauge Theory As A String Theory," IASSNS-HEP-92/45, to appear in the Floer Memorial Volume.
6. K. Li, "Topological Gravity With Minimal Matter," Nucl. Phys. **B354** (1991) 711, "Recursion Relations In Topological Gravity With Minimal Matter," Nucl. Phys. **B354** (1991) 725.
7. M. Bershadsky, W. Lerche, D. Nemeschansky, and N. P. Warner, "Extended $N = 2$ Superconformal Structure Of Gravity and W-Gravity Coupled to Matter," CALT-68-1832.
8. S. Mukhi and C. Vafa, "Two Dimensional Black Hole As A Topological Coset Of $c = 1$ String Theory," HUTP-93/A002.
9. S. Panda and S. Roy, "On The Twisted $N = 2$ Superconformal Algebra In 2D Gravity Coupled To Matter," IC/93/81.
10. M. Bershadsky, S. Cecotti, H. Ooguri, and C. Vafa, "Holomorphic Anomalies In Topological Field Theories," HUTP-93/A008,RIMS-915.
11. S. Cecotti, P. Fendley, K. Intrilligator, and C. Vafa, "A New Supersymmetric Index," Nucl. Phys. **B386** (1992) 405.
12. S. Cecotti and C. Vafa, "Ising Model And N=2 Supersymmetric Theories," HUTP-92/A044, SISSA-167/92/EP (1992).
13. S. Elitzur, G. Moore, A. Schwimmer, and N. Seiberg, "Remarks On The Canonical Quantization Of The Chern-Simons-Witten Theory," Nucl. Phys. **B326** (1989) 108.

14. S. Axelrod, S. DellaPietra, and E. Witten, "Geometric Quantization Of Chern-Simons Gauge Theory," J. Diff. Geom. **33** (1991) 787.

15. B. Kostant, *Orbits, Symplectic Geometry, and Representation Theory,* Proc. U.S.-Japan Seminar on Differential Geometry (Kyoto, 1965), *Quantization And Unitary Representations*, Lecture Notes In Mathematics, Vol. 170 (Springer, Berlin, 1970), "Line Bundles And The Prequantized Schrodinger Equation," Coll. Group Theoretical Methods in Physics, Marseille (1972) p. 81.

16. J.-M. Souriau, *Quantification Geometrique*, Comm. Math. Phys. **1** (1966) 374, *Structures Des Systems Dynamiques* (Dunod, Paris, 1970).

17. N. Woodhouse, *Geometric Quantization* (Oxford Univ. Press, Oxford, 1980).

18. J. Sniatycki, *Geometric Quantization And Quantum Mechanics* (Springer, New York, 1980).

19. P. Candelas and X. C. de la Ossa, "Moduli Space Of Calabi-Yau Manifolds," Nucl. Phys. **B355** (1991) 455.

20. S. Hawking, "The Cosmological Constant Is Probably Zero," Phys. Lett. **13B** (1984) 403.

21. E. Witten and B. Zwiebach, "Algebraic Structures And Differential Geometry In 2-D String Theory," Nucl. Phys. **B377** (1992) 55.

22. E. Verlinde, "The Master Equation of 2-D String Theory," Nucl. Phys. **B381** (1992) 141.

23. B. Zwiebach, "Closed String Field Theory: Quantum Action And The B-V Master Equation," Nucl. Phys. **B 390** (1993) 33.

POINT-LIKE SUBSTRUCTURE IN STRING THEORY

Michael B. Green
DAMTP, Silver Street, Cambridge CB3 9EW, UK

ABSTRACT

This talk reviews a class of string theories which possess point-like structure. These theories, which are defined in terms of a sum over world-sheets with boundaries on which coordinates satisfy fixed Dirichlet conditions, possess power-behaved high energy fixed-angle scattering amplitudes. Furthermore, the high-temperature behaviour of the free energy also indicates point-like short-distance properties. The effect of point-like structure on the low-energy spectrum is also discussed.

It is a privilege to be talking at this meeting in honour of Abdus Salam, a man greatly admired for his achievements not only in several key areas of theoretical physics but also as a tireless campaigner on behalf of science in the developing nations.

The origins of string theory lie in attempts to explain experimental data concerning the strong interactions in the late 1960's, before the advent of QCD. Although QCD provides a very compelling explanation of the short-distance properties of hadrons it is notoriously difficult to describe their large-scale features in terms of point-like constituent quarks and gluons despite valiant attempts based on large-scale numerical computations. Such features are intrinsically very complicated when described in terms of confined constituents and it might be that an alternative set of variables are more useful. Indeed, strings (colour electric flux tubes) arise as the natural degrees of freedom in the lattice formulation of QCD and the property of confinement suggests that this description should survive the continuum limit.

Although a string theory that is equivalent to Yang–Mills with an $SU(3)$ colour group might be very complicated, major simplifications occur if the group is taken to be $U(n)$ in the limit $n \to \infty$ with fixed $g_{eff}^2 \equiv g^2 n$ (where g is the Yang–Mills coupling constant[1)]). In this limit the perturbation expansion decomposes into a power series in n^{2-2h}, where h is the minimum genus of the surface on which a given Feynman diagram can be drawn. This topological expansion is reminiscent of the genus expansion in conventional perturbative string theory. Assuming that the large-n theory is equivalent to some kind of string theory, closed strings would correspond to glueballs while open strings (carrying flavour quantum numbers at their end-points) would correspond to mesons (baryons are solitonic states of large-n QCD[2)]).

⋆ Email: M.B.Green@strings.amtp.cam.ac.uk

A crucial feature of large-n QCD is that the sum of the leading diagrams in the $1/n$ expansion (diagrams with spherical topology) are known to exhibit aymptotic freedom characteristic of point-like short-distance behaviour (even though it is not possible to evaluate even the leading terms of four-dimensional QCD explicitly†). These diagrams correspond to *tree-level* string theory amplitudes (possessing poles and no cuts) which should therefore have point-like high-energy fixed-angle scattering behaviour, characterized by power decrease with energy at fixed angle. This is quite different from the tree amplitudes of conventional ('gravitational') string theory which decrease exponentially with energy at fixed angle.

Another contrast between the short-distance behaviour of QCD and gravitational string theory is provided by comparing the temperature dependence in the two cases. The Hagedorn temperature in string theory seems to be analogous to the deconfining temperature in Yang–Mills theory. However, the very high temperature behaviour of the confining phase of large-n $U(n)$ Yang–Mills theory (as deduced by Polchinski [5)]) is quite dissimilar from the behaviour of gravitational string theory.

In addition to possessing point-like substructure, a string theory of QCD must obviously not include a massless spin-2 state (a graviton) or a massless scalar state (a dilaton) in its spectrum – states that are so beloved of string theories of gravity. According to folklore the leading glueball (closed-string) states of QCD lie on the pomeron Regge trajectory which has intercept approximately equal to one. It is tempting to think that the absence of point-like structure in gravitational string theories is directly related to the presence of the graviton in the spectrum while the presence of hard partons in a putative QCD string is directly related to the absence of a spin-2 massless in that case.

If any of the issues to be discussed below have any connection at all with QCD they are concerned with the glueball sector, which arises in Yang–Mills theory in the absence of quarks. There are other necessary properties of a string theory of QCD that arise from the properties of the quarks that will not be discussed further here. For example, a QCD string theory should provide a natural coupling to external currents due to electro-weak sources. Furthermore, meson (and baryon) amplitudes should exhibit spontaneously broken chiral symmetry, with an accompanying massless Goldstone mode (the pion).

Dirichlet string theory

In conventional open-string perturbation theory a stringy Feynman diagram of a given order in the string coupling constant is expressed as a sum over Riemann surfaces with a given number of handles and boundaries‡ on which the coordinates $X^\mu(\sigma,\tau)$ satisfy Neumann boundary conditions, *i.e.*, $\partial_n X^\mu|_B = 0$ (where n denotes the normal at the boundary labelled B and $\mu = 0, 1 \ldots, D-1$ where the space-time dimension will often be set to its critical value, $D = 26$, below). Each boundary represents the trajectory of a free open-string end-point travelling at the speed of light and the theory describes the dynamics of both open-ended and closed strings interacting with each other.

† Two-dimensional QCD with quarks was solved by 't Hooft [3)] and recently pure Yang–Mills theory has been shown to be equivalent to a string theory in two dimensions of arbitrary topology [4)]).

‡ Non-orientable theories, which also involve cross-caps, will not be considered here.

An alternative theory (refs. [6,7] and references therein) is obtained by imposing fixed Dirichlet boundary conditions, $\partial_t X^\mu|_B = 0$ (where t denotes the tangent at the surface), on each boundary. In this case an entire boundary, B, is mapped into a particular space-time point, y^μ_B, which is to be integrated in the complete amplitude. Summing over all possible insertions of boundaries now leads to a modification of the purely closed-string theory (there are no physical open-string states). A theory combining both types of boundaries can also be considered[8)]. In this case the presence of Dirichlet boundaries modifies the conventional open-string theory. In theories with orientable world-sheets the Chan–Paton factor associated with a Neumann boundary is $U(n)$ (the 'flavour' group) which may be a distinct group from the $U(m)$ (or 'flavour' group) associated with a Dirichlet boundary. I will now review some properties of this kind of string theory.

(1). String theory with Dirichlet boundary conditions has very different short-distance properties from those associated with normal string theory. The fact that Dirichlet boundary conditions induce point-like short-distance properties can be deduced from the high energy fixed-angle behaviour of elastic scattering amplitudes. For example, the amplitude with four external massless spin-2 closed-string ground states (with polarization vectors $\zeta^{\mu\nu}$) on a world-sheet with the toplogy of a disc with a Dirichlet boundary has a term that behaves as $\zeta_1^{\mu\nu}\zeta_{2\nu\rho}\zeta_3^{\rho\omega}\zeta_{4\omega\mu}/(stu)^{1/2}$ at fixed angle as $s, -t, -u \to \infty$ (terms with other contractions of polarization vectors have divergences that will be described later). This contrasts with the exponential decrease characteristic of conventional string theory. In the theory with both Neumann and Dirichlet boundaries the lowest-order point-like contribution to the scattering of open strings comes from a world-sheet with the topology of an annulus with one Neumann and one Dirichlet boundary. With four external open-string ground states attached to the Neumann boundary the amplitude behaves as $(-s)^{-5/2}(-t)^{-5/2} + perms.$

The origin of this power behaviour can be seen most clearly by considering the process in a light-cone parametrization in which $\tau = X^+$ and the $+$ component of the momentum density on the string is constant, $P^+(\sigma) = 2\pi$ (where the $\pm$ components of any vector, V^μ, are defined by $V^\pm = (V^0 \pm V^{D-1})/\sqrt{2}$). A scattering process is described by a Mandelstam diagram[9)] in which the $\tau - \sigma$ plane is parameterized so that any string labelled i is represented by a horizontal strip of width p_i^+ in the σ direction (equal to the total $+$ momentum component carried by the string) and the edges of the strip are identified thereby imposing closed-string boundary conditions. The interactions are represented by points at which the strips join and split and the total width in σ of the diagram is fixed since p^+ is conserved through the diagram. Upon integrating over all possible positions for the joining and splitting points such a diagram represents a tree-level contribution to the amplitude given by the sum over string histories (all world-sheets) that join the incoming string states (at $\tau = -\infty$) to the outgoing states (at $\tau = \infty$).

Inserting a conventional Neumann boundary into this world-sheet gives a quantum correction to the tree diagram. The boundary is represented by a horizontal slit along which the transverse coordinates satisfy $\partial_\sigma X^i(\sigma,\tau) = 0$. The position of the slit in τ and σ and its length are to be integrated (these are moduli associated with the boundary insertion). It is easy to see that intermediate physical open-string states arise from the upper limit of the integral over the length.

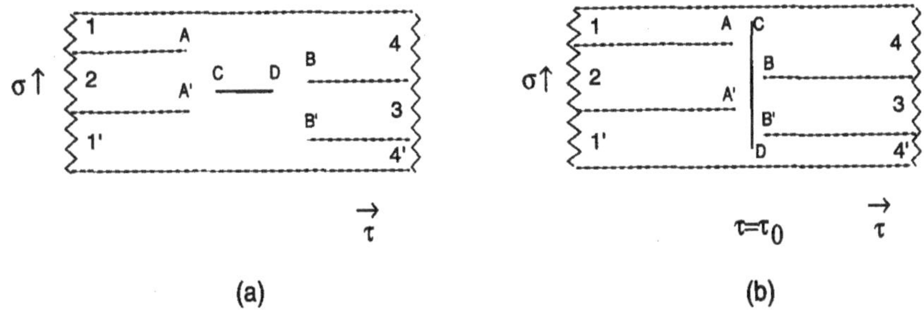

Fig. 1: (a) A world-sheet with a Neumann boundary correction to the amplitude for elastic scattering of closed strings in the light-cone parametrization in which τ is identified with the light-cone 'time', X^+. The width of any string is equal to its value of p^+. The incoming strings join at the interaction point A and the final strings are produced at the point B (the primed letters indicating points that are identified with the unprimed ones by the closed-string boundary condition that identifies the dotted boundaries on opposite sides of each strip). The horizontal slit between C and D represents the Neumann boundary and gives rise to an intermediate open-string state. (b) The same process as in (a) but with Dirichlet boundary conditions. The Dirichlet boundary is represented by the vertical slit at $\tau = \tau_0$ on which $X^i(\sigma, \tau_0) = y^i$ (where $i = 1, \ldots, D-2$ labels the transverse directions). In both cases the position and length of the slit (as well as the positions of the interaction points) are integrated in the expression for the amplitude.

A Dirichlet boundary is represented in this frame by a vertical slit in the world-sheet (*i.e.*, a slit at a fixed value of $\tau_0 = X_0^+$) along which the value of the transverse coordinates, $X^i(\sigma) = y^i$, is constant (which is then to be integrated). The two-dimensional position of the boundary in the string diagram as well as its length are also to be integrated. In the presence of this boundary the sum over histories that defines the amplitude is projected onto the subset of histories in which a finite fraction of the total P^+ carried by the string is at a single space-time point at $\tau = \tau_0$ (where the fraction and the space-time position are integrated). This is a subset of zero measure compared to the histories that contribute to scattering in the absence of Dirichlet boundaries. High-energy fixed-angle scattering is then dominated by configurations in which a finite fraction of the energy carried by each incoming string accumulates at a point at the collision time and these point-like densities scatter before dissipating in the outgoing strings[8)].

These fixed power contributions are modified by higher order effects due to the insertion of more boundaries which generate cuts in position space that give corrections to pure power behaviour in the deep euclidean region, which is a little reminiscent of conventional field theory. These corrections are logarithmic in four dimensions.

(2). Additional evidence of interesting point-like structure in the Dirichlet theory comes by considering its temperature dependence[10)]. Polchinski [5)] has argued that the analytic continuation to asymptotic temperatures ($\beta \to 0$) of the confining phase of large-n $U(n)$ Yang–Mills theory is quite different from the behaviour of any conventional string theory. In particular, he showed that the correlation function of two

Wilson–Polyakov loops (Wilson loops with unit winding number in the compact imaginary time direction*) at large spatial separation is dominated by the exchange of a glueball (closed-string) state with temperature-dependent mass $M_G(\beta)$ given by

$$M_G(\beta)^2 \sim -\frac{g^2 n}{\beta^2} \tag{1}$$

in the limit $\beta \to 0$. This is tachyonic because the confining phase is unstable at high temperature. In usual closed-string theory the lowest-mass closed-string state that contributes to the correlation function (a state with unit winding number) has a finite mass in the small β limit, which is very different from (1). However, the insertion of Dirichlet boundaries leads to a substantial renormalization of the small-β closed-string mass in accord with (1).

(3). Another striking feature of the zero-temperature Dirichlet theory is that the mass renormalization induced by the insertion of point-like boundaries is singular in certain processes. Such singularities indicate the presence of certain non-perturbative constraints that lead to a substantial shift in the bare closed-string spectrum that has not so far been fully elucidated[11)].

(4). Neumann and Dirichlet theories are related by duality of the target space. Target-space duality is a well-known discrete symmetry of closed-string theory that is simply illustrated by the example of a closed string moving in flat space-time with compact dimensions describing a torus of scale R, in which case the momentum modes are discrete and the string possesses winding numbers in the compact directions. Duality relates this closed-string theory to a closed-string theory in a space with scale $R' = \alpha'/R$ and with an interchange of the rôles of the discrete momenta and winding numbers. This duality transformation can be viewed as the transformation that changes the world-sheet coordinates X^μ into Y^μ, where

$$\partial_\alpha X^\mu = \epsilon_{\alpha\beta}\partial^\beta Y^\mu. \tag{2}$$

In theories in which the (orientable) world-sheets have boundaries (2) transforms a Neumann condition into a Dirichlet condition so that duality relates the usual Neumann theory at scale R to a Dirichlet theory at scale α'/R. Explicitly, the mode expansion for the open-string coordinates on a strip (with $0 \le \sigma \le \pi$) is

$$X^\mu = x^\mu + 2\alpha' p^\mu \tau + i\sqrt{2\alpha'}\sum_{n\neq 0}\frac{1}{n}\alpha_n^\mu \cos n\sigma e^{-in\tau}, \tag{3}$$

where $p^\mu = m^\mu/R^\mu + (y_2^\mu - y_1^\mu)/2\alpha'\pi$, while[7)]

$$Y^\mu = y_1^\mu + \frac{y^\mu}{\pi}\sigma + \sqrt{2\alpha'}\sum_{n\neq 0}\frac{1}{n}\alpha_n^\mu \sin n\sigma e^{-in\tau}, \tag{4}$$

where $y^\mu = 2\pi m^\mu R'^\mu + y_2^\mu - y_1^\mu$ and $y_2^\mu - y_1^\mu \le 2\pi R'^\mu$. The discrete momentum of the Neumann theory (labelled by the integers m^μ) is replaced in the Dirichlet theory by the

* The time direction is treated as imaginary and compactified on a circle of circumference $\beta = 1/kT$ so as to describe finite temperature.

winding number of an open string with fixed end-points. The y^μ_B are the positions of the end-points in the Dirichlet theory while they are identified in the Neumann theory with constant modes of gauge potentials coupling to the string end-points in the compact dimensions. In the $R \to \infty$ ($R' \to 0$) limit $y_2^\mu - y_1^\mu$ vanishes and $p^\mu = m^\mu / R^\mu$ becomes the continuous Minkowski-space momentum of the Neumann theory. In the $R \to 0$ ($R' \to \infty$) limit the only terms that survive are those with $m = 0$, so that $y^\mu = y_2^\mu - y_1^\mu$ and y^μ_B spans the D-dimensional dual Minkowski space-time. From here on D will be restricted to the critical dimension ($D = 26$) and only the $R \to \infty$ and $R \to 0$ theories will be discussed.

(5). World-sheets with boundaries may be assembled from open-string propagators joining three-string vertices (as, for example, in the framework of string field theory). The propagator has the form

$$\Delta = \frac{1}{L_0 - 1}, \tag{5}$$

where the zero-mode Virasoro generator, L_0, is given in the Neumann theory by

$$L_0^N = \alpha' p^2 + N, \tag{6}$$

where $N = \sum_{n \neq 0} (\alpha_{-n} \cdot \alpha_n + n c_{-n} b_n + n c_n b_{-n})$ and the fermionic ghost modes, b_n and c_n satisfy the usual anti-commutation relations. In the Dirichlet theory L_0 is given by

$$L_0^D = \frac{1}{4\alpha'} \left(\frac{y_2 - y_1}{\pi} \right)^2 + N. \tag{7}$$

At low energy ($\alpha' \to 0$) the $N \neq 0$ states of the Neumann theory become infinitely massive and decouple, leaving a theory with massless states only. Likewise, at high energy (the $\alpha' \to \infty$ limit) the position-space singularities away from the light-cone in the Dirichlet theory decouple, leaving a theory with only light-cone singularities[8),7)]. The expression for L_0^D is obtained from that for L_0^N by reinterpreting the (dimensionless) momentum of the Neumann theory, $\sqrt{2\alpha'} p^\mu$, as the separation of the end-points ($y^\mu / \sqrt{2\alpha'} \pi$) in the Dirichlet theory.

(6). In the theory with both types of boundary condition the general world-sheet also involves propagators for open strings with different boundary conditions at either end-point. The propagator for such strings has an expansion in 1/2-integral modes[12),13)]. Since the Neumann and Dirichlet boundaries are associated with Chan–Paton factors for different unitary groups such strings carry the defining representation of different Chan–Paton groups at either end-point.

Comments

The presence of world-sheet boundaries with constant Dirichlet conditions affects the short-distance properties of string theory quite dramatically. Amplitudes of the tree-level theory obtained by summing over world-sheets with arbitrary numbers of

Dirichlet boundary insertions have point-like fixed angle behaviour which is suggestive of the behaviour of the large-n limit of $U(n)$ Yang–Mills theory. Furthermore, the high temperature limit of the Dirichlet string theory also has features in common with large-n Yang–Mills theory.

The simple bosonic Dirichlet theory described in this talk has various problems, the most obvious one being the presence of the usual closed-string tachyons that indicate instabilities requiring a more sophisticated treatment which would take into account a condensate of Dirichlet boundaries as well as handles. In addition, string perturbation theory amplitudes with Dirichlet boundaries have divergences associated with the presence of intermediate gauge-singlet open strings with fixed endpoints that couple to closed strings[11)]. In particular, the open-string level-one vector state is a Lagrange multiplier field that should be integrated, leading to constraints on the closed-string spectrum that are not seen in string perturbation theory. The possibility that the presence of point-like structure may be related directly to the massless spectrum of the theory is most intriguing but as yet there is no evidence that the massless graviton (that seems to be an intrinsic feature of usual string theory) is absent in the Dirichlet theory as it must be in any realistic description of large-n Yang–Mills theory.

It is important to develop methods for summing over all possible insertions of boundaries. The induced metric on a Dirichlet boundary is singular since the whole boundary is mapped into a single point in the target space-time. Alternatively, by making a singular reparametrization the boundary can be collapsed to a point on the world-sheet so that the induced metric is well-behaved and the effect of the boundary is encoded into a local singularity of the intrinsic world-sheet metric. In this language the sum over boundary insertions is related to the problem of including singular metrics in the sum over histories that defines perturbative closed-string theory.

Generalizations of the boundary conditions are possible. For example, a theory with both Neumann and Dirichlet conditions describes open strings with end-points which may be fixed or free. Another possibly important variant is a theory which involves a sum over all possible ways of dividing each boundary into segments with Neumann or constant Dirichlet conditions where the target-space position and the length of each Dirichlet segment is integrated[7)] (which may be based on ideas in refs.[14, 12,13]). This introduces point-like structure at the string end-points.

REFERENCES

1. G. 't Hooft, *A planar diagram theory for strong interactions*, Nucl. Phys. **B72** (1974) 461.
2. E. Witten, *Baryons in the 1/N expansion*, Nucl. Phys. **B160** (1979) 57.
3. G. 't Hooft, *A two-dimensional model for mesons*, Nucl. Phys. **B75** (1974) 461.
4. D. J. Gross, *Two-dimensional QCD as a string theory*, Nucl. Phys. **B400** (1993) 161.
5. J. Polchinski, *High temperature limit of the confining phase*, Phys. Rev. Lett. **68** (1992) 1267.
6. M.B. Green, *The influence of world-sheet boundaries*, proceedings of the HARC Symposium, Houston (January, 1992).
7. M.B. Green, *Space-Time duality and Dirichlet string theory*, Phys. Lett. **266B** (1991) 325.
8. M. B. Green, *Point-like structure and off-shell dual strings*, Nucl. Phys. **B124** (1977) 461.
9. S. Mandelstam,*Interacting-string picture of dual resonance models* , Nucl. Phys. **B64**, 205 (1973).
10. M.B. Green, *Temperature Dependence of String Theory in the Presence of World-Sheet Boundaries*, Phys. Lett. **282B** (1992) 380 .
11. M.B. Green, *The influence of world-sheet boundaries on closed string theory*, Phys. Lett. **302B** (1993) 29.
12. E.F. Corrigan and D.B. Fairlie, *Off-shell states in dual resonance theory*, Nucl. Phys. **B91** (1975) 527.
13. M.B. Green, *Locality and currents for the dual string*, Nucl. Phys.**B103** (1976) 333.
14. J.H. Schwarz, *Off-shell dual amplitudes without ghosts*, Nucl. Phys. **B65** (1973) 131.

DIMENSIONAL REDUCTION in QUANTUM GRAVITY

G. 't Hooft
Institute for Theoretical Physics
Utrecht University
Postbox 80 006, 3508 TA Utrecht, the Netherlands

Abstract

The requirement that physical phenomena associated with gravitational collapse should be duly reconciled with the postulates of quantum mechanics implies that at a Planckian scale our world is not 3+1 dimensional. Rather, the observable degrees of freedom can best be described as if they were Boolean variables defined on a two- dimensional lattice, evolving with time. This observation, deduced from not much more than unitarity, entropy and counting arguments, implies severe restrictions on possible models of quantum gravity. Using cellular automata as an example it is argued that this dimensional reduction implies more constraints than the freedom we have in constructing models. This is the main reason why so-far no completely consistent mathematical models of quantum black holes have been found.

With the request to write a short paper in honor of Abdus Salam I am given the opportunity to contemplate some very deep questions concerning the ultimate unification that may perhaps be achieved when all aspects of quantum theory, particle theory and general relativity are combined. One of these questions is the dimensionality of space and time.

At first sight our world has three spacelike dimensions and one timelike. This was used a a starting point of all quantum field theories and indeed also all string theories as soon as they invoke the Kaluza Klein mechanism. Also when we quantize gravity perturbatively we start by postulating a Fock space in which basically free particles roam in a three plus one dimensional world. Naturally, when people discuss possible cut-off mechanisms, they think of some sort of lattice scheme either in 3+1 dimenisional Minkowski space or in 4 dimensional Euclidean space. The cut-off distance scale is then suspected to be the Planck scale.

Unfortunately any such lattice scheme seems to be in conflict with local Lorentz invariance or Euclidean invariance, as the case may be, and most of all also with coordinate reparametrization invariance. It seems to be virtually impossible to recover these symmetries at large distance scales, where we want them. So the details of the cut-off are kept necessarily vague.

The most direct and obvious *physical* cut-off does not come from non-renormalizability alone, but from the formation of microscopic black holes as soon as too much energy would be accumulated into too small a region. From a physical point of view it is the black holes that should provide for a natural cut-off all by themselves.

This has been this author's main subject of research for over a decade. A mathematically consistent formulation of the black hole cut-off turns out to be extremely difficult to find, and in this short note I will explain what may well be the main reason for this difficulty: nature is much more crazy at the Planck scale than even string theorists could have imagined.

One of my starting points has been that quantum mechanics itself is not at all a mystery to me. The emergence of a Hilbert space with a Copenhagen interpretation of its inner products is a quite natural feature of any theory with the following characteristics at a local scale: the system must have *discrete* degrees of freedom at tiny ditance scales, and the laws of evolution must be *reversible* in time. With discrete degrees of freedom one can construct Hilbert space in a quite natural way by postulating that any state of the physical degrees of freedom corresponds to an element of a basis of this Hilbert space[1]. Reversibility in time is required if we wish to see a quantum superposition principle: the norm of all states is then preserved, even if they are quantum superpositions of these basis elements.

Needless to say, one might suspect that some or all of the quantum mechanical postulates could break down at the Planck scale[2]. But then one might as well throw away anything we know about physics, and that is not the route I want to follow. I have never seen convincing models where ordinary quantum mechanics breaks down at a microscopical level but is somehow recovered at the atomic scale. Therefore I prefer not to speculate that quantum mechanics breaks down at the Planck scale, but in stead to suspect that quantum mechanics becomes trivial there: quantum superpositions are still allowed there but become irrelevant.

Black holes then present a major challenge. At first sight they render time reversibility impossible. Objects thrown into a black hole can never be retrieved[3]. Things are often presented as if black holes connect our world to other universes via wormholes[4], or, if one prefers not to refer to these other universes one says that information thrown in can not be retreived anymore[5]. According to this view information is preserved only if one considers multiply connected universes[6]. This is useless if one wishes to recover quantum mechanical behavior in our universe by itself. However, at closer inspection one finds that any object thrown into a black hole actually does leave some signals behind in our own world[7], and it is conceivable that a unitary theory for our own universe can be built using this as a starting point.[8]

The main difficulty then is to formulate exactly what our derees of feedom are. Remarkably, it is relatively easy to give a fairly precise estimate of *how many* degrees of freedom we have. This can be deduced in two equivalent ways[9, 10]. One is by considering capture and emission of objects by black holes as scattering experiments. If these are described quantum mechanically one can deduce information about the size of phase space. It must be finite, increasing exponentially with the surface area of the black hole horizon. The other way to deduce the same information is by using thermodynamics. One derives the entropy S of a black hole, finding

$$S = 4\pi M^2 + C\,, \tag{1}$$

in natural units. The constant C is not known (in fact there could be as yet unknown subdominant terms in this expresion, increasing slower than M^2 as the mass increases). In principle C could be infinite (even for small M), which basically corresponds to the remnant theory. I think an infinite C would induce major problems (again implying a deviation from ordinary quantum mechanical behavior) in a consistent theory, so I will henceforth assume C to be bounded.

This entropy is often attributed 'to the space-time metric itself', as if there would be yet another separate contribution from the quantized fields in this metric, such as the contribution to the entropy of objects outside the black hole. However, this would not be correct. First of all, the contribution of objects at some distance from the black hole will always be negligible unless they are really very far away (at $|x| \gg R_{\rm Schwarzschild}$). But most importantly, the contribution of quantum fields to the entropy *diverges* very near the horizon[10]. The reason for this divergence can be easily understood physically: arbitrary amounts of matter can be thrown in and arbitrary amounts are waiting to radiate out; they contribute infinitely to the total number of physical degrees of freedom.

Yet the black hole entropy had just been argued to be finite, *even with the quantum fields present!* We conclude that one has to attribute the black hole entropy not to the space-time metric itself but to the quantized fields present there (which of course does include the small quantum fluctuations of the metric itself), and then one must choose a cut-off sufficiently close to the horizon such that it exactly matches the known black hole entropy (1). In short, the black hole entropy *includes* the entropy of the quantized fields in its neighborhood[10].

In any quantum theory there is a 'third law of thermodynamics' relating the entropy to the total number of degrees of freedom: the dimension of the vector space describing all possible states our system can be in is the exponent of the entropy. For instance in a discrete theory described by n spins that can take only two values ('Boolean variables'), the dimension $\mathcal{N}$ of Hilbert space is

$$e^S = \mathcal{N} = 2^n \ , \tag{2}$$

Hence the entropy directly counts the number of Boolean degrees of freedom.

Considering the fact that at the Hawking temperature the contribution to the entropy of fields anywhere in the region $R < |x| \ll R^3$ pales compared to the black hole entropy itself, we can now make an important observation concerning the relevant degrees of freedom of a black hole:

The total number of Boolean degrees of freedom, n, in a region of space-time surrounding a black hole is

$$n = \frac{S}{\ln 2} = \frac{4\pi M^2}{\ln 2} = \frac{A}{4\ln 2} \ , \tag{3}$$

where A is the horizon area.

We can carry this argument one step further. Consider just any closed spacelike surface, with $S(2)$ topology and total surface area A. Consider all possible field and metric configurations inside this surface. We ask how many mutually orthogonal states there can be. If we want these states to be observable for the outside world we have to insist that the total energy inside the surface be less than $1/4$ times its linear dimensions, otherwise our surface would lie within the Schwarzschild radius. Let us first ask how many states would an ordinary quantum field theory allow us to have, given these limits on the volume V and the energy E. Now this is not hard. The most probable state would be a gas at some temperature $T = 1/\beta$. Its energy would be approximately

$$E = C_1 Z V T^4 \,, \tag{4}$$

where Z is the number of different fundamental particle types with mass less than T and C_1 a numerical constant of order one, all in natural units. The total entropy S is

$$S = C_2 Z V T^3 \,, \tag{5}$$

where C_2 is another dimensionless constant. Now the Schwarzschild limit requires that

$$2E < (V/\tfrac{4}{3}\pi)^{\frac{1}{3}} \,, \tag{6}$$

hence, with eq. (4),

$$T < C_3 Z^{-\frac{1}{4}} V^{-\frac{1}{6}} \,, \tag{7}$$

so that

$$S < C_4 Z^{\frac{1}{4}} V^{\frac{1}{2}} = C_5 Z^{\frac{1}{4}} A^{\frac{3}{4}} \,. \tag{8}$$

The C_i are all constants of order 1 in natural units. Since in quantum field theories, at sufficiently low temperatures, Z is limited by a dimensionless number we find that this entropy is small compared to that of a black hole, if the area A is sufficiently large.

Next consider a set of N black holes, with masses M_i. They contribute to the energy $\sum_i M_i$. So

$$\sum_i M_i < C_6 A^{\frac{1}{2}} \,, \tag{9}$$

while their total entropy is given by

$$S = C_7 \sum_i M_i^2 \,, \tag{10}$$

where we note that the contribution of their movements to the entropy is negligible. We see that ineq. (10) is saturated when one black hole has the largest possible size that still fits inside our area. Its entropy is

$$S_{max} = \tfrac{1}{4} A \,, \tag{11}$$

and this is as large as we can ever make it. This therefore answers our question concerning the total number of possible states. It is given by eqs. (2) and (3). The single black hole is the limit[11].

The importance of this result can hardly be overestimated. At first sight it is counterintuitive. One would have expected that the number of possible states would grow exponentially with the volume, not the area, as in any ordinary field theory with a cut-off.

So if one would take a regularized fermion theory with cut-off at the Planck scale one would get far too many states. But it is clear why our answer came out this way. Most of the states of a regularized quantum field theory would have so much energy that they would collapse into a black hole before they could dictate the further evolution of the system in time. If we want to avoid this a much more rigorous cut-off than a Planckian one must be called for. Note that if there are any fundamental bosons the number of possible states comes out to be strictly infinite.

But then one may come to appreciate this result after all. It means that, given any closed surface, we can represent all that happens inside it by degrees of freedom on this surface itself. This, one may argue, suggests that quantum gravity should be described entirely by a *topological* quantum field theory, in which all physical degrees of freedom can be projected onto the boundary. One Boolean variable per Planckian surface element should suffice. The fact that the total volume inside is irrelevant may be seen as a blessing since it implies that we do not have to worry about the *metric* inside. The inside metric could be so much curved that an entire universe could be squeezed inside our closed surface, regardless how small it is. Now we see that this possibility will not add to the number of allowed states at all. Our result suggests that one should not worry about creating universes inside test tubes.

The same can be said about wormholes. A wormhole with one end sprouting inside our closed volume and its other end somewhere else could connect the inside of our volume to the outside world, thus adding large quantities of possible states. We now believe that this is not allowed in a decent theory of quantum gravity. In a previous publication [8] it was explained why the functonal integral describing black holes probably has to be limited to topologically trivial field configurations only. We have detailed ideas concerning the consistency of such a requirement but will not elaborate on this here. So much for wormholes.

We would like to advocate here a somewhat extreme point of view. We suspect that there simply *are* not more degrees of freedom to talk about than the ones one can draw on a surface, as given by eq. (3) The situation can be compared with a hologram of a three dimensional image on a two-dimensional surface. The image is somewhat blurred because of limitations of the hologram technique, but the blurring is small compared to the uncertainties produced by the usual quantum mechanical fluctuations. The details of the hologram on the surface itself are intricate and contain as much information as is allowed by the finiteness of the wavelength of light - read the Planck length.

It is tempting to take the limit where the surface area goes to infinity, and the surface is locally approximately flat. Our variables on the surface then apparently determine all physical events at one side (the black hole side) of the surface. But since the entropy of a black hole also refers to all physical fields outside the horizon the *same* degrees of freedom determine what happens at this side. Apparently one must conclude that a two-dimensional surface drawn in a three-space can contain all information concerning the entire three-space. In fact, this should hold for *any* two-surface that ranges to infinity. This suggests that physical degrees of freedom in three-space are not independent but, if considered at Planckian scale, they must be infinitely correlated.

If one could determine the equation of motion of these variables on the two-plane one would also possess the remedy for the black hole information paradox. In a Rindler space we could put our information surface at the origin of the Rindler coordinate frame. The transformation rules for our variables under Lorentz transformations would correspond to the Rindler equations of motion, and pure states would evolve into pure states while the spectrum density of the black hole would continue to be the one dictated by its entropy.

The infinite correlations in three-space could also present a new starting point for resolving the Einstein-Rosen-Podolsky paradox. In terms of present day quantum field theoretical degrees of freedom it is not possible to interpret quantum mechanics deterministically. But certain cellular automaton models can give a faint resemblace to quantum behavior. The discreteness of our degrees of freedom on the two-surface strongly remind us of a 2+1 dimensional cellular automaton. But this would constitute a speculation on top of the previous one. There are various technical problems one has to face if such considerations were to be persecuted further.

In cellular automaton models for quantum mechanics the problem is not the Copenhagen interpretation[1]. This comes out quite naturally. The more tantalizing problem is how to understand the stability of our vacuum state[13]. Models can be constructed producing Hamiltonians that can be realistic but for which there are nearly always negative energy eigenstates. The absence of negative energy states in the real world is probably related to the presence of the gravitational force, for which the time coordinate is handled very differently from non- gravitational theories. But natural curvature of space and time is equally difficult to realize in cellular automaton models of this sort[12]; we just suspect that these problems are related but exactly how is not understood.

But there are other problems as well. One of these is the presence of the group of Lorentz transformations and the fact that this symmetry group is non-compact. This is of course at the heart of the black hole horizon problem. There, Lorentz transformations are playing the role of time translations. In any theory where the physical degrees of freedom are discrete it is extremely difficult to reproduce anything resembling Lorentz invariance [1]. As for other invariances such as rotational invariance, one can usually realize

[1] In principle one could think of realizing one of the non-trivial discrete subgroups of the Lorentz group, but in practice this seems to be impossible to reconcile with locality requirements.

symmetry under one of their finite subgroups and then it is not unnatural to suspect that complete invariance will be recovered in the thermodynamic limit as a consequence of renormalization group effects.

The evolution law of the physical degrees of freedom on a 2-surface is another mystery. Ultimately one wishes to recover full invariance with respect to the Poincaré group (which is a precisely defined invariance group for all states that have asymptotic in- and out-states, as the ones used in an S matrix formalism, but not in quantum cosmology). Now this implies that one not only needs an evolution law for translations in the time direction but also a quite similar looking law for translations in the direction orthogonal to the surface. These two evolution laws should commute with each other in spite of their complete independence. We will show how to reproduce such a feature in cellular automaton models, however, as we will see, requiring these models to possess physically interesting, that is, sufficiently non- trivial interactions, will remain a difficult obstacle.

The question whether models exist in 3+1 dimensions that are such that the data on a two-dimensional surface will determine all observables elsewhere is an extrelely intriguing one. For definiteness, let us consider a rectangular lattice in 4-space. Momentarily we will ignore requirements such as Lorentz covariance; it is sufficient to require that signals do not go faster than some limited speed c. The prototype of our models is a cellular automaton. The data f on every site on the lattice can be represented by an integer modulo some number p. This p will often be taken to be a prime number. The time evolution is defined by some local law. Conventionally, one imposes that the value of $f(\mathbf{x},t)$ be a given function of the values of $f(\mathbf{x}_i, t-1)$, where $\{\mathbf{x}_i\}$ are a finite set of nearest neighbors of $\mathbf{x}$.

Now such a model is deterministic in the classical sense. Quantization can be introduced either by allowing $f(\mathbf{x}_i, t)$ to be operators in Hilbert space, or even more simply, by declaring all states of the cellular automaton to be basis elements of Hilbert space. "Quantization" is then trivial. There are problems with this latter proposal in that the Hamiltonian then does not seem to possess a well-defined ground state. Again, let's not dwell on that.

In order to obtain "dimensional reduction" we will have to postulate a further constraint: the values of f on a sheet should fix the values elsewhere. This we can realize in principle by also postulating a law of evolution in the z direction. Since further discussion of this 4 dimensional problem becomes a bit intricate it is illustrative to remove one dimension and treat space as 2 dimensional, space-time as 3 dimensional. Here we will require that the data on a *line* should be sufficient to determine the data elsewhere. Part of the rectangular lattice is depicted in Fig. 1.

Suppose now that on every plaquette of the lattice a relation among the data f is imposed. Thus, in the figure there are constraints of the form

$$g_a\big(f(A),f(B),f(C),f(D)\big)=0\,, \tag{12a}$$
$$g_b\big(f(E),f(F),f(G),f(H)\big)=0\,, \tag{12b}$$
$$g_c\big(f(A),f(B),f(F),f(E)\big)=0\,, \tag{12c}$$
$$g_d\big(f(D),f(C),f(G),f(H)\big)=0\,, \tag{12d}$$
$$g_e\big(f(A),f(D),f(H),f(E)\big)=0\,, \tag{12e}$$
$$g_f\big(f(B),f(C),f(G),f(F)\big)=0\,. \tag{12f}$$

We require all these constraints to be such that if in any of these equations three of the four entries are given the fourth will be uniquely determined.

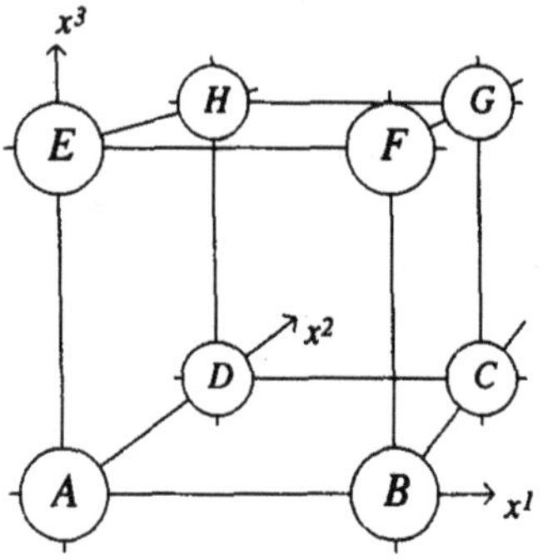

Fig. 1

The lattice sites in x^1, x^2, x^3 coordinates

If we choose the x, y, t coordinates in one of the principle directions of the lattice the situation becomes a bit singular. It is preferrable to define

$$\begin{aligned} t &= x^1 + x^2 + x^3\,, \\ x &= x^1 - x^2\,, \\ y &= x^3 - \tfrac{1}{2}(x^1 + x^2)\,. \end{aligned} \tag{13}$$

if on two nonsecutive t layers the data are known successive application of eqs (12a–f) will produce the data in all of space-time. But now consider the series of points $\mathbf{x}(n)$, $n \in \mathbb{Z}$, defined by

$$\begin{aligned} \mathbf{x}(3n+1) &= \mathbf{x}(3n) + \mathbf{e}^1\,, \\ \mathbf{x}(3n+2) &= \mathbf{x}(3n+1) - \mathbf{e}^2\,, \\ \mathbf{x}(3n+3) &= \mathbf{x}(3n+2) + \mathbf{e}^3\,, \end{aligned} \tag{14}$$

where $\mathbf{e}^{1,2,3}$ are the three unit vectors of the lattice (the signs in front of the $\mathbf{e}^i$ can actually be chosen at will). Suppose $f(\mathbf{x}(n))$ are given for all n. Successive application of the six identities (12 a- -f) then also fixes all data elsewhere.

However, one cannot choose the constraints (12a–f) any way one likes.

This is because the data elsewhere will be *over*-determined. Suppose that in the Figure the data are given at D, H, E and F. They form part of a series (14). By applying four of the six equations (12) the other data on the cube are determined. The remaining two must then be satisfied automatically. If these would not be satisfied our system would be constrained further, in such a way that practically no solutions survive at all. The point is that eqs (12a, b) generate translations along the x^1x^2 plane, eqs. (12c, d) translations along the x^1x^3 plane and (12e, f) translations along the x^2x^3 plane. These three translation operators, viewed as operators in Hilbert space, should commute with each other. Only when they are chosen very meticulously these commutation requirements can be met. One can also say that we have translation operators defined on the one dimensional series of data on the points (14). They define translations in the directions $\mathbf{e}^i \pm \mathbf{e}^j$, which should all commute with each other. Two linear combinations of these, $U_1(\delta t)$ and $U_2(\delta y)$, should generate *independent* translations in directions orthogonal to the line $\sigma(\mathbf{e}^1 - \mathbf{e}^2 + \mathbf{e}^3)$ and a third, U_3, corresponds to a translation in the direction of the line. The three independent translation operators one then has should commute with each other. Commutation with U_3 is usually easy to implement by choosing the evolution not to depend on the coordinate n, but commutation of U_1 with U_2 is as hard as finding two different yet commuting local Hamiltonians for a quantum system.

Let us phrase the constraints the following way. Consider Fig. 1. One is free to choose $f(A)$, $f(B)$, $f(D)$ and $f(E)$. Then the values of $f(C)$, $f(F)$ and $f(H)$ are uniquely determined by eqs (12a, c) and (e). But the value of $f(G)$ is overdetermined. The three equations (12b, d) and (f) should all yield the same value. Writing the solutions to eqs (12a–f) as

$$f(C) = h_a\big(f(A), f(B), f(D)\big) \,, \tag{15a}$$
$$f(G) = h_b\big(f(E), f(F), f(H)\big) \,, \tag{15b}$$
$$f(F) = h_c\big(f(A), f(B), f(E)\big) \,, \tag{15c}$$
$$f(G) = h_d\big(f(D), f(C), f(H)\big) \,, \tag{15d}$$
$$f(H) = h_e\big(f(A), f(D), f(E)\big) \,, \tag{15e}$$
$$f(G) = h_f\big(f(B), f(C), f(F)\big) \,, \tag{15f}$$

our requirement corresponds to

$$h_b\Big(f(E), h_c(f(A), f(B), f(E)), h_e(f(A), f(D), f(E))\Big) =$$
$$h_d\Big(f(D), h_c(f(A), f(B), f(D)), h_e(f(A), f(D), f(E))\Big) = \tag{16}$$
$$h_f\Big(f(B), h_c(f(A), f(B), f(D)), h_b(f(A), f(B), f(E))\Big).$$

An easy way to implement these commutation constraints is by choosing the functions g_i to be *linear in the functions* f_i *modulo* p :

$$g_i(\{f_j\}) = \sum_j A_{ij} f_j + B_i \mod p, \tag{17}$$

where the coefficients A_{ij} must all have an inverse modulo p. FFrom this one gets three equations for $f(G)$:

$$f(G) = K_{1,\alpha} f(A) + K_{2,\alpha} f(B) + K_{3,\alpha} f(D) + K_{4,\alpha} f(E) + K_{5,\alpha} \mod p, \quad \alpha = 1, 2, 3. \tag{18}$$

It is not hard to find sets of coefficients A_{ij}, B_i such that the 10 equations

$$K_{i,1} = K_{i,2} = K_{i,3} \tag{19}$$

are obeyed.

A next step is to attempt to find other realizations of our commutation requirement. Using a computer search program the author generated solutions which at first sight seemed to be quite different, but careful analysis reveiled that all solutions found were actually equivalent to the linear ones, eq. (17) after applying permutation operations on the numbers f. This could be seen as a disappointment because one might argue that a linear relation such as eq. (17) is too trivial to be of much physical interest. It implies that solutions can be superimposed onto each other, as if we were describing only "non-interacting" particles. Our challenge at present is to find any set of plaquette relations that does not allow a superposition procedure to obtain new solutions from old ones (note that superposition here means addition modulo a number, and is quite distinct from quantum mechanical superposition which is always allowed).

Our problem could be seen to simplify a bit by using different lattices. Basically what we want to achieve is a cellular automaton with two commuting but essentially different evolution laws. Suppose we have a triangular lattice in the $x-y$ direction as well as the $x-t$ direction, see Fig. 2. We may demand that the data on the x axis alone should determine all others. This implies that we have a relation U_1 determining how the data look in the y direction, and an evolution operator U_2 in the time direction. We have

$$f(F) = U_1\big(f(A), f(B)\big)\,, \tag{20a}$$
$$f(G) = U_1\big(f(B), f(C)\big)\,, \tag{20b}$$
$$f(P) = U_2\big(f(A), f(B)\big)\,, \tag{20c}$$
$$f(Q) = U_2\big(f(B), f(C)\big)\,, \tag{20d}$$
$$f(X) = U_2\big(f(F), f(G)\big)\,, \tag{20e}$$
$$f(X) = U_1\big(f(P), f(Q)\big)\,, \tag{20f}$$

and for all choices of $f(A)$, $f(B)$ and $f(C)$ the equations (12e) and (12f) should agree. These equations are easier to study than the equations (16). Again there are solutions where U_1 and U_2 are different, providing no direct relations between $f(F)$ and $f(P)$ or $f(G)$ and $f(Q)$, but all solutions we found are equivalent to linear ones modulo a prime number p.

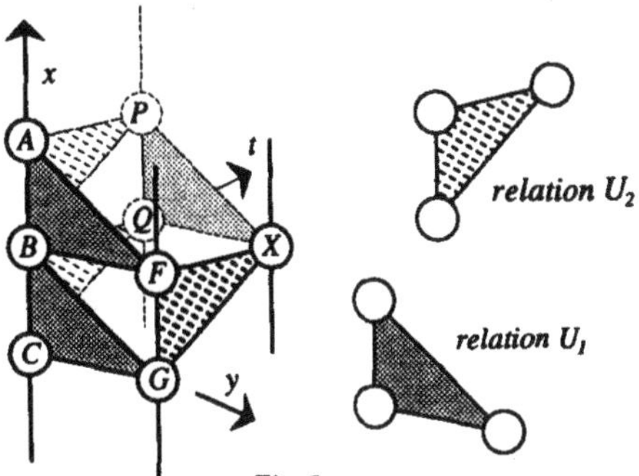

Fig. 2
Cellular automaton with two evolution laws on triangular lattices.

In the real world there is one more dimension. By choosing the initial data to be periodic in that one extra dimension the problem reduces to the three dimensional one. Ultimately one would also like to reobtain local Lorentz invariance and coordinate reparametrization invariance. These imply large symmetry groups that are as yet impossible to implement.

The reason why it is interesting to attampt to find an accidental non- trivial solution of the equations is that many cellular automata. defined on different lattice types, can be transformed into each other, as long as the evolution equations act upon nearest neighbors only. So we actually cover a large class of models. If the real world obeys comparable rules it could be equaivalent to such an automaton also, though, admittedly, we consider such a simple structure unlikely. Indeed, one can also devise interaction schemes among discrete variables that are essentially local but *not* equivalent to cellular automata of the kind considered here.

Even though the problems just mentioned are grave and conceptually difficult to disentangle, they do not seem to be insurmountable. Our basic problem is that there seem to be *too many* symmetry requirements and the Hilbert space of physically realizable states seems to be *too small*. The picture we sketched of the 2+1 (or perhaps it is better to say 2+2) dimensional nature of our micro- universe seems to follow from quite general arguments. Rejecting any of these arguments leads to quite different and perhaps equally if not more difficult problems, and one cannot help observing that one's preferences seem to be related to nearly religeous prejudices. Besides the author not many other physicists have tried this particular avenue. We advocate its further persuit.

References

1. G. 't Hooft, *Nucl. Phys.* **B 342** (1990) 471; *J. Stat. Phys.* **53** (1988) 323.
2. S.W. Hawking, *Phys. Rev.* **D 14** (1976) 2460.
3. S.W. Hawking and G.F.R. Ellis, "The Large Scale Structure of Space-time", Cambridge: Cambridge Univ. Press, 1973.
4. S.W. Hawking, *Phys. Rev.* **D 37** (1988) 904; S. Coleman, *Nucl. Phys.* **B 307** (1988) 864; *ibid.* **B 310** (1988) 643; S.B. Giddings and A. Strominger, *Nucl. Phys.* **B 307** (1988) 854.
5. C.G. Callen, S.B. Giddings, J.A. Harvey and A. Strominger, *Phys. Rev.* **D 45** (1992) R1005.
6. V.P. Frolov and I.G. Novikov, *Phys. Rev.* **D 42** (1990) 1057.
7. T. Dray and G. 't Hooft, *Nucl. Phys.* **B 253** (1985) 173; T. Dray and G. 't Hooft, *Commun. Math. Phys.* **99** (1985) 613; G.'t Hooft, *Nucl. Phys.* **B 335** (1990) 138.
8. C.R. Stephens, G. 't Hooft and B.F. Whiting, "Black hole evaporation without information loss", Utrecht/Gainesville prepr. THU-93/20; UF-RAP-93- 11; gr-qc/9310006.
9. G. 't Hooft, "On the Quantization of Space and Time", Proc. of the 4th Seminar on Quantum Gravity, May 25- 29, 1987, Moscow, USSR, ed. M.A. Markov et al (World Scientific 1988).
10. G. 't Hooft, *Nucl.Phys.* **B 256** (1985) 727.
11. G. 't Hooft, *Physica Scripta* **T 36** (1991) 247.
12. G. 't Hooft, "A Two-dimensional Model with Discrete General Coordinate-Invariance", in "The Gardener of Eden", Physicalia Magazine, vol **12**, in honour of R. Brout, eds. P. Nicoletopoulos and J. Orloff, Brussels, 1990.
13. G. 't Hooft, K. Isler and S. Kalitzin, *Nucl. Phys.* **B 386** (1992) 495

Strings and Singularities

Cumrun Vafa

Lyman Laboratory of Physics, Harvard University
Cambridge, MA 02138, USA

We review aspects of spacetime singularities from the view point of string theory. Examples considered include cosmological, cosmic string and black-hole singularities. We also discuss the consistency of viewing black-holes as excited states of fundamental strings.

It is a great pleasure to speak in this conference in honor of Professor Abdus Salam. I have a deep respect for him, not only because of his important contributions to theoretical physics but also for his complete devotion to the cause of advancing science in third world countries.

In this talk I will discuss aspects of space time singularities in the context of string theory.

One of the main hints about the incompleteness of classical general theory of relativity comes from the fact that almost all interesting solutions have physical singularities. Even if we start with a non-singular universe typically in the past or in the future there are times when we hit a singularity and do not know how to continue physics beyond. Let us consider three examples of singularities:

A-Cosmological

In the standard big bang cosmology, the universe is believed to have started in a situation of infinite curvature– this is also the reason that the time before big bang cannot be defined. Moreover at the singularity the temperature is infinite which is rather unphysical.

B-Cosmic strings

There are other kinds of examples of curvature singularities which are expected to appear in general relativity, the simplest example of which is that encountered for an infinitely thin cosmic string. In this case the geometry of space is singular only along the string itself, where the geometry is conical. This means that the transverse geometry looks flat except some piece of the angular region is missing.

C-Astrophysical Singularities: Black holes

In most astrophysical objects which are compact enough we encounter singularities. The most well known example is that of a black-hole singularity. These are very important to understand because star evolution could naturally lead to black holes, and thus their physical existence is an inescapable consequence of relativity combined with the the existence of compact objects.

One of the standard reactions to the existence of such singularities is to express a hope that quantum effects should naturally take care of these infinities. After all, quantum phenomena smooth things out, and so one should expect that smearing these infinities will be a quantum resolution to the paradox. Before the advent of string theories such hopes were expressed without a realistic hope of testing it, as we did not have any viable candidates of a quantum theory of gravity. With the emergence of string theory as essentially the only viable candidate of a quantum theory of gravity, we are thus forced to face the question: How does quantum strings resolve the singularities inherent in the general theory of relativity?

It turns out that string theory does have ways to cure these singularities at least in the context of simple examples. Though the complete story has not yet emerged we consider these examples as typical of how string deals with singularities. The way string theory resolves these singularities is not just by quantum smearing out but more surprisingly even *classically* strings do not see a singularity. We will now discuss these in turn in the context of concrete examples.

1. Strings and Cosmology

String cosmology has been modeled by putting strings in a box with periodic boundary conditions [1]. Consider a box of size L in all directions, measured in string (=planck) units (Fig. 1). Consider some matter in the box in an equilibrium state at temperature T. Assume that as universe evolves the temperature changes in an adiabatic way (i.e. to maintain constant entropy). We thus have a function $T(L)$. Then in point particle theory as $L \to 0$, the temperature diverges $T \sim 1/L$. Moreover the dynamics of relativity implies that (in a flat space with zero cosmological constant) $L \sim t^{1/2}$ for small time $t > 0$ after the big bang. The curvature of space time blows up as $t \to 0$.

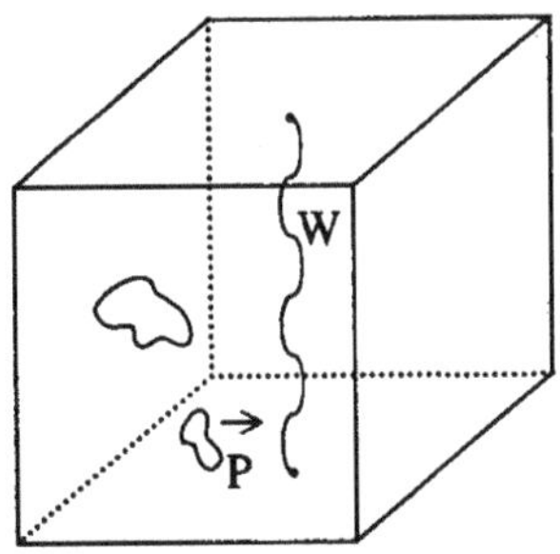

Fig. 1: The strings in a box can be in different types of states: They could be wound around the box (W) or oscillating inside the box, or moving from one side of the box to another (P), or some combination thereof.

The story is very different in the context of string theory: There are two major modifications: 1-There is a very important symmetry for a string in a box: Strings in a box of size L and in a box of size $1/L$ behave exactly the same [2][1]! There is no experiment we can perform do distinguish in which geometry the string is living in. This is amazingly strange. The point is that string states are sensitive to the size of the box L only in two ways: Their center of mass momentum is quantized in inverse units of L, i.e.

$$P = \frac{n}{L}$$

This is also true for point particles. However strings can also be wrapped around the box. Their energy is proportional to how many times m they wind around the box

$$W = mL$$

Note that if we exchange $L \leftrightarrow 1/L$, and if we relabel the momentum and winding states $P \leftrightarrow W$, (i.e. $n \leftrightarrow m$) we get exactly the same string spectrum. Not only the spectrum is the same but also all the interaction between states in the box of size L is the same as between the dual states in the box of size $1/L$. This example is the simplest example of a phenomenon known as mirror symmetry, in which a string propagating on two inequivalent manifolds behaves the same. In string theory such pairs of manifolds are thus indistinguishable. The duality symmetry in the context of the strings in a box is very stringy as it clearly depends on the existence of winding modes which are absent in point particle theories. It is also not a symmetry of p-brane theories as can be easily seen by looking at their spectra (for example for 2 dimensional extended objects we get states with energy $1/L, L$ and L^2 and the duality symmetry is thus lost). So if we insist on the existence of the duality symmetry $L \to 1/L$ we are forced to consider one dimensional extended objects, i.e. strings.

We thus see that in the context of string theory $L \to 0$ is no more singular than $L \to \infty$, which is known to be non-singular. The fact that there is this duality symmetry is nicely complemented by the appearance of a maximum temperature for stings, known as the Hagedorn temperature. Its existence is a deep fact of string theory and is dictated by the exponential growth in the number of string states with mass M, i.e.

$$d(M) \sim \exp(\beta_H M)$$

This gives rise to the limiting temperature known as the Hagedorn temperature $T_H = \beta_H^{-1}$.

The dynamical implementation of this duality in string theory requires the existence of a scalar field which is known as the dilaton which transforms non-trivialy under $L \to 1/L$. Its existence drastically modifies the dynamics of general relativity, as we will now briefly recall in the context of string cosmology [3]. The existence of the dilaton field restores some of the intuition one has with conservation of energy. For example, in the context of a flat universe with zero cosmological constant, having a constant energy density of matter results in inflation, which looks somewhat like the perpetual generation of energy. One's naive intuition would be that if there is a certain energy density, expansion of the universe while keeping the energy density constant should require 'work' and thus there should be an obstacle to it. The presence of dilaton restores this intuition in string theory! This suggests in particular that inflation is not a natural phenomenon in string theory (at least in the very early universe where dilaton field is not frozen).

In the context of string cosmology, if we plot the total energy E in the box as a function of log$L$ (see Fig. 2),

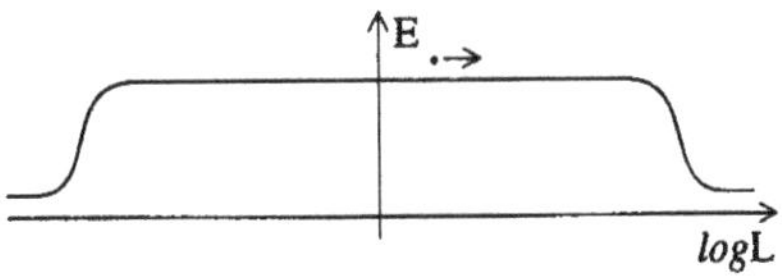

Fig. 2: The total energy E in the universe as a function of the (logarithmic) size of the universe. The state of the universe is analogous to that of a particle rolling on the potential E (with some damping and modulation of the potential). The fact that E is an even function of log$L$ is a manifestation of scale duality. Note that the flat region near L=1 is the Hagedorn phase of string, and seems not to display any particularly singular behaviour.

then the dependence of $L(t)$ can be inferred roughly by thinking of the state of the universe as a particle rolling on the potential E (with some damping and modulation of the potential–see [3]). Note that, as mentioned above there is no singularity as $L \to 0$. Moreover the Planckian region $L \sim 1$ (log$L \to 0$) is also a smooth region where the universe slowly rolls on the almost flat potential E, where the flatness is a consequence of the fact that in this region we are at the Hagedorn phase and the energy is rather insensitive to L. Also, it was using this model of string cosmology that it was suggested in [1] that long strings (or winding strings) if not decayed may be an obstacle for the expansion of the universe because E will then grow linearly with L in such cases and this may be a natural basis for the selection of 4 macroscopic dimensions for spacetime (two long strings have a hard time finding each other in more than 4 spacetime dimensions because 2+2=4)[2].

2. Cosmic Strings–Orbifold Singularities

Let us now move to the singularities that appear in the case of infinitely thin cosmic strings. Consider having a cosmic string with a deficit angle of $4\pi/3$ (see Fig. 3).

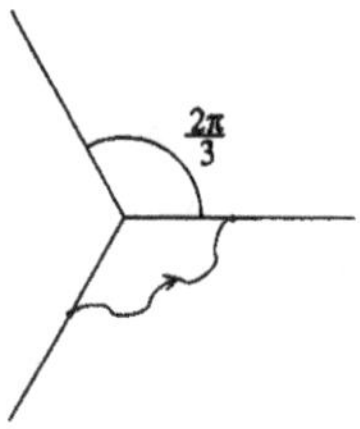

Fig. 3: The transverse plane to a cosmic string with deficit angle $4\pi/3$. Note that we can describe this geometry by identifying points on R^2 which differ by $2\pi/3$ rotation. In string theory we can get new states, twisted states, whose two endpoints differ by the Z_3 action.

This means that the angular variable goes from 0 to $2\pi/3$. It is more convenient to imagine the presence of the full 2π angular region but identify the three pieces each subtending a $2\pi/3$ angular region by dividing by a Z_3 symmetry, by identifying $z \rightarrow exp(2\pi i/3)z$ where the plane perpendicular to the direction of cosmic string is denoted by the complex parameter z. This way of viewing the cosmic string is also known as orbifolds, i.e., the two dimensional space of the cosmic string can be viewed as a quotient

$$R^2/Z_3$$

If one considers a field theory in this background one finds that unitarity is lost at the quantum level. The easiest way to see this is that one can start with the Hilbert space

[2] The dual picture of this is well known: Namely if we take $L \rightarrow 0$ then the momentum modes now play the role of massive modes. The momentum modes are just the excitation states for point particle theory. It is well known that for point particle field theories with more than 4 dimensions the theory is trivial (again the argument is based on $2 + 2 = 4$) and so thermal equilibrium is difficult to maintain.

of particles propagating on R^2 and project onto the Z_3 invariant subspace. However, this cannot possibly be consistent with unitarity because the Z_3 non-invariant states that we have projected out will be pair produced and so if we delete them from the spectrum we just get a non-unitary theory. If we include them we would be back to R^2 and not the cosmic string configuration.

Strings have a beautiful way of resolving this inconsistency [4]. In the context of strings projecting to the Z_3 invariant subspace of the Hilbert space is not the end of the story: We end up introducing two new sectors replacing the states we projected out. The reason for the presence of these sectors is that we can have string states which are closed up to Z_3 action (as depicted in Fig. 3). These states are closed strings when we recall that the points on the plane differing by Z_3 action are identified with each other. The two new sectors we get in this way are called the twisted sectors. The way string theory resolves the inconsistency of field theory with unitarity is that in the loop amplitudes instead of producing states which are not Z_3 invariant and have been projected out, we produce pairs of twisted string states. This turns out to restore the unitarity. Again we see it is the extended nature of strings that allows to have the extra sectors which restores unitarity, just as the extended nature of winding modes in the previous example was responsible for the duality invariance of strings.

3. Black-holes and Strings

Finally we would like to discuss how string theory comes to terms with black holes. This area needs to be further developed and better understood. However there is a two dimensional toy model for black holes which is very similar to the actual four dimensional theory [5] which can be discussed in the context of string theory. Actually the object of direct interest in string theory turns out to be a supersymmetric version of the black-hole [6] (supersymmetry here comes from the ghost degrees of freedom and is not part of the matter theory). The metric has the same singularity structure as the 4d black hole; it is given by

$$g = \frac{d\gamma d\bar{\gamma}}{1-\gamma\bar{\gamma}}$$

(with a dilaton field turned on). This back ground (with the fermionic degrees of freedom included) can be described as string propagation on the (supersymmetric) coset

$$SL(2)/U(1)$$

(with an appropriate 'twisting' [6]), where γ and $\bar{\gamma}$ can be viewed as particular (Gaussian) choice of coordinates on $SL(2)$. This theory is clearly singular at $\gamma\bar{\gamma} = 1$. However as shown in [6] this theory is equivalent to strings propagating on a circle of radius 1, which is not singular! In other words *the mirror theory to strings on black hole is strings on circle.* There is a way to see this [7] in the same explicit manner as in the duality for string in a box using the techniques in [8]. It is found that by going to the conjugate variables to the γ's, this can be accomplished. Let us briefly discuss how this comes about. The world sheet action for string propagating on $SL(2)/U(1)$ (ignoring the fermionic contributions for simplicity) can be written as

$$\int \beta\bar{\partial}\gamma + c.c. + \beta\overline{\beta}(1-\gamma\overline{\gamma}) \tag{1}$$

Upon integrating $\beta, \overline{\beta}$ out we get the usual sigma model action for the black hole with the metric written above.

$$S = \int -\frac{\overline{\partial}\gamma\partial\overline{\gamma}}{(1-\gamma\overline{\gamma})} \tag{2}$$

However we can also integrate out the γ's and we end up with

$$\int \frac{\overline{\partial}\beta\partial\overline{\beta}}{\beta\overline{\beta}} + \beta\overline{\beta} \tag{3}$$

Now rewriting $\beta \to \beta_0 exp(iX + \phi)$, with β_0 being a constant, leads to the standard action for string propagating on the circle (parametrized by X), where ϕ plays the role of the scale of the worldsheet metric (Liouville field), and $\beta_0\bar{\beta}_0 = \mu$ is identified as the cosmological constant on the worldsheet[3]. Note that the way string resolved this singularity was *not* by a change of coordinate. In other words β cannot be written as a function of γ alone. If this were the case we would simply have a change of coordinates in the same sense as in the

[3] It has been discovered [7] that the γ–description is related to a ground ring generator, whereas β is related to the tachyon vertex. So the two descriptions have complementary virtues. Note also that the tachyon states, which are given by $\beta^k\bar{\beta}$ and $\beta\bar{\beta}^k$ in the β–description are represented in the γ–description by $(\partial\bar{\gamma})^k\bar{\partial}\gamma/(1-\gamma\bar{\gamma})^{k+1}$ and $\partial\bar{\gamma}(\bar{\partial}\gamma)^k/(1-\gamma\bar{\gamma})^{k+1}$ which look like deforming the black hole geometry by higher spin gravity theories–i.e. W-gravities. In this connection it is natural to expect that what corresponds in the β picture to tachyon condensates corresponds in the the black-hole picture to turning on the higher spin W-gravity [7]. This suggests that self-dual geometries on the cotangent of the black-hole may be the natural setup to describe black-hole deformations.

classical theory of relativity. Rather, what we have here is $\beta(\gamma, \partial\gamma)$, and so this change of fields, does not translate to a geometrical operation on the black-hole background. Again we see that the sigma-model description is responsible for avoiding any singularity (a deeper understanding of this can be achieved by including the fermionic terms [7] which mix with γ, β fields in the action).

We thus see that the black-hole configuration is not singular: The γ-picture which looks singular is equivalent to its 'mirror' the β-picture, which is not singular. Having argued for the absence of physical singularity in this 2 dimensional model for black-hole it is natural to ask other questions relevant for d–dimensional black-holes. There is a lot more to be learned from string theory about black-holes. One such question is in regard to its thermal properties. Consider a highly excited string state (a string state with large oscillation number). Such a state has a mass much bigger than its size in Planck units (identifying string scale with Planck scale). Therefore it is a candidate for a black-hole. This is in line with 't Hooft's idea of viewing fundamental elementary states themselves as black-holes. In this connection one would be naturally led to ask questions relevant for black-hole dynamics from the view point of strings. One natural question is how long does it take for a black-hole to evaporate? This is thus translated to the question of how long it takes for the highly excited string state to decay. What is the spectrum of massless states one is left with? Does the decay product of a massive string state resemble a thermal distribution of massless string modes? If so how does the temperature of this distribution depend on the mass of the excited string state?

At first sight this correspondence may raise some paradoxes. The first question is that of entropy of a black-hole: It is natural to assume that the reason for the non-vanishing entropy of the black-hole is that there is a lack of information as to how the energy M is shared among various string states. However this runs to immediate problem: The entropy of the black-hole is dimension dependent and it goes as $exp(M^{d-2})$ for a d-dimensional spacetime, whereas the Hagedorn behaviour of strings gives $exp(M)$ for the entropy of a string state with mass M. However due to a very interesting observation of Susskind [9] this is not necessarily a contradiction as there may be a mass renormalization due to gravitational dressing, which effectively replaces $M \to M^{d-2}$. This may remove a fundamental obstacle in identifying a massive string state with a black-hole.

There are other interesting connections: If we compactify superstrings from 10 dimensions down to 4 on a six dimensional torus, and consider for a fixed point in space, a string state which winds around one of the cycles of the internal torus of length L. Let us further assume that $L >> 1$. Then viewed from the four dimensional view point we have a state whose mass is $M = L$ and whose extent is much less than L, and so is a black hole. But we know that in string theory such a winding state is absolutely stable (if it has no oscillations on it–otherwise it would decay to a winding state with no oscillations). Thus the question is why such a black-hole does not decay. The answer is simply that it also carries a $U(1)$ charge equal to $Q = L$. So we have a state with $Q = M$, which is the condition for extremality of a black-hole. Such a black-hole is not expected to radiate from four dimensional view point either. In fact the situation is more interesting, because string theory also predicts that for any string state $M \geq Q$ (which in the conformal language follows from the unitarity bound $h \geq Q^2/2$), which is precisely the bound expected for charged black-holes! This lends further support to the natural idea that string states themselves may be viewed as black-holes[4].

These points suggest that we should identify all the extremal black hole states with absolutely stable string states. We are thus naturally led to ask the question: how does one classify stable states in a string theory? Moreover, if we bring two such states together, do we get another stable state? These questions are being currently considered [10]. The results are in line with the intuition that if something is stable, there is some kind of 'charge' keeping it from absolute decay (i.e. the state is the minimum energy state in a given charge sector).

4. Other Examples of Singularity

We have shown that many interesting examples of singularities are harmless in string theory. One should be careful before making a sweeping generalization that all singularities are harmless. In fact there are examples of singular backgrounds (manifolds) which seem harmful in string theory. These manifolds can be reached with finite action from smooth ones. An important example of this happens for the quintic threefold at a particular point on moduli ($\psi = 1$) (see [11]). It would be very important to understand what happens to string theory in such a limit, and whether it signals an incompleteness of string theory.

I would like to thank K. Intriligator, L. Susskind and E. Verlinde for valuable discussions. This work was supported in part by NSF grants PHY-87-14654 and PHY-89-57162 and a Packard fellowship.

[4] The similar bound in the case of heterotic strings is $M \geq \sqrt{Q^2 - 1}$, which shows that for large enough Q we get the usual bound expected on classical grounds, but also there is no contradiction with the existence of massless charged states in string theory.

References

[1] R. Brandenberger and C. Vafa, Nucl. Phys. B316 (1988) 28.

[2] K. Kikkawa and M. Yamasaki, Phys. Lett. B149 (1984) 357;
N. Sakai and I. Senda, Prog. Theor. Phys. 75 (1986) 692;
V. Nair, A. Shapere, A. Strominger and F. Wilczek, Nucl. Phys. B287 (1987) 402;
B. Sathiapalan, Phys. Rev. Lett. 58 (1987) 1597;
E. Silverstein, *Large-Small Equivalence in String Theory*, hepth-9201009;
G. Moore, *Finite in All Directions*, hepth-9305139.

[3] A.A. Tseytlin and C. Vafa, Nucl. Phys. B372 (1992) 443.

[4] L. Dixon, J. Harvey, C. Vafa and E. Witten, Nucl. Phys. B274 (1986) 285.

[5] E. Witten, Phys. Rev. D44 (1991) 314.

[6] S. Mukhi and C. Vafa, *Two Dimensional Black-hole as a Topological Coset Model of c=1 String Theory*, 1993 preprint, HUTP-93/A002 , TIFR/TH/93-01.

[7] C. Vafa and E. Verlinde, unpublished notes.

[8] E. Martinec and S. Shatashvili Nucl. Phys. B368 (1992) 338.

[9] L. Susskind,*Some Speculations about Black Hole Entropy in String Theory*, 1993 preprint, hepth-9309145.

[10] K. Intriligator and C. Vafa, work in progress.

[11] P. Candelas, X.De le Ossa, P. Green and L. Parkes, *Essays on Mirror Manifolds*, ed. by S. T. Yau, International Press, 1992.

SUPERSTRINGS RECONCILE GRAVITY WITH QUANTUM MECHANICS

DANIELE AMATI
International School for Advanced Studies and INFN
Trieste, 34014 Italy

Many aspects of Einstein gravity seem hard to reconcile with quantum mechanics. Short distance quantum fluctuation – at the level of the Planck length – are too strong for the consistency of the theory. Furthermore, black hole solutions, classical unavoidable fate of some systems, seem to transform quantum mechanical pure states into mixed states thus destroying the information contained in their coherence.

A consistent theory of quantum gravity should clarify how these problems are solved: providing on one hand the short distance deviations to Einstein's description and showing on the other, how black holes operate in a coherent quantum treatment. This implies to avoid the hybrid description of a quantum state in a black hole classical background treating instead the whole process - that includes black hole formation and evaporation - in a quantum framework.

Superstrings, being a consistent quantum theory that contains gravity seem a very unique basis to answer these very fundamental questions. To start with, it is necessary to understand how to ask them: indeed the theory is well defined in a perturbative (topological) expansion and with a large - but limited - variety of realizations.

We will consider, as a realistic candidate, a (type II) superstring theory with four flat uncompactified dimensions and we will study[1] on shell scattering amplitudes for which the theory provides well defined perturbative expansions. It is obvious that through scattering amplitudes short distances may be explored as shown by the revelation of quasi point like structures in hadrons through deep inelastic scattering. In order to form a black hole, on the other hand, one has to concentrate energy (mass) in a small volume, with radius smaller than the Schwarschild radius of the energy poured into it. A realistic way would be to produce it by letting many particles impinge spherically, but we shall discuss the simpler case of a very energetic two particle collision with an impact parameter smaller than the Schwarschild radius of the process.

We shall analyse an elastic scattering in those conditions with the objective of identifying a black hole as an intermediate state produced in the process. It should be revealed by a singularity (much as a bound state or resonance) appearing as an infinite genus effect. This cannot be obtained from a perturbative development without a reorganization of contributions that allows to compute and re-sum all orders within the relevant contributions. In our treatment[1] this is provided by a high energy expansion valid in the asymptotic regime defined by

$$\frac{E^2}{M_p^2} > 1 \qquad g \ll 1 \tag{1}$$

where g is the string loop coupling parameter and E is the c.m. collision energy (let's say of two gravitons), M_p being the Planck mass. Recall that M_p and thus the Newton constant $G = \frac{\hbar}{M_p^2}$ are related to the string mass M_s and the string coupling g (the only two parameters of the theory) by $M_p = \frac{M_s}{g}$ so that $M_p \gg M_s$ in our asymptotic regime.

In this regime gravity is a strongly interacting theory. Indeed, every power of g^2 appearing in higher orders may be enhanced by a power of $\frac{E^2}{M_s^2}$ thus giving powers of

$$\frac{g^2 E^2}{M_s^2} = \frac{E^2}{M_p^2} = \frac{GE^2}{\hbar} > 1.$$

Our definition of leading contributions implies the consideration of only enhanced g^2, but those to all orders. Non-leading contributions will be given by the expansion in non-enhanced g^2.

For impact parameters b much larger than the string length $\lambda_s = \frac{\hbar}{M_s}$, the leading scattering amplitude turns out to be

$$S(b,E) = e^{2i\delta_0}, \quad \delta_0 = -\frac{GE^2}{\hbar} logb\Lambda \qquad (b \gg \lambda_s), \tag{2}$$

where Λ is the (infinite) irrelevant Coulomb phase. This leading eikonal expression stemming from string theory has also been obtained with several other methods[2]. The scattering angle is given by the saddle point of the eikonal phase in the Fourier transformation that brings from the impact parameter to the transfer momentum representation. The leading amplitude (2) gives[1]

$$\theta_{cl} = \frac{2\hbar}{E}\frac{\partial\delta_0}{\partial b} = \frac{8GE}{b} = \frac{2R}{b}; \quad R \equiv 4GE \tag{3}$$

result that coincides with the classical Einstein deflection angle.

Let us notice that R, as defined in Eq. (3), represents the Schwarschild radius of the process. In our asymptotic regime R may be smaller or larger than λ_s depending on how the small and large factors in the expression

$$\frac{R}{\lambda_s} = 4g\frac{E}{M_p} \tag{4}$$

are organized. These two possibilities give rise to very different behaviours for the scattering amplitude if b starts decreasing.

For $R < \lambda_s$ (this means large but fixed $\frac{E}{M_p}$ and $\frac{M_s}{M_p} \to 0$) string effects soften gravity at small distances, λ_s appears as a minimum explorable length as implied[1] by a modified uncertainty relation $x \sim \frac{\hbar}{q} + q\alpha'$.

For $R > \lambda_s$ new subleading terms proportional to powers of $\frac{R^2}{b^2}$ start counting for b approaching R. At the first non-leading order $(O(\frac{R^2}{b^2}))$ the elastic scattering amplitude is given by[3]

$$S_1 = exp \quad 2i(\delta_0 + \delta_1 + \delta_2 + \ldots) \tag{5}$$

$$\delta_1 = \frac{6}{\pi}\frac{G^2E^2}{b^2} logE^2$$

$$\delta_2 = \frac{2G^3E^4}{\hbar^2 b^2}[1 + \frac{2i}{\pi}(1 - logb\Lambda)logE^2] \tag{6}$$

δ_1 may be viewed as an i.r. finite real quantum effect anyhow irrelevant in our asymptotic regime. δ_2, instead, has an i.r. divergent imaginary part due to soft graviton bremsstrahlung. Its i.r. free – due to a Block-Nordsieck mechanism – real part, is of order $\frac{1}{\hbar}$ as the leading term δ_0 and represents indeed a classical looking correction to it of $O(\frac{R^2}{b^2})$ because

$$\frac{G^3E^4}{\hbar b^2} = \delta_0 \frac{R^2}{b^2} \log b$$

thus giving rise to a modification

$$sin\frac{\theta_{cl}}{2} = \frac{R}{b}(1 + \frac{R^2}{b^2} + \ldots) \qquad (7)$$

to the Einstein deflection angle.

The actual computation[3] of $Re\delta_2$ –identified with the first non-leading eikonal – was quite involved and implied complex cancellations among i.r. singular diagrams. The extension of this calculation to higher orders represents a hard task obviously inappropriate for the resummation needed in order to disclose a singularity in b^2/R^2. We have thus attempted[4] an effective action approach trying to learn from the first non leading computation, which are the relevant degrees of freedom responsible for the classical corrections we are looking for, and which is their dynamics. Conforted, of course, by the confidence gained that if the relevant contributions are well identified a cavalier treatment of i.r. singularities is inessential to the computation of the i.r. free real eikonal.

The form of the asymptotic graviton emission vertex[5] to obtain Eqs.(5)(6) was used to write an effective action in terms of the relevant gravitons emitted modes h_{TT} that have a transverse character. These transverse modes (dashed lines in Fig. 1) are emitted – and eventually rescattered – by light like longitudinal ones h_{++} and h_{--} (wavy lines in Fig. 1) that, besides, interact with the scattering particles (full lines in Fig. 1) acting as sources i.e.

$$L_{source} = \sqrt{G}Eh^{--}(x)\delta(x^-)\delta^2(x_T) + \sqrt{G}Eh^{++}(x)\delta(x^+)\delta^2(x_T - b) \qquad (8)$$

$x_\pm = x_0 \pm x_3$ being the light like longitudinal coordinates and $x_T = (x_1, x_2)$ the transverse ones.

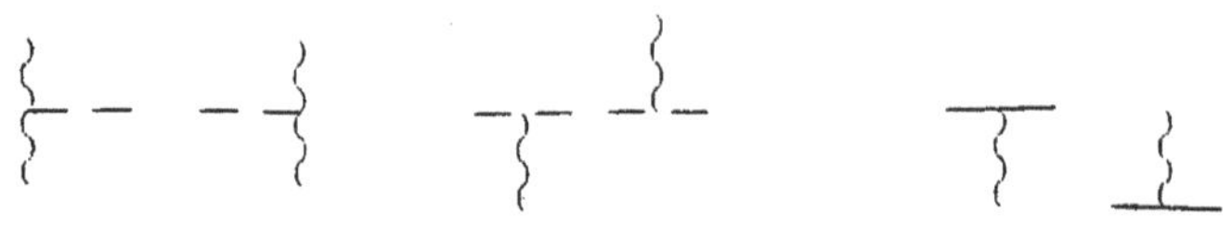

Fig. I - Interaction terms in the effective action

The emission vertex[5] was originally non local due to the different diagrams responsible for the emission. This non locality – represented by inverse square momenta – could be reabsorbed in a redefinition of fields that contain derivatives at the expense, however, of a higher derivative kinetic term of the mode representing the transverse degrees of freedom (dashed line propagator).

I will not reproduce explicitly in this lecture the actual form of the effective action. It coincides essentially with an asymptotic Regge–Gribov effective gravity action recently proposed by Lipatov[6] apart from total derivatives (surface terms) that are however important for a correct i.r. treatment.

The equations of motion define classical fields that, when introduced back into the action, give the scattering amplitude. Thus the problem is reduced to solve the equations of motion; the sources of Eq. (8) provide the inhomogeneous terms and the interactions of Fig. (1) provide the bilinear ones.

Furthermore, if rescattering is neglected it is possible to solve the longitudinal coordinate dependence of all fields. Rescattering is anyhow absent when the equations of motion are solved at the first iterative level where we have explicitly checked the results with the perturbative ones. Beyond that level there is no justification to neglect rescattering except for the fact that this slightly simplified problem is already able to show much of the richness of the whole approach and may allow us to identify the black hole dynamics we are after. As already said, without rescattering the longitudinal coordinate dependence may be solved. It shows classical metric fields that represent two shock waves for the longitudinal components, with a transverse component in between the two fronts, i.e.

$$h^{++} = \sqrt{G}E\delta(x_-)a(z,z^*)$$
$$h^{--} = \sqrt{G}E\delta(x_+)\overline{a}(z,z^*)$$

$$h_{TT} = \sqrt{G}GE^2\theta(x_+x_-)\frac{\partial^2}{\partial z^{*2}}\varphi(z,z^*)$$

with

$$a(z,z^*) = a^*(z,z^*) = \bar{a}(b-z,b-z^*)$$

$$\varphi^*(z,z^*) = \varphi(b-z,b-z^*)$$

$$z = x_1 + ix_2 \qquad z^* = x_1 - ix_2.$$

The profile functions $a, \bar{a}$ of the two shock waves and the function φ describing the sandwich transverse metric, are determined by two-dimensional equations of motion generated by a reduced transverse dynamics. The hint that the dynamics of high energy gravity could reduce to a $2-d$ theory has been met in a few other approaches[7]. The solutions of equations of motion are of course defined up to addition of solutions of the homogeneous system, freedom then used and exhausted, to implement regularity requirements as monodromy and correct u.v. and i.r. behaviours.

In order to check the general method, we have solved explicitly the $2-d$ equations at the first iterative expansion and shown that when introduced into the effective action they produce a scattering amplitude that exactly coincides with the first non-leading eikonal $exp2i(\delta_0 + Re\delta_2)$ with δ_0 and δ_2 given in Eqs. (2) (6) as obtained with the diagrammatic method.

But of course the objective is not to generate in a simpler way the eikonal as a series in b^2/R^2 but to have for it a non-perturbative approach so to investigate singularities that would reveal a black hole formation. The analysis of the infinite number of perturbative gravitons giving rise to the singularity should provide the spectrum of gravitons generated from the black hole evaporation. The elastic scattering amplitude obtained should be viewed as generated from how those evaporated gravitons merge then into a white hole that ends into the two final energetic particles.

The consistent quantum description of our approach could then answer other questions as - for instance - how unitarity is preserved when another particle participates in the process. The total process of black hole formation - in which the test particle and those giving rise to the black hole act together - and subsequent evaporation should show how the guaranteed agreement with basic axiom of quantum mechanics is implemented. We have not yet been able to solve the $2-d$ system of differential equations in a non-iterative way, so to detect and study the expected singularities. But a consistent method is at least clearly settled and should throw light on how a consistent quantum theory of gravity treats black hole formation and evaporation.

REFERENCES

1) D.Amati, M.Ciafaloni and G.Veneziano, *Int. J. Mod. Phys.* **A3** (1988) 1615, *Phys. Lett.* **B216** (1989) 41.
2) G.'t Hooft, *Phys. Lett.* **B198** (1987) 61; I.Muzinich and M.Soldate, *Ph. Rev.* **D37** (1988) 353; D.Kabat and M.Ortiz MIT-CTP 2069 (1992); E.Verlinde and H.Verlinde, *Nucl. Phys.* **B371** (1992) 246.
3) D.Amati, M.Ciafaloni and G.Veneziano, *Nucl. Phys.* **B347** (1990) 550.
4) D.Amati, M.Ciafaloni and G.Veneziano, SISSA 11/93/EP, *Nucl. Phys.* in press.
5) M.Ademollo, A.Bellini and M.Ciafaloni, *Nucl. Phys.* **B338**, (1990) 114.
6) L.N.Lipatov, *Nucl. Phys.* **B307** (1988) 705.
7) E.Verlinde and H.Verlinde, ref. 2) and PUPT-1319; M.Fabbrichesi, R.Pettorino, G.Veneziano and G.A.Vilkovisky, private communication.

V. Mathematical Physics and Condensed Matter

FULLERENES AND CARBON 60

Chen Ning Yang
Institute for Theoretical Physics
State University of New York
Stony Brook, New York 11794-3840

ABSTRACT

Some topological considerations are outlined for the structure of the fullerenes. Also the Hückel problem for Carbon 60 is solved.

It is a great pleasure to participate in this conference honoring Abdus Salam. We honor him for his sharp intellect, for his many beautiful achievements in fundamental physics, and for his tireless and dynamic leadership of this international center which has touched the lives of so many physicists all over the world.

What I shall discuss with you this afternoon concerns some new molecules: the fullerenes. They were discovered[1] in 1985 by Kroto, Smalley and their collaborators. Five years later, Krätschmer and Huffman[1] discovered a simple method of making these molecules and a new branch of chemistry was born. I shall discuss some topological and physical problems related to these molecules.

1. Topology

1. The fullerenes are molecules made of carbon atoms arranged on a shell, forming a polyhedron with pentagons and hexagons as its faces. Each carbon atom has three bonds which are the edges for the polyhadron. The standard soccer ball today has the shape of one fullerene, carbon 60, which is the most important and most prevalent of all the fullerenes. Kroto and Smalley had pointed out that because of Euler's theorem each fullerene molecule must have exactly twelve pentagons on its surface. To show this let p be the number of pentagons and h the number of hexagons. Then the number of faces is equal to

$$f = p + h. \tag{1}$$

Let n be the number of carbon atoms, i.e. the number of vertices. Then Euler's theorem states that

$$f - e + n = 2, \tag{2}$$

where e is the number of bonds. If we, for each bond, draw a small dot next to it on each face bordering it, we would have 2e dots. Now each pentagon has 5 dots and each hexagon 6. Thus

$$2e = 5p + 6h. \tag{3}$$

Similarly,

$$3n = 5p + 6h. \tag{4}$$

It follows from these equations that

$$p = 12 \quad \text{and} \quad n = 20 + 2h. \tag{5}$$

Thus n = even, a fact that had contributed to Kroto and Smalley's discovery of the structure of the fullerenes.

The smallest fullerene, n = 20 (h = 0), is the dodecahedron. One can prove without difficulty that it is topologically not possible to have n = 22. We shall now prove that for all larger even values of n, C_n is topologically possible. The proof follows that in Sah's paper[2].

Consider the four "caps" on fig. 1, each with 6 pentagons and 1, 2, 3 or 4 hexagons. We shall designate a carbon atom on the boundary of a cap with the number 2 or 3, depending on whether it has 2 or 3 bonds within the cap. It is obvious that each of the four caps has 12 boundary atoms, sequentially designated by 2, 3, 2, 3... . Thus if we project cap-a onto the upper hemisphere of a sphere and cap-b onto the lower hemisphere, the boundaries of the two caps can be made to mate, forming a fullerene for which h = a + b. Furthermore, we can insert the band shown in Figure 2 inbetween the two caps and make all three mate to form a fullerene for which h = a + b + 6. If m bands are inserted, we get h = a + b + 6m. Suitably choosing a, b and m we arrive at the conclusion that

The fullerene C_n is topologically possible for all even n ≥ 24.

2. In all methods of producing fullerenes, C_{60} is the most abundant, C_{70} the next. Is there a simple explanation of these facts?

It has been known to the chemists for decades that energetically it is not favourable to have two pentagons sharing the same C-C bond. The isolated pentagon rule (**IPR**) states that two pentagons are forbidden to share a C-C bond.

A fullerene satisfying **IPR** must have exactly 5 x 12 = 60 carbon atoms at the corners of the 12 isolated pentagons. *Thus the smallest fullerene with **IPR** is C_{60}.* It can further be shown[2] that C_{62}, C_{64}, C_{66} or C_{68} cannot satisfy **IPR**, and that the next bigger fullerene with **IPR** is C_{70}.

It may thus appear that the **IPR** is the proper explanation of the abundances of C_{60} and C_{70}. Unfortunately, C_{72} also has[2] an **IPR** structure, but is very nonabundant. Thus **IPR** alone is not enough to explain the great abundance of C_{60} and C_{70} and the rarity of C_{72}.

There are many solved and unsolved topological problems about the fullerenes. An interesting one which might be solvable, (though not useful), is the asymptotic form of the number of configurations for C_n as $n \to \infty$.

II. Hückel Problem for C_{60}

1. The C_{60} molecule has icosahedral symmetry, the rotational symmetry group I of the icosahedron, a group of 60 elements. It also has inversion symmetry. Thus it is invariant under $I \times Z_2$ when Z_2 is the two

element group consisting of the inversion operator P and the identity. We emphasize that C_{60} processes $I \times Z_2$ symmetry even though its **C-C** bonds are not all of the same length: those bordering two hexagons (h-h bonds) are different in length from those bordering a hexagon and a pentagon (h-p bonds).

To solve the Hückel problem for C_{60} we shall utilize[3] its $I \times Z_2$ symmetry.

The irreducible representations (IR) of $I \times Z_2$ are R_{1+}, R_{1-}, R_{3+}, R_{3-}, $R_{3'+}$, $R_{3'-}$, R_{4+}, R_{4-}, R_{5+} and R_{5-} where the indices represent the dimension and the parity of the IR. Now the 60 dimensional Hückel space is split into an even space and an odd space according to parity, each being 30 dimensional. Using standard character table we find that the even space is further split into IR's as follows:

$$R_{1+} + R_{3+} + R_{3'+} + 2R_{4+} + 3R_{5+}, \tag{6a}$$

and the odd space is split thus

$$2R_{3-} + 2R_{3'-} + 2R_{4-} + 2R_{5-}. \tag{6b}$$

2. The Hückel Hamiltonian is a 60 x 60 matrix. It depends on a parameter which characterizes the ratio of the couplings for the h-h and h-p bonds. We choose this parameter[3] to be α so that the h-h coupling has strength $2-\alpha$ and the h-p coupling has strength α. (Notice that when $\alpha = 1$

all couplings have strength 1.) The Hückel Hamiltonian **H** then has as matrix elements linear expressions in α. Its diagonal elements are all zero.

Each IR R_{ip} of $I \times Z_2$ can be further transformed according to the IR of the subgroup $\{\rho^o, \rho^1, \rho^2, \rho^3, \rho^4\}$ where ρ is an element of I representing a rotation through $2\pi/5$. The IR of this subgroup is one dimentional, and is characterized by an integer m, so that

$$\rho = \eta^m, \qquad [\eta = \exp(2\pi i/5)], \tag{7}$$

in exact analogy with the quantum number m for the rotational group SO_3 familiar in elementary quantum mechanics.

We shall call the eigenalue of P $p(=\pm 1)$ and refer to even and odd parities.

If we transform **H** into many blocks so that each block belongs to only one (mp) then all nondiagonal blocks vanish because of the $I \times Z_2$ symmetry of **H**. The diagonal blocks will be called $\mathbf{H}_{mp}$. The characteristic polynomial P_{mp} for each $\mathbf{H}_{mp}$ factorizes into characteristic polynomials Q_{ip}, each belonging to one IR of $I \times Z_2$. Using (6a) and (6b) we find

$$P_{o+} = Q_{1+}Q_{3+}Q_{3'+}Q_{5+} \tag{8}$$

$$P_{1+} = Q_{3+}Q_{4+}Q_{5+} \tag{9}$$

$$P_{2+} = Q_{3'+}Q_{4+}Q_{5+}. \tag{10}$$

The lack of $Q_{3'+}$ in (9) is because the representation $R_{3'+}$ does not have the state m=1. Similarly the lack of R_{4+} in (8) is because R_{4+} does not have the state m=0, etc..

Now each P_{mp} is a 6-th degree polynomial in the eigenvalue λ, because given (mp) it is only necessary[3] to assign wave function values to the 6 sites bordering one hexagon to be able to find the wave function on all 60 sites. This fact enables one to explicitly write down the polynomials P_{0+}, P_{1+}, and P_{2+}. The greatest common divisor of (8)–(10) is Q_{5+}. That of (9) and (10), $Q_{4+}Q_{5+}$. This way we find all five polynomials Q_{i+}:

$$Q_{5+} = \lambda^3+2\lambda^2+\lambda(-4+8\alpha-6\alpha^2)-8+16\alpha-2\alpha^2+\alpha^3,$$

$$Q_{4+} = \lambda^2-\alpha\lambda-4+4\alpha-2\alpha^2,$$

$$Q_{3'+} = \lambda-2+(1/2)(1-\sqrt{5})\alpha,$$

$$Q_{3+} = \lambda-2+(1/2)(1+\sqrt{5})\alpha,$$

$$Q_{1+} =\lambda+2+\alpha. \tag{11}$$

Similarly we find for odd parity

$$Q_{5-} = \lambda^2-\alpha\lambda-4+6\alpha-3\alpha^2,$$

$$Q_{4-} = \lambda^2-\alpha\lambda-4,$$

$$Q_{3'-} = \lambda^2+(1/2)\lambda\alpha(3+\sqrt{5})-4+\alpha(5-\sqrt{5})+(1/2)\alpha^2(-5+3\sqrt{5}),$$

$$Q_{3-} = \lambda^2+(1/2)\lambda\alpha(3-\sqrt{5})-4+\alpha(5+\sqrt{5})+(1/2)\alpha^2(-5-3\sqrt{5}). \tag{12}$$

Notice that in each of these nine polynomials, for each term the sum of the powers of λ and α is always $\leq$ the maximum power of λ in the polynomial. [This fact is obvious when we remember that **H** is linear in α.]

Putting (11) and (12) equal to zero we obtain all the eignevalues λ of the Hückel Hamiltonian for all α.

References:

1. See e.g. M. S. Dresselhaus, G. Dresselhaus and P. C. Ekluid, "Fullerenes", MIT preprint.
2. C. H. Sah, "Combinational Buckyballs", Stony Brook Mathematics Department preprint.
3. Yuefan Deng and Chen Ning Yang, Phys. Letters A170, 116 (1992).

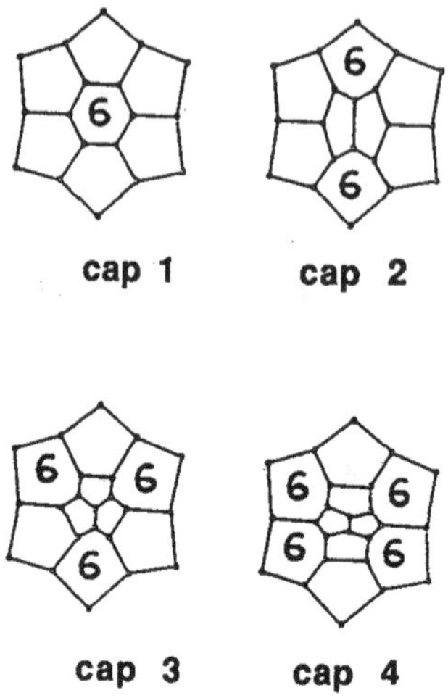

Fig. 1 Caps a. The number of hexagons in each cap is a. Each cap is bordered by 12 atoms to be designated by 2 or 3 according to whether the atom has 2 or 3 bonds within the cap. (From Sah's paper[2]).

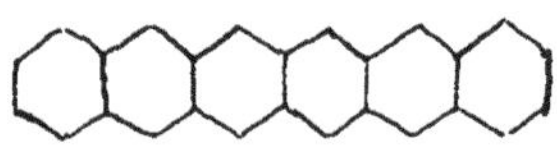

Fig. 2. Band of 6 hexagons. The leftmost bond should be made to coincide with the rightmost bond.

Translations, Integrals and Fourier Transforms In the Quantum Plane *

Chryssomalis Chryssomalakos and Bruno Zumino

Department of Physics
and
Theoretical Physics Group
Lawrence Berkeley Laboratory
University of California
Berkeley, California 94720

Abstract

We present a formulation of covariant translations in the quantum plane. We are led to an extension of the algebra of the coordinate functions and their dual derivatives by the quantum analogue of their eigenvalues. Jackson exponentials emerge as the corresponding eigenfunctions. An integral invariant under quantum translations is introduced and is used to define quantum Fourier transforms.

*This work was supported in part by the Director, Office of Energy Research, Office of High Energy and Nuclear Physics, Division of High Energy Physics of the U.S. Department of Energy under Contract DE-AC03-76SF00098 and in part by the National Science Foundation under grant PHY90-21139.

1 Introduction

Since its inception, the quantum plane has been envisioned by many as a paradigm for the general program of q-deformed physics. Such an endeavour presupposes the availability of adequate mathematical tools, integration being one of the most indispensable among them. The aim of this paper is to address aspects of the problem of integration in the quantum plane in a manner that will keep the results accessible to physicists.

We begin, in section 2, by specifying the meaning of a translation in the quantum plane. We find that consistency requires non-trivial commutation relations between the variables that describe translations and the coordinates on the plane as well as the derivatives with respect to these coordinates. We are then able to generate finite translations by q-exponentiating a translation generator. We introduce next, translations for the derivatives and interpret them, together with the ones introduced earlier, as (non-commuting) eigenvalues of the coordinate and derivative operators.

In section 3 we define, along the lines of [1], an integral on the plane essentially by the requirements of linearity and its vanishing when evaluated on a derivative of a function. We are then able to show, using a prescription about how to treat displacements in the integrand, that the above properties imply invariance under finite translations as well.

Section 4 contains a definition of the quantum Fourier transform which is seen to retain, in a q-deformed way, basic properties of its classical analogue. We end the paper by introducing, in section 5, "vacuum projectors". Using them we recover the integration prescription of section 3 in a constructive way.

2 Translations in the q-plane

We recall now the construction of the quantum plane [1] . We deal in the following with the (non-commutative) algebra of functions on the quantum plane enlarged so as to also include derivatives that operate on these functions. We choose as generators of $\mathcal{A}$ the coordinate functions x^i, $i = 1, \ldots, n$ (together with the unit function 1_x) and the derivatives dual to them, ∂_j, $j = 1, \ldots, n$ (together with the unit 1_∂). A set of consistent commutation relations among the above generators is known:

$$\begin{aligned} x^i x^j &= q^{-1} \hat{R}^{ij}_{kl} x^k x^l \\ \partial_l \partial_k &= q^{-1} \hat{R}^{ij}_{kl} \partial_j \partial_i \\ \partial_k x^i &= \delta^i_k + q \hat{R}^{ij}_{kl} x^l \partial_j . \end{aligned} \tag{1}$$

Here, $\hat{R}$ is an invertible solution of the quantum Yang-Baxter equation:

$$\hat{R}_{12}\hat{R}_{23}\hat{R}_{12} = \hat{R}_{23}\hat{R}_{12}\hat{R}_{23}$$

and satisfies the characteristic equation:

$$\hat{R}^2 - \lambda \hat{R} - 1 = 0, \qquad \lambda \equiv q - q^{-1} \tag{2}$$

Explicitly, $\hat{R}^{ij}_{kl} = R^{ji}_{kl}$ where R is the $GL_q(n)$ R- matrix of [2], given by:

$$R = q \sum_i e_{ii} \otimes e_{ii} + \sum_{i \neq j} e_{ii} \otimes e_{jj} + \lambda \sum_{i>j} e_{ij} \otimes e_{ji}$$

where $i, j = 1, \cdots, n$ and e_{ij} is the $n \times n$ matrix with single nonzero element (equal to 1) at (i, j). The above commutation relations permit unambiguous ordering of an arbitrary monomial in the x's and ∂'s into any desired order. One can now write down, if one wishes, differential equations for functions of the x's and study for example quantum mechanical systems by solving Schroedinger's equation in deformed space. In doing so, as well as in many other applications, one is sooner or later bound to be confronted with the problem of (spatially) translating functions of the x's. One place, in particular, where this question would certainly manifest itself, would be in the statement of finite translation invariance of any sort of integral one adopts for the quantum plane. One has then to first make more precise the notion of translation - it is natural, for example, to require a certain covariance. In its simplest form, that would be the requirement that the translated coordinates obey the same algebra as the original ones (we would also need, of course, reduction to the correct classical limit $x^i \mapsto x^i + a^i$ as the deformation parameter approaches its classical value). In bialgebra language, we are in search of a coproduct for the algebra of the x's, namely a map $\Delta : \mathcal{A}_x \to \mathcal{A}_x \otimes \mathcal{A}_x$ that is an algebra homomorphism: $\Delta(xy) = \Delta(x)\Delta(y), x, y \in \mathcal{A}_x$, and satisfies some standard additional conditions (see for example [3] - here $\mathcal{A}_x$ is the subalgebra of $\mathcal{A}$ generated by the coordinates). Such a map has been sought for for quite some time now, without success. Faced with this fact, we elect to pursue a different approach. We introduce a set of "displacements" a^i, $i = 1, \ldots, n$ and require that the $x + a$'s obey the same commutation relations as the x's. We would also like the displacements to be of "coordinate nature" *i.e.* we postulate $a - a$ commutation relations identical to those of the x's. We are then forced to introduce non-trivial $a - x$ commutation relations. The resulting algebra is:

$$\begin{aligned} a^i a^j &= q^{-1} \hat{R}^{ij}_{kl} a^k a^l \\ x^i a^j &= q \hat{R}^{ij}_{kl} a^k x^l. \end{aligned} \tag{3}$$

We check for covariance:

$$\begin{aligned} (x_1 + a_1)(x_2 + a_2) &= x_1 x_2 + a_1 x_2 + x_1 a_2 + a_1 a_2 \\ &= x_1 x_2 + (q^{-1} \hat{R}^{-1}_{12} + 1) x_1 a_2 + a_1 a_2. \end{aligned}$$

Also:

$$\begin{aligned} q^{-1} \hat{R}_{12} (x_1 + a_1)(x_2 + a_2) &= q^{-1} \hat{R}_{12} x_1 x_2 + q^{-2} x_1 a_2 \\ &\quad + q^{-1} \hat{R}_{12} x_1 a_2 + q^{-1} \hat{R}_{12} a_1 a_2. \end{aligned} \tag{4}$$

Using now the characteristic equation (2) in the form:

$$q^{-1} \hat{R}^{-1} + 1 = q^{-1} \hat{R} + q^{-2}$$

and comparing the two expressions above we find:

$$(x_1 + a_1)(x_2 + a_2) = q^{-1}\hat{R}_{12}(x_1 + a_1)(x_2 + a_2)$$

as desired. It is interesting to compare (3) with the commutation relations between coordinates and differentials introduced in [1] :

$$x_1\xi_2 = q\hat{R}_{12}\xi_1 x_2\ ;$$

the displacements a^i are the bosonic analogue of the ξ's (the algebra (3) has been introduced by Majid, in the context of braided Hopf algebras, in [4]). We can also give consistent $\partial - a$ commutation relations:

$$\partial_k a^i = q^{-1}(\hat{R}^{-1})^{ij}_{kl} a^l \partial_j \qquad (5)$$

(again, similar to the $\partial - \xi$ ones). Consider now the translation generator T defined by:

$$T \equiv a^i\partial_i \equiv a \cdot \partial.$$

Using (3), (5), we easily find:

$$[T, x^i] = a^i, \quad T\partial_i = q^2\partial_i T, \quad Ta^i = q^{-2}a^i T. \qquad (6)$$

These allow us to build a finite translation operator by "q-exponentiation". We have:

$$T^n x^i = x^i T^n + [n]_q T^{n-1} a^i \qquad (7)$$

where $[n]_q \equiv (1 - q^{2n})/(1 - q^2)$ and therefore:

$$x^i e_q(T) = e_q(T)(x^i - a^i) \qquad (8)$$

where:

$$e_q(T) \equiv \sum_{n=0}^{\infty} \frac{1}{[n]_q!} T^n, \quad [n]_q! \equiv [1]_q [2]_q \dots [n]_q$$

is the Jackson exponential (see for example [5] and references therein). Alternatively, we can write (7) in the form:

$$T^n x^i = x^i T^n + [n]_{q^{-1}} a^i T^{n-1}$$

which gives:

$$e_{q^{-1}}(T)x^i = (x^i + a^i)e_{q^{-1}}(T) \tag{9}$$

or, more generally:

$$e_{q^{-1}}(T)f(x) = f(x+a)e_{q^{-1}}(T). \tag{10}$$

One can regard (8) (or (9)) as an eigenvalue equation for the operator x^i. To make this more precise, we introduce coordinate and derivative vacua, denoted by $|\Omega_x\rangle$ and $|\Omega_\partial\rangle$ respectively, which satisfy:

$$x^i|\Omega_x\rangle = 0, \quad 1_x|\Omega_x\rangle = |\Omega_x\rangle, \quad \partial_i|\Omega_\partial\rangle = 0, \quad 1_\partial|\Omega_\partial\rangle = |\Omega_\partial\rangle$$

with similar relations for x, ∂ acting from the right:

$$\langle\Omega_x|x^i = 0, \quad \langle\Omega_x|1_x = \langle\Omega_x|, \quad \langle\Omega_\partial|\partial_i = 0, \quad \langle\Omega_\partial|1_\partial = \langle\Omega_\partial|.$$

The action of x^i on a function $f(\partial, a)$, denoted by $x^i(f(\partial, a))$, is expressed in terms of the coordinate vacuum as:

$$x^i(f(\partial, a))|\Omega_x\rangle = x^i f(\partial, a)|\Omega_x\rangle. \tag{11}$$

In words, to compute the left hand side of (11), we order it with all the x's on the right, where they anihilate the vacuum, and what remains is termed "the action of x^i on $f(\partial, a)$". We can define the (more familiar) action of the derivatives on functions of x, a in a similar manner. Actions from the right are also obviously defined via "left vacua" $\langle\Omega_x|$, $\langle\Omega_\partial|$. With these (standard) definitions, (8) gives:

$$(e_q(T))x^i = e_q(T)a^i \tag{12}$$

which suggests the interpretation of a^i as the eigenvalue of x^i (x^i acts here from the right). In the classical limit $q \to 1$, the a's commute with everything. Notice that $e_q(T)$ is a common eigenfunction for all x^i, the noncommutativity of the latter being reflected in the non-trivial $a - a$ commutation relations. It is interesting to note that one can interpret the derivatives ∂_i the way one does in classical analysis: $\partial_i(f(x))$ is the coefficient of a^i in the expansion of $f(x+a)$ around x [6]. Indeed, from (10) we have:

$$e_{q^{-1}}(T)(f(x)) = f(x+a) \tag{13}$$

which, by expanding the Jackson exponential, gives:

$$f(x+a) = f(x) + a^i\partial_i(f(x)) + \mathcal{O}(a^2), \tag{14}$$

the only difference in the quantum case being that one has to specify an ordering before identifying the derivative (above, we took the a's to stand to the left of the x's).

The interpretation given to the a^i above naturally leads to the question whether a similar construction is possible for the derivatives. To this end, we introduce the momentum-space analogue of the a's, which we call p_i, $i = 1,\ldots,n$ and find that the following commutation relations are consistent with the rest of the algebra:

$$\begin{aligned} p_l p_k &= q^{-1}\hat{R}^{ij}_{kl} p_j p_i \\ p_l \partial_k &= q\hat{R}^{ij}_{kl} \partial_j p_i \\ p_k x^i &= q^{-1}(\hat{R}^{-1})^{ij}_{kl} x^l p_j \\ p_k a^i &= q^{-1}(\hat{R}^{-1})^{ij}_{kl} a^l p_j . \end{aligned} \tag{15}$$

A host of useful identities can now be computed. We give a list involving invariant bilinears ($\alpha \cdot \beta \equiv \alpha^i \beta_i$):

$$\begin{aligned} Tp_i - p_i T &= 0 \\ x\cdot\partial\, x^i &= x^i + q^2 x^i\, x\cdot\partial \\ x\cdot\partial\, a^i - a^i\, x\cdot\partial &= 0 \\ \partial_i\, x\cdot\partial &= \partial_i + q^2 x\cdot\partial\,\partial_i \\ x\cdot\partial\, p_i - p_i\, x\cdot\partial &= 0 \\ x\cdot p\, x^i &= q^{-2} x^i\, x\cdot p \end{aligned}$$

$$
\begin{aligned}
x \cdot p\, a^i - a^i\, x \cdot p &= 0 \\
\partial_i\, x \cdot p &= x \cdot p\, \partial_i + p_i \\
x \cdot p\, p_i &= q^2 p_i\, x \cdot p \\
a \cdot p\, x^i &= q^{-2} x^i\, a \cdot p - q^{-1} \lambda a^i\, x \cdot p \\
a \cdot p\, a^i &= q^{-2} a^i\, a \cdot p \\
\partial_i\, a \cdot p &= q^{-2} a \cdot p\, \partial_i - q^{-1} \lambda T p_i \\
p_i\, a \cdot p &= q^{-2} a \cdot p\, p_i \\
(1 + q\lambda x \cdot \partial)\, x^i &= q^2 x^i (1 + q\lambda x \cdot \partial) \\
\partial_i (1 + q\lambda x \cdot \partial) &= q^2 (1 + q\lambda x \cdot \partial)\, \partial_i \\
(x \cdot p)(a \cdot p) &= q^2 (a \cdot p)(x \cdot p).
\end{aligned} \tag{16}
$$

We easily find now how ∂_i commutes with $e_q(x \cdot p)$:

$$\partial_i e_q(x \cdot p) = e_q(x \cdot p)\partial_i + p_i e_q(x \cdot p). \tag{17}$$

Also:

$$\partial_i e_{q^{-1}}(x \cdot p) = e_{q^{-1}}(x \cdot p)\partial_i + e_{q^{-1}}(x \cdot p) p_i. \tag{18}$$

We can therefore interpret the p's as (non-commuting) eigenvalues of the derivatives:

$$\partial_i (e_q(x \cdot p)) = p_i e_q(x \cdot p)\,, \tag{19}$$

Notice that being eigenvalues of derivatives, rather than momenta, the p's become real commuting quantities in the classical limit.

A couple of remarks are in order here. The first regards the covariance of the scheme described above under the coaction of $GL_q(n)$. The commutation relations given in (1), (3), (5) and (15) go into themselves when x, ∂, a and p transform according to:

$$
\begin{aligned}
x^i \mapsto (x')^i &= T^i_j x^j \\
a^i \mapsto (a')^i &= T^i_j a^j \\
\partial_i \mapsto (\partial')_i &= \partial_j M^j_i \\
p_i \mapsto (p')_i &= \partial_j M^j_i
\end{aligned}
$$

where T^i_j is a $GL_q(n)$ matrix, $M^t = (T^t)^{-1}$ (M^t denotes the transpose of M) and we take, as in [1], the elements of T to commute with all the variables and derivatives above. A second point that deserves attention is the fact that derivations do not commute with translations. In general:

$$\partial_i(f(x+a)) \neq \partial_i(f(x))|_{x\mapsto x+a}. \tag{20}$$

This can be traced to the fact that ∂_i does not commute with $a\cdot\partial$. Indeed, in order for (20) to be an equality, we would need (using (14)):

$$\begin{aligned} \partial_i(f(x)+a\cdot\partial(f(x))) &= \partial_i(f(x))+a\cdot\partial(\partial_i(f(x))) \Rightarrow \\ \Rightarrow \partial_i(a\cdot\partial(f(x))) &= a\cdot\partial(\partial_i(f(x))) \Rightarrow \\ \Rightarrow \partial_i\, a\cdot\partial &= a\cdot\partial\,\partial_i \end{aligned}$$

while, in our case, (6) holds: $a\cdot\partial\,\partial_i = q^2\partial_i\, a\cdot\partial$ (one can make a different choice of commutation relations which will make (20) into an equality but then (14) is not satisfied - we will not explore this further here).

3 Invariant Integration

We wish now to turn our attention to the problem of integration. The integral we are looking for is a linear map from functions on the quantum plane to complex numbers. Keeping in mind the classical limit, we expect it to be defined only for a class $\mathcal{A}^I_x$ of elements of $\mathcal{A}_x$ - we will use the notation $\langle f\rangle$ for the average, or integral, of f in that class. Such a map acquires interest when endowed with specific covariance properies. In our case, it is natural to require invariance under translations. This can be expressed in infinitesimal form as the requirement that the integral of a derivative vanish:

$$\langle\, \partial_i(f(x))\,\rangle = 0, \quad f\in\mathcal{A}^I_x. \tag{21}$$

A prescription for computing such an integral is known [1] . One first expands $f(x)$ in a sum of monomials in the x's and uses the commutation relations to bring each such monomial into some standard ordering (the same for all monomials). Then one performs the classical integral (from minus infinity to plus infinity) of the ordered function - the result is the quantum integral $\langle f \rangle$. Different standard orderings of the x's change only the overall normalization and the result satisfies (21)(notice that ∂_i is the *quantum* derivative). We would like though to be able to talk about finite translation invariance, *i.e.* we would like our integral to satisfy an equation like

$$\langle f(x+a) \rangle = \langle f(x) \rangle. \tag{22}$$

To make this precise, we ought to generalize the prescription for integration given above to the case of a function of x *and* a (since x, a do not commute, such a generalization is not trivial). Nevertheless, the natural approach works: to compute $\langle f(x+a) \rangle$, expand in monomials of x, a, use the commutation relations to move all the a's, say, to the left of each monomial and out of the integral, and then compute the quantum integral of the x's as before (notice that the a's need not be brought into any standard order). That such an integral satisfies (22) can easily be seen as follows. From (13) we have:

$$\begin{aligned} \langle f(x+a) \rangle &= \langle e_{q^{-1}}(a \cdot \partial)(f(x)) \rangle \\ &= \sum_{n=0}^{\infty} \frac{1}{[n]_{q^{-1}}!} \langle (a \cdot \partial)^n (f(x)) \rangle \,. \end{aligned}$$

We may now use the commutation relations given in (5) to move all the a's in $(a \cdot \partial)^n$ to the left (and then out of the integral). The form of (5) ensures that each term left in the integrand, except for $n = 0$, will be the derivative of some function of the x's and the integral of these vanishes by (21); (22) then follows. We should emphasize here that the integral of $f(x, a)$ is not, in general, translationally invariant (*i.e.* while $\langle f(x+a) \rangle = \langle f(x) \rangle$ holds, $\langle f(x+a, a) \rangle \neq \langle f(x, a) \rangle$ in general). In the same spirit, we define the integral $\langle f(x, a, p) \rangle$: we move a and p to the left and then perform quantum integration on the x's - we 'll need this in defining the Fourier transform in the next section.

We close this section with a formula for integration by parts, valid when the Jackson exponential of $x \cdot p$ is in the integrand. From (17) we get:

$$\partial_i(e_q(x \cdot p)f(x)) = e_q(x \cdot p)\partial_i(f(x)) + p_i e_q(x \cdot p)f(x)$$

which, upon integration, gives:

$$\begin{aligned} \langle e_q(x \cdot p)\partial_i(f(x))\rangle &= -p_i\langle e_q(x \cdot p)f(x)\rangle \\ &= -\langle \partial_i(e_q(x \cdot p))f(x)\rangle \end{aligned} \tag{23}$$

4 Fourier Transform

Armed with the tools developed in the previous section, we are now (almost) ready to discuss Fourier transforms in the quantum plane. The only ingredient missing is the observation that

$$\langle f(x+a,p)\rangle = \langle f(x,p)\rangle. \tag{24}$$

To see this, write $f(x,p)$ in the form:

$$f(x,p) = \sum_i g_i(p)f_i(x)$$

to get:

$$\begin{aligned} e_{q^{-1}}(T)(f(x,p)) &= \sum_i e_{q^{-1}}(T)(g_i(p)f_i(x)) \\ &= \sum_i g_i(p)e_{q^{-1}}(T)(f_i(x)) \\ &= \sum_i g_i(p)f_i(x+a) \end{aligned} \tag{25}$$

where, in the second line, we used the fact that p commutes with T. A look at (15) shows that the commutation relations of $p, x+a$ are identical to those of p, x which allows us to sum the last line of (25):

$$e_{q^{-1}}(T)(f(x,p)) = f(x+a,p). \tag{26}$$

Integrating both sides of (25) and using (26) we then get:

$$\begin{aligned}\langle f(x+a,p)\rangle &= \sum_i g_i(p)\langle f_i(x+a)\rangle \\ &= \sum_i g_i(p)\langle f_i(x)\rangle \\ &= \langle f(x,p)\rangle.\end{aligned}$$

We define now the Fourier transform $\tilde{f}(p)$ of a function $f(x)$ by:

$$\tilde{f}(p) \equiv \langle e_q(-ix\cdot p)f(x)\rangle\ . \tag{27}$$

We will need the following properties of the Jackson exponential

$$e_q(\alpha+\beta) = e_q(\beta)e_q(\alpha) \text{ for } \alpha\beta = q^2\beta\alpha \tag{28}$$

and

$$e_q(\alpha)e_{q^{-1}}(-\alpha) = 1 \tag{29}$$

to derive the analogue of a property of Fourier transforms, familiar in the classical case. Setting $f(x+a) \equiv f_a(x)$ we have:

$$\begin{aligned}\tilde{f}(p) &= \langle e_q(-ix\cdot p)f(x)\rangle \\ &= \langle e_q(-i(x+a)\cdot p)f_a(x)\rangle \\ &= \langle e_q(-ia\cdot p)e_q(-ix\cdot p)f_a(x)\rangle \\ &= e_q(-ia\cdot p)\tilde{f}_a(p) \Rightarrow \\ \Rightarrow \tilde{f}_a(p) &= e_{q^{-1}}(ia\cdot p)\tilde{f}(p)\end{aligned} \tag{30}$$

(where in the third line we used the last of (16)). Notice that, as in the classical case, the factor in front of $\tilde{f}(p)$ is actually a one dimensional representation of translations. Under $x \mapsto x+a$:

$$\begin{aligned}e_{q^{-1}}(ix\cdot p) &\mapsto e_{q^{-1}}(ix\cdot p + ia\cdot p) \\ &= e_{q^{-1}}(ix\cdot p)e_{q^{-1}}(ia\cdot p).\end{aligned}$$

5 Vacuum Projectors

In this section we introduce a "vacuum projector" E which realizes the operator $|\Omega_\partial\rangle\langle\Omega_x|$ (up to a possible normalization factor) in terms of coordinates and derivatives (a similar object, in a Hopf algebra context, has been introduced in [7]). It is given by the formal expansion:

$$E = \sum_{k=0}^{\infty} \frac{(-1)^k}{[k]_{q^{-1}}!} x^{i_1} \dots x^{i_k} \partial_{i_k} \dots \partial_{i_1}. \tag{31}$$

We scetch the proof. Setting:

$$E_k \equiv x^{i_1} \dots x^{i_k} \partial_{i_k} \dots \partial_{i_1}$$

we find:

$$E_k x^i = [k]_q x^i E_{k-1} + q^{2k} x^i E_k$$

(x^i is the i-th coordinate). We then have:

$$\begin{aligned} Ex^i &= \sum_{k=0}^{\infty} \frac{(-1)^k}{[k]_{q^{-1}}!} E_k x^i \\ &= \sum_{k=1}^{\infty} \frac{(-1)^k}{[k]_q!} q^{k(k-1)} [k]_q x^i E_{k-1} + \sum_{k=0}^{\infty} \frac{(-1)^k}{[k]_q!} q^{k(k-1)} q^{2k} x^i E_k \\ &= -\sum_{k=0}^{\infty} \frac{(-1)^k}{[k]_q!} q^{k(k+1)} x^i E_k + \sum_{k=0}^{\infty} \frac{(-1)^k}{[k]_q!} q^{k(k+1)} x^i E_k \\ &= 0. \end{aligned}$$

As a result, $E^2 = E$. One similarly checks that $\partial_i E = 0$.

We can now easily realize the projector $|\Omega_x\rangle\langle\Omega_\partial|$ as well. We know from [1] that $\mathcal{A}$ admits the $*$-involution (which we denote by a bar):

$$\bar{x^i} = x^i, \quad \bar{\partial_i} = -q^{2(n+1-i)} \partial_i, \quad \bar{q} = q^{-1}$$

(corresponding to a real quantum plane). It then follows immediately that $\bar{E}$, given explicitly by:

$$\bar{E} = \sum_{k=0}^{\infty} \frac{1}{[k]_{q^{-1}}!} q^{2k(n+1)} q^{-2(i_1+i_2+\dots+i_k)} \bar{E}_k \tag{32}$$

where $\bar{E}_k \equiv \partial_{i_1} \ldots \partial_{i_k} x^{i_k} \ldots x^{i_1}$, realizes the operator $|\Omega_x\rangle\langle\Omega_\partial|$. An alternative form for E_k, as a function of $x \cdot \partial$, is:

$$E_k = q^{-k(k-1)}(x \cdot \partial)(x \cdot \partial - [1]_q)(x \cdot \partial - [2]_q) \ldots (x \cdot \partial - [n-1]_q).$$

One can easily show that $\bar{E}_k$ can also be expressed in terms of $x \cdot \partial$.

The above objects allow us to approach the problem of integration from an alternative point of view. We can use the vacua introduced earlier to define the integral of a function $f(x)$ via:

$$\langle f(x)\rangle = \langle\Omega_\partial|f(x)|\Omega_\partial\rangle. \tag{33}$$

This definition automatically satisfies $\langle\partial_i(f(x))\rangle = 0$ and therefore it also satisfies $\langle f(x+a)\rangle = \langle f(x)\rangle$. Notice however that in deriving this last invariance property we do not need any ad-hoc rules about how to commute a's (or, for that matter, p's) through the "integral sign". Indeed, choosing the normalization $E = |\Omega_\partial\rangle\langle\Omega_x|$, $\bar{E} = |\Omega_x\rangle\langle\Omega_\partial|$ (which, in turn, implies $\langle\Omega_x|\Omega_\partial\rangle = 1$) (33) gives:

$$\begin{aligned}
\bar{E}f(x)E &= |\Omega_x\rangle\langle\Omega_\partial|f(x)|\Omega_\partial\rangle\langle\Omega_x| \\
&= \langle f(x)\rangle|\Omega_x\rangle\langle\Omega_x| \\
&\equiv \langle f(x)\rangle\delta(x).
\end{aligned}$$

However, as we have seen, E_k and $\bar{E}_k$ can be expressed as functions of $x \cdot \partial$ only. Refering back to the list given in (16), we see that a^i, p^j commute with $x \cdot \partial$ and this justifies postulating the integration procedure described in section 3.

Acknowledgments

We wish to thank Peter Schupp and Paul Watts for helpful discussions. The development of the subject in section 2 has benefited from early discussions one of the authors (BZ) had with Julius Wess - we thank him for them.

This work was supported in part by the Director, Office of Energy Research, Office of High Energy and Nuclear Physics, Division of High Energy Physics of the U.S. Department of Energy under Contract DE-AC03-76SF00098 and in part by the National Science Foundation under grant PHY90-21139.

It is a pleasure to dedicate this paper to Abdus Salam.

A Appendix

We give here explicitly, the full set of commutation relations for the two-dimensional case. The $\hat{R}$-matrix in this case is:

$$\hat{R} = \begin{pmatrix} q & 0 & 0 & 0 \\ 0 & \lambda & 1 & 0 \\ 0 & 1 & 0 & 0 \\ 0 & 0 & 0 & q \end{pmatrix}$$

from which we get, in obvious notation:

x − x

$$xy = qyx$$

x − ∂

$$\begin{aligned} \partial_x x &= 1 + q^2 x\partial_x + q\lambda y\partial_y \\ \partial_x y &= qy\partial_x \\ \partial_y x &= qx\partial_y \\ \partial_y y &= 1 + q^2 y\partial_y \end{aligned}$$

x − a

$$\begin{aligned} xa &= q^2 ax \\ xb &= qbx + q\lambda ay \\ ya &= qay \\ yb &= q^2 by \end{aligned}$$

x − p

$$\begin{aligned} p_x x &= q^{-2} xp_x \\ p_x y &= q^{-1} yp_x \\ p_y x &= q^{-1} xp_y \\ p_y y &= q^{-2} yp_y - q^{-1}\lambda xp_x \end{aligned}$$

∂ − ∂

$$\partial_x\partial_y = q^{-1}\partial_y\partial_x$$

$\partial - a$

$$\begin{aligned}
\partial_x a &= q^{-2} a \partial_x \\
\partial_x b &= q^{-1} b \partial_x \\
\partial_y a &= q^{-1} a \partial_y \\
\partial_y b &= q^{-2} b \partial_y - q^{-1} \lambda a \partial_x
\end{aligned}$$

$\partial - \mathbf{p}$

$$\begin{aligned}
p_x \partial_x &= q^2 \partial_x p_x \\
p_x \partial_y &= q \partial_y p_x \\
p_y \partial_x &= q \partial_x p_y + q \lambda \partial_y p_x \\
p_y \partial_y &= q^2 \partial_y p_y
\end{aligned}$$

$\mathbf{a} - \mathbf{a}$

$$ab = qba$$

$\mathbf{a} - \mathbf{p}$

$$\begin{aligned}
p_x a &= q^{-2} a p_x \\
p_x b &= q^{-1} b p_x \\
p_y a &= q^{-1} a p_y \\
p_y b &= q^{-2} b p_y - q^{-1} \lambda a p_x
\end{aligned}$$

$\mathbf{p} - \mathbf{p}$

$$p_x p_y = q^{-1} p_y p_x$$

References

[1] J. Wess and B. Zumino, "Covariant Differential Calculus On the Quantum Hyperplane", *Nucl. Phys. B (Proc. Suppl.)* **18B** (1990)

[2] N. Yu. Reshetikhin, L. A. Takhtadzhyan and L. D. Fadeev, "Quantization of Lie Groups and Lie Algebras" *Leningrad Math. J.* **1** 193 (1990)

[3] E. Abe, "Hopf Algebras", Cambridge Univ. Press, 1980

[4] S. Majid, "Braided Momentum In the q-Poincare Group", preprint DAMTP/92-65

[5] H. Exton, "q-Hypergeometric Functions And Applications", Ellis Horwood, 1983

[6] S. Majid, private communication in Ixtapa, Mexico

[7] C. Chryssomalakos, P. Schupp and P. Watts, "The Role Of the Canonical Element In the Quantized Algebra Of Differential Operators $\mathcal{A}\rtimes\mathcal{U}$", preprint LBL-33274, UCB-PTH-92/42

VI. Salam's Collaborators and Students

Unified Models based on Grassmann Coordinates

R. Delbourgo

Department of Physics, University of Tasmania
Hobart, Tasmania, AUSTRALIA 7005

Abstract

Lorentz scalar Grassmann coordinates may be used for classifying elementary particles and their interactions. Internal symmetry schemes based on this concept are reviewed with particular emphasis on the case of 5 or 4 anti-commuting variables.

1 INTRODUCTION

My close association with Abdus Salam, spanning a period of some seventeen years, has been a happy and fruitful one. It began at Imperial College in 1959, when he lectured to me as a final year Physics undergraduate; it developed into a supervisory role for him during my PhD studies over the next three years and it finally blossomed into a full working relationship until about 1976, when I emigrated to Australia. This talk is a tribute to him for the help, guidance and friendship which he so generously offered throughout those years. Naturally his influence over my development has been enormous and, as you will see in this lecture, his many and varied physics contributions have shaped the way I think and act today. If you find anything attractive about what I shall describe presently, you may attribute it to the influence of Abdus Salam.

I have many fond memories of those times of course; in an attempt to give a more concrete expression to these feelings, I rummaged through my university memorabilia and came across some lecture material, which I have brought with me. Although Salam's notes are merged with those from other lecturers, it is not hard to pick out his individual lessons: his presentation of 2 subjects (Quantum Mechanics à la Dirac, and Statistical Mechanics) stands out as logical, purposeful and woffle-free. Many students did not like the stark, no-nonsense delivery, but I personally enjoyed it a lot. When therefore I decided to embark on a postgraduate career and scoured various Imperial College departments on the lookout for something interesting, I quickly ditched optics and got sucked into theoretical physics. I can well remember Salam's animated discourse that afternoon in his office in the Huxley building about his most recent research enterprise, namely the 2-component neutrino theory and his recent discovery of γ_5 invariance. Of course, I was too fresh to absorb very much at the time, but I distinctly recall being swept by his nervous excitement and enthusiasm. It was the catalyst I needed for settling on my future career. The following year we moved from the soon-to-be-demolished Royal College of Science into the new Physics building and I began PhD studies starting with the compulsory diploma coursework. As it happened, Salam was again one of our lecturers; on this occasion he explained the mysteries of Symmetries and Fields to the class. Tom Kibble had been my official supervisor in 1960/61 but Salam took over the responsibility the following year, initiating me into the intricacies of the "gauge technique"—a brand new way he'd found of handling the Green functions in gauge field theories. That work was written up in early 1964, by which time I was undergoing my postdoctoral apprenticeship at Wisconsin.

Salam may have sensed a keenness on my part to continue working with him and perhaps he found my assistance to be of some value. Anyway, when he offered me a job at the newly founded International Centre for Theoretical Physics (located in Piazza Oberdan then), I jumped at the chance of returning to Europe. The opportunity of collaborating with John Strathdee, my contemporary at Imperial, was an added inducement. Those early days at the Centre were rather exciting; it had just commenced operation and Salam fought extremely hard to make it a successful venture, something of which the world (and especially developing countries) could be proud. How he managed to carry out research at the same time, in such trying circumstances, I shall never know. John and I used to joke that Abdus did most of his work at airports or on airplanes since he was constantly on the move, fighting tooth and nail to set the Centre on a firm financial footing. The fact that we are all housed in this magnificent complex is a monument to his energy and tenacity during those years.

In 1966 I moved back to Imperial as a lecturer and this allowed me to continue collaborating with Abdus, picking his brains and trying in some small way to learn the tricks of the trade. I hope that some of his insights and instincts have rubbed off on me. In any event I regard Salam as a true mentor and friend and I am especially honoured to have been invited for such an important occasion. Except perhaps for John Strathdee, it is unlikely that anyone present can claim to cover the range of interactions that I have experienced with him.

Now to business. I would like to tell you about our efforts in Tasmania at trying to build a unified model of elementary particles, using a somewhat bizarre approach, based on Lorentz *scalar* Grassmann coordinates. In this research I have been aided and abetted by Ruibin Zhang, Martin White, Simon Twisk, Lorella Jones and latterly by Peter Jarvis and Roland Warner. Others, who will be duly referenced† have applied similar sorts of ideas in the past but none with our accent nor with our obstinacy.

2 GRASSMANN VARIABLES IN PHYSICS

Anticommuting *operators* were invented by Dirac in 1926 in order to account automatically for the curious antisymmetric states needed to describe spin 1/2 particles. We come across them early in our quantum mechanics education today but they must have caused quite a stir at the time. Disregarding arguments, Dirac enunciated the anticommutators,

$$\{a, a^\dagger\} = 1, \qquad \{a, a\} = \{a^\dagger, a^\dagger\} = 0. \tag{1}$$

During the course of the next decade, when field theory came into prominence, it was a natural and logical step for Jordan, Wigner, Pauli and Fock to invoke the concept of equal time anticommutators for electron *field operators*:

$$\{\psi(\mathbf{x}), \psi^\dagger(\mathbf{y})\} = \delta(\mathbf{x} - \mathbf{y}), \qquad \{\psi, \psi\} = \{\psi^\dagger, \psi^\dagger\} = 0. \tag{2}$$

The idea that there might exist fundamental anticommuting *coordinates*—call them θ—came much later. Those ideas are nicely summarised in reviews[1)] by Berezin and DeWitt. Writing anticommutators for such coordinates in the form

$$\{\theta_I, \partial^J\} = \delta_I^J, \qquad \{\theta_I, \theta_J\} = \{\partial^I, \partial^J\} = 0 \tag{3}$$

one sees the close resemblance with the Fock formalism and the Heisenberg commutation relations, if one regards $\partial^J \equiv \partial/\partial\theta_J$ as the conjugate momentum to θ_J. An important and more concrete extension of the idea was made[2)] by Salam & Strathdee and by Ferrara, Wess & Zumino. They suggested adding 4 spinorial coordinates θ_α

†and if I have omitted anyone inadvertently, I offer my humble apologies.

(and replicas if necessary) to space-time so as to connect them with generators Q_α of supersymmetric transformations, with

$$\{Q_\alpha, Q_\beta\} = (\gamma.PC)_{\alpha\beta}; \tag{4}$$

Supersymmetry is alive and kicking as a theoretical idea, but nature seems to have shunned it in the low energy regime, as far as we can tell.

A return to the original concept of Lorentz scalar coordinates occurred with the development of quantized gauge theories. You will hardly need reminding that unitarity requires gauge fields in non-Abelian theories to be accompanied by ghost fields C (and an auxiliary scalar field B, for reasons of elegance). The local gauge invariance of the theory gets replaced by the global BRST symmetry in the quantised version[3]. A nice way to show this is by going to an OSp(4/2) formalism whereby two Grassmann parameters $\theta^1, \theta^2 \leftrightarrow \theta, \bar{\theta}$ are appended[4] to space-time, whereupon the BRST transformations are neatly expressed in an Sp(2) invariant manner[5]:

$$\begin{aligned} \delta A_\mu &= \theta^I D_\mu C_I, \\ \delta C_J &= \theta^I(\epsilon_{IJ}B - gC_I \times C_J/2), \\ \delta B &= \theta^I(gC_I \times B/2 - g^2 C^J \times (C_I \times C_J)/12). \end{aligned} \tag{5}$$

Gravity can be treated in the same vein and the same sorts of ideas can be applied to Hamiltonian systems for enforcing constraints.

3 GRASSMANN SPACE

My own motivation for exploring the concept of Grassmann coordinates stems from Kaluza-Klein (K-K) theory. As you are familiar, the traditional K-K scheme relies on adding one bosonic, compactified extra coordinate which somehow curls up into a small circle of Planckian scale. All fields are to be regarded as functions of the 5 coordinates of an enlarged space-time, $\Phi(x^\mu, x^5) \equiv \Phi(X^M)$. Because the dependence on the fifth coordinate is periodic over a circle of radius R, it is customary to expand the fields in an infinite number of (discrete) normal modes:

$$\Phi(X) = \sum_n \phi_n(x^\mu) \exp(inx^5/R). \tag{6}$$

Typically, such fields are acted upon by a five-dimensional d'Alembertian operator,

$$\partial^\mu \partial_\mu + \partial^5 \partial_5 \rightarrow \partial^\mu \partial_\mu + n^2/R^2, \tag{7}$$

showing us that there exists an infinite tower of particles with increasingly enormous masses n/R. Of course one of the nice things about the K-K theory is that gravity and electromagnetism are unified within an extended metric structure; in the original ansatz (no extra scalar fields), the gravitational field and the em field appear as components of a five-dimensional metric tensor ($\kappa \propto \sqrt{G_{Newton}}$),

$$G_{MN} \equiv \begin{pmatrix} G_{\mu\nu} & G_{\mu 5} \\ G_{5\nu} & G_{55} \end{pmatrix} \equiv \begin{pmatrix} g_{\mu\nu} - \kappa^2 A_\mu A_\nu & -\kappa A_\mu \\ -\kappa A_\nu & -1 \end{pmatrix} \tag{8}$$

What is more, this concept can be generalised to more complicated internal symmetry groups; Salam and Strathdee have played a key role in such developments[6)].

However contemplating an infinite sum of superheavy states brings on distinct feelings of unease, if not for the question of renormalisability (a lost cause for gravity anyhow), then at least for the cosmological consequences (production and survival of these excitations). I was therefore led to exploring an alternative version, in which the extra coordinates are taken as anticommuting. This is hardly a radical idea, since anticommuting particle fields abound in nature. The first problem was to see if one could mimic conventional K-K constructions. Zhang and I managed to solve this poser[7], as did Ellicott and Toms. For QED all one need do is attach two real fermionic coordinates, θ^1 and θ^2 (or a complex θ and its conjugate $\bar{\theta}$), to the four bosonic coordinates x^μ. Flat space then becomes amenable to an OSp(4/2) invariant treatment in which the extended distance is given by

$$ds^2 = dx^\mu \eta_{\mu\nu} dx^\nu + d\theta^m \eta_{mn} d\theta^n, \tag{9}$$

where the fermionic "Minkowski" metric η_{mn} has to be antisymmetric in its indices. Turning to curved space, $X^M = (x^\mu, \theta^m)$, the analogue of K-K comes out to be

$$G_{MN} \equiv \begin{pmatrix} G_{\mu\nu} & G_{\mu n} \\ G_{m\nu} & G_{mn} \end{pmatrix} \equiv \begin{pmatrix} g_{\mu\nu} - \kappa^2 \bar{\theta} A_\mu A_\nu \theta & \kappa(A_\mu \theta)_n \\ \kappa(\bar{\theta} A_\nu)_m & \eta_{mn} \end{pmatrix}. \tag{10}$$

Indeed one can go further and generalise the construction to internal groups like SU(N) by simply invoking an extended set of complex Grassmann coordinates θ_I; $I = 1 \ldots N$. It is not very difficult to reproduce the standard model, using an ansatz such as above[8], and even a unified gauge theory like Georgi-Glashow[9] SU(5). The virtue of such an approach is it that we partially sidestep the question "Where are these coordinates?"—only fermion bilinears have physical significance; also we are straightaway restricted to a finite number of states, owing to the terminating nature of fermionic expansions. The group theory and the representations are correspondingly circumscribed and the theory is easily verified or dismissed, i.e. it has the good feature of being strongly predictive!

4 GRASSMANN SPIN

This is a good place to mention that Grassmann scalars provide an amusing way of imagining spin[10)]. If one invokes two[‡] real θ's, obeying the conjugation property

$$\theta_1^* = \theta^1 = \epsilon^{12}\theta_2 = \theta_2, \qquad \theta_2^* = \theta^2 = \epsilon^{21}\theta_1 = -\theta_1, \tag{11}$$

the spin operators are then construed as rotation operators in the internal θ-space. They correspond to the generators L of an Sp(2) group,

$$L_{IJ} = \theta_I \partial_J + \theta_J \partial_I; \qquad \partial_J \equiv \epsilon_{JI}\partial^I, \tag{12}$$

since they are *symmetric* in their indices. In fact if we identify the three (hermitian) combinations,

$$S_1 = (L_{11} + L_{22})/4, \quad S_2 = -i(L_{11} - L_{22})/4, \quad S_3 = -iL_{12}/2, \tag{13}$$

it is easy to establish that $\mathbf{S} \times \mathbf{S} = i\mathbf{S}$. As far as the θ monomials are concerned, it is readily ascertained that (θ_1, θ_2) constitute a spin 1/2 doublet and $(1 \pm \theta_1\theta_2)/2$ are two spin 0 singlets. Nor will it surprise you to hear that one can represent relativistic spin by progressing to a complex two-component θ-space (or four real θ), wherein the 4×4 Dirac matrices can be expressed as suitable bilinear combinations of θ_α and $\partial^\beta = \partial/\partial\theta_\beta$. This approach must be contrasted with that of Berezin and Marinov[11)], who rely on a five-dimensional Grassmannian variable[§].

5 INTERNAL GRASSMANN SYMMETRY

I'd now like to turn to the prime motivation for introducing anticommuting coordinates, namely the desire to gain a deeper understanding of "internal symmetries". The Grassmannian K-K theory suggests that the appendage of fermionic coordinates

[‡]one Grassmann coordinate produces no structures of interest. At least a pair is needed to incorporate symplectic forms.

[§]Casalbuoni and Gatto[12)] offer an alternative route by considering two θ but also allowing the bilinears $\theta_1\partial_2 - \theta_2\partial_1, \theta_1\theta_2, \partial_1\partial_2$.

provides a setting for the marriage of gravitation and internal gauge symmetry. The extended space-time $X = x, \theta$, not only specifies the *where* (= t) and *when* (= $\mathbf{x}$) of an event but also the *what* (= property, θ) of it. Without getting too enmeshed in the fermion-boson sector $G_{\mu n}$ it pays to take a closer look at this internal Grassmann space (associated with the fermion-fermion sector G_{mn}), to see if leads to other interesting consequences. In what follows I am going to suggest to you that it provides an elegant framework for comprehending the "generation problem" or family replication phenomenon.

Presuming, as I am, that nature contains hidden anticommuting coordinates, the first question is "How many?" At the end of the day we know that last unbroken gauge group is $\mathrm{SU}(3)_{QCD} \times \mathrm{U}(1)_{QED}$ so it is sensible to include at least four complex Grassmann θ. Anything less will not be able to account for the properties of charge and colours. The fact that these four θ can form the basis of an SU(4) group, à la Pati-Salam[13)], is no coincidence, since the fourth coordinate is a neutral colour, in a sense. One can go the extra mile and ask that the electroweak gauge group $SU(2)_{EW}$ be also accommodated directly. In that case one is led to enlarge the Grassmann manifold to 5 complex θ in the manner of Georgi and Glashow. Let me begin then with the straightforward unified Grassmann model based on an SU(5) group. Later on I shall return to the even more economic version based on SU(4).

Start therefore with the five coordinates bearing the Georgi-Glashow charge (Q) assignments:

$$(\theta_1, \theta_2, \theta_3, \theta_4, \theta_5) \leftrightarrow (1/3, 1/3, 1/3, 0, -1)_Q \tag{14}$$

so that we may associate them with the generic states $(\bar{D}_{Red}, \bar{D}_{Green}, \bar{D}_{Blue}, \nu_\ell, \ell^-)$. Clearly the conjugate coordinates carry the opposite quantum numbers. Borrowing the Salam-Strathdee superfield idea[2)], we can imagine extended fields which are functions of these internal coordinates $\Phi(\theta, \bar{\theta})$ and can be expanded in a Taylor series in θ & $\bar{\theta}$. The simplest sort of superfield (the analogue of an analytic function of a complex variable) will depend on $\bar{\theta}$ alone and its expansion consists of six terms terminating in $(\bar{\theta})^5$,

$$\Phi = N + \bar{\theta}^I e_I + \bar{\theta}^I \bar{\theta}^J U_{IJ}/2 + \bar{\theta}^I \bar{\theta}^J \bar{\theta}^K u_{IJK}/6 + \bar{\theta}^I \bar{\theta}^J \bar{\theta}^K \bar{\theta}^L E_{IJKL}/24 + \bar{\theta}^1 \bar{\theta}^2 \bar{\theta}^3 \bar{\theta}^4 \bar{\theta}^5 n. \tag{15}$$

A full generation of leptons is held in either the odd or the even components of the above expansion; for instance, if you inspect the even component fields, you will discern the full complement of quarks and leptons (including the antineutrino!):

$$\begin{aligned} N &= \nu_e \\ U_{IJ} &= (U_{ij}, U_{i4}, U_{i5}, U_{45}) = (\bar{u}, \bar{d}, u, \ell^-) \\ E_{IJKL} &= (E_{ij45}, E_{1234}, E_{1235}) = (d, e^+, \bar{\nu}_e) \end{aligned} \tag{16}$$

where the colour indices i, j range from $1\ldots3$. This is hardly a surprise because the terms of the expansion consist of SU(5) representations, 1, 5, 10, $\overline{10}$, $\bar{5}$ & 1, grouping themselves nicely into two 16's of SO(10). That SO(10) should feature naturally is also obvious from the fact that we are dealing with 10 real Grassmann variables. More striking is the fact that the correct antisymmetric physical representations arise at this level and no others.

Even more intriguing is the ability of the scheme to accommodate generations of particles[14)]. The crucial point is that in general we ought to be expanding the superfield Φ in both θ and $\bar{\theta}$. Since $(\bar{\theta}\theta)$ is an SU(5) scalar we can contemplate further components of Φ which are associated with such scalar products. One is led to speculate that powers of $(\bar{\theta}\theta)$ may be connected with generation number. Taking this line and specifically considering an odd-component superfield, to which we successively attach $(\bar{\theta}\theta)$, we obtain

$$[\bar{\theta}^I e_I + \bar{\theta}^I\bar{\theta}^J\bar{\theta}^K u_{IJK}/6 + \bar{\theta}^1\bar{\theta}^2\bar{\theta}^3\bar{\theta}^4\bar{\theta}^5\bar{n}_e] \quad \leftrightarrow \quad \text{Generation 1?} \tag{17}$$

$$(\bar{\theta}\theta)[\bar{\theta}^I m_I + \bar{\theta}^I\bar{\theta}^J\bar{\theta}^K c_{IJK}/6 + \bar{\theta}^1\bar{\theta}^2\bar{\theta}^3\bar{\theta}^4\bar{\theta}^5\bar{n}_\mu] \quad \leftrightarrow \quad \text{Generation 2?} \tag{18}$$

$$(\bar{\theta}\theta)^2[\bar{\theta}^I \tau_I + \bar{\theta}^I\bar{\theta}^J\bar{\theta}^K t_{IJK}/6 + \bar{\theta}^1\bar{\theta}^2\bar{\theta}^3\bar{\theta}^4\bar{\theta}^5\bar{n}_\tau] \quad \leftrightarrow \quad \text{Generation 3?} \tag{19}$$

Although this superficial approach has its attractions, it comes unstuck when one considers the (unwanted) residual set of 5's, viz. $(\bar{\theta}\theta)^3[\bar{\theta}^I\psi_I]$ and $(\bar{\theta}\theta)^4[\bar{\theta}^I\chi_I]$. We surely do not welcome these *incomplete* 4'th and 5'th generations, which will

Table 1: The full set of $SU(5)$ multiplets contained in the coordinate expansion of a superfield Ψ. p and q, the column and row numbers, correspond to the term $(\bar{\theta})^p(\theta)^q$ in the Taylor series for Ψ.

$q\backslash p$	0	1	2	3	4	5
0	1	5	10	$\overline{10}$	$\overline{5}$	1
1	$\overline{5}$	$1+24$	$5+\overline{45}$	$10+\overline{40}$	$\overline{10}+\overline{15}$	$\overline{5}$
2	$\overline{10}$	$\overline{5}+45$	$1+24+75$	$5+\overline{45}+\overline{50}$	$10+\overline{40}$	$\overline{10}$
3	10	$\overline{10}+40$	$\overline{5}+45+50$	$1+24+75$	$5+\overline{45}$	10
4	5	$10+15$	$\overline{10}+40$	$\overline{5}+45$	$1+24$	5
5	1	5	10	$\overline{10}$	$\overline{5}$	1

endanger anomaly cancellation, whatever else they may do! The difficulty suggests that a fuller investigation is warranted. Let us therefore expand the Ψ superfield completely:

$$\Psi(\theta,\bar{\theta}) = \sum_{p,q=0}^{5} (\bar{\theta})^p(\theta)^q\psi(p,q), \tag{20}$$

The components $\psi(p,q)$ can be easily expressed as linear combinations of SU(5) multiplets and we arrive at the Table 1:

We discover an embarrassment of entries, $2^{10} = 1024$ in all, as this is the fundamental representation of the superalgebra gl(512,512). These states can be decomposed under a variety of subalgebras, such as the chains

$$so(20) \oplus so(20) \rightarrow so(20) \rightarrow u(10) \rightarrow u(5) \oplus u(5) \rightarrow su(5) \oplus u(1) \oplus u(1),$$
$$so(20) \rightarrow sp(10) \oplus sp(2) \rightarrow su(5) \oplus u(1) \oplus u(1);$$

nevertheless there are too many representations for comfort and whether we like it or not, we must consider ways of cutting them down to size. One obvious symmetry is the reflection property about the main diagonal which can be made to correspond to charge conjugation. This suggests that we can halve the number of states by requiring Ψ to be antisymmetric about the diagonal, reflected states being identified as charge conjugates and carrying opposite chirality. Unfortunately that still leaves an excess of states and further pruning is necessary. Another possible way to halve the system again is to use reflections about the cross diagonal: $\psi(p,q) \rightarrow \psi(5-q, 5-p)$, as embodied in the "Grassmannian duality" rule

$$\breve{\psi}^{J_1..J_q}_{I_1..I_p} \equiv \epsilon_{I_1..I_5}\epsilon^{J_1..J_5}\psi^{I_{p+1}..I_5}_{J_{q+1}..J_5}/(5-p)!(5-q)! \tag{21}$$

By identifying $\breve{\psi}$ with ψ, up to a sensibly chosen phase factor (depending on the number of indices), we can finish up with a model[15)] summarised in Table 2, where chirality assignments alternate between the even and odd pieces.

In this arrangement we obtain three standard families of $(5+\bar{10})_R$ as well as one exotic family $(\bar{10}+\bar{15}+40+\bar{45})_R$, which is anomaly-free as one can check from the anomaly coefficients listed in the next Table (3). Evidently too many fermions spell trouble for asymptotic freedom unless the gauge group is enlarged beyond SU(5). However, if these exotic extras are very heavy one can probably get by in the low energy ($\sim$ 1 TeV) region.

Table 2: Surviving multiplets of a unified SU(5) Grassmannian Model. States marked by $*$'s are conjugates of their reflections about the principal diagonal or are repetitions of their reflections about the cross-diagonal.

$p\backslash q$	0	1	2	3	4	5
0	0	5_R	10_L	$\bar{10}_R$	$\bar{5}_L$	0
1	$*$	0	$5_R + \bar{45}_R$	$10_L + \bar{40}_L$	$\bar{10}_R + \bar{15}_R$	$*$
2	$*$	$*$	0	0	$*$	$*$
3	$*$	$*$	0	0	$*$	$*$
4	$*$	$*$	$*$	$*$	0	$*$
5	0	$*$	$*$	$*$	$*$	0

Table 3: Anomaly coefficients for $SU(5)$ representations of fixed chirality.

$SU(5)$ Representation	1,24,75	5,10	15	40	45	50
Anomaly coefficient	0	1	9	16	6	15

6 AN SU(4) GRASSMANNIAN MODEL

It is possible to construct other models along these lines by varying the duality phase relations or by imposing other suitable restrictions, but more often than not a proliferation of fermion states seems to ensue. In view of that propensity it is natural to look for more economical schemes. To that end let me retrace my steps to the SU(4) scheme, by dropping the uncharged θ coordinate that was used to implement $SU(2)_{EW}$ painlessly. This tack was advocated some time back by Casalbuoni and Gatto[12], though in a Fock space setting and with its origin in *preons*[13] — their scheme made it difficult to argue for generations without enlarging the group structure by adding more Grassmann creation and annihilation operators. In our own (complex) coordinate setting, where families are more natural, we shall hang on to the charge assignments,

$$(\theta_1, \theta_2, \theta_3, \theta_4) \leftrightarrow (1/3, 1/3, 1/3, -1)_Q$$

so that Q remains a (traceless) generator of the overall SU(4). Now superfields which are purely functions of $\bar{\theta}$ contain exactly one generation of particles,

$$\Phi = \nu + \bar{\theta}^I d_I + \bar{\theta}^I \bar{\theta}^J u_{IJ}/2 + \bar{\theta}^I \bar{\theta}^J \bar{\theta}^K e_{IJK}/6 + \bar{\theta}^1 \bar{\theta}^2 \bar{\theta}^3 \bar{\theta}^4 n, \tag{22}$$

without repetition¶. Written out, the component fields are

$$\nu_e, \quad \begin{pmatrix} \bar{d} \\ e^- \end{pmatrix}, \quad \begin{pmatrix} \bar{u} \\ u \end{pmatrix}, \quad \begin{pmatrix} d \\ e^+ \end{pmatrix}, \quad \bar{\nu}_e$$

As we expect to find other families (associated with dependence on θ), let us perform a complete expansion of the superfield, $\Phi(\theta, \bar{\theta}) = \sum_{p,q=0}^{4} \phi(p,q)(\bar{\theta})^p(\theta)^q$, writing out the SU(4) multiplets $\phi(p,q)$ in a magic square as we did before with SU(5). See Table 4.

There are many fewer states now, only $2^8 = 256$, giving grounds for optimism. What is more, special pleading for chirality is no longer needed. One simply takes the entire Φ to be left-handed[12] in order that all helicity states are properly accounted

¶Note, however that the statistics alternates as the powers of θ increase, pointing to a grading of generators.

Table 4: The full set of $SU(4)$ multiplets contained in the coordinate expansion of a superfield Φ. p and q, the column and row numbers, correspond to the term $(\bar{\theta})^p(\theta)^q$ in the Taylor series for Φ.

$q\backslash p$	0	1	2	3	4
0	1	4	6	$\bar{4}$	1
1	$\bar{4}$	$1+15$	$4+20$	$6+\overline{10}$	$\bar{4}$
2	$\bar{6}$	$\bar{4}+\overline{20}$	$1+15+20'$	$4+20$	6
3	4	$\bar{6}+10$	$\bar{4}+\overline{20}$	$1+15$	4
4	1	4	$\bar{6}$	$\bar{4}$	1

for, just as in SO(10) models. For example the monomial corresponding to $q = 1, p = 0$ is a $\bar{4}_L$ and can be construed as the charge conjugate of a 4_R: it is the opposite helicity partner to the 4_L lying at $q = 0, p = 1$. Invoking Grassmann duality conditions (reflections about the cross-diagonal) in order to cut down the content further reaps dividends. As before, the dual components $\breve{\phi}$ of ϕ come about by defining

$$\breve{\phi}^{J_1..J_q}_{I_1..I_p} \equiv \epsilon^{I_1..I_4}\epsilon_{J_1..J_4}\phi^{I_4..I_{p+1}}_{J_4..J_{q+1}}/(4-p)!(4-q)!. \tag{23}$$

By imposing phase relations such as $\breve{\phi} = (-1)^{pq}\phi$, we may eliminate the 10-fold representation at $q = 1, p = 3$ — a dangerous multiplet containing a charge two lepton ϕ^4_{123} amongst other states — as well as the $p = 2, q = 2$ 15-fold. This pares down the entries to Table 5.

By focussing on the three standard families, we can better appreciate how they fit into the magic square. This is shown in the next Table (6).

The entries also highlight the fact that $SU(2)_{EW}$ must act rather strangely on the states in question. In particular, the weak isospin lowering operator T_L^- must only act on the even components, removing a θ_4 or adding a $\bar{\theta}^4$. Conversely, the raising operator T_L^+ acts on the odd components, creating a θ_4 or destroying a $\bar{\theta}^4$,

Table 5: A reduced set of left-handed $SU(4)$ multiplets obtained after imposing Grassmannian duality conditions. $*$ signify states related by conjugation or duality.

$q\backslash p$	0	1	2	3	4
0	1	4	6	$\overline{4}$	1
1	$*$	$1+15$	$4+20$	6	$*$
2	$*$	$*$	$1+20'$	$*$	$*$
3	$*$	$*$	$*$	$*$	$*$
4	$*$	$*$	$*$	$*$	$*$

Table 6: Locations of the 3 basic generations within the magic square. Such states may well mix through the mass matrix induced Higgs Yukawa interaction, so e, μ, τ represent *generic* names at this stage.

$q\backslash p$	0	1	2	3	4
0	ν_e	$e^-, \bar{d}$	$\bar{u}, c$	μ^+, s	$\bar{\nu}_\mu$
1	e^+, d	ν_e, ν_τ	$\tau_-, \bar{b}$	$\bar{t}, c$	μ^+, s
2	$\bar{c}, u$	τ^+, b	ν_τ	$\tau_-, \bar{b}$	$\bar{u}, c$
3	$\mu^-, \bar{s}$	$\bar{c}, t$	τ^+, b	ν_e, ν_τ	$e^-, \bar{d}$
4	ν_μ	$\mu^-, \bar{s}$	$\bar{c}, u$	e^+, d	ν_e

Table 7: Action of left-handed weak isospin lowering and raising operators on the field components of the 4×4 square.

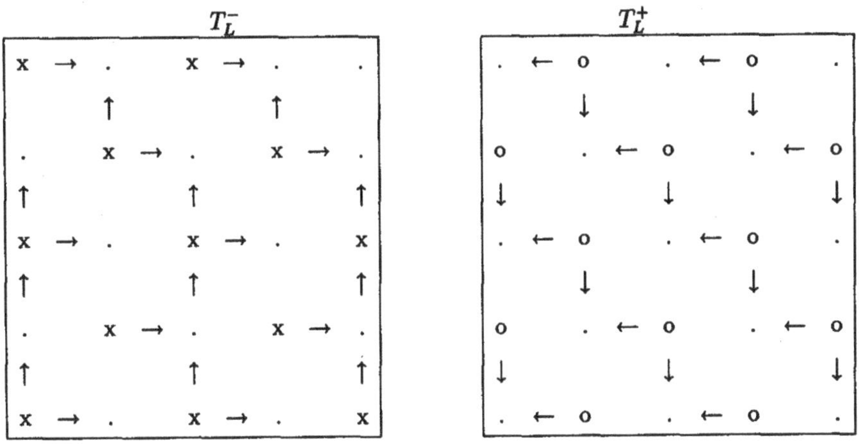

as depicted in Table 7. This can all be expressed algebraically:

$$T_L^- = (\bar{\theta}^4 + \partial^4)\mathcal{P}_e/\sqrt{2}, \quad \partial^4 \equiv \partial/\partial\theta_4; \qquad T_L^+ = (\theta_4 + \bar{\partial}_4)\mathcal{P}_o/\sqrt{2}, \quad \bar{\partial}_4 \equiv \partial/\partial\bar{\theta}^4; \tag{24}$$

whereupon the third component of left isospin is obtained via

$$T_L^3 = [T_L^+, T_L^-]/2 = [\mathcal{P}_e - \mathcal{P}_o + (\bar{\partial}_4 + \theta_4)\partial^4 - \bar{\theta}^4(\bar{\partial}_4 + \theta_4)]/4. \tag{25}$$

In these equations $\mathcal{P}_e$ and $\mathcal{P}_o$ stand for the projectors onto the even and odd monomials. It is quite easy to verify that, in spite of the grading, one does indeed have an $SU(2)_L$ Lie algebra,

$$[T_L^3, T_L^\pm] = \pm T_L^\pm. \tag{26}$$

Furthermore one readily finds that the monomials $\theta_4 \sim e_L^+$, $(1 + \bar{\theta}^4\theta_4)/\sqrt{2} \sim \bar{\nu}_{eL}$ behave as singlets, while the doublet comprises $((1 - \bar{\theta}^4\theta_4)/\sqrt{2}, \bar{\theta}^4) \sim (\nu_{eL}, e_L^-)$, in agreement with the standard assignments[16].

It will not have escaped your notice that the system admits a right-handed weak isospin in which the role of the even and odd projectors is interchanged:

$$T_R^- = (\bar{\theta}^4 + \partial^4)\mathcal{P}_o/\sqrt{2}, \qquad T_R^+ = (\theta_4 + \bar{\partial}_4)\mathcal{P}_e/\sqrt{2},$$

$$T_R^3 = [T_R^+, T_R^-]/2 = [\mathcal{P}_o - \mathcal{P}_e + (\bar{\partial}_4 + \theta_4)\partial^4 - \bar{\theta}^4(\bar{\partial}_4 + \theta_4)]/4. \tag{27}$$

It goes without saying that the right and left-handed groups commute, $[\mathbf{T}_R, \mathbf{T}_L] = 0$. That being so, it is a complete mystery of this formalism why nature has chosen to break the handedness and gauge left isospin rather than right. Of course it is within the realms of possibility that left-right mirror symmetry is broken spontaneously[17] and that right-handed bosons acquire superheavy masses by a yet-to-be-confirmed Higgs mechanism.

If you still feel uncomfortable about the notion of graded weak isospin (the raising and lowering operators being 'different' from the third component) one remedy is to append another (real) Grassmann coordinate, call it η. Now take the superfields to be functions of the θ and of this extra η. Let the overall Bose superfield be written as

$$\Phi(\theta,\bar{\theta},\eta) = B(\theta,\bar{\theta}) + \eta F(\theta,\bar{\theta}),$$

guaranteeing that B is uniformly Bose and F uniformly Fermi. In this picture B contains part of the fermion multiplets in its odd sector while F contains the remainder in its even sector‖. The isospin operators have uniform (Bose) grading in this description and we may perhaps happier about the spin-statistics connection:

$$T_L^- = (\bar{\theta}^4 + \partial^4)\eta/\sqrt{2}, \qquad T_L^+ = (\theta_4 + \bar{\partial}_4)\partial_\eta/\sqrt{2},$$

$$T_L^3 = [[\partial_\eta, \eta] + (\bar{\partial}_4 + \theta_4)\partial^4 - \bar{\theta}^4(\bar{\partial}_4 + \theta_4)]/4. \tag{28}$$

$$T_R^- = (\bar{\theta}^4 + \partial^4)\partial_\eta/\sqrt{2}, \qquad T_R^+ = (\theta_4 + \bar{\partial}_4)\eta/\sqrt{2},$$

$$T_R^3 = [-[\partial_\eta, \eta] + (\bar{\partial}_4 + \theta_4)\partial^4 - \bar{\theta}^4(\bar{\partial}_4 + \theta_4)]/4. \tag{29}$$

You should have recognised by now that the gauge fields are associated with bilinear operators formed out of θ and ∂. Specifically the SU(N) generators are connected with the combination

$$F_I^J = \bar{\theta}^J\bar{\partial}_I - \theta_I\partial^J; \qquad [F_I^J, F_K^L] = \delta_I^L F_K^J - \delta_K^J F_I^L, \tag{30}$$

hence the covariant derivative of Ψ is constructed as ever,

‖Conversely the Higgs fields are assumed to be embedded in even sectors of B and odd sectors of F.

$$\partial_\mu \to D_\mu \equiv \partial_\mu - A^I_{\mu J} F^J_I / \sqrt{2},$$

and actions arise as 8-fold Berezin integrals,

$$\mathcal{L} = \int d^4\theta d^4\bar{\theta} \bar{\Phi} i\gamma.D\check{\Phi} + g \int d^4\theta d^4\bar{\theta} \Phi^c \check{\Phi} H. \tag{31}$$

The place and status of the Higgs superfield H is more problematic. In the absence of any direct experimental evidence one way or the other, we might as well assume that the Higgs bosons are held in the same magic squares (with different duality phases maybe) and that they enjoy normal Φ^4 interactions, producing zero-charge expectation values. A preliminary investigation[14)] of this idea for the SU(5) scheme showed that a seesaw[18)] mechanism is feasible (with Fritsch-like mass matrices) because of the properties of Grassmann integration. The case of SU(4) remains to be worked out fully and this is something which is currently engaging our attention. One thing we can be certain of is that Grassmannian SU(4) models do not readily admit $SU(2)_{EW}$ triplets of Higgs, but they do allow several copies of doublets, probably as many as three generations. At the moment we are sufficiently encouraged by our superficial successes at classification and interaction to pursue these concepts.

7 THE LAST WORD

Throughout this talk you have heard me make liberal reference to Salam's published papers. This is not because I deliberately inserted them out of a sense of duty and respect. On the contrary, the references have been included because they are a true measure of his towering contributions to particle physics and any omission of them would have been a gross oversight.

ACKNOWLEDGEMENTS

I am grateful to Dr. R. C. Warner for helping me elucidate the role of left-right groups within the SU(4) Grassmannian scheme and for tracing relevant references, not to mention his other insights.

REFERENCES

1. Berezin, F. A., "The Method of Second Quantization", (Acad. Press, NY, 1966); DeWitt, B., "Supermanifolds", (Cambridge University Press, Cambridge, 1984).
2. Salam, A. and Strathdee, J., Nucl. Phys. B76, 477 (1974); Ferrara, S., Wess, J., and Zumino, B., Phys. Lett. 51B, 239 (1974).
3. Taylor, J. C., Nucl. Phys. B33, 436 (1971); Slavnov, A., Theor. and Math. Phys. 13, 174 (1972); Becchi, C., Rouet, A. and Stora, R., Phys. Lett. 52B, 344 (1974); Tyutin, I. V., Int. Report FIAN 39 (1975); Nakanishi, N., Prog. Theor. Phys. 35, 111 (1976).
4. Curci, G. and Ferrari, R., Phys. Lett. 63B, 91 (1976); Ojima, I., Prog. Theor. Phys. 64, 625 (1980); Bonora, L. and Tonin, M., Phys. Lett. 98B, 48 (1981); Bonora, L., Pasti, P. and Tonin, M., Nuovo Cim. 63A, 353 (1981).
5. Delbourgo, R. and Jarvis, P. D., J. Phys. A15, 611 (1982); Delbourgo R., Jarvis, P. D. and Thompson, G., Phys. Lett. 109B, 25 (1982).
6. Cho, Y. M., J. Math. Phys. 16, 2029 (1975); Mecklenburg, W., Phys. Rev. D21, 2149 (1980); Salam, A. and Strathdee, J., Ann. Phys. 141, 316 (1982) Witten, E., in Proceedings of the Shelter Island Conference (MIT Press, 1985); Randjbar-Daemi, S., Salam, A. and Strathdee, J., Nucl. Phys. B214, 491 (1983).
7. Delbourgo, R. and Zhang, R. B., Phys. Lett. 202B, 296 (1988); Ellicott, P. and Toms, D. J., Class. Quant. Grav. 6, 1033 (1989).
8. Delbourgo R. and Zhang, R. B., Phys. Rev. D38, 2470 (1988); Delbourgo R., Twisk, S. and Zhang, R. B., Mod. Phys. Lett. A3, 1073 (1988).
9. Georgi, H. and Glashow, S., Phys. Rev. Lett. 32, 438 (1974); Georgi, H., Quinn, H. and Weinberg, S., Phys. Rev. Lett. 33, 451 (1974).
10. Delbourgo R., Mod. Phys. Lett. 4A, 1381 (1989); Delbourgo, R., Jones, L. M. and White, Phys. Rev. D41, 679 (1990).

11. Berezin, F. A. and Marinov, M., Ann. Phys. 104, 336 (1977);
Casalbuoni, R., Nuovo. Cim. 33A, 389 (1976).

12. Casalbuoni, R. and Gatto, R., Phys. Lett. B88, 306 (1979); B90, 81 (1980); see also Krolikoswki, W., Acta Phys. Polon. B19, 599 (1989).

13. Pati, J. C. and Salam, A., Phys. Rev. D8, 1240 (1973);
and Nucl. Phys. B214, 109 (1983); *ibid* B234, 223 (1984).

14. Delbourgo, R., Mod. Phys. Lett. 4A, 1381 (1989);
Delbourgo, R. and White, M., Mod. Phys. Lett. 5A, 355 (1990).

15. Delbourgo, R., Jarvis, P. D. and Warner, R. C., Aust. J. Phys. 44, 135 (1991); and "Grassmann Coordinates and Unified Models" to appear in J. Math. Phys.

16. Glashow, S., Nucl. Phys. 22, 579 (1961);
Salam, A. and Ward, J. C., Phys. Lett. 13, 168 (1964);
Weinberg, S., Phys. Rev. Lett. 19, 1264 (1967);
Salam, A. in Nobel Symposium (Wiley Interscience, New York, 1968).

17. Pati, J. C. and Salam, A., Phys. Rev. D10, 275 (1974);
Mohapatra, R. N., and Pati, J. C., Phys. Rev. D11, 566, 2558 (1975);
Senjanovic, G. and Mohapatra, R. N., Phys. Rev. D12, 1502 (1975).

18. Gell-Mann, M., Ramond, P. and Slansky, R., in "Supergravity" (North-Holland, Amsterdam, 1979).

TWENTY YEARS LATER: WHY I STILL BELIEVE IN SU(4)-COLOR

JOGESH C. PATI
Department of Physics, University of Maryland
College Park, Maryland 20742, USA

Abstract

Following some preliminary remarks of a historical nature, the recent success of the conventional approach to grand unification as regards the meeting of the coupling constants is summarized. Its shortcomings as regards the arbitrariness in the Higgs sector are noted and a case is made for an alternative approach to unification based on a purely gauge origin of the fundamental forces. This seems to call for the ideas of local supersymmetry and preons. Preonic ideas in turn seem to require SU(4)-color symmetry. The advantages of the marriage of these ideas as regards an understanding of the origins of (a) diverse mass-scales, (b) family-replication and (c) inter-family mass-hierarchy are discussed.

1. Preliminary Remarks

Professor Abdus Salam is a great scientist, a humanitarian and a symbol for the third world. I wish to dedicate this talk with pleasure and gratitude to him on the occasion of his sixty-seventh birthday.

Before I enter into the scientific part of my talk, I thought I would say a few words, befitting the occasion of this meeting, first about some rare qualities of Professor Salam and then about the way in which our collaboration started and grew.

It hardly needs to be said that Professor Salam has made some landmark contributions to physics, spanning more than three decades. This in itself is rather unusual. But I believe his unique and most valuable contribution to science and humanity, one that is perhaps unparalleled in the world, is the sacrifice he has made of his time, energy, and personal comfort in promoting the cause of science in different corners of the globe, in particular the third world. It literally pains Professor Salam to see that talented young people in the third world simply do not get opportunities to pursue the path of science to the best of their abilities. The efforts he made to resolve this problem has led to the creation of many new research centres throughout the world.

A symbol of his efforts is the creation of the International Centre, in this beautiful city of Trieste. The Centre stands as a monument of learning providing opportunities for the East to meet the West and the North to meet the South. The two goals which I believe the Centre has achieved remarkably well is the production of high-quality science on the one hand and the provision of meaningful opportunities to the third-world scientists for exposure to frontier developments in science on the other hand. As a frequent visitor and organizer of many activities at the Centre, I have always been impressed by (a) the total number of visitors whom the Centre invites per year from different countries in the third world; (b) the high quality of the symposia and the workshops which the Centre arranges throughout the year for the benefit of these visitors, and above all (c) the fruitful research which evolves owing to interactions between the visitors and the lecturers for the workshops. The success of each of these phases is in large part due to the insistence by Professor Salam on high quality. I have also often been touched by the extreme generosity of Professor Salam in dealing with the problems of each individual visitor. His door is always open to one and all.

As a passing remark, I should add that if the Centre has been a success, it owes this both to the dedicated efforts of Professor Salam and to the whole-hearted support and a true appreciation of the goals of the Centre by the Government of Italy, the people of Italy and the city of Trieste. In this sense, I believe that the Centre has been fortunate to have the best of both worlds: The most able leader and the most receptive country. It has thereby emerged as an example for other individuals and countries to follow.

I will now like to talk about my collaboration with Professor Salam.

2. The Strength in Our Collaboration

My collaboration with Professor Salam started in the summer of 1972. It lasted for over a decade and involved many exciting and memorable episodes. During this period, I used to come to Trieste for about two months every summer. The Adriatic Sea, the surrounding small mountains and the daily view of the setting sun have always been a source of inspiration to me. In part because of this but especially because of the warmth which I received from Professor Salam as well as from all the people at the Centre I have always loved to come back here again and again.

During the first six years, which was the most intense phase of our collaboration, Professor Salam also used to visit the University of Maryland during the academic year, often twice a year, for about a week each time. During these visits, the two of us would hide in the hotel room run by the University's Center for Adult Education, so that we would not be disturbed. There, we would spend the first few days in discussing physics and the last few days in writing papers. This was indeed a very productive period of our collaboration, marked with the joy of creativity and mutual appreciation.

During our collaboration there have been many letters, arguments, and even heated exchanges about taste and judgement in physics, but always in a good-natured spirit.

My great respect for Professor Salam is, of course, natural, in part because of his pre-eminence, but in part also because of our common Eastern tradition, which teaches us to respect the older, who is automatically taken to be the wiser. Because of this tradition, I still address him as "Professor Salam" rather than by his first name "Abdus". But what has touched me immensely over the years is that he also treated me throughout our collaboration with respect and affection[1]. This served to establish a mutual bond between us which has lasted until today.

In our conversations, Professor Salam had some favorite phrases. For example, he would sometimes come up with an idea and if I expressed that I did not particularly like it for such and such reason, he would get impatient and say to me: "My Dear Sir, what do you want: Blood?" I would smile and say "No Professor Salam! I would like something better".

Just as he would sometimes become impatient with me, there would be occasions when I would also get impatient with him and even say to him "Professor Salam! You just don't understand!" But, whether I was right or wrong, he never took it ill. In fact, because of these free exchanges, our collaboration became all the more strong.

3. The Origin of SU(4)-Color

Due to lack of time, I cannot possibly cover the many episodes and the different stages of our collaboration. Let me therefore focus only on its initial phase: how it started and how it evolved in the first month, leading to the idea of SU(4)-color. This will allow me to illustrate a rare quality of Professor Salam, that is bringing out the best in a collaboration.

I met Professor Salam first as a student-participant in the Colorado Summer School of 1959 where he lectured on "Symmetries and Group Theory". But my close association with him began only in the summer of 1972, here at Trieste. Just before coming to Trieste that summer, I had spent six months at Delhi University, while I was on sabbatical leave from the University of Maryland. During those six months, I got somehow out of phase with the developments which were taking place in the USA and Europe. When I reached Trieste in May 1972, I found that everyone was talking about 't Hooft's proof of renormalizability of spontaneously broken gauge theories and the SU(2) × U(1) electroweak model[2]. Many papers appeared in 1972 which tried to build variants of the SU(2) × U(1) model. For example, there was a model proposed by Georgi and Glashow[3] which got rid of the neutral current weak interactions altogether because experiments were hinting that they may be absent. Then, there was the model of Bars, Halpern and Yoshimura[4] which proposed a gauge theory of the electroweak and the strong forces, but the idea underlying the model is very different from what prevails today. I realized that I had a lot to catch up on.

However, as I was trying to do this, I felt that something basic was missing from the main thrust of activities. It appeared to me that the heart of the matter lay not in building variants of the SU(2) × U(1) model but in removing the arbitrariness of the model first in the gauge sector and then in the Higgs sector, involving the scalar mass, the quartic and the Yukawa coupling parameters of the Higgs boson.

One aspect of the gauge-sector of the model that bothered me from the beginning was the assignment of the abelian weak hypercharge to the different SU(2)-doublets and the singlets. These assignments did not follow from an underlying principle and thus did not have an *apriori* meaning. They were simply put in by hand to fit the known charges.

Moreover, it seemed unsatisfactory that quarks and leptons are introduced in the model without a rationale for their co-existence and for the asymmetry that is built in between these two kinds of matter. A third ingredient that was also bothersome was the disparity with which the model treated the left and the right chiral fermions. This amounted to putting in non-conservation of parity by hand. Short of a compelling reason why (in Pauli's words) God should be weakly left-handed, it seemed that the introduction of such a disparity is arbitrary and thus unaesthetic.

Last but not least, it appeared that if the weak and the electromagnetic interactions are generated by a gauge principle, the fundamental strong interactions must also have their origin in the same principle. For these reasons, I felt then that the SU(2) $\times$ U(1) model, even it were to be eventually borne out by experiments, which it was, can not constitute a fundamental theory by itself. It must emerge from within a bigger structure which should remove these shortcomings.

In those days, there used to be a tradition at the Centre of having an informal seminar once a week, followed by a tea. In one of those tea gatherings, I casually mentioned to Professor Salam that there ought be a *compelling reason* for the way the weak hypercharge is assigned to the various fields in the SU(2) $\times$ U(1) model – as compelling as the eigenvalues of I_3 in an SU(2) symmetry and those of I_3 and F_8 in an SU(3)-symmetry. An underlying theory should also provide a reason for the co-existence of quarks and leptons and simultaneously for their hypercharge assignments, such that the charges of the electron and the proton are equal and opposite. I mentioned that in order to implement these features, one would need to put quarks and leptons in one common multiplet of a higher gauge symmetry group, which must be spontaneously broken at a high energy scale to account for the observed disparity between quarks and leptons.

Now, the idea of putting quarks and leptons in the same multiplet was rather unconventional at that time. Rather than expressing any reservation about it, as some others did, Professor Salam responded immediately by saying "That seems like an excellent idea! Let us develop it together". It is this sort of spontaneous appreciation and encouragement – a quality that is rare – that has helped to enrich our collaboration at every step. Given the high esteem that I had for him, I, of course, welcomed wholeheartedly his participation. Thus, from that tea conversation began our collaboration.

After a few days, Professor Salam made the important remark that if one could understand why the electron and the proton have equal and opposite charges, one would then have an answer to Feynman's question as to why the electron and the proton, rather than the positron and the proton, exhibit the same sign of longitudinal polarization in β-decay. This made the search for such an answer even more pressing.

Over the next two weeks, we had many discussions at the blackboard and several false starts. One Sunday afternoon, as I was working in my apartment in Sistiana (a sea-side resort near Trieste), it suddenly appeared as a flash that quarks and leptons can be united in a very elegant manner by using the symmetry SU(4)-color. With SU(4)-color the whole spectrum of quarks and leptons fitted beautifully into a 4 $\times$ 4-structure of a symmetry group $SU(4)^{flavor} \times SU(4)^{color}$ operating on four flavors (u, d, c, s) and four colors (r, y, b, ℓ)[5]. Such a structure accounted naturally for the vanishing of the sum of the quark and the lepton charges and thereby for the vanishing of $(Q_{e^-} + Q_p)$. Within it, the neutrino and the electron are interpreted simply as the up and the down quarks of lepton color.

I was naturally elated by the picture which had thus emerged and called Professor Salam over the phone to convey my feelings. As usual, he was working at his desk and asked me to come over to the Centre. I caught a bus and reached the Centre in late afternoon. As I described the incomplete idea, it became clear to both of us that a gauging of SU(4)-color would provide a compelling reason, based on the requirement of renormalizability of the full set of interactions, for why the weak interactions must utilize the flavor degree of freedom including isospin in the first place any why *they must be universal with respect to quarks and leptons.* Because of these features, we both felt that SU(4)-color may well turn out to be fundamental. Professor Salam stressed to me that we must now find the right Higgs system which would break the relevant gauge symmetry appropriately. That Sunday afternoon thus marked the origin of SU(4)-color.

It became clear to us in the next few days that if quarks and leptons have to be united in SU(4)-color, *one important by-product of such a unification is that the known strong interactions must be generated entirely by the SU(3)-color gauge symmetry,* so that the effective gauge symmetry at low energies must be $SU(2)_L \times U(1)_Y \times SU(3)^c$[6, 7]. This became known eventually as the standard model symmetry.

We wrote up this aspect of our thinking in a short draft which we submitted to J.D. Bjorken for presentation at the 1972 Batavia Conference and then in a paper which appeared in the Physical Review[7]. In this paper, we initiated the concepts of quark-lepton unification in terms of SU(4)-color as well as the idea of unification of forces in terms of a single gauge coupling constant, without exhibiting explicitly a symmetry to implement this idea. In detail, we stressed in this paper the desirability of the combined gauge symmetry $SU(2)_L \times SU(1)_Y \times SU(3)^c$ as an effective low-energy symmetry and left the development of suitable Higgs system, for more extended symmetries including SU(4)-color, to two later papers[8]. The discovery of asymptotic freedom, which took place about a year later in May 1973[9], gave a final boost to such a gauge structure, in particular to $SU(3)^c$.

The idea of SU(4)-color in turn generated within the next year and a half other ideas such as those of (i) the gauge symmetry $SU(2)_L \times SU(2)_R \times SU(4)^C$[8], (ii) left-right symmetry and right-handed neutrinos implying $m_\nu \neq 0$[8,10], (iii) non-conservations of baryon and/or lepton numbers, owing to spontaneous breaking of the gauged generator B-L, leading to phenomena such as proton decay which has emerged as one of the hall marks of quark-lepton unification[8,11], (iv) explicit construction of grand unification models such as those of SU(5)[11], SO(10), $[SU(4)]^4$ E_6 and SU(16)[12]. These finally led to the demonstration of the meeting of the coupling constants with a prediction for $\sin^2\theta_W$ in the context of such models.[13] All of these ideas, except for minimal non-supersymmetric SU(5), are very much alive today.

I now turn to the scientific part of my talk, where I explain why, twenty years later, I still believe in SU(4)-color. In this context, I discuss (i) the advantages of the symmetry group $SU(2)_L\times SU(2)_R\times SU(4)^c$ as a stepping stone to higher unification and (ii) the current status of the conventional approach to grand unification with and without supersymmetry, in particular its recent success with the meeting of the coupling constants, as well as its shortcomings as regards the arbitrariness in the Higgs sector. A case is made for an alternative approach to unification based on *a purely gauge-origin* of the fundamental forces and thus for a theory which is devoid of the Higgs sector at a fundamental level. It is suggested that this idea is best realized by combining the idea of local supersymmetry with that of preons, which treats the Higgs bosons as well as the quarks and the leptons as composites. The advantages of a recently developed preonic approach[14] with regard to economy in the building blocks and the parameters, as well as an understanding of (i) the origin of diverse mass scales, (ii) family replication[15] and (iii) inter-family mass-hierarchy[16] are presented. It is stressed that the preonic approach lends itself to many tests by which it can be falsified. Finally, a recent work which demonstrates that the unity of forces may well occur at the level of the preons, rather than quarks and leptons, near the Planck scale, is presented[17]. As it turns out, the preonic approach needs and thus preserves intact the idea of SU(4)-color.

4. SU(4)-Color and Higher Symmetries

Standard model gauge symmetry may be embedded within a higher symmetry containing SU(4)-color, which unites quarks and leptons by assuming that lepton number represents the fourth color[7]. This naturally *requires* the existence of ν_R as the fourth color partner of the u_R's. The minimal gauge symmetry which contains SU(4)-color and ensures quantization of electric charge is[8]

$$\mathcal{G}_0 = SU(2)_L \times SU(2)_R \times SU(4)^c$$

The representations of the left and the right-handed fermions of the electron-family with respect to the symmetry group $\mathcal{G}_o$ are shown below:

$$F_L^e = \begin{pmatrix} u_r & u_y & u_b & u_\ell & = & \nu^e \\ d_r & d_y & d_b & d_\ell & = & e^- \end{pmatrix}_L = (2,1,4)$$

$$F_R^e = \begin{pmatrix} u_r & u_y & u_b & u_\ell & = & \nu^e \\ d_r & d_y & d_b & d_\ell & = & e^- \end{pmatrix}_R = (1,2,4)$$

The μ and τ-family fermions have identical representations. The symmetry $\mathcal{G}_0$ naturally suggests that the basic laws of nature are chiral and yet left-right symmetric (parity conserving), and that parity violation is a low-energy phenomenon brought about by spontaneous symmetry breaking which makes $m_{W_R} \gg m_{W_L}$[8,10].

The minimal symmetry $\mathcal{G}_0$ can, of course, be embedded further within simple groups such as SO(10), E_6 and SU(16)[12], which ensure one gauge coupling constant. Viewed as a part of such a bigger symmetry or even otherwise, $\mathcal{G}_0$ is assumed to break spontaneously into the SM symmetry $SU(2)_L\times U(1)_Y\times SU(3)^C$ at a scale $M_0 \gg 1$ TeV, where $Y = I_{3R} + \frac{B-L}{2}$. A particularly desirable Higgs multiplet, which implements this breaking is $\underline{126}$ of SO(10) which contains the multiplet $\Delta_R \sim (1, 3_R, 10^C)$ of $\mathcal{G}_0$. The VEV of $<\Delta_R>= v_R = M_0 \gg 1$ TeV not only makes W_R's and the leptoquark gauge bosons heavy, but it gives a heavy Majorana mass $M_R = h_M v_R$ to ν_R's which breaks L and $B-L$ by two units. Here, h_m denotes the relevant Yukawa coupling. This heavy Majorana mass, combined with the much lighter Dirac masses of the neutrinos m_D^i, which are family-dependent, leads to the light left-handed neutrinos via the standard see-saw mechanism[18]

$$m(\nu_L^i) \sim (m_D^i)^2/M_R \ll m_D^i$$

Now, a number of symmetry breaking patterns, e.g. two-step breaking of SO(10), leads via RGE, to $v_R \simeq 10^{12}$ GeV and thus $M_R \approx 10^{12\pm1}$ GeV for $h_M = 1/10$ to 10. This in turn leads to:

$$m(\nu_L^e) \sim 10^{-8} \text{ eV}, \; m(\nu_L^\mu) \sim 10^{-3} \text{ eV}, \text{ and } m(\nu_L^\tau) \sim 10 \text{ eV},$$

corresponding to $m_D^i \approx 2$ MeV, 500 MeV and 50 GeV for ν_e, ν_μ and ν_τ respectively. This mass pattern and the associated mixings for the $(\nu_e - \nu_\mu)$ combination matches well the MSW solution[19] for a resolution of the solar neutrino puzzle. Furthermore, the mass of ν_τ has the right magnitude for ν_τ to serve as hot dark matter, which also seems to be needed, together with about a 70% mixture of cold dark matter, to account for the COBE anisotropy and structure formation[20]. In other words, any higher symmetry containing $\mathcal{G}_0$ as a subgroup and breaking in two steps via $\mathcal{G}_0$ to the SM symmetry or any underlying theory (see discussions later) in which $\mathcal{G}_0$ breaks at about $10^{11} - 10^{12}$ GeV to the SM via $<\Delta_R>$ yields a pattern for neutrino masses and mixings which go well with the MSW solution and the COBE data together with models for structure formation. Thus, if the MSW solution and the current interpretation of the COBE data are reconfirmed, they would together provide a strong motivation for the existence of new physics at an intermediate scale of order 10^{11} GeV and also a hint for left-right symmetry.[21]

The planned sharpening of the solar neutrino studies and of the COBE data can thus turn out to be most revealing.

Advantages of $\mathcal{G}_0 = SU(2)_L\times SU(2)_R\times SU(4)^C$: To summarize, the symmetry group $\mathcal{G}_0$ *even without being embedded in a simple group*, brings in a number of attractive features which are worth noting:

(i) quark-lepton unification through SU(4)-color.
(ii) Quantization of electric charge leading to $Q_p + Q_{e^-} = 0$.
(iii) Left-right symmetry, with the associated concept of spontaneous violation of parity.
(iv) Naturally massive neutrinos, with the right masses and mixings which may be needed for a resolution of the solar neutrino puzzle and also to provide hot dark matter.
(v) Finally, $B - L$ as a local gauge symmetry. Following the old arguments given by Lee and Yang, based on the Eötvos-type experiments, it follows that the massless gauge particle coupled to $B - L$ must acquire a mass through SSB and, thereby, $B - L$ must be violated spontaneously (e.g. $< \Delta_R >$ violates L and $B - L$ by two units). Such $B - L$ violation may well be necessary to implement baryogenesis in the presence of electroweak effects which erase $B - L$ conserving baryon-excess generated at high temperatures[22].

Because of these special and most desirable features, I believe that the symmetry group $\mathcal{G}_0$ is likely to be part of a fundamental theory. In particular, the new concepts brought in by it are likely to survive one way or another. Any higher symmetry such as SO(10) or E_6 or SU(16) which contains $\mathcal{G}_0$ as a subgroup would, of course, naturally retain all the advantages of $\mathcal{G}_0$ listed above.

5. Unity Within the Conventional Approach: Proton Decay and $\sin^2 \hat{\theta}_W$:

By "conventional" approach to grand unification, I mean the view that the unity of the electroweak and the strong forces occur at some ultrahigh momentum scale $M_U \sim 10^{15\pm1}$ GeV (say), where the quarks and the leptons exhibit their unity as well[7,8,11,13]. This amounts to assuming, of course, that quarks and leptons are elementary.

It has been known for sometime that both the dedicated proton decay searches especially at the IMB and the Kamiokande detectors, and more recently the precision measurements of the SM coupling constants (in particular $\sin^2 \hat{\theta}_W$) at LEP, put severe constraints on grand unification models without supersymmetry. Owing to such constraints, the non-SUSY minimal SU(5) and, for similar reasons, the one-step breaking non-SUSY SO(10)-model, as well, are now excluded beyond a shadow of doubt.

But the idea of the union of the coupling constants g_1, g_2, and g_3 can well materialize in accord with the LEP data, if one either invokes supersymmetry[23,24] into minimal SU(5) (or SO(10)) or assumes a two-step breaking of a higher symmetry like SO(10) into the SM, with or without SUSY. See Fig. 1, which shows the *impressive meeting* of the three coupling constants of the minimal supersymmetric standard model (MSSM) with an assumed SUSY-threshold around 1 TeV. Such a model can, of course, be embedded within a minimal SUSY SU(5) or SO(10) model, which would provide the rationale for the meeting of the coupling constants at a scale $M_U \approx 10^{16}$ GeV. In SUSY SU(5) or SO(10), dimension 5 operators do in general pose problems for proton decay. But the relevant parameters can be arranged to avoid conflict with experiments. This, together with the requirement that the relic neutralino density be consistent with cosmology, turns out to restrict severely the SUSY-parameter space and thereby leads to predictions for the mass-spectrum of the SUSY particles, which can provide a test of the idea.

It is encouraging that the super-Kamiokande (to be completed in April 1996) is expected to be sensitive to the $e^+\pi^0$ mode up to partial lifetimes of few $\times 10^{34}$ years, to the $\overline{\nu}K^+$ and $\overline{\nu}K^0$ modes with partial lifetimes $\leq$ few $\times 10^{33}$ years and to the non-canonical $n \to e^- e^+ \nu_e$ and $p \to e^- \pi^+ \pi^+$ modes with partial lifetimes $< 10^{33}$ years. Thus the super-Kamiokande with a fiducial mass of 22,000 tons of water, together with other forthcoming facilities, in particular, ICARUS with a sensitive mass of 4700 tons of liquid argon per module (three to be constructed), provide a *big ray of hope* that first of all (a) one will be able to probe much deeper into neutrino physics in the near future and second of all (b) proton-decay may even be discovered within the twentieth century, following the completion of these detectors.

6. A Perspective: Unity with Quarks or Preons?

Talking of a perspective of the field in the future, it is good to focus attention on the meeting of the coupling constants for the minimal SUSY SU(5) or SO(10) models, as exhibited in Fig. 1. The relevant question is: *Is this meeting a reflection of the "truth" or is it somehow deceptive?*

On the one hand, the manner in which the union occurs is certainly impressive and has prompted some to express the opinion that this union confirms, though indirectly, that both SUSY and grand unification have been discovered. On the other hand, many, myself included, feel that such a view is not warranted in part because the meeting of the coupling constants can occur in alternate ways in accord with the LEP data and τ_p. I shall give one such example at the end. The main reason why I feel that such a view is premature at present is because I believe that *a fundamental theory should exhibit not only a unity of the basic particles and of their forces but also should be devoid of arbitrariness in the Higgs sector*, if the Higgs bosons are elementary.

In this sense, neither minimal SUSY SU(5) nor SUSY SO(10) is likely to constitute a fundamental theory *by itself*, because in either scheme, the lagrangian contains two parts:

$$\mathcal{L}((SU(5) \text{ or } SO(10)) = \mathcal{L}_{gauge} + \mathcal{L}_{non-gauge}(m_i^2, h_{ab}^i, \lambda_{ij})$$

While the gauge-part is governed by a single universal gauge coupling, the parameters of the non-gauge part consisting of the scalar masses m_i^2, the Yukawa couplings h_{ab}^i and the quartic self couplings of the scalars λ_{ij} are arbitrary. In short, the unity in the context of these theories is only *partial*, comprising just the gauge forces, but leaving out the forces mediated by the Higgs scalars.

Furthermore, neither SUSY SU(5) nor SUSY SO(10) by itself has the scope of explaining the origins of (a) the three chiral families, (b) inter-family mass-hierarchy and (c) the hierarchical mass-scales which span from M_{Planck} to m_ν. The superstring theories, despite the multitude of solutions, do not seem to be capable of yielding a conventional grand unification model like SUSY SU(5) or SUSY SO(10), with the desired spectrum and parameters (even nearly), and thereby removing the arbitrariness of such a model in the Higgs sector.

Towards a Pure Gauge Origin of the Fundamental Forces: Since it is the Higgs-sector which is the major source of the drawbacks noted above, it seems appropriate to explore an alternative approach to unification in which all the forces of nature have a purely gauge-origin and the Higgs-sector is dispensed with at the elementary level. In this case, the Higgs-force should arise effectively only at a composite level. Notwithstanding the goal of unification, the technicolor idea which treats the Higgs bosons as composite, but quarks and leptons as elementary, is in fact an attempt precisely in this direction. But this idea encounters several difficulties which are first of all practical in nature, because of conflicts with flavor-changing processes and oblique electroweak parameters. Second, the difficulties of technicolor are also aesthetic in nature because, as a generic feature, it needs excessive proliferation in elementary constituents but does not produce any accompanying benefits as regards a synthesis or an understanding of the basic issues. This second difficulty remains even within the variant of walking technicolor. I therefore believe that if the Higgs bosons are composite, quarks and leptons must also be composites of a *common* set of constituents – generically called "preons".

One prima facie *advantage* of the preonic idea compared to that of technicolor is obvious economy in the building blocks. Because quarks, leptons and Higgs bosons share common constituents, there is no need for technifermions and extended technicolor in a preonic theory. Also, the familiar difficulties of technicolor as regards flavor-changing processes and the oblique electroweak parameters can be avoided at least within a class of preonic theories (the model to be presented here is in fact one of this nature). Furthermore, one major attraction of the preonic idea, which incidentally has kept me occupied in the development of this idea well over a decade, is that it can be built as a *pure gauge theory* (like QCD with massless quarks), so that there are no non-gauge mass, quartic and Yukawa coupling parameters of scalars at the preon level.[14] This leads to real economy in the parameters as well.

Inspite of these advantages, it is well known that the preonic idea brings with it new major hurdles, which need to be overcome if the idea has to get off the ground. First, one needs to find a mechanism which would adequately protect the masses of composite quarks and leptons compared to their compositeness scale. Second, one needs to understand family-replication. Third, considering that all three families are presumably made of the same type of constituents, bound by the same force, one needs to understand why there is such a large hierarchy between the masses of the three families – in particular, why $(m_e/m_t) \sim 10^{-5}$ – in a simple and natural way.

Overcoming the Hurdles:
Problems of the type noted above, have plagued composite models from the time the idea was initiated in 1973-74[25]. If at all they may be resolved, they would need some novel features in the dynamics.

Precisely such novel features seem to emerge[26,14], however, by combining the ideas of local supersymmetry with that of preons and by recognizing that a dynamical breaking of supersymmetry is forbidden in a class of theories due to the Witten index theorem[27], except for the presence of gravity. Thus SUSY-breaking is induced in this class of theories, if at all, only by the team-effort of the strong non-perturbative metacolor force *and* the weak perturbative gravitational force. Now for the case of massless preons, the preonic fermion condensates $< \overline{\psi}\psi >$ (see notation below) as well as the metagaugino condensate $< \lambda \cdot \lambda >$ break SUSY. As a result, assuming that these condensates form, each of them must be damped by one (or higher) power of $(\Lambda_M/M_{p\ell})$, corresponding to the effect of one (or multiple) graviton exchange, where Λ_M denotes the scale-parameter of the metacolor force. Barring unexpected vanishing of the leading term, we thus expect $\langle\overline{\psi}^a\psi^a\rangle = a_{\psi_a}\Lambda_M^3(\Lambda_M/M_{P\ell})$, $\langle\lambda\cdot\lambda\rangle = a_\lambda\Lambda_M^3(\Lambda_M/M_{P\ell})$. Here a_{ψ_a} and a_λ denote dimensionless numbers of order unity and the subscript a denotes preon flavor or color-index (see later). We expect a_{ψ_a} to be typically smaller than a_λ by factors of 5 to 10 because ψ's are in the fundamental and λ's in the adjoint representation of the metacolor gauge symmetry.

A number of independent consistency arguments, based on (a) the idea of the unity of forces (see remarks below), (b) the value of m_W, and (c) the masses of the light neutrinos suggest that $\Lambda_M \sim 10^{11}$ GeV[14].

The $< \lambda\cdot\lambda >$ and $< \overline{\psi}\psi >$-condensates thus induce SUSY-breaking mass-splittings δm_S of order $a_\lambda\Lambda_M(\Lambda_M/M_{P\ell}) \sim 1$ TeV $\ll \Lambda_M$. Furthermore, since $< \overline{\psi}\psi >$ is responsible for breaking $SU(2)_L \times U(1)_Y$ as well as for giving masses to the composite quarks and leptons, one naturally obtains[14]:

$$(m_W, m_Z) \sim (1/10)\Lambda_M(\Lambda_M/M_{P\ell}) \sim 100 \text{ GeV};$$

$$(m_q, m_\ell) \leq (1/10)\Lambda_M(\Lambda_M/M_{P\ell}) \sim 100 \text{ GeV}.$$

The factor of (1/10) arises on dynamical ground, where use has been made of the fact that we expect a_ψ to be of order (1/5 to 1/10) of a_λ. Thus, one sees the reason why quark and lepton masses are *necessarily* protected compared to their inverse size Λ_M and also why the mass-scales span over such a wide range: from $m_{P\ell}$ to Λ_M to m_W. To see, how families replicate and how the scales descend from m_W to m_e to m_ν one needs to enter, at least briefly, into the details of the model.

7. A Model with a Unification of Scales – From $M_{P\ell}$ to m_ν

The model[14] assumes that the effective lagrangian just below the Planck scale possesses $N = 1$ local supersymmetry and a gauge symmetry of the form $G_M \times G_{fc}$.

Here $G_M = SU(N)_M$ (or SO(N)$_M$) generates an asymptotically free metacolor gauge force which binds preons and G_{fc} denotes the flavor-color gauge symmetry, which is assumed to be either $\mathcal{G}_0 = SU(2)_L \times SU(2)_R \times SU(4)^c$[8], or a subgroup of $\mathcal{G}_0$ containing SU(2)$_L\times$ U(1)$_Y\times$ SU(3)^c. The gauge symmetry $G_M \times G_{fc}$ operates on a set of six positive and six negative *massless* chiral preonic superfields $\Phi_\pm^{a,\sigma} = (\varphi, \psi, F)^{a,\sigma}_{L,R}$, each belonging to the fundamental representation N of SU(N). Thus "a" runs over six values: $(x, y); (r, y, b, \ell)$, where (x, y) denote the two basic flavor-attributes (u, d) and (r, y, b, ℓ) the four basic color-attributes of a quark-lepton family[8]. The index σ runs over metacolor quantum numbers. The metacolor gauge multiplet is made of v_μ, λ and D, which denote the metagluon, the metagaugino and the auxiliary component respectively. Likewise the gauge multiplets of the other groups. The representation content of the preonic superfields under the gauge symmetry is shown below:

		$SU(2)_L\ \times$	$SU(2)_R\ \times$	$SU(4)^C_{L+R}\ \times$	$SU(N)_{L+R}$
$\Phi_+^{f,\sigma} = (\varphi_L^f, \psi_L^f, F_L^f)^\sigma$	$\sim$	2_L,	1,	1,	N
$\Phi_-^{f,\sigma} = (\varphi_R^f, \psi_R^f, F_R^f)^\sigma$	$\sim$	1,	2_R,	1,	N
$\Phi_+^{c,\sigma} = (\varphi_L^c, \psi_L^c, F_L^c)^\sigma$	$\sim$	1,	1,	4_c,	N
$\Phi_-^{c,\sigma} = (\varphi_R^c, \psi_R^c, F_R^c)^\sigma$	$\sim$	1,	1,	4_c,	N

Here f stands for two flavors (x and y) and c four colors (r, y, b, and ℓ). Since quarks and leptons carry flavor and color, they must be metacolor singlet composites consisting of at least two preons, one carrying flavor and the other carrying color[25].

We see that the preonic content presented above is indeed the *minimum*, providing the attributes of just one family. *In particular, note that there is no repetition of any entity at the preonic level unlike the case of the quark-lepton families.* One then asks: Are there natural reasons within this minimal system for the origins of (a) family replication and (b) mass-hierarchy?

Before entering into this discussion, we note that the effective lagrangian of this model turns out to possess, in accord with the motivations mentioned earlier, only gauge and gravitational interactions. No mass, no non-gauge Yukawa and quartic couplings are even allowed by the gauge symmetry, supersymmetry and the non-anomalous R-symmetry U(1)$_X$. As a result, the model possesses *at most only three or four parameters* (see below) corresponding to the coupling constants of the gauge symmetry $G_M \times G_{fc}$. Even these few would merge into one near the Planck scale if there is an underlying unity of forces-in particular through superstrings (see remarks later).

Such a model has not yet been derived from a superstring theory, although there does not appear to be any bar, in principle, in this regard especially in the context of four-dimensional construction[28]. Even without such a derivation, however, if one introduces the preon-picture mentioned above through an effective lagrangian just below the Planck scale, the model is already most economical and one is able to derive a number of advantages. These include:

(i) **Origin of Diverse Mass Scales:** The model provides a reason[14] for the origin of all the diverse scales – from M_{Planck} to m_ν – and thereby of the small numbers such as $(m_W/M_{pl}) \sim (m_t/M_{pl}) \sim 10^{-17}, (m_c/M_{pl}) \sim 10^{-19}, (m_e/M_{pl}) \sim 10^{-22}$, and $(m_\nu/M_{pl}) < 10^{-27}$ – *by introducing just one fundamental input parameter*: the coupling constant α_M associated with the metacolor force. Briefly speaking, this comes about as follows. Corresponding to an input value $\overline{\alpha}_M \approx 1/27$ to $1/32$ at $M_{Pl}/10$, which incidentally is suggested independently by the idea of the unity of forces (see remarks below), the metacolor force generated by $SU(N)_M$ becomes strong at a scale $\Lambda_M \approx 10^{11} GeV$ for N=5 to 6. Thus the first big step in the hierarchical ladder leading to the small number $(\Lambda_M/M_{Pl}) \sim 10^{-8}$ arises naturally through renormalization group equations due to the slow logarithmic growth of $\overline{\alpha}_M$ and its perturbative input value at $M_{Pl}/10$.

The next step arises due to the Witten-index theorem which, as mentioned above, leads to a damping of SUSY-breaking condensates by the factor $(\Lambda_M/M_{P\ell})$. This provides a reason why $\delta m_S \sim \Lambda_M(\Lambda_M/M_{P\ell}) \sim 1$ TeV and $m_W \sim m_t \sim (1/10)\Lambda_M(\Lambda_M/M_{P\ell})$ 100 GeV $\ll \Lambda_M \ll M_{P\ell}$. One special feature of the model is that, while $m_t \sim 100$ GeV it naturally provides extra symmetries in the fermion mass-matrix (see below) which protect $m_{u,d,e}$, barring corrections of order 1 MeV. There is also a good reason why the masses of the fermions belonging to the muon-family are intermediate between those belonging to the e and the τ-families. Finally, the familiar see-saw mechanism for neutrinos with $m(\nu_R^i) \sim \Lambda_M \sim 10^{11}$ GeV and $m(\nu^i)_{Dirac} \propto \Lambda_M(\Lambda_M/M_{Pl})$ yields $m(\nu_L^i) \leq 10^{-3}M_{Pl}(\Lambda_M/M_{Pl})^3 \sim 10^{-27}M_{Pl}$. In this way, the model provides a *common origin* of all the diverse scales – from M_{Pl} to m_ν, as desired. Supersymmetry and compositeness play an essential role in achieving this result.

Note that electroweak symmetry breaking in the preon-picture is dynamical, like technicolor. *But unlike technicolor, there is no strong force at the TeV-scale.* The intrinsic scale of the new dynamics, responsible for electroweak symmetry-breaking, is the metacolor scale of order 10^{11} GeV. This scale incidentally goes well not only with the see-saw neutrino masses that are needed for the MSW solution of the solar neutrino puzzle but also with the desired scale for a breaking of the Peccei-Quinn symmetry.

(ii) **Family Replication:** Although a two-body metacolor singlet fermion-boson composite of the form $\psi^a\phi^{b*}$ can yield the desired quantum numbers of quarks and leptons[25], given that the masses of quarks and leptons are protected for reasons mentioned above, one can argue plausibly that such a "massless" spin-1/2 composite made of a "massless" spin-1/2 and a spin-0 constituent is dynamically inconsistent unless it consists of at least one additional third ingredient, which is necessarily neutral with respect to flavor and color. Thus, this third constituent is the metagluon (v). In short, the minimum dimensional composite operator representing quarks and leptons corresponds to a three-body composite of the form $\psi\phi^*v$ and its SUSY-transforms.

Recognizing that in a SUSY theory, fermionic constituents can be interchanged by their boson partners (i.e. $\psi \leftrightarrow \varphi$ and $v_\mu \leftrightarrow \lambda$ or $\overline{\lambda}$ etc.), there exist several alternative three-particle combinations with identical quantum numbers, which can make a left-chiral $SU(2)_L$-doublet family q_L^i – e.g. (i) $\sigma_{\mu\nu}\psi_L^f \varphi_R^{c*} v_{\mu\nu}$, (ii) $\sigma^{\mu\nu}\varphi_L^f \psi_R^{c*} v_{\mu\nu}$, (iii) $\psi_L^f \psi_R^{c*}\lambda$ and (iv) $\varphi_L^f(\sigma^\mu\overline{\lambda})\partial_\mu\varphi_R^{c*}$. Here $f = x$ or y corresponds to up or down flavors and $c = (r, y, b)$ or ℓ corresponds to the four colors. *The plurality of these combinations, which stems because of SUSY, is in essence the origin of family-replication.* By constructing composite superfields, it is found[15] curiously enough, that at the level of minimum dimensional composite operators (somewhat analogous to qqq rather than $qqqq\overline{q}$ for QCD) there are just three linearly independent chiral families $q_{L,R}^i$, and, in addition, two *vector-like families* $Q_{L,R}$ and $Q'_{L,R}$, which couple vectorially to W_L's and W_R's respectively. Each of these composite families is, of course, accompanied by its scalar superpartners. To sum up, we see that owing to fermion-boson pairing in SUSY, the model provides a good reason for replication (i.e. number of chiral families being more than one) and at least some rationale, subject to the assumption of saturation at the level of minimum dimensional composite operators, as to why the number of chiral families is three. Clearly this last assumption needs further justification. Pending such a justification, however, it turns out, as noted below, that the mass-matrix of the fermions belonging to this system of *five families* – three chiral and two vector-like-provides a compelling reason for inter-family hierarchy.

(iii) **Inter-family Mass-Hierarchy**: Owing to symmetries of the underlying theory and the quantum numbers of the composites, it turns out[14,16] that the 5×5 mass-matrix of the three chiral and two vector-like families acquire relatively heavy masses through the metagaugino condensate of order $\Lambda_M(\Lambda_M/M_{P\ell}) \sim 1$ TeV, but the direct mass-terms of the three chiral families vanish barring corrections of order (at most) 1 MeV. The latter acquire masses primarily through their mixings with the two vector-like families. Such mixings are induced through the preonic $\langle\overline{\psi}\psi\rangle$-condensates. Although all three chiral families are on a par in that their mixings with the two vector-like families are comparable to each other, *the rank of such a 5×5 see-saw type mass-matrix automatically guarantees one linear combination of the chiral families to be massless*, apart from corrections of order 1 MeV. This combination may be identified with the electron family. At the same time, the heaviest fermion (top) acquires a mass of order 100-150 GeV and the members of the muon-family acquire masses which are naturally intermediate between those of the electron and the tau family.

In this way, one obtains a simple reason for the inter-family mass-hierarchy – i.e. why $m_{u,d,e} \ll m_{c,s,\mu} \ll m_{t,b,\tau}$[16]. In particular, one understands both why $m_e \sim O(1$ MeV) and the heaviest chiral quark (top) is around 100-150 GeV, so that $(m_e/m_t) \sim 10^{-5}$. It needs to be stressed that all this is realized, even though the model started with no fundamental input parameters except for the three gauge couplings. The simplicity of this explanation seems to provide support for the preonic approach. It also leads one to expect that there ought to exist two vector-like families in the SSC/LHC-range, since they are crucial to the explanation.

(iv) **CP Violation:** The model provides an elegant mechanism for spontaneous CP violation, which is shown to vanish (for the observed processes), if the masses of the electron family were set to zero[16], – i.e. if the direct mass terms of the chiral families $m_{(0)}(q_L^i \to q_R^j)$ in (1) are put strictly to zero. Allowing for these direct mass-terms the model predicts $|\epsilon| \sim (m_d/m_b)\sin\xi \sim 2 \times 10^{-3}$ for a maximal CP violating phase. It also predicts an electric dipole moment for the neutron $\approx$ (1 to $\frac{1}{20}$) $\times 10^{-25} ecm$, which is observable.

(v) **Crucial Tests:** One distinguishing feature of the preon model is that it leads to several crucial predictions[16] by which it can really be falsified if it is wrong. These include:

(1) $m_t \leq 150$ GeV.

(2) $K^0 \leftrightarrow \overline{K}^0$ and $K_L \to \overline{\mu}e$ are normal as in SM, in so far as contributions from box graphs involving quarks and W's are concerned. But diagrams containing squarks and gluinos need, as in all SUSY theories, that the squarks of different generations have to be degenerate to better than about 5% if $m_{\tilde{q}} \sim O$ (3 TeV). Such a degeneracy in principle can arise depending upon the relative values of certain mass matrix elements, but it is not apriori guaranteed by the model.

(3) $A(``Z" \to t\overline{c}) \approx \left(\frac{g_2}{\cos\theta_W}\right)$ (1/6 to 1)%

(4) $A(``Z" \to c\overline{u})$ which gives $\Delta m(D - \overline{D})_{New} \approx (10 - 1) \times 10^{-14}$ GeV, while $\Delta m_{expt} \leq 1.3 \times 10^{-13}$ GeV and $\Delta m(D - \overline{D})_{SM} \approx (1 - 4) \times 10^{-15}$ GeV.

(5) $A(Z \to \overline{\mu}e)$ which gives $B(\mu \to 3e) \approx$ (1 to 5) $\times 10^{-13}$

(6) $\tau_\tau = (\tau_\tau)_{SM}(1 + \epsilon) = (\tau_\tau)_{SM}[1 + (1 to 4)\%]$; N_ν (LEP) $= 3 - 2\epsilon = 3 -$ (.02to.08). Note that corrections to τ_τ and N_ν are correlated[29].

(7) Prominent $\nu_\mu - \nu_\tau$ mixing with $m(\nu_\tau) \approx$ (1 to 50) eV, $m(\nu_\mu) \approx (10^{-3} - 10^{-1})$ eV, $m(\nu_e) \approx 10^{-8}$ eV, and $\nu_\mu - \nu_e$ mixing consistent with the MSW solution.

(8) $(edm)_n \approx$ (10 to 1/2) $\times 10^{-26}$ ecm.

(9) Existence of SUSY partners with masses $\sim$ (100 GeV - few TeV)

(10) 2 Higgs doublets with familiar SUSY charged and neutral members corresponding to $\tan\beta \approx$ (30-40).

(11) Finally, as mentioned before, the model predicts two vector-like families, which are the hall-marks of the model. Their masses including QCD-renormalization effects, are given by[16]:

$$m_{E,E'} \approx 200 - 700 \text{ GeV}; \; m_{U,D,U',D'} \approx (600 - 3000) \text{ GeV}$$

For a variety of reasons, including renormalization group analysis, it turns out that the number of these vector-like families must be precisely two – no more, no less; one of these couples to W_L and the other to W_R.

(vi) **Supersymmetry Breaking:** The preonic theory requires the presence of a new metacolor force in the *observable sector* which becomes strong at a superheavy scale $\Lambda_M >> 1\ TeV$. Such a force would not, however, be permissible if quarks and leptons were elementary. The existence of such a force, in the presence of gravity, allows the possibility of a dynamical breaking of supersymmetry directly in the observable sector (albeit with a damping by the Planck mass), which thus transmits efficiently into the masses of the squarks, the gluinos and the winos. This may well be an advantage over a dynamical breaking of supersymmetry occurring entirely in the hidden sector of a superstring theory, which seems to be the only possibility in such a theory if quarks and leptons are elementary.

8. Unity of Forces at the Preon Level

Although the derivation of a preon model resembling that proposed in Ref. 14 from a superstring theory is still awaited, it is intriguing to ask whether the coupling constants g_1, g_2 and g_3 extrapolated in the context of the preon model from their measured values at low energies do indeed meet with each other as well as with g_M near the Planck scale, for any reasonable choice of the metacolor gauge symmetry G_M and the flavor-color gauge symmetry G_{fc} which operates near the Planck scale. The precise nature of G_M and G_{fc} may hopefully get determined ultimately by an underlying superstring theory if preons have their origin from such a theory. In this case, despite the non-unifying appearance of the effective symmetry $G_M \times G_{fc}$, the constraints of grand unification including the quantization of electric charge and the familiar equality of the coupling constants near the unification scale M_{SU} would still hold, especially for $k = 1$ Kac-Moody algebra, barring, of course, Planck-scale threshold effects.

Note that the extrapolation based on renormalization group equations is fully determined in regions I and II(see Fig. 2) because the spectrum and the gauge symmetry are fixed, while in region III, there is only a few discrete choices which can be made as regards the metacolor gauge symmetry G_M and the flavor-color gauge symmetry G_{fc}. The scale Λ_M is fixed at about 10^{11} GeV (within a factor of 3, say) by requiring consistency with the hierarchy of scales, in particular by the observed mass of m_W.

In a recent work, Babu and I found[17] that with the measured low-energy values of g_1, g_2 and g_3 at LEP and thus with $\sin^2\theta_W \simeq .2333, \alpha = 1/127.9$ and $\alpha_3 = .118$ at m_Z as inputs, the coupling constants including g_M, show a clear tendency to converge to a common value within a few percent of each other[30] at a scale $M_U \approx (2-5)\times 10^{18}$ GeV, for example for the choice $G_M = SU(5)_M$ and $G_{fc} = SU(2)_L \times U(1)_{I3R} \times SU(4)^c$, (see Fig. 2). It demonstrates that the unity of forces may well occur at the level of preons in a manner that is truly novel compared to the conventional approach of elementary quarks and leptons.

9. Concluding Remarks

To conclude, let me compare the two alternative approaches to unification: One based on the idea of elementary quarks and leptons and the other on the idea of preons.

On the negative side, the preon picture possesses a certain arbitrariness with regard to the choice in the pattern of symmetry breaking and correspondingly in the precise set of condensates which form near Λ_M. While this type of arbitrariness is present in one form or another in alternative approaches as well (compare e.g. with the needed VEVs of the scalar partners of ν_R and N in the three-generation Calabi-Yau models), it is clearly desirable to understand the dynamics of locally supersymmetric gauge theories to shed some light on these issues. Of special interest is the question *whether vectorial global symmetries such as isospin and parity can break dynamically in a SUSY QCD-type theory* to which the premises of the Vafa-Witten theorem do not apply. Such a breaking has been presumed within the preonic approach. If it can be derived on dynamical grounds, one would have a compelling reason for the breakdown of isospin, parity, baryon number and perhaps other global symmetries in SUSY QCD-type theories. It also remains to be seen whether the preon-content and the gauge symmetry of the type suggested in Ref. 14 can be derived from an underlying superstring theory. On the practical side, as mentioned above, squarks of different generations have to be fairly degenerate (as in all SUSY theories) to avoid excessive flavor-changing neutral-current (FCNC) processes. Such a degeneracy could in principle arise in the model if certain mass matrix elements dominate, but there does not appear to be any obvious reason for the same. A natural resolution of this as well as of the other issues mentioned above are the challenges confronting the preon picture.

On the positive side, while the pattern of the condensates needs to be assumed within the preonic approach, the scales of the condensates including those with a damping by (Λ_M/M_{Pl}) are motivated on general grounds. Furthermore, even without being derived from a superstring theory as yet, the preonic approach shows many attractive features:

(a) It is economical in parameters and building blocks and thus is devoid of much of the arbitrariness that goes with the conventional approach to grand unification involving elementary quarks, leptons and Higgs bosons.

(b) It is viable with regard to all precision measurements of the electroweak parameters, unlike standard technicolor. The issue with regard to FCNC processes is, however, still open (see remarks above).

(c) The preonic approach seems capable of addressing successfully the three basic issues pertaining to the origins of (1) the diverse mass-scales from M_{Pl} to m_ν, (2) family replication and (3) inter-family mass-hierarchy. None of these issues can be addressed within the conventional approach to grand unification with elementary quarks, leptons and Higgs bosons, with or without supersymmetry. And, as I mentioned before, the hope of addressing these issues by deriving a conventional supersymmetric grand unification model from a superstring theory seems to be marred because the existing vast set of solutions including Calabi-Yau, Orbifold and four-dimensional constructions do not show promise in this regard.

(d) The recent work of Ref. 17 opens up the possibility that the unity of forces can occur near the Planck scale through preons rather than quarks and leptons, in accord with the recent LEP data.

(e) The *real advantage* of the preonic approach, it seems to me, is that while it is capable of addressing some very basic issues, it also offers some crucial tests listed above, by which it can be excluded, if it is wrong. This is so, inspite of the fact that the compositeness scale Λ_M is found to be very high, of order 10^{11} GeV.

Thus, the preonic approach to unification, seems to offer some *distinct advantages* and as such it deserves further theoretical study in all its aspects involving dynamics of locally supersymmetric QCD-type theories on the one hand and a derivation of a model of the type proposed in Ref. 14 from a superstring theory on the other hand.

In this connection, it is interesting to recall that the idea of $SU(3)$-flavor as an approximate symmetry, suggested by Ikeda, Ogawa and Ohnuki; Wess; and Yamaguchi, turned out to be correct, but the identification of the fundamental triplet with the observed $(p, n$, and $\Lambda)$ rather than with their constituents – i.e. the quarks – turned out to be wrong. Even earlier, the idea of Yang and Mills for generating strong interactions by a non-abelian gauge symmetry turned out to be right, but its association with the isospin symmetry of protons and neutrons, rather than with the color symmetry of their constituent – the quarks, also proved to be wrong. One wonders whether history is being repeated now, by adherence to the common notion that the unity of forces should occur at the level of the observed quarks and leptons. The idea of the superstring theories and therefore of the unity of forces at a fundamental level may well be right. But it remains to be seen, and here experiments of the type mentioned above can help, whether the fundamental fields emerging from such a theory correspond to the observed quarks and leptons or possibly to a new layer of constituents – the preons.

By raising the alternative of preons, I am, of course, not suggesting that the search for an ultimate layer of constituents would necessarily continue for ever. Sooner or later, assuming that adequate funding of high energy experiments is not halted prematurely, we may well be able to see clear marks of such a layer and thereby also of the "final theory". I believe that this could be the case even if, on energetic grounds, one can not *directly* observe the physics at ultrahigh energies approaching the Planck scale and thereby "see" the ultimate building blocks. This is because, on the basis of concrete cases at hand discussed in the text, it seems likely that such an ultimate theory, although defined at ultrahigh energies, could well cast clear shadows at low energies by which it can be probed with a reasonable degree of confidence. In addition to compatibility with observation, *the two criteria which one may use for deciding whether or not one has arrived at such an "ultimate theory" are: (a) it should be able to provide simple answers to all the basic questions which one would be capable of raising and (b) it should somehow be singled out by its uniqueness and thus by its lack of arbitrary parameters.*

The question is: Have we arrived at such a theory already? My answer is that we have not. I say this because there are still many loose ends and gaps. Among others, we do not yet have (i) a principle to derive a superstring theory, (ii) a reason for the four-dimensional nature of space-time in which we live, (iii) an insight into the non-perturbative aspect of the string theories so as to *understand* the choice of the ground state and supersymmetry breaking and (iv) a good reason for the exact or the approximate vanishing of the cosmological constant. Optimistically, speaking, however, there are reasons to believe that we may not be far from an "ultimate theory" in that quite a few of its major ingredients are in. The basic premises of the superstring theories and thereby of grand unification in a broader perspective are very likely among these.

However, using the criteria mentioned above, I believe that the common view in this regard that quarks and leptons are the ultimate building blocks is likely to be false. As mentioned above, the main reason why I prefer the alternative of the supersymmetric preonic theory over that of elementary quarks, leptons and Higgs bosons is that the former dictates one to write the fundamental lagrangian as a *pure gauge theory* in terms of massless fields with no non-gauge couplings.

Fortunately, the two alternatives can be distinguished from each other by pre-SSC and especially post-SSC/LHC experiments of the type mentioned here.

If, on the basis of these experiments, preonic ideas turn out to be right, what would that mean for the idea of SU(4)-color? The above discussion shows that this symmetry has played a crucial role even in the preonic approach. In fact, if ν_R's are relevant, I can not think of any elegant way to construct a composite model without this symmetry. This is why, twenty years later, I still believe in SU(4)-color.

Finally, I must say that I feel especially fortunate to have had the privilege of a most intense and fruitful collaboration with Professor Salam, marked by many moments of shared joy and happiness. For this and for the warmth and friendship which he has extended to me, I wish to express once again my deepest gratitude, together with my very best wishes for the future.

10. Acknowledgements

The research described in this talk is supported in part by a grant from the National Science Foundation. It is a pleasure to thank Qaisar Shafi and Joseph Sucher for many valuable discussions and for their help in the preparation of this manuscript.

I also would like to thank Seifallah Randjbar-Daemi and the other organizers of the Salamfest for arranging an excellent meeting and for inviting me to give this talk.

References

[1] To illustrate this, I digress for a moment. I come from the state of Orissa lying in the east coast of India - about 200 miles to the south of Calcutta. Orissa possesses one of finest temples in India, called Konarak, built about 800 years ago, with superb sculpture, engraved on the stone. Professor Salam visited this temple and was much impressed by its beauty. He described it to me as the Taj Mahal of the Hindus. In the early stages of our collaboration, he once told me "Jogesh: since you come from the land of the Konarak, I expect nothing less than a Konarak in physics from you". This remark, coming from him, has always been a source of inspiration for me.

[2] S. Weinberg, *Phys. Rev. Lett.* **19** (1967) 1269; Abdus Salam, in Elementary Particle Theory, Nobel Symposium, ed. by N.Svartholm (Almqvist, Stockholm, 1968), p. 367; S.L. Glashow, *Nucl. Phys.* **22** (1961) 57a .

[3] H. Georgi and S.L. Glashow, *Phys. Rev. Lett.* **28**, 1494 (1972).

[4] I. Bars, M.B. Halpern and Y. Yoshimura, *Phys. Rev. Lett.* **29**, 969 (1972).

[5] Recall that in 1972 there was no need to consider more than four flavors. Effectively, such a 4×4-structure is to be viewed as a merger of two families, each being a 2×4. In reality, one must, of course gauge a suitable chiral counterpart of $SU(4)^{flavor}$ together with $SU(4)^{color}$, which could be vectorial or chiral. This is what we did by gauging $SU(2)_L \times SU(2)_R \times SU(4)^c$ or $[SU(4)]^4$ with mirrors (see later). But for the purposes of classification and assignment of electric charge, which is vectorial, it suffices to use the non-abelian vectorial $SU(4)^f \times SU(4)^c$.

[6] To clarify the background of this suggestion, the idea of $SU(3)^c$ as a global symmetry was introduced implicitly through paraquarks by O.W. Greenberg, *Phys. Rev. Lett.* **13**, 598 (1964). M. Han and Y. Nambu (*Phys. Rev.* **139**, B1006 (1965)) introduced the same symmetry explicitly and made the important suggestion of gauging $SU(3)^c$, but in a manner that violated $SU(3)^c$ explicitly through the coupling of the photon to integer charge quarks (icq); this spoilt the renormalizability of the theory. They also introduced medium strong interactions utilizing flavor symmetry. These two shortcomings were removed and the advantages of the combined gauge structure $SU(2)_L \times SU(1)_Y \times SU(3)^c$ were noted in two papers by Salam and myself starting with a contribution to the 1972 Batavia Conference (Ref. 7). H. Fritzsch and M. Gell-Mann also promoted $SU(3)^c$ gauging at the same conference. They, and many others, favored the alternative of the fractionally charged quarks (fcq) with confinement. To Salam and myself, this was only an alternative along with the one of liberated gauged icq; the two alternatives arise within the same theory depending upon the nature of SSB. We therefore proposed, over the next few years, some key experiments which could help distinguish between the two alternatives. Eventually, by the mid 1980's the two-photon experiments seemed to favor clearly fcq over gauged icq. The two central ideas of quark-lepton unification and the consequent non-conservations of baryon and lepton numbers, introduced in Refs. 7 and 8, are of course, independent of the nature of quark-charges.

[7] J.C. Pati and Abdus Salam; Proc. 15th High Energy Conference, Batavia, reported by J.D. Bjorken, Vol. 2, p. 301 (1972); *Phys. Rev.* **8** (1973) 1240.

[8] J.C. Pati and Abdus Salam; *Phys. Rev. Lett.* **31** (1973) 661; *Phys. Rev.* **D10** (1974) 275.

[9] D. Gross and F. Wilczek, *Phys. Rev. Lett.* **30**, 1343 (1973); H.D. Politzer, *Phys. Rev. Lett.*, 1343 (1973).

[10] R.N. Mohapatra and J.C. Pati, *Phys. Rev.* **D11**, 566 (1975); *Phys. Rev.* **D11**, 2558 (1975); G. Senjanovic and R.N. Mohapatra, *Phys. Rev.* **D12**, 1502 (1975).

[11] H. Georgi and S.L. Glashow, *Phys. Rev. Lett.* **32** (1974) 438.

[12] **SO(10)**: H. Georgi, Proc. AIP Conf. Williamsburg (1974); H. Fritzsch and P. Minkowski, *Ann. Phys.* (NY) **93** (1975) 193. $[\mathbf{SU(4)}]^4$: J.C. Pati and Abdus Salam, *Phys. Lett.* **58B**, 333 (1975). **SU(16)**: J.C. Pati, A. Salam and J. Strathdee, *Il. Nuov. Cim.* **26A** (1975) 77; *Nucl. Phys.* **B185** (1981) 445. **E(6)**: F. Gursey, P. Ramond and P. Sikivie, *Phys. Lett.* **60B** (1976) 177.

[13] H. Georgi, H. Quinn and S. Weinberg, *Phys. Rev. Lett.* **33** (1974) 451.

[14] J.C. Pati, *Phys. Lett* **B228** (1989) 228.

[15] K.S. Babu, J.C. Pati and H. Stremnitzer, *Phys. Lett.* **B256** (1991) 206.

[16] K.S. Babu, J.C. Pati and H. Stremnitzer, *Phys. Rev. Lett.* **67** (1991) 1688.

[17] K.S. Babu and J.C. Pati, *Phys. Rev. D. Rap. Comm.* **48**, R1921 (1993).

[18] M. Gell-Mann, P. Ramond and R. Slansky, in Supergravity, ed. by D. Freedman and P. Van Nieuwenhuizen (North Holland, 1979), p. 315; T. Yanagida, Proc. Workshop on "Unified Theory and Baryon Number in the Universe," eds. O. Sawata and A. Sugamoto, KEK, Japan (1979); R.N. Mohapatra and G. Senjanovic, *Phys. Rev. Lett.* **44** (1980) 912.

[19] S.P. Mikheyev and Y. Yu Smirnov, *Nuovo Cim* **9C**, 17 (1986); L. Wolfenstein, *Phys. Rev.* **D17**, 2369 (1978).

[20] G. Smoot et al., *Ast. J.* **396** (1992) L1; E.L. Wright et al., *Ast. J.* **396** (1992) L13; R.K. Schaefer and Q. Shafi, *Nature* **359** (1992) 199; A. Klypin et al., Santacruz SCIPP-92-52; R.L. Davis et al., *Phys. Rev. Lett.* **69** (1992) 1856.

[21] With only ν_L's and Higgs doublet ϕ (as in SM and SU(5)), higher dimensional operators such as $\nu_L \nu_L \langle\phi\rangle^2/M$ can give Majorana mass to ν_L. But if these are induced by gravity with $M \sim M_{P\ell}$, one would get $m(\nu_L) \sim 10^{-5}$ eV, which is too small to be relevant to the MSW solution.

[22] V. Kuzmin, V. Rubakov and M. Shaposhnikov, *Phys. Lett.* **155B** (1985) 36.

[23] S. Dimopoulos and H. Georgi, *Nucl. Phys.* **B193** (1981) 150; N. Sakai, *Z. Phys.* **C11** (1982) 153.

[24] P. Langacker and M. Luo, *Phys. Rev.* **D 44** (1991) 817; U. Amaldi, W. de Boer and H. Furstenau, *Phys. Lett.* **B260** (1991) 447; J. Ellis, S. Kelley and D.V. Nanopoulos, *Phys. Lett.* **B260** (1991) 131. F. Anselmo, L. Cifarelli, A. Peterman and A. Zichichi, *Nuovo Cim.* **A104** (1991) 1817.

[25] Flavon-chromon type composite models of quarks and leptons were initiated in J.C. Pati and Abdus Salam, *Phys. Rev.* **D10** (1974) 275 (footnote 7). This idea was subsequently considered by them and several other authors, see e.g. H. Terezawa, *Prog. Theor. Phys.* **64** (1980) 1763 and O.W. Greenberg and J. Sucher, *Phys. Lett.* **99B** (1981) 339. Supersymmetric Versions of this idea were proposed by J.C. Pati and A. Salam, *Nucl. Phys.* **B214** (1983) 109; ibid **B234** (1984) 223 and R. Barbieri, *Phys. Lett.* **121B** (1983) 43. In these attempts, there did not, however, exist a reason for the protection of the masses of the composite quarks and leptons. Subsequently, the idea of quasi-Nambu-Goldstone fermions, suggested by W. Buchmüller, R.D. Peccei and T. Yanagida, *Phys. Lett.* **124B** (1983) 67, and developed by several authors, did provide a reason in a limited

sense, but had problems of internal consistency in SUSY-breaking. These types of problems are evaded in Ref. 26 and 14.

[26] J.C. Pati, M. Cvetic and H. Sharatchandra, *Phys. Rev. Lett.* **58** (1987) 851.

[27] E. Witten, *Nucl. Phys.* **B 185** (1981) 513.

[28] H. Kawai, D. Lewellen and S. Tye, *Phys. Rev. Lett.* **57**, (1986) 1832; I. Antoniadis, C. Bachas and C. Kounnas, *Nucl. Phys.* **B289** (1987) 187.

[29] For a discussion of this and other cases see e.g. K. Babu, J.C. Pati and X.M. Zhang, *Phys. Rev.* **46** (1992) 2190.

[30] Non-renormalizable interactions induced by Planck-scale physics and threshold effects at M_{Pl} and Λ_M could, in general, account for such differences.

[31] Other cases including some threshold effects are being considered by K. Babu, M. Parida and J.C. Pati. G_M gets restricted between $SU(4)_M$ and $SU(6)_M$.

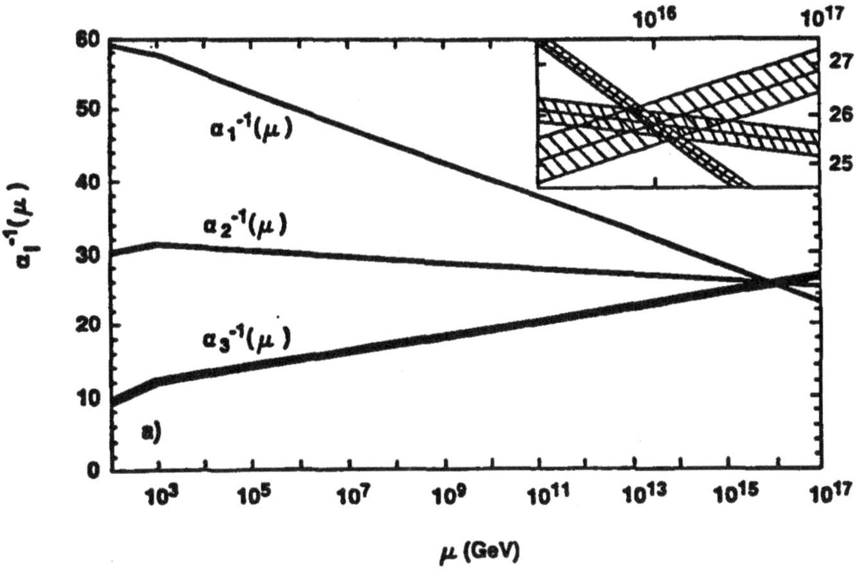

Figure 1: Running of coupling constants in SUSY minimal $SU(5)$ (Ref. 24).

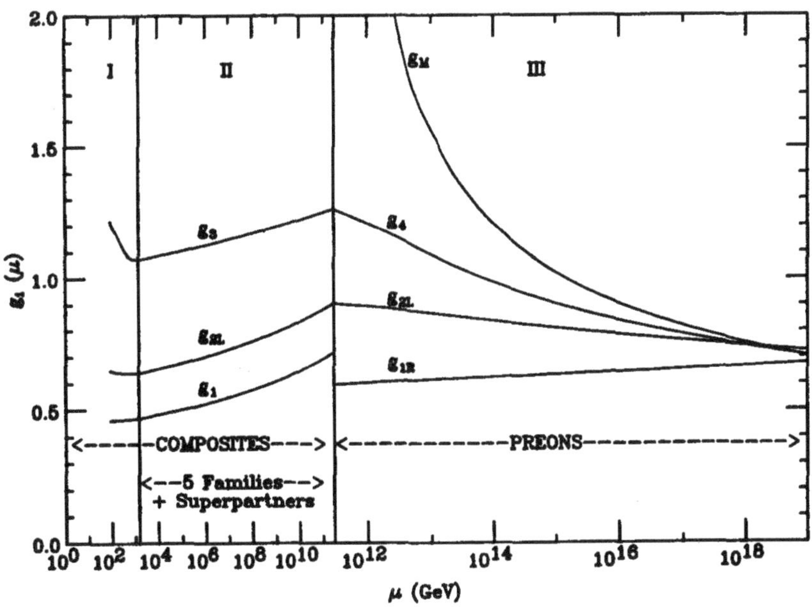

Figure 2: Running of coupling constants in the preon model (see text and Ref. 17).

Towards A 'Standard Model' of Large Scale Structure Formation *

Q. Shafi
Bartol Research Institute
University of Delaware, Newark, DE 19716

Abstract

We explore constraints on inflationary models employing data on large scale structure mainly from COBE temperature anisotropies and IRAS selected galaxy surveys. In models where the tensor contribution to the COBE signal is negligible, we find that the spectral index of density fluctuations n must exceed 0.7. Furthermore the COBE signal cannot be dominated by the tensor component, implying $n > 0.85$ in such models. The data favors cold plus hot dark matter models with n equal or close to unity and $\Omega_{HDM} \sim 0.2 - 0.35$. Realistic grand unified theories, including supersymmetric versions, which produce inflation with these properties are presented.

*Updated version of talk presented at the 'Salamfest', Trieste, March '93.

1 Introduction

The inflationary universe scenario [1] provides an attractive resolution of the well known "horizon", "flatness", and "causality" puzzles encountered in the standard big bang cosmology. This scenario is most easily realized within the framework of both ordinary and supersymmetric grand unified theories (GUTS). In the simplest realizations of inflation, the density fluctuations are Gaussian and close to the Harrison-Zeldovich form, and the background density equals the critical density ρ_c. Primordial nucleosynthesis implies that less than 10% of the background density is composed of baryons, and so the bulk of matter is "dark", presumably in the form of relic elementary particles.

Almost a decade ago it was pointed out by Shafi and Stecker [2] that two types of dark matter, cold and hot, would provide, within the inflationary context, an elegant way of understanding observations which indicated a surprising amount of clustering on larger than cluster scales. Since only the cold component clustered on smaller scales, and the hot component only clusters on larger scales, such a universe would have an enhanced large scale clustering power. Examples of GUTS containing cold plus hot dark matter (C+HDM) were also presented. The implications of this picture for microwave background anisotropies were worked out in 1989 [3,4]. Mass functions [5], bulk streaming motions [3,4], cluster number densities and correlation functions, as well as "great attractors" were also found to be compatible [6,7] with data.

Recent data on large scale structure from a variety of sources, particularly COBE and IRAS selected galaxy surveys, provide additional strong support to this remarkably "simple" scenario for structure formation. The COBE group [8,9] found an anisotropy amplitude which they characterized by an averaged quadrupole moment amplitude $\delta T/T = Q_{rms-PS}/T_0 = 6.2 \pm 1.5 \times 10^{-6}$. As we have emphasized [10], the earlier predictions [3,4] for C+HDM models (3/4 CDM, 1/4 HDM) with negligible baryonic matter content were $7.8 \times 10^{-6}/b_8$, while for CDM alone they were significantly smaller ($4.7 \times 10^{-6}/b$), where b_8 =(rms mass fluctuation on the scale 8 h^{-1}Mpc)$^{-1}$. Including a 5% baryon density (implied by primordial nucleosynthesis) modifies these predictions to $8.2 \times 10^{-6}/b_8$ for this C+HDM mixture and $5.1 \times 10^{-6}/b_8$ for CDM. Observationally derived values of b_8 seem to fall in the range $1.3 < b_8 < 2.5$, meaning the COBE result was in remarkable agreement with C+HDM models. COBE also lent some weak support for the inflationary power spectrum; the data are consistent with the $P(k) \propto k^n$ spectrum with $n = 1.1 \pm 0.5$ [8,9]. Following the COBE result other pieces of evidence pointing to the C+HDM model were discovered: compatibility with early quasar formation [11,12], compatibility with galactic correlations and velocities, [13,14,15],

APM correlations [15,16], quiet local Hubble flow [17], "cosmic mach number" [18], and counts in cells [19,20]. Thus, there seems no doubt that the C+HDM model is a candidate worthy of serious consideration.

While we know that the primordial density power spectrum $P(k) \propto k$ with C+HDM provides a good fit to the data, it has long been known that the inflationary scenario does not quite yield this simple form. In recent studies, these correction factors have been exploited to yield power spectra $P(k) \propto k^n$ with n significantly less than 1 as a means to repair the relative large/small scale problems of the pure CDM model. In this paper, one of our tasks will be to undertake a study of the range of allowed n. Since we do not precisely know the HDM fraction of the universe, we will explore the two dimensional parameter space $n - \Omega_{HDM}$, and examine the constraints imposed by the data.

We do this by comparing the model predictions with data from COBE, the power spectrum of IRAS galaxies, IRAS counts-in cells, and bulk velocities from the POTENT analysis. We will also consider requirements for the early formation of quasars. After fitting the normalization and the bias factor for IRAS galaxies, we calculate χ^2 for each model. We then compare the results of the models and present χ^2 contour plots in the $n - \Omega_{HDM}$ fraction plane. We also consider the possibility that some of the COBE signal was produced by the long wavelength gravity waves generated during inflation.

Finally, if we are to take seriously the whole picture of inflation and C+HDM we need to identify plausible models of inflation which are compatible with our analysis of the large scale data. Here we will consider only "realistic" models of inflation. By this, we mean models in which the inflating field is a part of a larger theory which contains the following elements:

- neutrinos with masses in the eV range.
- a cold dark matter candidate.
- a high energy particle physics structure which is compatible with low energy physics and its hints about higher symmetries.
- a successful baryogenesis following inflation.

Without these elements we find the inflation model to be somewhat *ad hoc.*

2 Constraints from Large Scale Structure Data in the Linear Regime

In this section, we will describe data and our χ^2 analysis of theoretical predictions for these observations, but first we will discuss a few general issues concerning the models and the testing strategy employed here.

First of all, the problem with drawing strong conclusions about any given model of structure formation is that there are so many parameters to vary that we have a multi-dimensional parameter space to explore. We will choose to take best guess values for three of them, the baryonic fraction Ω_{baryon}, the Hubble constant, $H_0 = 100h\ km\ \sec^{-1}$, and the cosmological constant Λ. In the present analysis we will use only the central value, $h = 0.5$. Primordial nucleosynthesis strongly constrains the baryon density [21] as $0.010 \leq \Omega_{baryon}h^2 \leq 0.015$. Using $h = 0.5$, we can express this 95% confidence constraint as

$$\Omega_{baryon} = 0.05 \pm 0.01 \tag{2.1}$$

The hot dark matter fraction (combined with the hubble constant) also specifies the neutrino mass. If we have one flavor of neutrino whose mass is in the eV range, usually taken to be ν_τ, then

$$\Omega_{HDM} = \left(\frac{m_{\nu_\tau}}{23\ \text{eV}}\right)\left(\frac{0.5}{h}\right)^2 . \tag{2.2}$$

Finally, we will take the cosmological constant Λ to be zero. We find it hard to understand how a cosmological constant corresponding to an energy density of order $\sim (10^{-3}eV)^4$ can be generated. Furthermore, there is some weak observational evidence against such a model: too few gravitational quasar lensing events, and very small streaming velocities are generated [19].

The growth of density fluctuations is affected by the dynamics of the matter content, producing a scale dependent modification of the density fluctuation power spectrum. The relative growth as a function of scale is summarized in Figure 1, where we present the transfer functions in Fourier space as a function of Fourier wavenumber $k = 2\pi/\lambda$. In figure 1, we see that increasing the neutrino fraction decreases the amount of growth, and hence the amplitude, on small scales (large k). We also note that there are some modifications of the shape of the spectrum on quite large scales. For example the transfer function with $\Omega_{HDM} = 0.3$ at $k \sim 0.07h$ Mpc^{-1} is slightly steeper than that of either $\Omega_{HDM} = 0.0$ (CDM) or $\Omega_{HDM} = 0.5$.

In order to do our comparison in the most unambiguous way, we will confine our attention mostly to the regime where linear theory is appropriate. In the past, it was common to normalize linear power spectra by use of the fact that the rms optically selected galaxy density fluctuations are $\delta N/N = 1$ on a scale of 8 h^{-1} Mpc. In the most naive case, where one assumes that galaxies (light) trace the mass distribution, one would set the rms mass fluctuation also equal to one on this scale. More recently, however, it has become apparent that the galaxies are more strongly clustered than the mass, i.e., they are biased tracers of the mass. The usual method for dealing with this complication is to assume that there is a linear relation between the optical galactic density and mass fluctuations using a bias parameter b, which is independent of scale $b = (\delta N/N)/(\delta M/M)$. The mass fluctuation $\sigma(R)$ in a sphere of radius R is calculated in linear theory by

$$\sigma^2(R) \equiv \frac{1}{2\pi^2}\int_0^\infty dk\; k^2 P_{th}(k)\left[3\frac{j_1(kR)}{kR}\right]^2, \tag{2.3}$$

where $j_1(x)$ is the first spherical Bessel function and the term in brackets is the Fourier transform of a sharp edged sphere of radius R. Clearly, when a perturbation amplitude approaches unity, linear theory is no longer appropriate. Even with b as large as 2, this implies $\sigma(8\ h^{-1}\ \mathrm{Mpc}) = 0.5$, which is still quite non-linear. Using the spherical collapse approximation we estimate that for $\sigma = 0.5$, we are already highly contaminated by non-linear effects. If we consider scales for which $\sigma(R) \leq 0.4$ we can limit the non-linear corrections to $\lesssim 30\%$.

A look at figure 1 shows that the greatest difference between the models considered here occurs on the small scales, and limiting our testing to $k \leq 0.3h$ Mpc^{-1} implies that our testing will have a somewhat weakened ability to discriminate between models with different HDM fractions. In section III we will consider some constraints from non-linear structures to help us pin down the dark matter fraction.

We will first give a brief description of the large scale structure data, followed by a description of our calculations for different models.

2.1 Large Scale Structure Data

Here we will discuss the particulars of the data we are using. We explain our reasons for choosing the data and method of interpretation of this data.

2.1.1 COBE data

The large amplitude of the COBE measured temperature fluctuations is characterized by the extrapolated quadrupole moment Q_{rms-PS}. This amplitude was a factor of 2-3 larger than that predicted in the usual CDM models fit to galactic structure, and so gave strong support for the

C+HDM models [10]. However, various authors seem not content to use the COBE analysis of the amplitude. Instead they choose to use the sky variance at 10° which, when used with a Gaussian beam, implies a somewhat smaller amplitude for the fluctuations (*e. g.*, [16,22] corresponding to $Q_{rms-PS} = 15.3\mu$K). The COBE beam pattern, however, is only approximately a Gaussian shape. More careful analysis of the COBE results [23] using the actual beam pattern seems to confirm the original higher value of the amplitude. In addition, use of the correct beam pattern reduces the variance of the fit amplitude, so the best fit $n = 1$ COBE amplitude now corresponds to $Q_{rms-PS} = 17.1 \pm 2.9\ \mu$K. Although it is customary to quote the results in terms of the quadrupole moment, this corresponds to only the very small wavenumber end of the spectrum. The best fit amplitude of the quadrupole moment will be somewhat dependent on the value of n used in the analysis. In order to find a better quantity than the quadrupole, Wright *et al.*[23] recommend normalizing the amplitude to the hexadecupole $\delta T_4 = 12.8 \pm 2.3\ \mu$K, when $n \neq 1$, as in the inflationary models we consider here. This yields a best fit amplitude less dependent on the value of n.

2.1.2 The IRAS Large Scale Survey of Galaxies

The IRAS survey of galaxies done by the QDOT (Queen Mary-Durham-Oxford-Toronto) collaboration [24], extends as deep as several hundred Mpc, approaching the smallest scales observable by the COBE satellite. Combining the COBE data with the IRAS survey of galaxies thus covers the whole range of scales where the fluctuations can be described by the linear theory. The QDOT IRAS survey has measured redshifts for 1 - in - 6 IRAS galactic sources (1824 galaxies) with IRAS fluxes > 0.6 Jy. Using the redshift of the source as a distance indicator, combined with angular position data, one has a three dimensional picture of the galaxy distribution. This distribution can be analyzed to directly extract the power spectrum of density fluctuations [25,26]. The results of the Feldman, et al. [26] analysis is shown as the power spectrum data in Figure 2.

We will use this QDOT IRAS data to test models with theoretical power spectra given by

$$P_{th}(k) = Ak^n[T(k)]^2. \tag{2.4}$$

For $n \sim 1$ and the transfer functions $T(k)$ in figure 1, we can see that the power spectra go like $P(k) \sim k$ on very large scales and like $P(k) \sim k^{-3} - k^{-4}$ on very small scales, with a peak somewhere around $k \sim 0.02 - 0.10$. In order to get the proper normalization for the amount of structure on scales up to the power spectrum peak scale, one must measure the power on scales larger than the power spectrum peak scale. In the mass fluctuation integral (equation 2.3) there are significant contributions to the mass fluctuations in 20 Mpc spheres coming from 100 Mpc scales ($k \sim 0.03h/\mathrm{Mpc}$). Thus we feel that to test power spectra of the type considered here, we require a survey out to a depth at least as large as the QDOT survey. A comparison of the QDOT and 1.2 Jy Berkeley power spectra (see Ref. [26]) suggests that the Berkely survey may still be too small to be seeing all of the large scale power.

Since the largest differences between the models under consideration here occurs on smaller scales, we would like to supplement our analysis with a different estimator of power on small scales. To this end we use the "counts - in - cells" statistic from Ref. [24]. Here the redshift space of IRAS galaxies is cut up into near cubical cells of side length ℓ and the number of galaxies in each cell is counted. The cell to cell variance of galactic number is a direct indication of the underlying density fluctuations of length $\sim \ell$ in which the galaxies reside. The QDOT counts in cells data is presented in figure 3a.

2.1.3 POTENT bulk flow velocities

We also use the bulk flow velocities from the POTENT analysis [27]. They represent the rms velocities of spherical regions of radius R. The velocity has first been filtered with a Gaussian of width $R = 12h^{-1}$ Mpc to reduce noise. This data is shown in figure 3b. The POTENT analysis has been done using the IRAS galaxies of the Berkeley survey, and so suffers from the problem that enough of the universe has not been sampled to get a fair estimate of the velocities. This is exacerbated by the fact that velocities are even more sensitive to the very large scale power than the mass fluctuation. However, we would still like to find a way to use these velocities because they do not depend on the bias. (Velocities are generated by the gravity of density fluctuations.) Since we have velocities from one local patch we will include the cosmic variance of the predicted velocity for any given patch of the universe in our analysis. Our treatment will be described in the next section.

2.2 Description of χ^2 Analysis

The reduced χ^2 statistics is calculated by comparing the set of N predictions y_i^{th} and observations y_i^{obs} according to the following formula

$$\chi^2 = \frac{1}{N_{d.o.f.}} \sum_{i=1}^{N} \left(\frac{y_i^{th} - y_i^{obs}}{\sigma_i^{obs}} \right)^2 , \tag{2.5}$$

where σ_i^{obs} are the standard observational errors and $N_{d.o.f.}$ equals the number of observations N minus the number of fitted theoretical parameters in the model. In our analysis we use the 36 values of the IRAS $P(k)$, the 4 IRAS counts-in-cells, the 5 POTENT velocity values, the b_I determination, and the COBE amplitude, which we will count as two points. In the χ^2 analysis of $n = 1$ models in ref. [19], the COBE data is counted as two points, one for the sky noise at 10° and another for the value of Q_{rms-PS}. Because of the complications of the non-gaussian beam pattern we have avoided this procedure, but we still choose to weight the hexadecupole amplitude as if it were two points in the analysis. This forces the amplitude of density perturbations to be slightly closer to the COBE normalized amplitude than by weighting it as only one point. The results of the analysis is quite similar if we were to weight COBE as only one data point. Thus we have a total of $N = 48$ data points which contribute to the χ^2 sum.

For each value of n and Ω_{HDM} we find the values of b_8 and b_I which minimize $\chi^2(b_8, b_I)$ for each model. This is effectively a least squares fitting of b_8 and b_I for each model, and so the number of degrees of freedom is reduced by 2 to 46 in our χ^2 formula. The confidence levels of rejecting the hypothesis that the data fits the model are found from integrating the normalized χ^2 distribution. The probability of getting $\chi^2 > 1.38$ is $< 5\%$, so we can reject such models with 95% confidence. We will now proceed to discuss how we compute the y_i^{th}.

2.3 Model Predictions from Linear Theory

Here we will explain how we compare our theoretical power spectra $P_{th}(k)$ to the data presented in the previous section. First of all, we can calculate the hexadecupole δT_4 (measured by COBE) with the formula given in Ref. [28] for the coefficient of the fourth spherical harmonic for each of the power spectra we are considering. However, we can find an effective wavenumber k_{eff} for which the amplitude of $P(k_{eff})$ is directly proportional to the amplitude of the hexadecupole. We have found that the value $k_{eff} = 1.05 \times 10^{-3} h/\mathrm{Mpc}$ accurately characterizes the hexadecupole moment for the limited range of n we are studying ($0.70 < n < 1.00$).

The COBE experiment cannot distinguish between temperature fluctuations generated by the Sachs-Wolfe effect of scalar (density) fluctuations and tensor (gravity wave) fluctuations, both of which are products of inflation. This signal confusion is worsened by the fact that the ratio of the moments generated (at least on COBE scales) for scalar and tensor contributions are nearly independent of scale [29,30,31]. The overall amplitudes of the moments for scalar and tensor modes, however, can be quite different, and one must construct a specific model of inflation and then evaluate the density perturbation and gravity wave amplitudes during inflation. In some models of inflation, such as "new" inflation [1] the contribution of gravity waves to the COBE signal typically is negligible, and the COBE signal relates only to the scalar density perturbations.

Models of inflation which produce a significant amount of gravity waves cannot be summarized with a universal formula. However, there is a relation derived from the toy model "power law inflation" which relates the amount of gravity wave contribution in the COBE signal to the density power spectrum exponent n [22]

$$\frac{(\Delta T/T_0)^2_T}{(\Delta T/T_0)^2_S} \approx 7(1-n). \tag{2.6}$$

This relation also is reasonably accurate for chaotic inflation models. Since the COBE signal is a quadrature sum of the multipole moments, decreasing n decreases the fraction of COBE signal due to density waves, and thus implies a smaller amplitude for $P_{th}(k)$. Thus the amplitude of the COBE quadrupole moment is $\sqrt{8-7n}$ times larger than would be expected from the density component alone. We will consider models with gravity waves produced according to this formula.

To simulate the IRAS power spectrum $P_I(k)$ we have to apply a couple of correction factors to our theoretical $P_{th}(k)$. First of all the distribution of galaxies in redshift space on large scales appears more clustered because of the doppler contribution to the redshift from velocity perturbations [32]. With a biased galaxy distribution (b_I) this correction can be made by multiplying with the factor

$$P(k) \to \left[1 + \frac{2}{3b_I} + \frac{1}{5b_I^2}\right] P(k) \tag{2.7}$$

However, on smaller scales there is the opposite effect; the doppler shifts from the peculiar velocities of galaxies become large in comparison to the Hubble velocities and in fact wash out this redshift clustering effect. This effect can be described with a velocity dependent factor [33]

$$P(k) \to P(k)\frac{\sqrt{\pi}}{2}\frac{\mathrm{erf}(kR_v)}{kR_v} \tag{2.8}$$

where $R_v = 4.4\ h^{-1}$ Mpc for IRAS galaxies.

Applying these corrections to a linear power spectrum seems to give reasonable agreement with the power spectrum of "galaxies" in N-body simulations.

To do the testing, we first must calculate the linear theory power spectra for all models with a primordial power spectral index $0.70 < n < 1.00$ and $0 < \Omega_\nu < 0.5$. We calculate the power spectra for these models in steps of $\delta n = 0.02$ and $\delta\Omega_\nu = 0.05$ using our form for $P(k)$:

$$P_{th}(k) = Ak^n [T(k)]^2 , \tag{2.9}$$

We will vary the amplitude A by a factor of $2^{\pm 1}$ away from the COBE implied central value in 20 logarithmically spaced steps. However, since the definition of A is author dependent, we will discuss our results in terms of the parameter b_8 for each model. Thus we are effectively varying b_8 in a range centered on the COBE best fit b_8.

The value of b_I is somewhat more constrained than values of b_8. For the IRAS galaxies, separate determinations of b_I have been made by comparing their velocities and distributions. These dynamical tests yield the 95% confidence values $b_I = 1.16 \pm 0.42$ (Ref. [34]) and $b_I = 1.23 \pm 0.46$ (ref. [35]). However, the POTENT analysis [36] of the Berkeley IRAS galaxies finds that the 95% confidence interval for $b_I = 0.5 - 1.3$. We therefore combine these measurements to say that $b_I = 1.1 \pm 0.3$ to cover the 95% confidence overlap region of the b_I determinations. We consider the measurement of $b_I = 1.1 \pm 0.3$ to be a *bona fide* data point which we add to our χ^2 test. As is usual, we will assume a constant bias factor independent of scale, and we will allow b_I to vary in 7 linearly spaced steps between 0.8 and 1.4. We calculate the χ^2 for each of the values of b_I and b_8, and find the set (b_I, b_8) which gives the lowest value of χ^2, and plot this minimum χ^2 in our $\Omega_\nu - n$ contour plots. Thus we allow each model to put its best face forward in our test. Some of these models with $b_I = 1.00$ are shown in figure 4.

We calculate the POTENT rms velocities by the same procedure as described in ref. [11]. In our comparison of the rms POTENT velocities to theoretical rms velocities, we will incorporate the cosmic variance of the velocity field. The idea is that with a Gaussian density field, the velocity field will also be Gaussian. The magnitude of the velocity vector will have a χ^2 distribution with 3 degrees of freedom. The variance of the rms velocity is much larger than the POTENT velocity errors. For example the 68% theoretical confidence range on the average predicted velocity magnitude $< v >$ corresponds to ($< v > -0.48 < v >$, $< v > +0.32 < v >$), while the POTENT errors are less than $\sim$ 15 %. This is plotted in figure 3b for the $n = 0.96$, 25% HDM model. If we normalize an $n = 1.00$ power spectrum to COBE then the predicted velocities are within the POTENT 1 σ error bars, regardless of the HDM fraction (see e.g., $n = 1.00$, 50% HDM model in figure 3b). As we decrease n below 1.00, we will decrease the predicted velocities well below the POTENT values. Thus the upper limit on the predicted velocity will be the most relevant quantity for comparing theory to observations. We combine this upper limit in quadrature with the POTENT 1 σ error, and use this as a fairer estimate of the error bar in our χ^2 analysis.

While adding errors in quadrature is strictly correct only for Gaussian errors, the velocity distribution is not too far from a Gaussian. We plot several theoretical predictions for $< v >$ against the POTENT velocities in figure 3b. Even with these huge error bars, we find that the predicted velocities for models with significant gravity wave contributions will still have trouble matching the POTENT derived velocities.

2.4 Results of Linear Data Analysis

The results of our χ^2 test are presented in figures 5 and 6 for models without and with gravity waves, respectively. Starting with our absolute best fitting model $n = 1.00$, 30% HDM and working outward, we plot 9 concentric curves with confidence levels of .5, 1, 5, 10, 25, 50, 68, and 95 %. When one reaches the level of 1% areas where the theory does not match the data begins to become discernible in the plots. The first thing to note about the graphs is that the overall level of χ^2 for models with $n \sim 1.00$ is quite low. This result supports earlier claims that $n = 1$ models normalized to COBE have sufficient power to explain "large scale power" apparent in galactic clustering measures. If the measurement errors were smaller our test would be a much better discriminator between models.

The χ^2 contours for models with no gravity waves are shown in figure 5. As can be seen we can rule out models with $n \lesssim 0.7$ at 95% confidence. This is easily understood since models with a lot of "tilt" already have little galaxy scale power, and large HDM fractions exaggerate this behavior.

The χ^2 contours for models with gravity wave contributions according to equation 2.6 are shown in figure 6. It is readily apparent that the allowed parameter space is much smaller. Here the 95% confidence limit on n rules out models with $n \lesssim 0.85$. We note that for our 95% confidence limit an amount $\gtrsim$ 50 % of the COBE signal is attributed to the effect of density fluctuations. This tells us that to fit large scale structure one requires that the COBE signal cannot be dominated by inflation generated gravity waves. Our best fit models are those in which at least 80% of the COBE signal is due to density fluctuations.

Our two conclusions from this analysis are that the power spectrum must be close to the Harrison-Zeldovich form ($n = 1.00$) and that the COBE signal must be mostly due to density fluctuations.

3 Constraints from Data on Non-Linear Structures

As noted in the previous section, we find the best models are those which have $\Omega_{HDM} \gtrsim 0.1$. The statistical confidence of this conclusion is not very high, on the order of 10% for a given value of n. This is not surprising, since we have confined ourselves to only the largest scales, $k \lesssim 0.2h$ Mpc^{-1}, where the transfer functions for all the models do not differ too much (see figure 1). The best place to discriminate between these models is on smaller scales, where we have data only on the non-linear part of the power spectrum.

3.1 The Data

The relation between non-linear structures and the amplitude of the linear power spectrum is quite complicated, and it is non-trivial to extract strong conclusions from this data. We will consider two such constraints which we feel can be used relatively safely - constraints on the amplitude of $\sigma(8h^{-1}$ Mpc) or equivalently b_8, and the requirement that quasars form early enough to be compatible with observations. The quasar constraint is a lower bound on the power spectrum amplitude while $\sigma(8h^{-1}$ Mpc) is an upper bound (at a somewhat larger scale).

3.1.1 High redshift quasars

The discovery of quasars with high redshifts (about 20 with $z \geq 4$) was a direct challenge to theories of structure formation. One needs a minimum amplitude for density fluctuation on galactic scales to account for the quasar population. Efstathiou and Rees (ref. [37]) considered the formation of quasars in a highly biased ($b_8 = 2.5$) $n = 1$ CDM model. The basic strategy is that in order for a massive black hole to form and power the quasar emission, one first requires a host dark matter halo to supply the gravitational potential to induce baryonic infall. The number density of structures for a given power spectrum can be computed using the Press-Schecter or BBKS techniques. The number density of structures depends exponentially on a parameter ν given by

$$\nu = \frac{\sigma_c}{\sigma(R)} \tag{3.1}$$

where R is the radius of the initial collapsing region appropriate for the objects in question and σ_c is the linear theory value of $\sigma(R)$ which corresponds to the gravitational collapse and vitalization of the object. Using the spherical collapse approximation $\sigma_c = 3(12\pi)^{2/3}/20 = 1.69$. In this application we use the Gaussian filtered mass fluctuation $\sigma_G(R)$

$$\sigma_G(R)^2 = \frac{1}{2\pi^2}\int dk k^2 P(k) e^{-k^2R^2} \tag{3.2}$$

where the subscript "G" is to remind us that we are using a Gaussian filtering function. The parameter ν then has the physical meaning of the ratio of the overdensity in a collapsed object to the ambient rms overdensities. Thus specifying ν tells us the amplitude of the density fluctuations on the scale R. Also, since the calculated number density depends exponentially on ν, it is hoped that errors in the measured number density will not significantly change the value of ν.

Ref. [37]) estimated that the minimum mass for a quasar halo was $\geq 2\times 10^{12} M_\odot$. They found that a $b_8 = 2.5$, $n = 1$ CDM model could account for the number density of quasars at least out to a redshift of $z = 5$. Ref. [38]) improved on this treatment by showing that the $b = 2.5$ CDM could fit the quasar luminosity function at a variety of redshifts. Since both increasing the HDM fraction and reducing n lessen the amount of power on small scales, it is obvious to ask whether the amplitude of quasar scale fluctuations has decreased below that required for making quasars. Haehneldt (ref. [12] used the techniques of ref. [38] to limit the HDM fraction in $n = 1$ models and n in CDM models, finding that $n > 0.75$ in CDM models and the HDM fraction must be $\Omega_{HDM} \leq .3$. Schaefer and Shafi (ref. [11]) showed that 25% HDM and $n = 0.94 - 0.97$ is compatible with the quasar density out to a redshift of $z \sim 5$. Due to uncertainties in the theoretical parameters for calculating quasar number densities (see ref. [22]), we take a more generous limit than Haehrelt [ref. 12].

3.1.2 σ(8h^{-1} Mpc)

Since it has become traditional to specify the amplitude of linear theory by b_8, some attention has been paid to determining the value of b_8 from observations. There are several ways of getting at this quantity which seem to be converging on the range $b_8 = 1.5 - 2.0$ for the $\Omega = 1$ models. We will consider some of the more recent attempts to constrain this parameter.

The easiest way to estimate the density fluctuations is by finding the variance of the galactic number density. For the IRAS galaxies we have two estimates of the number density fluctuations: one from the QDOT survey [39] which is in perfect agreement with the estimate from the Berkeley survey [40]

$$b_I \sigma^{non-linear}(8h^{-1}\mathrm{Mpc}) = 0.69 \pm 0.04 \tag{3.3}$$

The 95% confidence upper limit is then $b_I\sigma^{non-linear}(8h^{-1}\mathrm{Mpc}) \leq 0.77$ If we take an extremely small value for $b_I = 0.5$ which is the POTENT 95% confidence lower limit, this would imply that $\sigma^{non-linear}(8h^{-1}\mathrm{Mpc}) \leq 1.54$. If we use the spherical collapse model to determine what that means in terms of the linear density fluctuation, this implies

$$\sigma^{linear}(8h^{-1}\mathrm{Mpc}) \leq 0.71, \text{ or } b_8 \geq 1.40. \tag{3.4}$$

To be extra cautious, we will adopt the constraint

$$\sigma^{linear}(8h^{-1}\mathrm{Mpc}) \leq 0.80, \text{ or } b_8 \geq 1.25 \tag{3.5}$$

for a new round of χ^2 testing.

3.2 Results of Non-Linear Analysis

In Figures 7 and 8 we again plot the χ^2 contours for models with and without significant gravity wave temperature anisotropies. The effect of our non-linear constraints is clear. The restriction $b_8 \geq 1.25$ forces the normalization of small Ω_{HDM}, $n \sim 1$ models to be too low to match the large scale structure data. This is symptomatic of the CDM models which, when normalized to COBE, have too much small scale power. The effect of enforcing the lower limit on the amplitude from quasars is to cut out a triangle of parameter space corresponding to large Ω_{HDM} and small n. This is symptomatic of $\Omega_{HDM} \sim 1$ model problems, there being not enough small scale power to explain the early epoch of quasar formation.

What we are left with is a patch of parameter space which has $n \sim 1$ and $\Omega_{HDM} = 0.3 \pm 0.2$. This is in agreement with earlier studies and shows that models with $\Omega_{HDM} \sim 1/4$ are significantly better fits to the data than with no HDM. The range of n is roughly the same as in the linear analysis, although the non-linear quasar constraint has effectively chopped off the low n, high HDM fraction corner of parameter space of the previous best fits. Our allowed region of parameter space overlaps the allowed region found by Liddle and Lyth, (ref. [41]). However, their analysis took the non-linear constraints at face value so their allowed region is somewhat smaller than ours. They noted however, that their allowed region was meant to be suggestive of trends in the data and should not be taken literally. On the other hand we are taking pains to be overconservative with non-linear constraints in the hope that our limits will be firmer.

Thus we find the following properties which it is desirable for inflation to have. We require a density perturbation spectrum which is quite close to a Harrison-Zeldovich spectrum as the data do not seem to favor much "tilting. For models with a negligible gravity wave contribution to the COBE signal, the best fits occur for $0.9 \lesssim n \lesssim 1.0$. In models with some gravity wave contribution, we find an even tighter range of best fit n values, namely $0.94 \lesssim n \lesssim 1.00$. Since we used equation 2.5 to determine this, we see that for $n = 0.94$ the gravity wave contribution to COBE is only 20%. Thus the data favor models for which density perturbations are $\gtrsim 80\%$ responsible for the COBE signal. We also find $0.15 \lesssim \Omega_{HDM} \lesssim 0.45$ gives the closest fits to the data, which implies we would nominally like one flavor of neutrino with a mass $\sim= 3 - 10$ eV. [The best fits imply a narrower range $\Omega_{HDM} \simeq 0.2 - 0.3$.] With these attributes in mind we proceed to explore a new model for inflation based on supersymmetric grand unification.

3.3 Supersymmetric Inflation

The presently measured gauge couplings of the standard model, when extrapolated to higher energies with supersymmetry (SUSY) broken at scales around 10^3 GeV [42], appear to merge at scales around 10^{16} GeV. This is a boost for supersymmetric GUTS, with $SU(5)$ or $SO(10)$ being the obvious gauge groups. In the presence of unbroken matter parity, either of them can provide a cold dark matter candidate in the form of LSP (lightest supersymmetric particle). However, hot dark matter in the form of massive neutrinos most naturally appear in the $SO(10)$ model. The supersymmetric $SO(10)$ scheme has some additional features which make it attractive from the particle physics viewpoint. For instance,

(a) A Z_2 subgroup of the center of $SO(10)$ (more precisely Spin (10)) is left unbroken if tensor representations are employed to do the symmetry breaking. This Z_2 symmetry [43], which is not contained in $SU(3)_c \times SU(2)_L \times U(1)_Y$, acts precisely as matter parity!

(b) In some versions of SUSY $SO(10)$, the important parameter $\tan\beta(\equiv \phi^u/\phi^d$, the ratio of the two vevs which provide masses to 'up' type and 'down' type quarks) is predicted to lie close to m_t/m_b [44]. One consequence of this is the identification of the 'bino' (the supersymmetric partner of the $U(1)_Y$ gauge boson) as the LSP, with mass $\sim 200 - 300$ GeV.

(c) Fermion Mass Ansatzes have recently attracted a fair amount of attention and are most simply realized within the framework of SO(10) [45].

To summarize, particle physics considerations as well as observations of large scale structure which favor a cold plus hot dark matter scenario, together suggest SUSY $SO(10)$ as an attractive way to proceed. Inflation, either 'new' or 'chaotic', can be implemented by introducing a suitable singlet superfield. Remarkably enough, singlets are typically employed to achieve the breaking of the GUT symmetry, (without breaking SUSY) and we exploit one of them to induce inflation!

Let Φ denote the $SO(10)$ singlet (inflaton) superfield, $\chi(\bar{\chi})$ are the higgs superfields in the $126(\overline{126})$ representations whose vevs provide Majorana masses to the right handed neutrino, and $16_i(i = 1, 2, 3)$ are the matter superfields. To simplify the discussion, we restrict attention to the sector involving an interplay only between these superfields. This allows us to discuss the salient features of the (chaotic) inflationary scenario including baryogenesis. Consider the renormalizable superpotential

$$\begin{aligned} W = \alpha\Phi(\chi\bar{\chi} - M_X^2) \quad & + \tfrac{M}{2}\Phi^2 + \tfrac{\beta}{3}\Phi^3 \\ & + \gamma_{ij} 16_i 16_j \bar{\chi} \end{aligned} \tag{3.6}$$

Note that $\Phi \to \Phi$ under the matter parity contained in $SO(10)$ (similarly $\chi, \bar{\chi} \to \chi, \bar{\chi}$, and $16_i \to -16_i$).

The superpotential W gives rise to a supersymmetric ground state in which (vevs refer to the scalar components of the superfields)

$$\begin{aligned} <\phi> &= 0, <16_i>= 0, \\ <\chi> &= <\bar{\chi}>^* \neq 0 \end{aligned} \tag{3.7}$$

It is clear that matter parity is unbroken and we expect the LSP to contribute to the cold dark matter component.

Even though $B - L$ is now broken at $M_X \sim 10^{16}$ GeV, the right handed 'tau' neutrino mass must be of order $10^{12} - 10^{13}$ GeV, if the 'light' tau is to be the dark matter component. The inflaton must be at least twice as heavy, and one simple way to have $m_\phi \sim 10^{13}$ GeV is to arrange the coefficients α and M in (4.8) to be of order m_ϕ/M_X and m_ϕ respectively. The decay rate of the inflaton into right handed neutrinos is given by

$$\Gamma \sim \frac{1}{4\pi}\left(\frac{m_\phi}{M_X}\right)^6 m_\phi \; GeV \tag{3.8}$$

With $m_\phi \sim 10^{13}\ GeV$, the reheat temperature T_r is of order $10^6\ GeV$. Baryogenesis via leptogenesis now proceeds along the lines given in ref. [46]. Note that because of the relatively low T_r, the otherwise vexing gravitino problem is neatly avoided in this approach.

Depending on the details, the spectral index n lies between 0.94 (if the quartic potential dominates) and 0.96 (with a quadratic potential dominant during the chaotic inflationary phase). The ratio $(\Delta T/T)^2_T/(\Delta T/T)^2_S \simeq 0.22(0.11)$, respectively.

4 Conclusions

We have performed a χ^2 goodness of fit test of the predicted linear theory power spectra against data on scales ranging from 1 to 10^4 Mpc, mainly from the COBE satellite and the QDOT IRAS survey of galaxies. We find that the inflation based scenario of large scale structure formation, in which the dark matter consists of cold plus hot components, can provide a good fit to large scale structure data.

Taking the primordial power spectrum to have spectral exponent n, we find with 95% confidence, respectively, that $n \gtrsim 0.7, (n \gtrsim 0.85)$, in models with (without) significant gravity wave contributions to the COBE anisotropy. The precise bound depends slightly on the HDM fraction. We find in models with a significant gravity wave anisotropy, that the COBE signal must not be dominated by the gravity wave contribution.

The best fit region for all data, including some constraints from non-linear structure, is roughly an ellipse (see figures 7 and 8). For models with small amplitude gravity wave anisotropies, the focii of the ellipse are approximately at ($\Omega_{HDM} = 0.20, n = 0.92$) and ($\Omega_{HDM} = 0.35, n = 0.98$). For models with a large tensor COBE anisotropy, the ellipse is more eccentric, with the focii at roughly ($\Omega_{HDM} = 0.15, n = 0.95$) and ($\Omega_{HDM} = 0.35, n = 0.98$). Thus, the best fits occur for $\Omega_{HDM} \sim 0.2 - 0.35$, and n very close to unity. Models of this type were proposed almost a decade ago by Shafi and Stecker [2].

Just as the seventies saw the rise of the standard model in particle physics, the nineties should see the emergence of a 'standard' model for large scale structure formation. Hopefully the scenario described here is a step in the right direction!

In conclusion, I would like to thank the organizers for arranging this memorable meeting and for inviting me to take part in it. Abdus Salam's health during the last few years has been a source of constant challenge which he has faced with great fortitude. We all wish him an early recovery and many happy years ahead of him.

Acknowledgement:

The work reported here has been carried out in collaboration with Dr. Robert K. Schaefer. I want to thank him for sharing over the years my conviction in the cold plus hot dark matter scenario.

References

[1] For reviews of inflation see A. Linde, Particle Physics and Inflationary Cosmology, Harwood Academic, NY (1990);
E. Kolb and M. Turner, The Early Universe, Addison-Wesley, Reading, MA (1990).

[2] Q. Shafi and F. W. Stecker, PRL, **53**, 1292, (1984).

[3] R.K. Schaefer, Q. Shafi, and F.W. Stecker, *Astrophys. J.*, **347**, 575 (1989).

[4] J. Holtzman, *Astrophys. J., Suppl.*, **74**, 1 (1989).

[5] F. Occhionero and R. Scaramella, AA, **204**, 3 (1988).

[6] A. van Dalen, and R.K. Schaefer, *Astrophys. J.*, **398**, 33 (1992).

[7] J.A. Holtzman and J.R. Primack, *Astrophys. J.*, **405**, 428 (1993).

[8] Smoot, *et al.*, 1992, apjl, **396**, L1 (1992).

[9] Wright, *et al.*, 1992, apjl, **396**, L5 (1992).

[10] R.K. Schaefer and Q. Shafi, *Nature*, **359**, 199 (1992).

[11] R.K. Schaefer and Q. Shafi, *PRD*, **47**, 1333 (1993).

[12] M.G. Haehnelt, *MNRAS*, in press, (1993).

[13] M. Davis, F.J. Summers, and D. Schlegel, *Nature*, **359**, 393 (1992).

[14] A. Klypin, J. Holtzman, J. Primack, and E. Regos, *ApJ*, (in press), (1993).

[15] Y.P. Jing, H.J. Mo, G. Börner, and L.Z. Fang, *AA* (in press) (1993).

[16] G. Efstathiou, J.R. Bond, and S.D.M. White, *MNRAS*, **258**, 1P (1992).

[17] D. Schlegel, M. Davis, F. Summers, and J. A. Holtzman, UCB preprint, (1993).

[18] S. Xiang and T. Kang, *MNRAS*, **259**, 761 (1992).

[19] A.N. Taylor and M. Rowan-Robinson, *Nature*, **359**, 396 (1992).

[20] D. Yu., Pogosyan and A.A. Starobinski, DAMTP preprint, (1993).

[21] T.P. Walker, G. Steigman, D.N. Schramm, K.A. Olive, and H.-S. Kang, *ApJ*, **51**, 376 (1991); M.S. Smith, L.H. Kawano, and R.A. Malaney, *ApJS*, **85**, 219 (1993).

[22] A.R. Liddle and D.H. Lyth, *Phys. Rept.*, (in press) (1993).

[23] Wright, *et al.*, *Apj* (1993).

[24] G. Efstathiou, N. Kaiser, W. Saunders, A. Lawrence, M. Rowan-Robinson, R.S. Ellis, and C.S. Frenk, *MNRAS*, **247**, 10P (1990).

[25] K.B. Fisher, M. Davis, M.A. Strauss, A. Yahil, and J.P. Huchra, *ApJ*, **389** 188 (1992).

[26] H. Feldman, N. Kaiser, and J.A. Peacock, preprint (1993).

[27] A. Dekel, in *Observational Cosmology*, Milan, ed. G. Chincarini, *et al.*, (in press) (1992).

[28] L.F. Abbott and M. Wise, *Nucl. Phys.*, **B244**, 541 (1984).

[29] L.F. Abbott and R.K. Schaefer, *Astrophys. J*, **308**, 546 (1986).

[30] R. Crittenden, J.R. Bond, R.L. Davis, G. Efstathiou, and P.J. Steinhardt, *PRL*, to appear.

[31] M. Turner, *Phys. Rev. D* (in press).

[32] N. Kaiser, *MNRAS*, **227**, 1 (1987).

[33] J.A. Peacock, *MNRAS*, **259**, 494 (1992).

[34] N. Kaiser, G. Efstathiou, R. Ellis, C.S. Frenk, A. Lawrence, M. Rowan-Robinson, and W. Saunders, *MNRAS*, **252**, 1 (1991).

[35] M. Rowan-Robinson, W. Saunders, A. Lawrence, and K.L. Leech, *MNRAS*, **253**, 485 (1991).

[36] A. Dekel, E. Bertschinger, A. Yahil, M.A. Strauss, M. Davis, and J.P. Huchra, *ApJ*, **412**, 1 (1993).

[37] G. Efstathiou and M.J. Rees, *MNRAS*, **230**, 5P (1988).

[38] M.G. Haehnelt and M.J. Rees, *MNRAS*, **263**, 168, (1993).

[39] W. Saunders, M. Rowan-Robinson, and A. Lawrence, *MNRAS*, **258**, 134 (1992).

[40] K.B. Fisher, M. Davis, M.A. Strauss, A. Yahil, and J.P. Huchra, *MNRAS* (in press) (1993).

[41] A.R. Liddle and D.H. Lyth, Univ. of Sussex preprint, AST/4-2 (1993).

[42] U. Amaldi, W. deBoer, and H. Furstenau, *Phys. Lett. B*, **260**, 447 (1991);
J. Ellis, S. Kelley, and D.V. Nanaopoulos, *Phys. Lett. B*, **249**, 441 (1990);
P. Langacker and M. Luo, *Phys. Rev. D*, 44, 817 (1991).

[43] T.W.B. Kibble, G. Lazarides, and Q. Shafi, *Phys. Lett.*, **B113** 237 (1982);
S.P. Martin, *Phys. Rev.*, **D46**, 2769 (1992).

[44] B. Ananthanarayan, G. Lazarides, and Q. Shafi, *Phys. Rev. D*, **44**, 1613 (1991);
ibid, *Phys. Lett. B*, **300**, 245 (1993).

[45] G. Lazarides and Q. Shafi, *Nucl. Phys.*, **B350**, 179 (1991);
K. Babu and Q. Shafi, *Phys. Rev. D*, **47**, 5004 (1993);
G. Anderson, *et al.*, Berkeley Preprint UCB-PTH-93/03 (1993) and references therein.

[46] G. Lazarides and Q. Shafi, *Phys. Lett. B*, **258**, 305 (1991);
M. Fukugita and T. Yanagitda, *Phys. Lett. B*, **174**, 45 (1986).

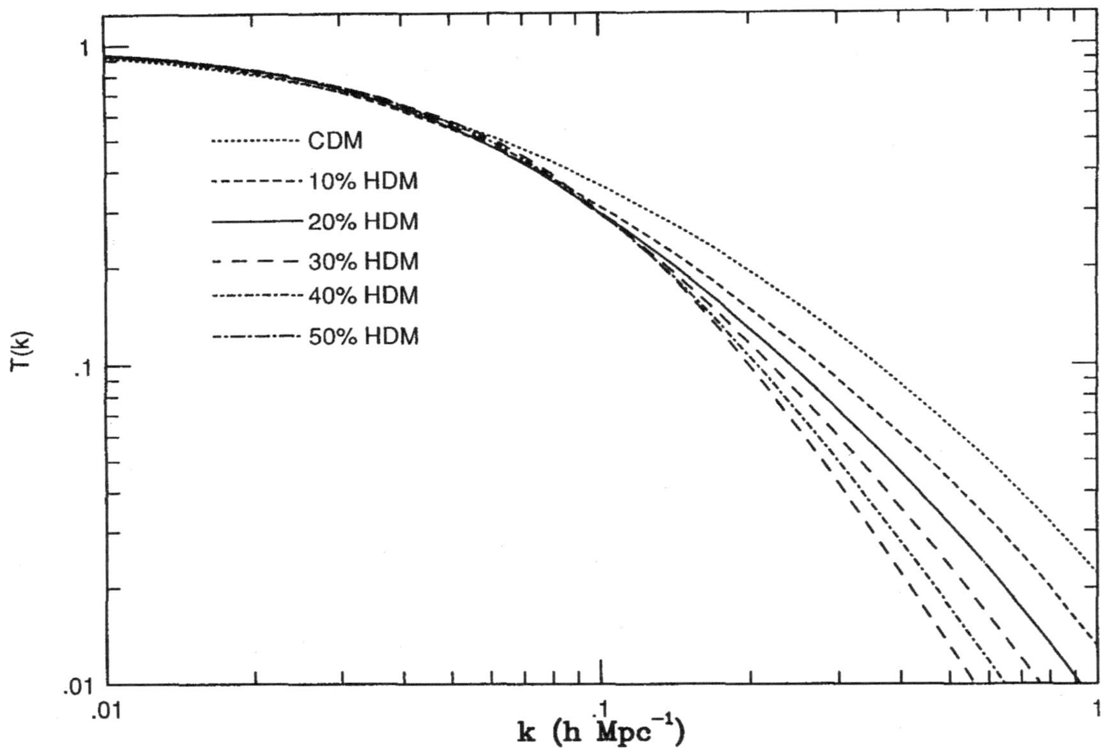

Fig. 1. Transfer functions for the cold plus hot dark matter models. The curves represent the relative growth of density fluctuations as a function of scale. The curves represent the present time (z=0) transfer functions and in a universe with $\Omega_{CDM} + \Omega_{HDM} + \Omega_{baryon} = 1$, where we have taken the canonical value of $\Omega_{baryon} =$ 0.05 from primordial nucleosynthesis.

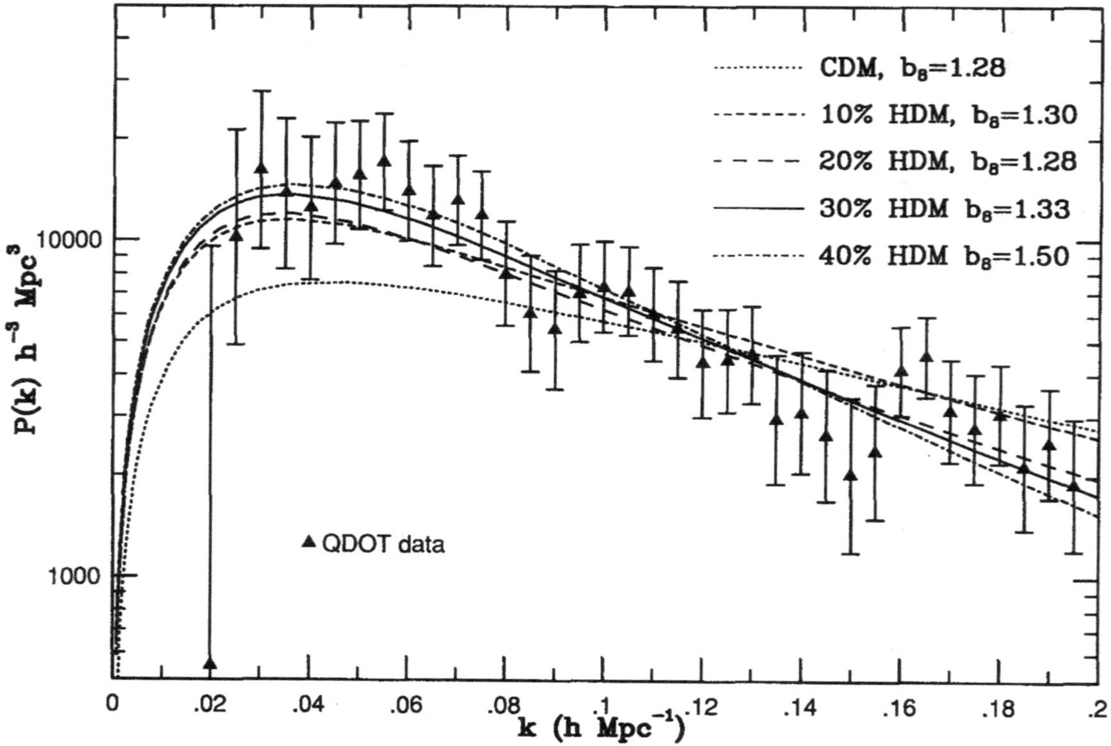

Fig. 2. Redshift space power spectra for IRAS galaxies (with $b_I = 1$) for a few models to illustrate how the fits are made. Here we see power spectra for 3 models with different n and HDM fractions normalized for minimum χ^2 to all data. Here we show the IRAS QDOT power spectrum data [27]. The models shown have COBE normalizations $b_I = 1$.

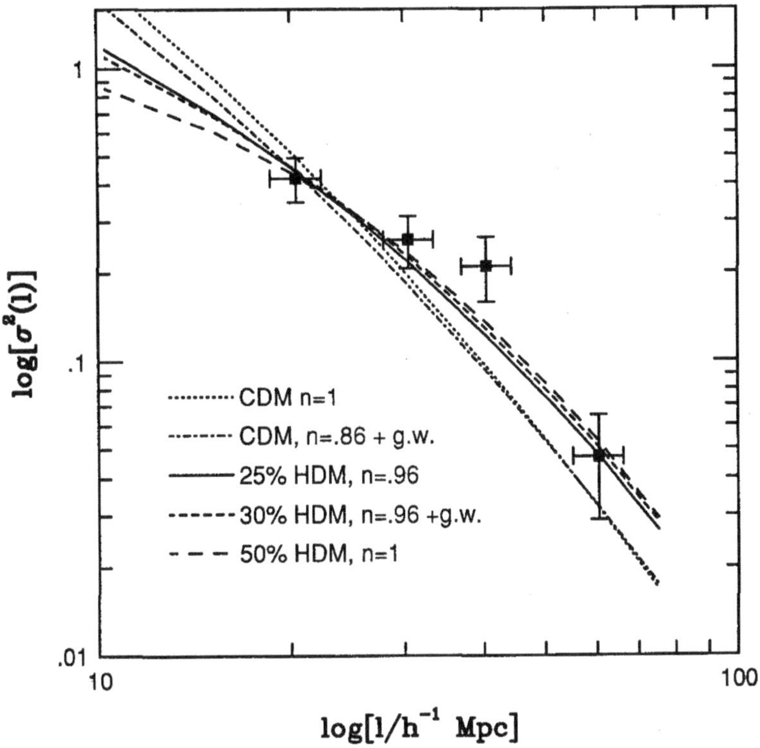

Fig. 3a. The counts in redshift space cells data from the IRAS QDOT survey [25]. Illustrated are the curves for some selected models. Low mass fluctuation amplitudes of some models are compensated by using high values of the IRAS bias b_I as can be seen here.

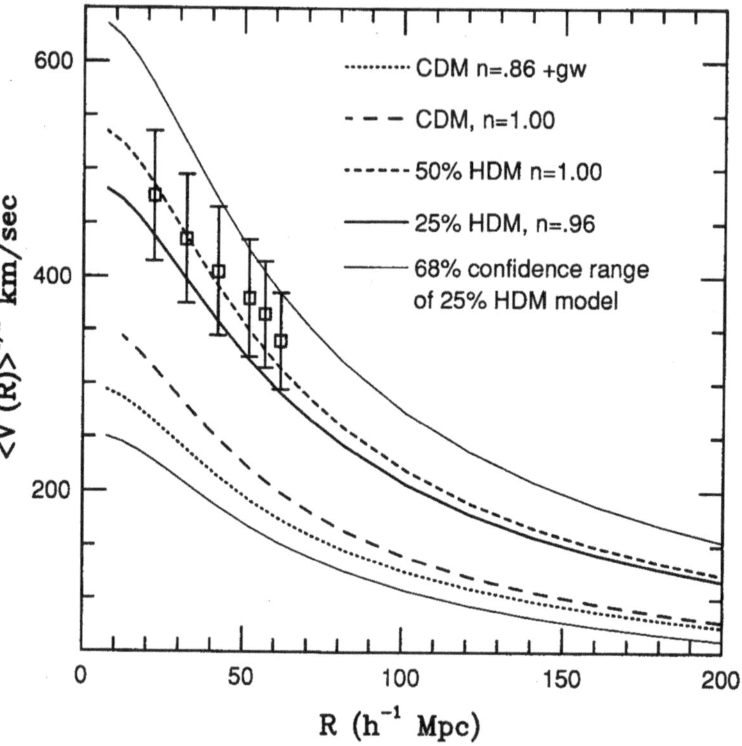

Fig. 3b. The values of the bulk streaming velocities extracted by the POTENT analysis program with 1 σ error bars [29]. Also shown are predicted velocity curves for selected models. The models with low n and gravity waves are not ruled out by this data because the cosmic variance for the velocity predictions in a single region are quite large. To illustrate this we have plotted the 68% confidence limits on the prediction of large scale streaming velocities for the 25% HDM model.

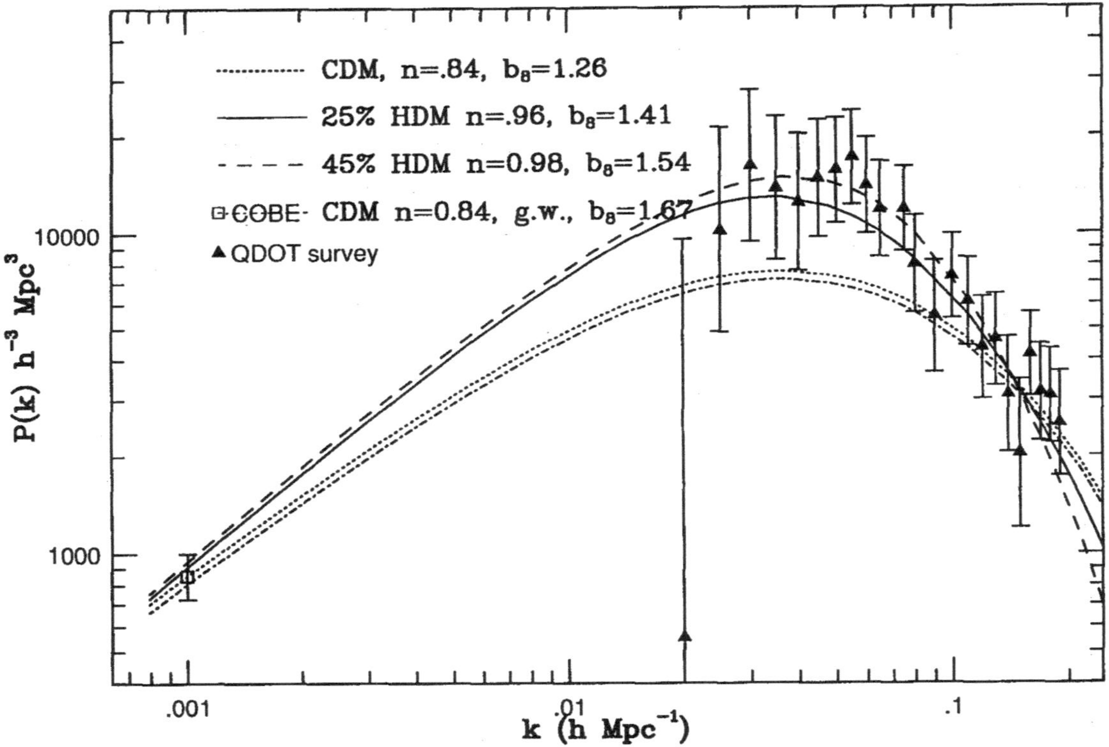

Fig. 4. Here we show the IRAS QDOT power spectrum data [27] and the COBE amplitude constraint [23] (converted to a redshift space power spectrum constraint for $b_I = 1.0$). The models shown have normalizations $b_8 = 1.28, 1.48$, and 1.64 for the models shown with 0%, 25%, and 50% HDM, respectively. We have shown only part of the QDOT data so the plot is easier to read.

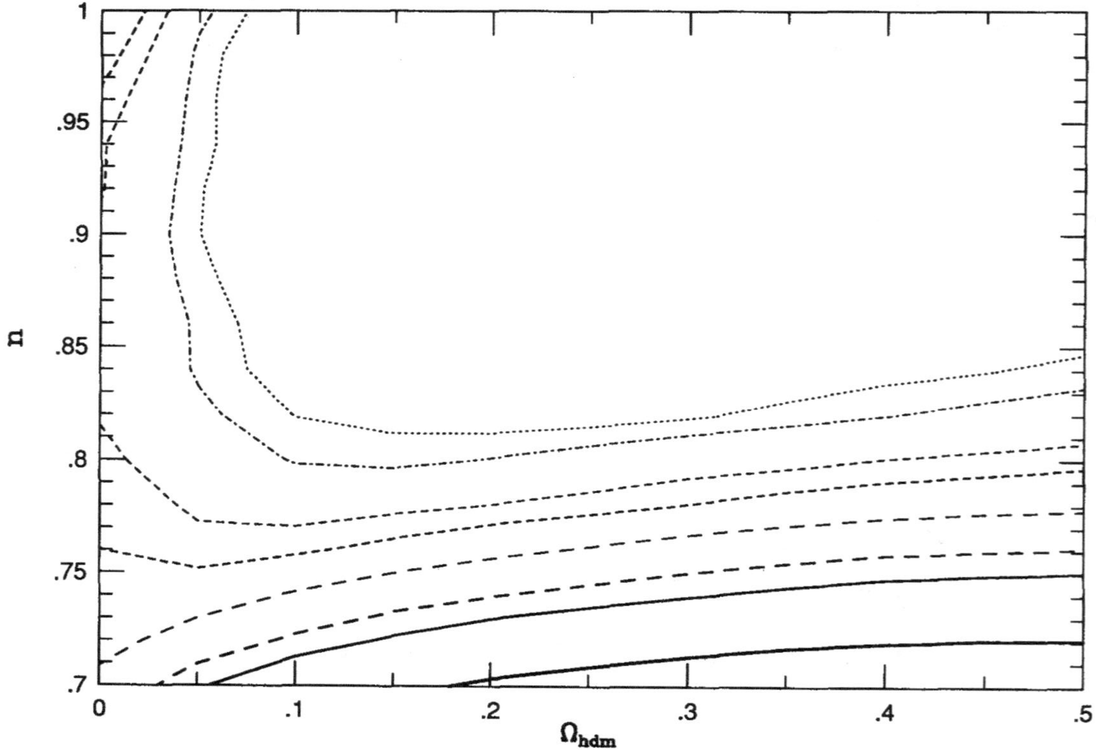

Fig. 5. Contour plot for χ^2 — No gravity waves χ^2 contours in the n-Ω_{HDM} plane for models with a negligible gravity wave content. Moving outward from the center of the graph are contours corresponding to .5% (dotted), 1% (dash-dotted), 5% (dashed), 10% (heavy dashed) confidence levels.

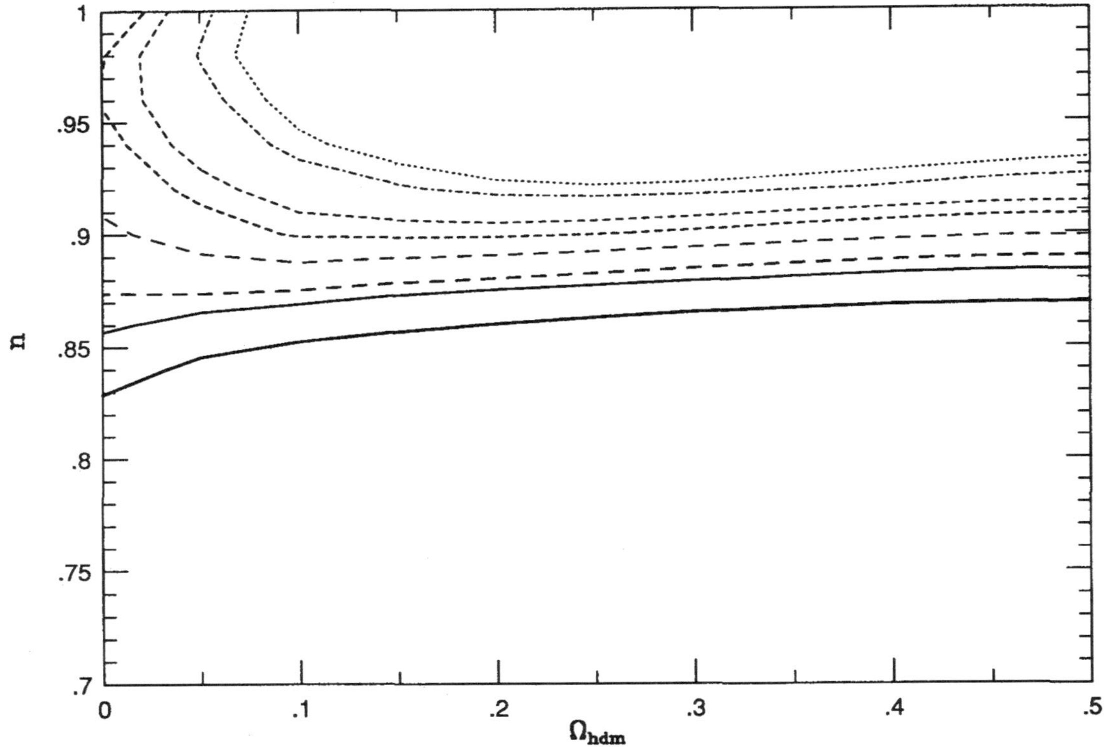

Fig. 6. Contour plot for χ^2 — With gravity waves χ^2 contours in the n-Ω_{HDM} plane for models with gravity waves. Moving outward from the center of the graph are contours corresponding to .5% (dotted), 1% (dash-dotted), 5% (dashed), 10% (heavy dashed), 25% (long dashed), 50% (heavy long dashed), 68% (solid), 95% (solid), 99% (heavy solid) confidence levels.

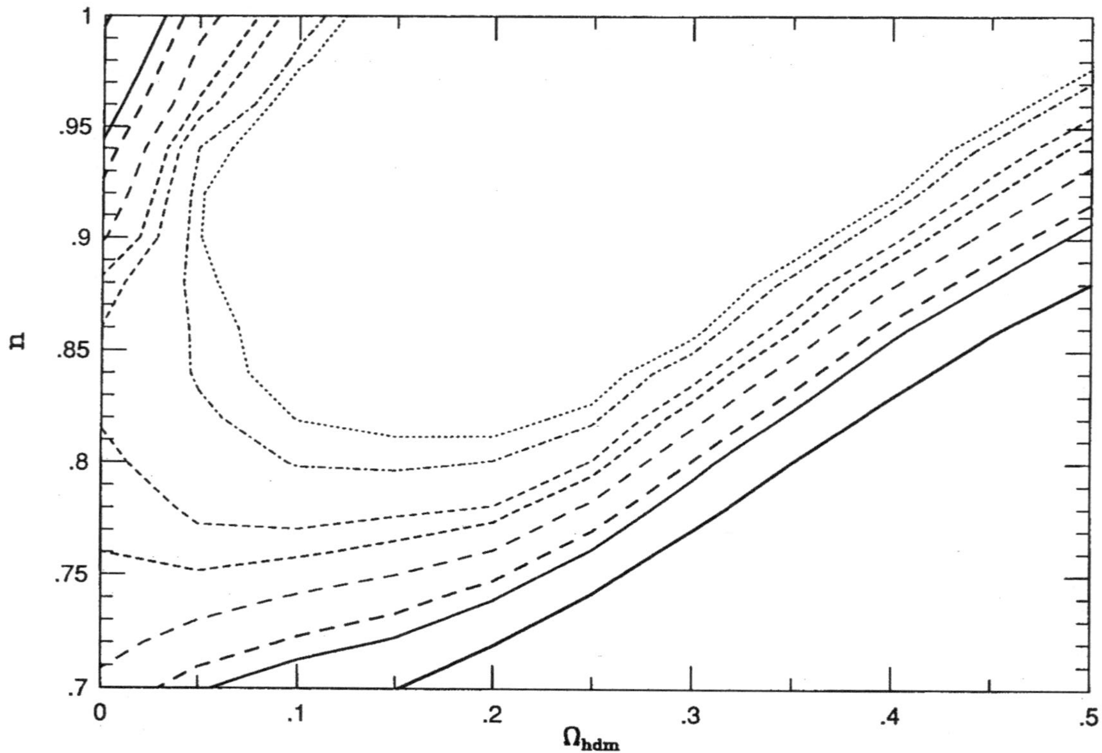

Fig. 7. Same as figure 5, but we have added the constraints from non-linear data. We have added the restriction that $b_8 \geq 1.25$ and that we have sufficient power for quasar formation. The procedure for adding these constraints in a χ^2 analysis is described in the text. The significance of the contours is here not well defined because of the way we have added the non-linear constraints.

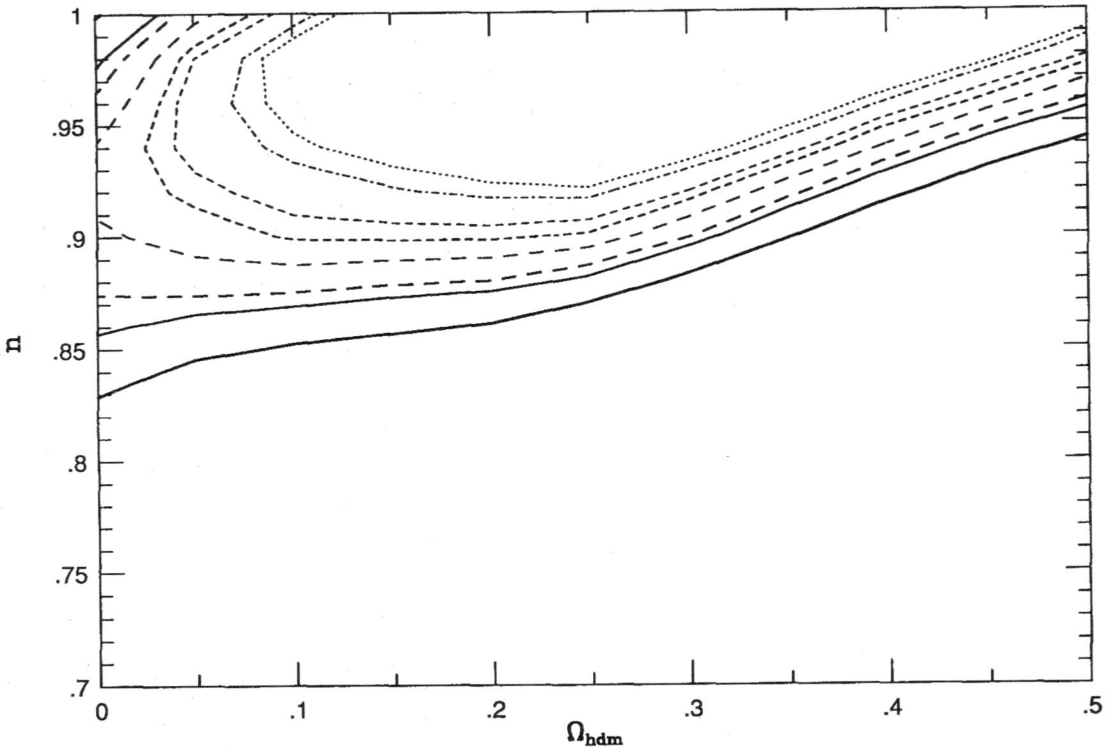

Fig. 8. Same as figure 6, but we have added the constraints from non-linear data. We have added the restriction that $b_8 \geq 1.25$ and that we have sufficient power for quasar formation. The significance of the contours is here not precise because of the way we have added the non-linear constraints.

Heavy Quark Spin Symmetry and $\tilde{U}(12)$ and $SU(4)$ Current Algebra

Riazuddin

Physics Department, King Fahd University of Petroleum and Minerals,
Dhahran 31261, Saudi Arabia

Fayyazuddin

Physics Department, King Abdulaziz University
P.O.Box 9028, Jeddah 21413, Saudi Arabia

Heavy quark spin symmetry in the context of old formalism of $\tilde{U}(12)$ and $SU(4)$ current algebra is reviewed.

1. Introduction

"Human existence depends upon compassion and curiosity. Curiosity without compassion is inhuman, compassion without curiosity is ineffectual". So said Prof. V. Weisskopf. It was curiosity which led Prof. Abdus Salam to seek " unity in Nature", the crowning achievement here was his unified theory of electroweak interactions for which he shared the 1979 Nobel Prize. It was compassion which led him to seek unity in " Mankind", the crowning achievement here was the creation and subsequent development of International Center for Theoretical Physics at Trieste for which he was justly awarded the Atom's for peace prize in 1967. Thus, Prof. Abdus Salam has the rare distinction of being an outstanding physicist and a brilliant administrator, concerned with the organization and development of science, especially in the third world countries. His concern for human welfare through science and technology is as great as his love for physics. He is equally at home with solving intricate problems of physics and those of isolation of scientists working in developing countries.

In 1960's, there was a great deal of interest in spin-flavor symmetry of light quarks [1], the so called static $SU(6)$. However, the derivation of several "no-go" theorems showed that non-trivial extensions of mixing a Lie global internal symmetry (e.g. flavor $SU(3)$) with the Poincare's group strong interaction were impossible [2]. The spin-flavor symmetry of light quarks was recently extended [3] to heavy quarks interacting through gluons in the limit of infinite heavy quark mass. In this limit the dynamics of heavy quarks interacting with the light degrees of freedom, consisting of gluons, light quarks and quark-antiquark pairs, is simplified by the heavy quark spin symmetry of QCD. The symmetry underlying this so called heavy quark effective theory, $HQET$, has allowed to derive model independent predictions for the form factors of semi-leptonic decays of B and D mesons near the zero recoil point [4].

The techniques used in mid 1960's by Salam, Delbourgo and Strathede [5] and others [6], in finding a connection of static $SU(6)$ with its covariant extension $\tilde{U}(12)$, where the latter symmetry is to be interpreted as holding for effective tree level vertices and *not* as a symmetry of the entire theory of a non-compact group, give an alternative way of realizing the heavy quark spin symmetry and studying its consequences. We shall come to this point later.

In order to see how the dynamics of heavy quarks interacting through gluons is simplified in the limit $M \to \infty$, where M is the mass of the heavy quark, and to find the symmetry underlying this theory, let us consider the Pauli form of Dirac equation [7]:

$$\left[-D_\mu D_\mu + \frac{g_s}{2}\sigma_{\mu\nu}G_{\mu\nu}\right]\Psi = -M^2\Psi, \tag{1}$$

where $G_{\mu\nu} = G^A_{\mu\nu}\frac{\lambda_A}{2}$ [λ_A being Gell-Mann's 3×3 matrices in color space], g_s is the gluon field strength and D_μ is the covariant derivative

$$D_\mu = \frac{\partial}{\partial x_\mu} + ig_s G_\mu. \tag{2}$$

The mass dependent term in Eq.(1) can be removed by the re-definition of the field Ψ [8]:

$$h_v = S\Psi, \tag{3}$$

where

$$S = \exp\left(-M\gamma \cdot v\, v \cdot x\right) \tag{4}$$

with $\gamma \cdot v = \gamma_\mu v_\mu, v_\mu$ being the four velocity of the quark [$v^2 = -1$]. This transformation leads to

$$\left[-\gamma \cdot v(v \cdot D) - \frac{D^2}{2M} + \frac{g_s}{4M}\sigma_{\mu\nu}G_{\mu\nu}\right] h_v = 0. \tag{5}$$

The corresponding Lagrangian is

$$\mathcal{L} = \bar{h}_v\left[i \not{v} v \cdot D + \frac{D^2}{2M} - \frac{g_s}{4M}\sigma_{\mu\nu}G_{\mu\nu}\right] h_v, \tag{6}$$

where we have defined $\not{v} = -i\gamma \cdot v$ so that $\not{v}^2 = -v^2 = 1$.

Since in the rest frame $\frac{1}{2}(1 \pm \gamma_4)$ projects out the upper and lower two components of the Dirac spinor, it follows that in an arbitrary frame, they are given by

$$h_\pm = \frac{1}{2}(1 \pm \not{v})h_v \tag{7}$$

with the properties

$$\not{v}h_\pm = \pm h_\pm. \tag{8}$$

When we substitute Eq.(7) in Eq.(5) and make use of Eq.(8), we obtain

$$\begin{aligned}\{(\mp iv \cdot D - \frac{D^2}{2M} + \frac{g_s}{4M}\sigma_{\mu\nu}G_{\mu\nu})h_\pm \\ \mp \frac{g_s}{4M}(v_\mu\gamma_\nu - v_\nu\gamma_\mu)G_{\mu\nu}h_v\} = 0.\end{aligned} \tag{9}$$

Due to the last term, the equations of motion for h_+ and h_- would be coupled. However, it is easy to see that the last term is $O(1/M^2)$. This is obvious in the frame $\mathbf{v} = 0$, since then the last term involves "relativistic" matrices γ_i. Thus, to order $1/M$ the equations of motion for $h_\pm$ are

$$\left[\mp iv \cdot D - \frac{D^2}{2M} + \frac{g_s}{4M}\sigma_{\mu\nu}G_{\mu\nu}\right] h_\pm = 0. \tag{10}$$

Here h_+ annihilates a heavy quark, while h_- creates a heavy antiquark.

The corresponding Lagrangian for $h_\pm$ is [9]:

$$\mathcal{L}_\pm = \bar{h}_\pm \left(\pm iv \cdot D + \frac{D^2}{2M} + \frac{g_s}{4M}\sigma_{\mu\nu}G_{\mu\nu}\right) h_\pm\,. \tag{11}$$

Note that the Lagrangian (6) or (11) contains leading $1/M$ corrections to the kinetic energy term, namely the first term.

Now from the relation

$$i\partial_\mu\Psi(x) = [\hat{P}_\mu, \Psi(x)], \tag{12}$$

it follows through the transformation (3) that

$$i\partial_\mu h_v = M\,\not{v}v_\mu h_v + [\hat{P}_\mu, h_v], \tag{13}$$

so that

$$i\partial_\mu h_\pm = \pm M v_\mu h_\pm + [\hat{P}_\mu, h_\pm]. \tag{14}$$

This shows that a derivative acting on h_+ corresponds to a factor of the residual momentum k_μ carried by the heavy quark. Thus one can write

$$-k_\mu = M v_\mu - p_\mu$$

or

$$p_\mu = M v_\mu + k_\mu \tag{15}$$

so that k_μ indicates how much heavy quark is off-mass shell. In the limit $M \to \infty$ (the no recoil limit) with v_μ and k_μ fixed,

$$v_\mu^{quark} \equiv \frac{p_\mu}{M} = v_\mu + \frac{k_\mu}{M} \to v_\mu. \tag{16}$$

Now one would expect the heavy quark to carry most of the momentum of the $\bar{q}Q$ bound state, but not all:

$$P_\mu = p_\mu + l_\mu = M v_\mu^{quark} + l_\mu, \tag{17}$$

where $P_\mu = M_B v_\mu$ is the momentum of the bound state and l_μ is that which is carried by the light degree of freedom. Now $M_B = M + m - B$, where B is binding energy supplied by the interaction through gluons. Thus from Eq.(17)

$$v_\mu^{quark} = \frac{M_B}{M} v_\mu - \frac{l_\mu}{M} \tag{18}$$

so that again $v_\mu^{quark} \to v_\mu$ as $M \to \infty$ and a comparison of Eq.(18) with Eq.(16) shows that in the limit $M \to \infty$, the interaction with gluons can change k_μ but not $v_\mu^{quark} = v_\mu$; the velocity of the heavy quark can be altered only by an external current which absorbs "infinite momentum".

Now from the relation

$$[\Psi(x), M_{\mu\nu}] = (\frac{1}{2}\sigma_{\mu\nu} + l_{\mu\nu})\Psi(x) \tag{19a}$$

and Eqs.(3), (4), (7), and (8), we obtain

$$[h_+, M_{\mu\nu}] = \big[\frac{1}{2}\sigma_{\mu\nu} + l_{\mu\nu} + M(x_\mu v_\nu - x_\nu v_\mu)\big] h_+. \tag{19b}$$

It follows from Eq.(19b), using the argument given above that the residual orbital angular momentum of the heavy quark goes to zero as $v_\mu^{quark} \to v_\mu$ in the heavy quark limit. Thus the spin of heavy quark is decoupled.

The third term in the Lagrangian (6) or (11) undergoes short-distance QCD corrections and as such is multiplied by the renormalization factor

$$Z_Q(\mu) = [\alpha_s(\mu)/\alpha_s(M_Q)]^{-9/25}$$

with $Z_Q(\mu = M_Q) = 1$.

2. Spin Symmetry of Heavy Quark

In the limit $M \to \infty$, the Lagrangian $\mathcal{L}_+$ (or $\mathcal{L}_-$) given in Eq.(11) has additional symmetries not present in the full QCD Lagrangian. One such symmetry, namely, the spin symmetry of heavy quark, is reflected in the fact that the first term in the Lagrangian (11) giving the gluon interaction vertex as $\pm i g_s \frac{\lambda_A}{2} v_\mu$, makes no reference to the Dirac structure at all which can couple to the spin degrees of freedom of $h_\pm$. More explicitly, we can define the spin operator

$$S_{\pm i}(v) = \int h_\pm^\dagger s_i h_\pm d^3x, \tag{20a}$$

where

$$s_i = -i\gamma_5 \not{v} \gamma \cdot \epsilon_i \tag{20b}$$

with $\epsilon_i \cdot v = \epsilon_{i\mu} v_\mu = 0$, $\epsilon_i^2 = \epsilon_{i\mu}\epsilon_{i\mu}$, $\epsilon_{j\mu}\epsilon_{k\mu} = \delta_{jk}$ [note that in the frame $\mathbf{v} = 0$, $\epsilon_{i\mu} = \delta_{i\mu}$, $s_i = \sigma_i$]. We note that

$$[S_{+i}, h_+] = -s_i h_+. \tag{21}$$

Hence

$$\begin{aligned} \delta h_+ &= i\boldsymbol{\theta} \cdot \mathbf{s}\, h_+ \\ \delta \bar{h}_+ &= -i\bar{h}_+ \boldsymbol{\theta} \cdot \mathbf{s} \end{aligned} \tag{22}$$

($\boldsymbol{\theta}$ being an infinitesimal parameter), showing that the Lagrangian $\mathcal{L}_+$ in Eq.(11) is invariant under $SU(2)_+$ of heavy quark spin symmetry. Similarly $\mathcal{L}_-$ is invariant under $SU(2)_-$ of heavy antiquark spin symmetry. It means that pseudoscalar and vector meson states $|P(v)\rangle$ and $|V(v,\epsilon)\rangle$, containing the same heavy quark, with momentum $P_\mu^B = M_B v_\mu$ can be related to each other

$$S_+(v)|P(v)\rangle = \eta |V(v,\epsilon)\rangle, \tag{23}$$

where η is a phase factor (which in our metric and convention of γ-matrices, can be conveniently chosen to be $\eta = -i$). Their masses are degenerate in this limit. This degeneracy in masses is lifted by the third term in the Lagrangian (11) giving, e.g. for $B^*(1^-)$ and $B(0^-)$ mesons, the mass difference $\delta_b \equiv (m_B^* - m_B)$ which scales like $1/M_b$.

The second symmetry of the Lagrangian (11) in the limit $M \to \infty$ arises when we introduce two distinct flavors $h_1(e.g.b)$ and $h_2(e.g.c)$. Since the first term in Eq.(11) makes no reference to masses $M_i(i = 1,2)$, and since mass is the only property which can distinguish between quarks of different flavors in QCD, the effective theory possesses a symmetry under which $h_1(v) \leftrightarrow h_2(v)$. It may be emphasized that this symmetry does not in any way depend on $M_1 = M_2$ but only on $M_i \gg \Lambda$, where Λ is a scale parameter such that $1/\Lambda$ determines the size of light degrees of freedom in the bound state and is a few hundred MeV, it may vary from process to process. Note also that the flavor symmetry holds between heavy quark fields of same velocity and not with the same momentum. This flavor symmetry together with the spin symmetry mentioned above gives rise to $SU(4)$ symmetry for the system $\begin{pmatrix} h_1(v) \\ h_2(v) \end{pmatrix}$ and has been used [3,4] to relate the matrix elements of the flavor changing effective currents which mediate weak decays of mesons containing heavy quarks. To see this consider the following matrix elements

$$\begin{aligned} \langle D^0(p')|i\bar{c}\gamma_\mu(1+\gamma_5)b|B^-(p)\rangle &= \langle D^0(p')|i\bar{c}\gamma_\mu b|B^-(p)\rangle \\ &= \frac{1}{\sqrt{4p_0p_0'}}\left[f_+(t)(p+p')_\mu + f_-(t)(p-p')_\mu\right], \end{aligned} \tag{24}$$

where

$$t = -q^2 = -(p-p')^2 = m_B^2 + m_D^2 + 2p\cdot p' = m_B^2 + m_D^2 + 2m_B m_D v\cdot v'.$$

Since $i\bar{c}\gamma_\mu b$ is a symmetry current in the above formalism, it follows that the above matrix is described by a single form factor

$$\begin{aligned} &\langle D^0(v')|i\bar{c}\gamma_\mu(1+\gamma_5)b|B^-(p)\rangle \\ &\quad = \langle D^0(v')|ic\gamma_\mu b|B^-(v)\rangle \\ &\quad = \frac{1}{\sqrt{4v_0v_0'}}\xi_0(-v\cdot v')(v+v')_\mu, \end{aligned} \tag{25}$$

where the function ξ_0 is normalized at the kinematic point corresponding to $v = v'$ so that

$$\xi_0(-v^2) = \xi_0(1) = 1. \tag{26}$$

Thus in $HQET$,

$$f_- = 0, \quad f_+ = \xi_0(-v.v') \tag{27}$$

with

$$f_+[(m_B - m_D)^2] = 1. \tag{28}$$

Now using the relations (20), (21) and (23) for b quark we have

$$\begin{aligned} &\langle D^0(v')|i\bar{c}\gamma_\mu(1+\gamma_5)i\gamma_5 \ \not{v}\gamma\cdot\epsilon b|B^-(v)\rangle \\ &\quad = \langle D^0(v')|i\bar{c}\gamma_\mu(1+\gamma_5)[S_+, b_+]vertB^-(v)\rangle \\ &\quad = i\langle D^0(v')|i\bar{c}\gamma_\mu(1+\gamma_5)b|B^{*-}(v)\rangle. \end{aligned} \tag{29}$$

By the use of the identity

$$\gamma_\mu\gamma_\nu\gamma_\lambda = \delta_{\mu\nu}\gamma_\lambda - \delta_{\mu\lambda}\gamma_\nu + \delta_{\lambda\nu}\gamma_\mu + \epsilon_{\mu\nu\lambda\rho}\gamma_5\gamma_\rho, \tag{30}$$

it is easy to see that the term involving Dirac matrices on the left hand side of Eq.(29) can be written as

$$-B_{\mu\nu}\gamma_\nu(1+\gamma_5), \tag{31}$$

where

$$B_{\mu\nu} = v_\mu\epsilon_\nu - \epsilon_\mu v_\nu + \epsilon_{\mu\nu\rho\lambda}v_\rho\epsilon_\lambda. \tag{32}$$

Then using Eq. (25), we obtain

$$\begin{aligned} &\langle D^0(v')|i\bar{c}\gamma_\mu(1+\gamma_5)b|B^{*-}(v)\rangle \\ &\quad = -iB_{\mu\nu}\langle D^0(v')|i\bar{c}\gamma_\nu(1+\gamma_5)b|B^-(v)\rangle \\ &\quad = \frac{-i}{\sqrt{v_0 v_0'}} B_{\mu\nu}\xi_0(-v.v')(v+v')_\nu. \end{aligned} \tag{33}$$

Thus the four form factors in the matrix elements $\langle D^0(v')|i\bar{c}\gamma_\mu(1+\gamma_5)b|B^{*-}(v)\rangle$ get related to the one universal form factor $\xi_0(-v\cdot v')$.

As another application, if we replace in Eq.(29) $|D^0\rangle$ by $|0\rangle$ and c by $q = u, d, s$, then we obtain on using Eqs.(31) and (32) and the first part of Eq.(33):

$$\langle 0|i\bar{q}\gamma_\mu(1+\gamma_5)b|B_q^*(v)\rangle = -iB_{\mu\nu}\langle 0|i\bar{q}\gamma_\nu(1+\gamma_5)b|B_q(v)\rangle. \tag{34}$$

With the usual definitions

$$\langle 0|i\bar{q}\gamma_\nu(1+\gamma_5)b|B_q(p)\rangle = \frac{1}{\sqrt{2p_0}} ip_\nu f_B = \sqrt{\frac{m_B}{2v_0}} f_B(iv_\nu),$$

$$\langle 0|i\bar{q}\gamma_\nu(1+\gamma_5)b|B_q^*(p)\rangle = \frac{1}{\sqrt{2p_0}}\epsilon_\mu f_{B^*} = \frac{1}{\sqrt{2m_{B^*}v_0}}\epsilon_\mu f_{B^*}, \tag{35}$$

Eq.(34) gives (noting $v^2 = -1$, $v \cdot \epsilon = 0$, *etc.*):

$$\sqrt{m_B} f_B = \frac{1}{\sqrt{m_{B^*}}} f_{B^*}$$

i.e. on using $m_{B^*} = m_B + \delta_B$, where $\delta_B \sim O(1/m_B)$, we obtain in the heavy quark limit

$$f_{B^*} = m_B f_B. \tag{36}$$

Since $B_0(j = 0^+)$ and $B_1^*(1^+)$ and $B_1(1^+)$ and $B_2^*(2^+)$ form two separate spin doublets, we also obtain

$$f_{B_1^*} = m_{B_0} f_{B_0} \tag{37}$$

and

$$f_{B_1} = 0 = f_{B_2^*}. \tag{38}$$

The second result follows from the fact that the matrix element of vector or axial vector currents between the vacuum and B_2^* vanishes. All the above results hold if B is replaced by D and the charm quark c is regarded as heavy enough, $m_c \gg \Lambda$.

There is another way to see the heavy quark spin symmetry of QCD. We go back to the work of Salam, Delbourgo and Strathdee [5] mentioned earlier. The meson field as a composite of quark- antiquark fields was written down in 1964:

$$\Phi_\beta^\alpha = \bar{q}^\alpha(x) q_\beta(x). \tag{39}$$

Now using the completeness of the Dirac γ matrices

$$\sum_A (\gamma_A)_\sigma^\rho (\gamma_A)_\beta^\alpha = 4\delta_\beta^\rho \delta_\sigma^\alpha, \tag{40}$$

one can express

$$\Phi_\beta^\alpha = \frac{1}{4}\sum_A (\gamma_A)_\beta^\alpha \phi_A. \tag{41}$$

Here α, β etc. are spinor indices $\gamma_A = 1, \gamma_5, \gamma_\mu, i\gamma_\mu\gamma_5, \sigma_{\mu\sigma}$ and ϕ_A are meson fields. As we have discussed earlier, for mesons with one heavy quark one can replace h_v by $h_+ = \frac{1+\not{v}}{2} h_v$ (we will not write the suffix v explicitly). hence in this case using

$$\sum_A (\gamma_A)^\rho_\sigma (1+\not{v})^\alpha_\gamma (\gamma_A)^\gamma_\beta = 4(1+\not{v})^\alpha_\gamma \delta^\rho_\beta \delta^\gamma_\sigma, \tag{42}$$

we can write for the heavy vector and pseudoscalar meson fields [c.f.Eq(41)] [10]:

$$H_a = \frac{1+\not{v}}{2}\left[\gamma_\mu P^*_{a\mu} + \gamma_5 P_a\right], \tag{43}$$

where $a = 1, 2, 3$ for u, d, and s quarks and $P^*_{a\mu}$ and P_a are annihilation operators normalized as

$$\langle 0|P^*_{a\mu}|Q\bar{q}_a(1^-)\rangle = \epsilon_\mu \tag{44}$$

$$\langle 0|P_a|Q\bar{q}_a(0^-)\rangle = 1. \tag{45}$$

We define the adjoint field

$$\bar{H}_a = \gamma_4 H^\dagger_a \gamma_4 = \left[-\bar{P}^*_{a\mu}\gamma_\mu - \bar{P}_a\gamma_5\right]\frac{1+\not{v}}{2}. \tag{46}$$

We note that

$$\not{v} H_a = H_a$$

$$H_a \not{v} = -H_a. \tag{47}$$

The Lorentz symmetry and $SU(2)$ spin symmetry of heavy quark are now obvious in view of Eqs.(20) and (21). The spin symmetry which relates P_a and $P^*_{a\mu}$ is automatically incorporated in Eq.(43).

In case of spinor field $\Psi(x)$, the wave function can be written

$$\langle 0|\Psi(x)|p\rangle \sim e^{ipx} u(p). \tag{48}$$

Thus, in view of Eqs.(43), (44) and (45) we can write for mesons containing a heavy quark ($P = B$ or D)

$$\langle 0|H_a|P_a(v)\rangle = \frac{1+\not{v}}{2}\gamma_5$$

$$\langle 0|h_a|P^*_a(v,\epsilon)\rangle = \frac{1+\not{v}}{2}\gamma\cdot\epsilon. \tag{49}$$

As an application of the above considerations, consider the matrix elements

$$\begin{aligned}&\langle B^+(p')|i\bar{u}\gamma_\mu d|B^0(p)\rangle\\&\quad=\frac{1}{\sqrt{4p_0p_0'}}F(q^2)(p+p')_\mu=\frac{1}{m_B\sqrt{4v_0v_0'}}F(t)(v+v')_\mu,\end{aligned}\tag{50}$$

where B^+ and B^0 form an isospin multiplet and due to the conservation of the vector current, we have only one form factor with the normalization $F(0)=1$. Using the wave function (49) in $HQET$, we have ($t=2m_B^2(1-v\cdot v')$):

$$\begin{aligned}&\langle B^+(v')|i\bar{u}\gamma_\mu d|B^0(v)\rangle\\&\quad=\frac{1}{\sqrt{4v_0v_0'}}A(-v.v')Tr(-\gamma_5\frac{1+\not{v}'}{2}i\gamma_\mu\frac{1+\not{v}}{2}\gamma_5)\\&\quad=\frac{-1}{\sqrt{4v_0v'}}A(-v\cdot v')(v+v')_\mu,\end{aligned}\tag{51}$$

where the comparison with Eq.(50) gives $A(-v^2)=A(1)=-1$. Let us now define

$$\begin{aligned}&\langle B^+(p')|i\bar{u}\gamma_\mu\gamma_5 d|B^{*0}(p)\rangle\\&\quad=\frac{i}{\sqrt{4p_0p_0'}}\left[F_0(t)\epsilon_\mu+F_+(t)p'\cdot\epsilon(p+p')_\mu+F_-(t)p'\cdot\epsilon(p-p')_\mu\right]\\&\quad=\frac{i}{\sqrt{4v_0v_0'}}\frac{1}{m_B}\left[F_0\epsilon_\mu+F_+m_B^2v'\cdot\epsilon(v+v')_\mu+F_-v'\cdot\epsilon m_B^2(v-v')_\mu\right].\end{aligned}\tag{52}$$

We obtain on using the wave functions (49),

$$\begin{aligned}&\langle B^+(v')|i\bar{u}\gamma_\mu\gamma_5 d|B^{*0}(v)\rangle\\&\quad=\frac{1}{\sqrt{4v_0v_0'}}A_1(-v.v')Tr(-\gamma_5\frac{1+\not{v}'}{2}i\gamma_\mu\gamma_5\frac{1+\not{v}}{2}\gamma\cdot\epsilon)\\&\quad=\frac{1}{\sqrt{4v_0v_0'}}A_1(-v.v')i\left[\epsilon_\mu(1-v\cdot v')+v'\cdot\epsilon v_\mu\right],\end{aligned}\tag{53}$$

where we have used $v\cdot\epsilon=0$. Note that this is only possible non-zero Lorentz covariant as the other possible terms $v_\mu\gamma_5$ or $v'_\mu\gamma_5$ give zero when the trace is taken. The comparison with Eq.(52) gives

$$F_0(t)=m_B(1-v\cdot v')A_1(-v.v');\quad m_BF_+(t)=m_BF_-(t)=\frac{1}{2}A_1(-v\cdot v').\tag{54}$$

Thus for $v \cdot v' = -1$ which corresponds to $t = (m_{B^*} - m_B)^2 = \delta_B^2 \to 0$, we get in the heavy quark limit

$$F_0(0) = 4m_B^2 F_+(0) = 4m_B^2 F_-(0) = 2m_B A_1(1). \tag{55}$$

Note that unlike $A(1)$ the normalization for $A_1(1)$ is not fixed. We shall make use of the result (55) in the next section.

3. Heavy Quark Limit and Chiral Symmetry

There are some interesting consequences of heavy quark limit when it is used in conjunction with the chiral symmetry consisting of partial conservation of axial vector current ($PCAC$) and current algebra. As an example, let us consider the $SU(4)$ algebra generated by the axial charges [11,12]

$$[S_i^{(+)}, S_j^{(-)}] = 2\delta_{ij} I_3, \tag{56}$$

where

$$S_i^{(+)} = -\int A_i^{(+)}(\mathbf{x}, 0) d^3x, \tag{57}$$

$$A_\mu^{(+)} = i\bar{u}\gamma_\mu\gamma_5 d, \quad A_i^{(+)} = -u^\dagger \sigma_i d, \quad \sigma_i = -i\gamma_4\gamma_i\gamma_5 \tag{58}$$

and

$$I_3 = -i\int V_{34}(\mathbf{x}, 0) d^3x, \quad V_{i4} = \frac{1}{2} q^\dagger \tau_3 q. \tag{59}$$

Taking the matrix elements of Eq.(56) between $|B^+(p)>$, we get

$$\begin{aligned} &\langle B^+(p)|[S_i^{(+)}, S_j^{(-)}]|B^+(p)\rangle \\ &\quad = 2\delta_{ij}\langle B^+(p)|I_3|B^+(p)\rangle \\ &\quad = \delta_{ij}. \end{aligned} \tag{60}$$

In order to evaluate the left hand side, we introduce a complete set of allowed states, the lightest of which is 1^- state $|B^{*0}(p')\rangle$, a member of the isopin multiplet (B^{*+}, B^{*0}). We take $i = j = 3$ and obtain

$$\int d^3p' |\langle B^+(p)|S_3^{(+)}|B^{*0}(p')\rangle|^2 + \sum_{n \neq \vec{B}^0} |\langle B^+(p)|S_3^{(+)}|n\rangle|^2 = 1\,. \tag{61}$$

We have neglected the contribution from the second term in the commutator in Eq.(60) since in a saturation by resonances it is "exotic" and as such is expected to be negligible compared to the first term. Note that the left hand side of Eq.(60)is positive definite. We work in the rest frame of B^+ meson, $\mathbf{p} = 0$. Then using Eq.(52), we obtain ($t = \delta_B^2 \to 0$)

$$\frac{1}{4m_B m_{B^*}} F_0^2(0) = 1 - \delta^2, \tag{62}$$

where δ^2 is the contribution from the second term on the left hand side of Eq.(61). In this connection it may be noted that the possible contributions of p-wave states $B_J(J = 0,1,2)$ to δ^2 vanish in the static limit in the frame $\mathbf{p} = 0$, the couplings involved being momentum dependent. PCAC applied to the axial vector current appearing on the left hand side of Eq.(52) gives

$$2f_\pi g_{B^*B\pi} = \left[F_0(0) + (m_{B^*}^2 - m_B^2)F_+(0)\right], \tag{63}$$

where $g_{B^*B\pi}$ is defined by

$$\langle B^+(p')|j_\pi|B^{*0}(p)\rangle = \frac{1}{\sqrt{4p_0 p_0'}} g_{B^*B\pi}(2p' - p)\cdot\epsilon \tag{64}$$

In the normalization used above $f_\pi = 130\ MeV$. In the heavy quark limit using Eq.(55):

$$\frac{F_+(0)}{F_0(0)} \sim \frac{1}{m_B^2}, \quad (m_{B^*}^2 - m_B^2) \sim 2m_B\delta_B,$$

i.e. is a constant. In this limit, Eqs.(62) and (63) give [12]:

$$\begin{aligned} g_{B^*B\pi} &\simeq \frac{(m_B m_{B^*})^{1/2}}{f_\pi}(1-\delta^2)^{1/2} \\ &\le \frac{(m_B m_{B^*})^{1/2}}{f_\pi}. \end{aligned} \tag{65}$$

Thus we can write

$$g_{B^*B\pi} = \lambda \frac{m_B}{f_\pi}, \tag{66}$$

where $\lambda = (1-\delta^2)^{1/2}[1 + O(\delta_B/m_B)] \le 1$. In other words

$$g_{B^*B\pi} = \frac{m_B}{f_\pi} \tag{67}$$

gives an upper limit on $g_{B^*B\pi}$. The following remarks are in order. For ρ and K^* mesons, the $SU(6)$ relation $g_{\rho\pi\pi} = \sqrt{2}m_\rho/f_\pi$, $g_{K^*K\pi} = m_{K^*}/\sqrt{2}f_\pi$ were known since 1966 [13,14]. These relations differ from $KSRF$ value [15], $g_{\rho\pi\pi} = m_\rho/f_\pi$ and $g_{K^*K\pi} = m_{K^*}/2f_\pi$ by a factor of $\sqrt{2}$. Experimentally $SU(6)$ relations for ρ and K^* are not good. But for the heavy quarks the relation (67) is expected to hold for the following reasons: (i) For $SU(4)$ algebra considered here, the natural frame is the $\mathbf{p} = 0$ frame, a good frame for heavy mesons, for which the static limit is a good approximation. The spin of B or D meson is flipped with the emission of a pion with negligible recoil (ii) The $PCAC$ relation (63) reduces to $F_0 = 2f_\pi g_{B^*B\pi}$ in the heavy quark limit. This relation is exact in this limit ($m_B \to \infty$) and is crucial in deriving the relation (48). This is not the case for ρ and K^* mesons (iii) Eq.(67) was derived in Ref.[16] using $PCAC$ and the non-relativistic bound state picture for mesons with a heavy quark. The above derivation of Eq.(67) thus shows a close relation between bound state picture and the algebraic approach followed above. The relation (66) was also obtained in Ref.[17], using Adler-Weisberger type sum rule with only pole contribution in the sum rule. Note for this type of algebra, the natural frame is the $\mathbf{p} \to \infty$ frame. How does the relation (66) compare with experiment? There is no experimental information on $g_{B^*B\pi}$. However, for D meson there are estimates from the analysis of D^* radiative decay [18] and give $\lambda \sim 0.5 - 0.8$, consistent with $\lambda \leq 1$ found in Eq.(66).

ACKNOWLEDGEMENTS

We were first introduced to particle physics by Prof. Abdus Salam and we take this opportunity to express our deep sense of gratitude to him for his encouragement throughout our work in this field. One of the authors (Riazuddin) would like to acknowledge the support of King Fahd University of Petroleum and Minerals for this work.

REFERENCES

1. F. Gursey and L. A. Radicati, Phys. Rev. Lett. **13** (1964) 173; B. Sakita, Phys. Rev. **136B** (1964) 1756.
2. See, for example, S. Coleman and J. Mandula, Phys. Rev. **159** (1967) 1251.
3. N. Isgur and M. B. Wise, Phys. Lett. **B232** (1989) 113 and **B237** (1990) 527; F. Hussain, J. G. Korner, K. Schilcher, G. Thomson and Y. L. Wu, Phys. Lett. **B249** (1990) 295; H. Georgi, Phys. Lett. **B240** (1990) 447; J. G. Korner and G. Thomson, Phys. Lett. **B264** (1991) 185. For a review, see H. Georgi,

"Heavy Quark Effective Field Theory" in Proc. Theoretical Advanced Study Institute (1991) p. 589, eds R. K.Ellis, C. T. Hill and J. D. Lykken (World Scientific, Singapore, 1992).

4. N. Isgur and M. B. Wise, Phys. Rev. Lett. **66** (1991) 1130; Ming-L-Lu, M. B. Wise and N. Isgur, Phys. Rev. D**45** (1992) 1553.
5. A. Salam, R. Delbourgo and J. Strathede, Proc. Roy. Soc. (London) A**284**, (1965) 146.
6. B. Sakita and K. Wali, Phys. Rev. Lett.**14** (1965) 404; Phys. Rev.**139**B (1965) 1355.
7. See for example, Fayyazuddin and Riazuddin, Quantum Mechanics (World Scientific, Singapore, 1990) p. 444. Eq.(1) follows from Eq.(18.108) on p. 444 with $\hbar = c = 1, e \to -g_s, A_\mu \to G_\mu$ and $F_{\mu\nu} \to G_{\mu\nu}$.
8. See for example H. Georgi, reference 3
9. See for example A. F. Falk, B. Grinstein and M. E Luke, Nucl. Phys. **B357** 185 (1991).
10. J.D.Bjorken, invited talk given at " Les Recontre de la Valle d'Aosta La Thuile", Editions Frontieres, Gif-Sur-Yvette (1990) 583; and H.Georgi, reference 3.
11. B. W. Lee, Phys. Rev. Lett. **14** (1965) 676.
12. Fayyazuddin, Phys. Rev. *D*1 (1993) Oct. In order to conform to the usual convention used in the present paper, $\sqrt{2}g_{B^*B\pi}$ and $\sqrt{2}f_P$ should be changed to $-g_{B^*B\pi}$ and f_P in this reference. However, $g_{B^*B\pi} = (m_B m_{B^*})^{1/2}/2f_\pi$ goes over to $(m_B m_{B^*})^{1/2}/f_\pi$.
13. Fayyazuddin, Riazuddin and M. S. K. Razmi, Phys. Rev **141** (1966) 1509.
14. J. J. Sakurai, Phys. Rev. Lett. **17** (1966) 552.
15. K. Kawarabyashi and M. Suzuki, Phys. ReV. Lett. **16** (1966) 255; Riazuddin and Fayyazuddin, Phys. Rev. **147** (1966) 1071.
16. M. Suzuki, Phys. Rev. D**37** (1988) 239.
17. C. A. Dominguez and N. Paver, Z. Phys. C**41**(1988) 270.
18. P. Cho and H. Georgi, HUTP- 92/ AO 43, Sep. (1992).

Aspects of σ Models

S. Randjbar-Daemi
International Centre for Theoretical Physics, Trieste 34100, **Italy**
and

J. Strathdee
International Centre for Theoretical Physics, Trieste 34100, **Italy**

Abstract

Some aspects and applications of the σ-models in particle and condensed matter physics are briefly reviewed.

1 Introduction

σ-models appear in many places in particle and condensed matter physics. The $O(4)$ invariant linear and non-linear σ-models were originally introduced by Gell-Man and Levy as a simple framework within which the $SU(2)_L \times SU(2)_R$ chiral symmetry and the partial conservation of the axial vector current were realized [1]. The linear model is also a good testing ground for some of the general ideas concerning renormalizability, symmetries and their spontaneous breakdown. The non-linear models on the other hand, incorporate the non-linear realization of symmetries and are used as phenomenological models whose tree graphs generate the soft pion amplitudes [2]. In general the dynamical fields are maps from the space-time to some target manifold. When the target space itself is a Lie group, the corresponding σ model is also called a *principal chiral model.* In this case the coordinates of the group manifold are identified with the Goldstone bosons of the spontaneous breakdown of a $G \times G$ symmetry to the subgroup G.

In string theory applications the target manifolds have a more complicated structure. Their geometry is fixed by some consistency requirements of the string theory itself [3].

The σ models are also encountered in the studies of the magnetic properties of matter in condensed matter physics. In this case the coordinates of the target space are somehow related to the magnetization order field [4]. Some of the behaviour of the magnetic systems near their phase transition points are understood by applying the renormalization group techniques to its non-linear σ-model description [5].

Finally one should also mention the unique perspective offered by these models in the study of the role of geometry and topology in the theory of the quantized fields.

The aim of this contribution is to briefly review some of these topics. No derivations of the results will be presented. The plan of the paper is as follows: in section 2 we mention some of the places where the principal chiral models and their various decendents appear. In section 3 we recount some condensed matter physics applications. Finally in section 4 we summarize some of the results contained in ref [6] and [7].

2 The principal chiral model and its generalizations

The principal chiral models are effective field theories which encapsulate and systematize the results of the current algebras on soft meson physics. At the moment there is no rigourous derivation of these models from an underlying QCD Lagrangian. However the following argument makes their formulation plausible [2],[8].

In the limit of vanishing (light) quark masses the QCD Lagrangian has a global symmetry group which acts independently on the left and the right handed massless fermions. For N flavours of massless quarks this group is $U(N)_L \times U(N)_R$. However the hadronic spectrum shows only an approximate $SU(N)$ flavour symmetry and there is no parity doubling of the meson mutiplets . Therefore this symmetry must be broken. The breakdown takes place by two different mechanisms. Firstly a $U(1)$ subgroup is broken by chiral anomalies. Then it is assumed that the remaining $SU(N)_L \times SU(N)_R \times U(1)_V$ subgroup breaks down spontaneously to $SU(N)_V \times U(1)_V$. The Goldstone bosons produced by this symmetry breaking are identified with the light scalar pseudo mesons. The symmetry in this phase is assumed to be realized non-linearly, namely the pseudo scalar Goldstone bososns are described by the coordinates of the manifold of the $SU(N)$ group. Together with some physical assumptions such as the *pion pole dominance* one lands almost uniquely at an action of the form

$$S = \frac{1}{4f^2}\int d^4x \, tr(\partial_\mu U \partial_\mu U^{-1}) + ... \tag{1}$$

where f is the pion decay constant and $U \in SU(N)$. The ... in equ. 1 indicate all G-invariant terms involving more than two derivatives of U. For the moment we shall ignore these extra terms.

To develop a perturbation theory one writes

$$\begin{aligned} U(x) &= e^{if\pi(x)} \\ \pi(x) &= \pi(x)^a Q_a \end{aligned} \tag{2}$$

where Q_a are the hermitian generators of $SU(N)$. By substituting from equ. 2 into equ.

1 and expanding in powers of π we obtain

$$S = \int d^4x \left[-\frac{1}{4} tr(\partial_\mu \pi \partial^\mu \pi) + \frac{f^2}{48} tr([\partial_\mu \pi, \pi]^2) + O(f^4) \right]$$

The theory obtained in this way is not power counting renormalizable in 4-dimensions. This should not come as a surprise, if one remembers that it is only the tree diagrams of the action which should be considered for the description of the low energy meson physics. The action of equ.(1) displays only the leading term in an expansion in powers of derivative. In principle it is possible to add infinite number of terms involving higher derivatives of U, compatible with the $SU(N) \times SU(N)$ symmetry. The resulting Lagrangian will contain an infinite number of phenomenological constants and may be renormalizable in some generalized sense. The action of equ. (1) also does not generate all the amplitudes of the fundamental QCD Lagrangian. For example to reproduce the processes which are mediated through the anomalous triangular diagrams of QCD we need to include a term which involves four derivatives of U. This is the the Wess-Zumino term [9] which is also needed if one requires that the discrete symmetries of the effective theory described by equ. (1) to coincide exactly with those of the fundamental QCD Lagrangian [10].

The Wess-Zumino term, Γ can not be written as a manifestly $SU(N)\times SU(N)$ invariant contribution to equ.(1) in 4-dimensions. To put it in a manifestly invariant form one needs to embed the Euclidean compactified 4-dimensional space time manifold as the boundary of a 5-dimensional space in the group manifold [10]. One then writes $\Gamma = k \int \omega$, where ω is a closed 5-form on the group manifold. To extend the group valued function $U(x)$ from the 4-dimensional boundary to the 5-dimensional space creates some ambiguity in physics which can be removed by assuming that the coupling constant k is an integer. Witten has argued that this integer should equal the n of the colour $SU(n)$ group. The addition of this term has interesting consequences both in 4-as well as in 2-dimensional space-times. For example in the 4-dimensional case, Witten has shown that the solitons of the σ-model will behave as fermions (bosons) if $k = n$ is odd (even) [11].

In 2-space-time dimensions the theory defined by equ.(1) is renormalizable. There is a logarithmic divergence at the 1-loop order, which can be defined away by a simple coupling constant renormalization

$$f = f_0 + \frac{Nf_0^3}{4\pi} ln\left(\frac{\mu_0}{\mu}\right)$$

where μ is the scale of definition of f. The running of the coupling is therefore governed by the renormalization group equation

$$\mu \frac{df}{d\mu} = \beta(f) = -\frac{Nf^3}{4\pi}$$

The $\beta(f)$ is negative and $f = 0$ is an UV stable fixed point. The asymptotic freedom of the model makes it an interesting analogue of more realistic models like QCD in $D = 4$.

The inclusion of the WZ-term to the 2-dimensional action changes the picture dramatically [12]. Indeed starting from

$$S = \frac{1}{4f^2} \int d^2x \; tr(\partial_\mu U \partial^\mu U) + k\Gamma_{WZW} \tag{3}$$

Witten has shown that

$$\beta(f) = -\frac{Nf^3}{4\pi} \left[1 - \left(\frac{f^2 k}{2\pi}\right)^2\right] \tag{4}$$

In equ.(3) Γ_{WZW} is the Wess-Zumino-Witten term which is defined in a manner analogous to the 4-dimensional case, namely as an integral over a 3-dimensional ball whose boundary is identified with the compactified 2-dimensional space-time [10]

$$\Gamma_{WZW} = \frac{1}{24\pi} \int_{B_3} d^2x \; \varepsilon^{\mu\nu\lambda} tr(g^{-1}\partial_\mu g g^{-1}\partial_\nu g g^{-1}\partial_\lambda g) \tag{5}$$

For the same reason as in 4-dimensions, the constant k must assume integer values only.

Now returning to equ.(4) we see that in addition to the origin the β function vanishes also at a non-zero value of the coupling constant, namely at $f_c^2 = 2\pi/|k|$ Unlike the origin, the new fixed point f_c is an IR stable point. The theory defined at f_c is conformaly invariant and at this value of f the physics can be described in terms of the representation theory of two copies of commuting Kac-Moody algebras. These algebras are generated by the separately conserved left and right moving Noether currents derived from the action integral of equ. (3) [12].

An obvious generalization of the model defined by equ.(1) is obtained by replacing the target space G by one of the factor spaces G/H of G. For example for $G/H = SU(2)/U(1) = CP^1$ the Lagrangian may be defined by

$$L = \frac{1}{2f^2}\partial_\mu \underline{n}.\partial^\mu \underline{n} \tag{6}$$

where $\underline{n}^2 = 1$ is a vector describing a $S^2 = CP^1$. The Theory is renormalizable at $D = 2$, where it is also asymptotically free. For $D > 2$ it has been shown that [5].

$$\beta(f) = (D-2)f - \frac{1}{2\pi}f^2 \tag{7}$$

Thus for $D > 2$ the asymptotic freedom is lost and a non-trivial ultraviolet stable fixed point is generated at $f_c = 2\pi(D-2)$. If the Euclidean Lagrangian of equ.(6) is interpreted as the Hamiltonian of a classical Heisenberg ferromagnet, then f_c will correspond to a critical temperature for the transition to a disordered phase.

Of course the most ambitious generalization of the action of equ.(1) is constructed by replacing the target space by an arbitrary manifold M with coordinates ϕ^a, $a = 1, 2...dimM$, and by writing the most general action

$$S = \int d^D x \left(\frac{1}{2} g_{ab}(\phi)\partial_\mu \phi^a \partial^\mu \phi^b + A_{ab}(\phi)\partial_\mu \phi^a \partial_\nu \phi^b \varepsilon^{\mu\nu} + ...\right) \tag{8}$$

where $g_{ab}(\phi)$, $A_{ab}(\phi)$, ... are tensor fields on M. The $D = 2$ version of equ.(8) has been used to obtain effective field theoretical description of the low energy dynamics of the massless modes of the string theory. In this case the field equations determining the dynamics of g_{ab}, A_{ab},.. are obtained by the requirement of the conformal invariance of the theory defined by equ. (8) [3].

3 $SU(2)$- Quantum Magnets and the $O(3)$ invariant σ-model

Now we turn to a very brief discussion of the magnetic systems. The analysis of the spin Hamiltonians has been a source of many interesting and important results in physics and mathematics. Physical concepts like spin waves, magnons and mathematical ideas like

Yang-Baxter equations and the quantum groups infact have their roots in the studies of such systems.

Consider a lattice of points $\underline{r} \in Z^n$ and attach the $SU(2)$ generators $S^{\alpha}_{\underline{r}}, \alpha = 1,2,3$ to the sites $\underline{r}$ of the lattice. The Heisenberg Hamiltonian is defined by

$$H = \frac{1}{2}\sum_{\underline{r},\underline{r}'} J_{\underline{r},\underline{r}'} S_{\underline{r}}.S_{\underline{r}'} \tag{9}$$

where $J_{\underline{r},\underline{r}'}$ are assumed to be translationally invariant exchange energies. The generators of course satisfy the usual $SU(2)$ Lie algebra at each site, viz,

$$[S^{\alpha}_{\underline{r}}, S^{\beta}_{\underline{r}'}] = i\varepsilon^{\alpha\beta\gamma} S_{\underline{r}}{}^{\gamma} \delta_{\underline{r},\underline{r}'} \tag{10}$$

Exact solutions of the Hamiltonian of equ.(9) are known only in $D = 1$ [13]. Therefore to understand the physics in higher dimensions the best we can do at present is to resort to approximation schemes. The spin wave analysis is one such method which can be applied in any number of dimensions to understand the longwavelength physics of the fluctuations around a semi-classical ground state [14].

Let us consider an isotropic $D = 1$ magnet with a nearest neighbour interaction,

$$H = \frac{J}{2}\sum_{r} S_r.S_{r+1} \tag{11}$$

Assume that at each site the S_r are given by a set of Pauli matrices. Then for J negative, the ground state will be given by a configuration in which all the spins are parallel,

$$\uparrow\uparrow\uparrow\uparrow\uparrow\uparrow\uparrow\uparrow \tag{12}$$

The state is an exact (ferromagnetic)-eigenstate of the Hamiltonian of equ.(11). For J positive on the other hand, the construction of an exact ground state is too complicated. However a semi-classical (anti-ferromagnetic) ground state, called the Neel state, can be envisaged in which the pairwise neighbouring spins point in opposite directions.

$$\uparrow\downarrow\uparrow\downarrow\uparrow\downarrow\uparrow\downarrow \tag{13}$$

Such configurations can be constructed on any bi-partite lattices. One can also generalize them to the spin chains with arbitrary values of the spin s.

To study the effects of the thermal and quantum fluctuations on the magnetic order, one defines an order parameter n, the so called staggered magnetization

$$n = \sum_{\underline{r}} (-)^{\sum r_i} S_{\underline{r}}.\hat{n} \tag{14}$$

Clearly in the semi-classical ground state we have

$$< 0|n|0 > \neq 0 \tag{15}$$

This could indicate the spontaneous breakdown of the rotational symmetry of the system. The corresponding Goldstone bosons are the spin waves or the magnons[4].

So far we have only summarized the standard lore. However in 1983 there was an interesting observation by Haldane [15] that the long wave-length fluctuations of n in the $s \to \infty$ limit of the Heisenberg chain equ.(11), can be described by an $O(3)$ invariant σ-model with a topological term. The corresponding Lgrangian is given by

$$L = \frac{1}{2f^2}(\partial_\mu \underline{n})^2 + \theta \varepsilon^{\mu\nu} \underline{n}.\partial_\mu \underline{n} \times \partial_\nu \underline{n} \tag{16}$$

together with $\underline{n}^2 = 1$. Here $f^2 = 2/s$ and $\theta = s/4$. Noting that

$$\int d^2x \; \varepsilon^{\mu\nu} \underline{n}.\partial_\mu \underline{n} \times \partial_\nu \underline{n} = 8\pi.integer \tag{17}$$

we see that the θ term in equ.(16) can be relevant only if s is a half integer number. It is known that for $\theta = 0$ the model described by equ. (16) is always disordered [5]. The spectrum consists of a triplet of massive states. It was conjectured by Haldane that the $s =$ half integer chain is gapless. Furthermore Affleck and Haldane argued [16] that the universality class of the critical chain is a $SU(2)$ Wess-Zumino-Witten model of the type given by equ. (3) with $k = 1$.

One might expect that starting from a 2-dimensional lattice rather than a chain, one should obtain a (2+1)- dimensional field theory analogous to the one defined by equ. (16).

In this case it is better to introduce a complex 2-vector

$$z = \begin{pmatrix} z_1 \\ z_2 \end{pmatrix}$$

which is related to $\underline{n}$ by $n^\alpha = z^\dagger \sigma^\alpha z$ where $\sigma^\alpha, \alpha = 1, 2, 3$ are the Pauli spin matrices. It is also assumed that $z^\dagger z = 1$. Thus a phase redefinition of z will leave $\underline{n}$ invariant. Therefore z can be regarded as the homogeous coordinate of a point on CP^1 which has the same topology as a 2-dimensional sphere S^2, described by the unit vector $\underline{n}$. It can be shown by direct calculation that

$$\frac{1}{4}(\partial_\mu \underline{n})^2 = |(\partial_\mu - iA_\mu)z|^2 \tag{18}$$

where

$$A_\mu = -\frac{i}{2}(z^\dagger \partial_\mu z - \partial_\mu z^\dagger z) \tag{19}$$

In 2+1 dimensions there is a topologically conserved current, viz,

$$J^\mu = \frac{1}{8\pi}\varepsilon^{\mu\nu\lambda}\underline{n}.\partial_\nu \underline{n} \times \partial_\lambda \underline{n} \tag{20}$$

$$= \frac{1}{4\pi}\varepsilon^{\mu\nu\lambda}F_{\nu\lambda} \tag{21}$$

where

$$\begin{aligned} F_{\nu\lambda} &= \partial_\nu A_\lambda - \partial_\lambda A_\nu \\ &= -i(\partial_\mu z^\dagger \partial_\nu z - \partial_\nu z^\dagger \partial_\mu z) \end{aligned}$$

Therefore a natural (2+1)-dimensional generalization of equ.(16) would be

$$L = \frac{1}{2f^2}(\partial_\mu n)^2 + \theta J_\mu A^\mu \tag{22}$$

$$= \frac{2}{f^2}|(\partial_\mu - iA_\mu)z|^2 + \frac{\theta}{4\pi}\varepsilon^{\mu\nu\lambda}A_\mu F_{\nu\lambda} \tag{23}$$

This model has soliton like solutions [17] and it has been shown by Wilczk and Zee [18] that they behave like bososns for $\theta = 0$ and like fermions for $\theta = \pi$. The cause of this phenomenon is basically the non-local interaction which is obtained by solving the algebraic field equations of the A-field and substituting the result back in equ. (22).

For arbitrary values of θ the statistics of the solitons should interpolate between the ones of Bose-Einstein and Fermi-Dirac, which is also called anyon statistics [19]. The mechanism of statistics change could induce interesting physical effects such as superconductivity. Unfortunately, however, unlike equ. (16), the term in equ. (20) does not seem to be present in the large s-field theoretical description of the planer magnets [20].

It is interesting to note that the conserved charge Q of the topological current J^μ of equ. (20) is given by,

$$Q = \frac{1}{8\pi} \int d^2x \, \varepsilon^{0\mu\nu} n.\partial_\mu n \times \partial_\nu n$$

By virtue of equ.(17) this is an integer.

4 Some generalizations

Affleck [21] and independently Read and Sachdev [22] have generalized the group $SU(2)$ to $SU(N)$ by simply regarding the $S_r{}^\alpha$ in equ.(11) as the $SU(N)$ generators. In this case of course the range of the index α will be from 1 to $N^2 - 1$. It is also assumed that the representation content of the generators is independent from the site index $\underline{r}$.

This generalization opens up interesting new possibilities. For example by regarding N as a new independent parameter one can now apply the well known method of $1/N$ expansion to study the system's behaviour for large values of N. In some sense this is similar to the large s approximation of the ordinary $SU(2)$ model. Indeed interesting results have been obtained in this way about the critical behaviour of the system by Affleck and Marston [23] and, by Read and Sachdev [22].

It is possible to give a generalization of the concepts of ferromagnetic as well as of the anti-ferromagnetic orders in this broader context. The dynamics of the long-wavelength fluctuations will still be governed by a field theory for which the dynamical variables are maps from the space-time manifold to a factor space G/H of G. The subgroup H is the invariance group of the generalized magnetic order. Here we shall summarize some of the results of this study as carried out in [6] and [7]. To this end the coordinates of G/H will be denoted by ϕ^μ where $\mu = 1, 2, ...dim\frac{G}{H}$. It will always be assumed that H contains the maximal Abelian subgroup of G. The coset corresponding to the point $\phi \in G/H$ can be represented by a group element $L(\phi)$. This representation is of course not unique. The action of $g \in G$ on G/H which maps $\phi \to \phi' = \phi'(\phi, g)$ is defined by $gL(\phi) = L(\phi')h$ where $h = h(\phi, g) \in H$. The detailed form of the functions $\phi'(\phi, g)$ and $h(\phi, g)$ depend on the representative element $L(\phi)$. The G-invariant geometry of G/H can be constructed from the pull-back of the Mauer-Cartan forms $L(\phi)^{-1}dL(\phi)$. These forms take their values in the Lie algebra of G. Thus, they may be expanded on a basis Q^α of this algebra

$$L(\phi)^{-1}dL(\phi) = e^\alpha Q_\alpha = -A^j H_j + e^{\bar{\alpha}} Q_{\bar{\alpha}}$$

where the generators Q_α satisfy the commutation relations

$$[Q_\alpha, Q_\beta] = c_{\alpha\beta}{}^\gamma Q_\gamma$$

The generators of the Cartan subalgebra are denoted be H_j and the $Q_{\bar{\alpha}}$ are the remaining generators.

Now to each site r of the lattice attach the weight vectors Λ_r. These are eigenvectors of the generators H_j ,viz,

$$H_j|\Lambda_{\underline{r}}>=|\Lambda_{\underline{r}}>\Lambda_{\underline{r}j}$$

The semi classical ground states of the system are defined in terms of a distributions of these vectors on the lattice. We assume that in the ground states the lattice can be decomposed into translationally invariant sublattices such that in each one of them the distribution of the Λ's is site independent. If there is only one translationally invariant sublattice we call the corresponding ground state ferromagnetic. The generalized anti-ferromagnetic ordering corresponds to the presence of more than one translationally invariant sublattice.

With this notational background we are now in a position to state some of our results.

1) For the generalized ferromagnetic ordering the dynamics of the long-wavelength fluctuations are governed by a set of non-relativistic field equations which reduce to the

Landau-Lifschitz equations in the special case of $G = SU(2)$ and $H = U(1)$. These equations have the following form:

$$\frac{\hbar}{i} F_{\mu\nu} \partial_t \phi^\nu = \frac{J}{2} k_{\mu\nu} \Delta \phi^\nu$$

where we have assumed an isotropic coupling $J < 0$ and where the "Laplacian" Δ is defined by

$$\Delta \phi^\nu = \partial_i{}^2 \phi^\nu + k^{\nu\rho} (\partial_\lambda k_{\rho\mu} + \partial_\mu k_{\rho\lambda} - \partial_\rho k_{\lambda\mu}) \partial_i \phi^\lambda \partial_i \phi^\mu$$

Here $k_{\mu\nu}$ is a G invariant tensor on G/H and $F_{\mu\nu} = \partial_\mu A_\nu - \partial_\nu A_\mu$, where $A_\mu = A_\mu{}^j \Lambda_j$ Thus it is seen that the equations are intrinsically non-relativistic, i.e. they involve a first order time - and second order space derivatives.

It is not difficult to verify that for $G/H = SU(2)/U(1)$ these equations reduce to the phenomenological Landau-Lifschitz equations,

$$\hbar \partial_t \underline{n} = -\frac{J}{2} s \underline{n} \times \nabla^2 \underline{n}$$

where s is the spin and $\underline{n}$ is a unit vector describing the target space S^2.

2) For the generalized anti-ferromagnetic order we also in general obtain non-relativistic low energy effective field equations. However in this case, unlike the ferromagnetic ordering, the low energy dispersion relations of the small oscillations have a linear form, viz, $\omega \sim |k|$ as $|k| \to 0$. These models become relativistic only if G/H is a symmetric homogenous space, such as the Grassmaniann s $\frac{SU(N+M)}{SU(N)\times SU(M)\times U(1)}$.

With two translationaly invariant sublattice in $D = 1$, the dynamics of the long wavelength fluctuations of the staggered magnetization is governed by

$$S = \int d^2x \left[\frac{1}{2} g_{ab} \partial_t \phi^a \partial_t \phi^b + k_{ab} \partial_x \phi^a \partial_x \phi^b \right]$$

where $g_{ab}(\phi)$ is another G invariant tensor field on G/H. If the space is symmetric, then there will be -up to a normalization- a unique second rank symmetric G-invariant tensor field on G/H. This means that the tensors g_{ab} and k_{ab} will be proportional and therefore the action will be relativistic. In general however there are more than one such fields and the model -as stated above- is not Lorentz invariant.

The tensors g_{ab} and k_{ab} are the coupling constants of our system and one may ask about their running under the renormalization group transformations. Indeed the 1-loop RG equations can be set up in quite general terms. As an example we have examined these equations for the flag manifolds of the form $\frac{SU(N)}{U(1)\times....\times U(1)}$ and shown that in $D = 2+\varepsilon$ the equations admit a non-trivial fixed point at which the theory becomes relativistic, i.e. the tensors g_{ab} and k_{ab} become proportional. This point is a generalization of the Polyakov fixed point given by equ.(7). At this point there is an enlargement of the global symmetries. One can find new discrete coordinate transformations on G/H which leave the effective fixed point theory invariant.

3) The θ term is also generated for the generalized 1-dimensional antiferromagnetic systems. This term is rather like the θ term in QCD and for a system with two translationally invariant sublattices it is given by $\frac{1}{2\pi}\int F$, where F is as defined above the pull-back of the canonical $U(1)$ connections. The weight vector defining the representation content of semi-classical antiferromagnetic ground state entering into the definition of F is $\Lambda = \Lambda_1 = -\Lambda_2$,where Λ_1 and Λ_2 are the two weight vectors associated with the sublattices 1 and 2 respectively.

4) The models admit finite action solution of Euclidean field equations with non-zero values of $\int F$. These solutions can be regarded as the generalizations of the soliton solutions of the CP^N models.

Acknowledgment : We would like to express our thanks to Anwar Shiekh for helping us with type setting and proof reading.

References

[1] M. Gell-Mann and M. Levy, Nuov.Cim. **16** (1960) 705

[2] S. Weinberg, Phys.Rev.Lett. **18** (1967) 188, 507 and Phy.Rev. **116** (1968) 1568

[3] E.S. Fradkin and A.A. Tseytlin, Phys.Lett. **158B** (1985) 316, ibid (1985) 69 and Nucl.Phys. **B261** (1985) 1
C.G. Callan, D. Friedan, E.J. Martinec and M.J. Perry, Nucl.Phys. **B262** (1985) 593

[4] For a recent review see, for example I. Affleck in *Strings, Fields and Critical Phenomena* Les Houches Summer School 1988, session XLIX, edited by E. Brezin and J. Zinn-Justin, (North Holand, Amsterdam,1990),
E. Fradkin, *Field Theories of Condensed Matter Systems* (Addison-Wesley, Redwood City, CA,1991)
E. Manousakis, Rev.Mod.Phys **63** (1991) 1

[5] A.M. Polyakov, Phys.Lett. **B59** (1975) 79
E. Brezin and J. Zinn-Justin, Phys.Rev.Lett. **36** (1976) 691
D.R. Nelson and R.A. Pelcovits, Phys.Rev. **B16** (1977) 2191

[6] S. Randjbar-Daemi, Abdus Salam and J. Strathdee, Phys.Rev **B48**(1993) 3190

[7] S. Randjbar-Daemi and J. Strathdee, Int.J. Mod.Phys. **A8**(1993) 3509

[8] See e.g. S. Weinberg, Physica **A96** (1979) 327
H. Leutwyler, *On the Foundations of Chiral Perturbation Theory* BUTP- 93/24

[9] J. Wess and B. Zumino, Phys.Lett. **37B** (1971) 95

[10] E. Witten, Nucl.Phys. **B223** (1983) 422

[11] E. Witten, Nucl.Phys. **B223** (1983) 433

[12] E. Witten, Comm.Math.Phys. **92** (1984) 455

[13] H.A. Bethe, Zeit.f.Phys. **71** (1931) 205

R.L. Orbach, Phys.Rev. **112** (1958) 309

R. Baxter, *Exactly Solved Models in Statistical Mechanics*, Academic Press 1982

[14] C. Kittel, *Quantum Theory of Solids*, Wiley, New York (1963)

[15] F.D.M. Haldane, Phys.Lett. **93A** (1983) 464, Phys.Rev.Lett. **50** (1983) 1153

[16] I. Affleck and F.D.M. Haldane, Phys.Rev. **B36** (1987) 5291

[17] A.A. Belavin and A.M. Polyakov, JETP Lett. **22** (1975) 245

[18] F. Wilczek and A. Zee, Phys.Rev.Lett. **51** (1983) 2250

[19] F. Wilczek, Phys.Rev.Lett. **48** (1982) 1144 and **49** (1982) 957

[20] E. Fradkin and M. Stone Phys.Rev. **B38** (1988) 7215

F.D.M. Haldane, Phys.Rev.Lett. **61** (1988) 1029

X.G. Wen and A. Zee, ibid **61** (1988) 1025

D. Domber and N. Read, Phys.Rev. **B38** (1988) 7181

[21] I. Affleck, Nucl.Phys. **B257** (1985) 397, ibid **B265** (1986) 409 and **B305** (1988) 582

[22] N. Read and S. Sachdev, Nucl.Phys. **B316** (1989) 609 and Phys.Rev. **B42** (1990) 4568

[23] I. Affleck and B. Marston, Phys.Rev. **B37** (1989) 3774

A REVIEW OF W STRINGS

Peter West

Department of Mathematics
King's College London
Strand, London WC2R 2LS
&
Institute of Theoretical Physics
Chalmers University of Technology
S-412 96 Göteborg SWEDEN

Abstract

Recent progress on the physical states and scattering amplitudes of the W_3 string is reviewed with particular emphasis on the relation between this string theory and the Ising model.

1. Introduction

I carried out my doctorate studies at Imperial College and had the good fortune to have Abdus Salam as my supervisor. When I began research, the paper of Wess and Zumino [33] had induced many of Europe's leading physicists to work on supersymmetry, and Abdus Salam and John Strathdee [34] had just written their classic paper discovering superspace and super-Feynman rules. With such rapid progress being made it was not easy for a graduate student to achieve anything of significance, but despite his many commitments Abdus Salam was always ready to give helpful advice and encouragement. It was impossible not to be infected by his great enthusiasm for new ideas and the enjoyment he derived from doing physics. One came away from his office feeling that all was possible and that failure was only a temporary phenomenon.

Although it had been understood [35] how to break supersymmetry using the classical potential, it was thought to be more desirable if it could be broken using radiative corrections. Abdus Salam characteristically encouraged Bob Delbourgo and myself to systematically examine every possibility. He also, however, advocated that if all else failed one could always tell the truth. In this case, as I eventually found, the truth was that if supersymmetry was preserved classically then the effective potential vanished [36]. It was this theorem which allowed others [37] to observe that supersymmetry solved the technical hierarchy problem.

The subject of my talk is W_3 strings. Paul Howe and myself, and also Bilal and Gervais, began studying this subject in 1988. The realisation that the W_3 algebra could be used to construct a new string theory was rather exciting, since we thought that a much stronger algebra than the Virasoro algebra must lead to a more exciting string than the bosonic string. After finding some interesting results, we became despondent. There were two reasons for this; firstly the only realisation known consisted of two scalars leaving no room for space-time. Secondly, the construction of the BRST charge by Thierry-Mieg had an intercept 4 which suggested the possibility of massless particles of spin greater than 2 in the theory. We realised, however, that tachyonic particles with spin necessarily have negative norm states, because we can choose our frame of reference such that the non-zero components carry a time-like index. We could not see how to eliminate such particles. No doubt, had we talked to Abdus Salam we would have continued, but in the event we gave up.

We now know that W_3 strings do not involve massless higher spin particles and have available representations involving more than 2 scalars. The message of my talk is that despite their, perhaps, rather artificial appearance W_3 strings show all the magic of ordinary strings, they obey a no ghost theorem, have scattering amplitudes which obey duality and factorisation and they are modular invariant. Also one finds that the Ising model pervades all aspects of the theory.

The discovery, by Zamolodchikov [1], of two dimensional algebras that contain currents of spins greater than two has led to many interesting new developments. The simplest such algebra, W_3 is generated by the spin 2 energy-momentum tensor $T(z) = \sum L_n z^{-n-2}$ and the spin 3 current $W(z) = \sum W_n z^{-n-3}$; the algebra being

$$\begin{aligned}
[L_n, L_m] &= (n-m)L_{n+m} + \frac{c}{12}n(n^2-1)\delta_{n+m,0} , \\
[L_n, W_m] &= (2n-m)W_{n+m} , \\
[W_n, W_m] &= \frac{16}{22+5c}(n-m)\Lambda_{n+m} \\
&\quad + (n-m)\Big[\frac{1}{15}(n+m+2)(n+m+3) - \frac{1}{6}(n+2)(m+2)\Big]L_{m+n} \\
&\quad + \frac{c}{360}n(n^2-1)(n^2-4)\delta_{n+m,0} ,
\end{aligned} \tag{1.1}$$

where

$$\Lambda_n = \sum_m : L_{n+m}L_{-m} : -\frac{1}{20}\Big[n^2 - 4 - \frac{5}{2}(1-(-1)^n)\Big]L_n \tag{1.2}$$

and the normal ordering means that $: L_pL_q := L_qL_p$ if $p > q$.

This algebra occurs naturally in the minimal model with central charge 4/5 since this theory contains a spin 3 primary field. The above algebra, however, is consistent for all values the central charge c. The most important new feature of these new algebras is that they are not Lie algebras; their commutators are expressed as a sum of terms some of which are quadratic in the generators.

An important consequence of this fact is that, unlike the Virasoro algebra, two representations of a W_3 algebra cannot, in general, be added to form a third. The earliest known realisation of the W_3 algebra is in terms of 2 scalars [2]; it is a generalisation of the classical Miura transformation that takes the mKdV equation into the KdV equation. It was found [3], however that one of the scalars only occurred through its energy-momentum tensor and consequently, one could replace it by any energy momentum tensor with the same central charge. In particular, one could use any number of scalar fields, say x^μ for $\mu = 0, 1, ..., D-1$ with a suitable background charge. Denoting the other of the two original scalars by φ, the W_3 generators, which carry a superscript m, corresponding to matter, are given by

$$T^m = T^\varphi + T^x \tag{1.3}$$

where

$$T^\varphi = -\frac{1}{2}(\partial\varphi)^2 - Q\partial^2\varphi, \tag{1.4}$$

$$T^x = -\frac{1}{2}\partial x^\mu \partial x_\mu - \alpha_\mu \partial^2 x^\mu, \tag{1.5}$$

and

$$W^m = -\frac{2i}{\sqrt{261}}\left[\frac{1}{3}(\partial\varphi)^3 + Q\partial\varphi\partial^2\varphi + \frac{1}{3}Q^2\partial^3\varphi + 2\partial\varphi T^x + Q\partial T^x\right], \tag{1.6}$$

It was natural given the existence of the W_3 algebra to try to build new string theories. Although there are, with hindsight, a number of ways of constructing the bosonic string, the path most suited to our present knowledge of the W_3 algebra is as follows: starting from the Virasoro algebra we construct the BRST charge Q and demand its square vanish. This condition implies that c = 26 and the intercept is one [4]. We then find a realization of the Virasoro algebra with c = 26, for example 26 free scalars $x^\mu, \mu = 0, 1, ..., 25$. The physical states are the non-trivial cohomology classes of Q which are also subject to a ghost constraint [4,5,15]. Given the physical states, one can then construct the scattering amplitudes.

The construction of a W_3 string along these lines was first discussed in reference [6] and the W_3 string with the field content given here was first formulated in reference [7]. Fortunately, the BRST charge for the W_3 algebra had previously been constructed [8] and found to square to zero if c = 100 and the intercept was 4. To build this operator one must introduce the usual reparameterization ghosts b and c, and W_3 transformation ghosts d and e which have spins 3 and -2 respectively. The total energy-momentum tensor T^{tot} and spin 3 current W^{tot} of the combined ghost and matter system are of the form

$$T^{tot} = T^m + T^{gh}, \quad W^{tot} = W^m + W^{gh} \tag{1.7}$$

where

$$T^{gh} = -2b\,\partial c - \partial b\,c - 3d\,\partial e - 2\partial d\,e, \tag{1.8}$$

$$\begin{aligned} W^{gh} &= -\partial d\,c - 3d\,\partial c - \frac{8}{9.27}[\partial(b\,e\,T^m) + b\,\partial e\,T^m] \\ &+ \frac{25}{6.9.27}(2e\,\partial^3 b + 9\partial e\,\partial^2 b + 15\partial^2 e\,\partial b + 10\partial^3 e\,b). \end{aligned} \tag{1.9}$$

Here the background charge Q is given by $Q^2 = 49/8$, and α is such that T^x has central charge $51/2$. The BRST charge Q is given by [8]

$$Q = \int dz\, j^{BRST}, \tag{1.10}$$

where

$$j^{BRST} = c(T^m + \frac{1}{2}T^{gh}) + e(W^m + \frac{1}{2}W^{gh}). \tag{1.11}$$

The reader will be aware of the difference between the background charge Q and the BRST charge Q from the different contexts in which they are used. Some useful relations are

$$T^{tot}(z) = \{Q, b(z)\}, \quad W^{tot}(z) = \{Q, d(z)\}, \tag{1.12}$$

as a consequence of which

$$[Q, T^{tot}(z)] = [Q, W^{tot}(z)] = 0. \tag{1.13}$$

It will be useful to discuss the various possible vacua associated with the ghosts. The natural vacuum, $|0\rangle$, of the ghost system is that for which $e(z) = \sum_n e_{-n} z^{n+2}$ and $d(z) = \sum_n d_{-n} z^{n-3}$ are well defined at $z = 0$. This requires [9]

$$\begin{aligned} e_n|0\rangle &= 0, \quad n \geq 3, \quad d_n|0\rangle = 0, \quad n \geq -2 \\ c_n|0\rangle &= 0, \quad n \geq 2, \quad b_n|0\rangle = 0, \quad n \geq -1. \end{aligned} \tag{1.14}$$

We can construct other vacua by acting on $|0\rangle$ with e_n for $n = 0, \pm 1, \pm 2$ and with c_n for $n = 0, \pm 1$. One of the most useful is

$$c_1 e_1 e_2 |0\rangle \equiv |\downarrow\rangle, \tag{1.15}$$

which is annihilated by e_n, c_n for $n \geq 1$ and b_n, d_n for $n \geq 0$. In terms of the conformal fields we may express the relation between the two states as

$$c\,\partial e\, e|0\rangle = |\downarrow\rangle, \tag{1.16}$$

where $c\,\partial e\, e$ is understood to be evaluated at $z = 0$. Similar formulae hold for the other vacuum states.

In order to gain a non-zero vacuum expectation value with respect to the state $|0\rangle$ we must have 3 factors of c and 5 of e. We set

$$\langle 0|c_{-1}c_0c_1e_{-2}e_{-1}e_0e_1e_2|0\rangle = \frac{1}{576}\langle 0|\partial^2 c\,\partial c\, c\partial^4 e\,\partial^3 e\,\partial^2 e\,\partial e\, e|0\rangle = 1 \tag{1.17}$$

A correlator will also vanish unless the sum of the momenta of all the vertex operators in the correlator is $2i$ times background charge.

2. States in effective bosonic strings

In W_3 and probably W_N $N \geq 4$ strings one finds that the physical states belong to only a subspace of the full Fock space of the theory that is similar to the bosonic string. In this section, we will consider what are the physical states that can arise in such effective bosonic string sectors, but rather than base our discussion specifically within the context of W strings we will consider bosonic string-like sectors in the abstract, so making the discussion clearly applicable to any string theory in which they may arise. By an effective bosonic string sector we mean a string, or sector of a string theory, that has D scalar fields x^μ which possess a background charge α^μ. The physical states of the sector belong to the Fock space subspace $\tilde{H}$ generated by α_n^μ and are taken to be subject to the Virasoro-like conditions

$$\tilde{L}_n|\tilde{\psi}\rangle = 0,\ n \geq 1\ \ ,(\tilde{L}_0 - a)|\tilde{\psi}\rangle = 0 \tag{2.1}$$

where the intercept a is to be specified and

$$\tilde{T} = -\frac{1}{2}\partial x^\mu \partial x^\nu \eta_{\mu\nu} - \alpha_\mu \partial^2 x^\mu \tag{2.2}$$

$$\tilde{T}(z) = \sum_n \tilde{L}_n z^{-n-2},\ \ i\partial x^\mu(z) = \sum_n \alpha_n^\mu z^{-n-1} \tag{2.3}$$

We note that the background charge need not, be such that $c = D + 12\alpha^2$ is equal to the critical value of 26.

The physical states in such a theory were found in reference [10] by using an extension of the methods of reference [11] and we now summarise these arguments. The first step is to introduce a set of operators that span $\tilde{H}$, but have sufficiently simple commutation relation properties with $\tilde{L}_n$ as to allow us to solve the constraints of equation (2.1). We begin by defining the operators

$$\mathcal{A}_n^\mu = A_n^\mu - \frac{nk^\mu - 2i\alpha^\mu}{2} F_n \tag{2.4}$$

where

$$A_n^\mu = \oint dz : i\partial x^\mu e^{ink\cdot x} :$$

$$F_n = \oint dz : \frac{k\cdot\partial\partial x}{k\cdot\partial x} e^{ink\cdot x} : \tag{2.5}$$

These will commute with $\tilde{L}_n$ provided $k\cdot(k-2i\alpha) = 0$. If we further take $k^2 = 0$, then they obey the algebra.

$$[\mathcal{A}_n^\mu, \mathcal{A}_m^\nu] = \left(n\eta^{\mu\nu} + 2n^3 k^\mu k^\nu\right) p.k\delta_{n+m,0} + mk^\mu \mathcal{A}_{n+m}^\nu - nk^\nu \mathcal{A}_{n+m}^\mu$$
$$-i\alpha^\mu k^\nu m^2 \delta_{n+m,0} p\cdot k + i\alpha^\nu k^\mu m^2 \delta_{n+m,0} p\cdot k. \tag{2.6}$$

We next define the operators

$$C_n = -\mathcal{A}_n^- - \left\{\frac{1}{2}\sum_p \sum_i : C_{n-p}^i C_p^i : -i(n+1)\alpha\cdot C_n\right\} + 1$$

$$C_n^i = A_n^i + i\alpha^i \delta_{n,0} \tag{2.7}$$

which of course also commute with $\tilde{L}_n$. These operators obey the algebra

$$[C_n^i, C_m^j] = \delta_{n+m,0}\delta_{i,j},\ [C_n^i, C_m] = 0,$$

$$[C_n, C_m] = (n-m)C_{n+m} + (26 - D - 12\alpha^2)\frac{n(n^2-1)}{12}\delta_{n+m,0} \tag{2.8}$$

Since C_n and C_n^i commute with $\tilde{L}_n$, we conclude that they generate physical states which are of the form

$$C_{-m_1}.....C_{-m_q} C_{-n_1}^{i_1}....C_{-n_p}^{i_p}|0,p\rangle \tag{2.9}$$

where $|0,p\rangle$ is the tachyon state and is annihilated by α_n^μ, $n \geq 1$ and p^μ satisfies $p\cdot k = 1$, $\frac{1}{2}p\cdot(p-2i\alpha) = a$.

We add to our collection of oscillators the

$$\phi_n = \oint dzz^{-1}e^{ink.x} \quad . \tag{2.10}$$

These have the relations

$$[\tilde{L}_p, \phi_m] = -p\phi_m \quad , \quad [C_p, \phi_m] = m\phi_{m+p}$$

$$[\phi_n, \phi_m] = 0 \quad , \quad [C_n^i, \phi_p] = 0 \; . \tag{2.11}$$

One can show that the oscillators C_n^i, C_n and ϕ_n do span the same Hilbert space as the original oscillators α_n^μ , $\mu = 0, 1, \ldots, D-1$. It also follows, from the above commutation relations, that the physical states do not contain ϕ_n and so are none other than those of equation (2.9)

We now wish to analyse the physical states. The operators C_{-n} obey a Virasoro algebra which has a central charge $\tilde{c} = 26 - D - 12\alpha^2$ and they act on highest weight states with weight $h = 1 - a$. The norms of these states is therefore controlled by the Kac determinant for the Virasoro-like operators C_n with the above central charge and weight. If $\tilde{c}$ is less than or equal to 1 then the states will only have positive norm if and only if $\tilde{c}$ is a member of the minimal unitary series $\tilde{c} = 1 - \frac{6}{(n+1)(n+2)}$ $n = 3, 4, \ldots$ and the intercepts $a_{r,s}$ are related to the weights of the corresponding minimal unitary series by $a_{r,s} = 1 - h_{r,s}$. The number of physical degrees of freedom is found by discarding those states of zero norm. This is the same as discarding those states which are descendant, but yet also highest weight with respect to the C_n. Consequently, the number of physical degrees of freedom c_n at level n is given by

$$\sum c_n x^n = \prod_{n=1}^{\infty} \frac{1}{(1-x^n)^{D-2}} \hat{\chi}_{h_i}(x). \tag{2.12}$$

The character χ_h is that corresponding to the above central charge and weight and is defined by

$$\chi_h(z) = \sum_{n=0}^{\infty} z^{h-\frac{1}{48}} \mathrm{dim} V_{n+h} = z^{h-\frac{1}{48}} \hat{\chi}_h(z) \tag{2.13}$$

where h is the weight of the highest weight state and V_q is the dimension of the space with weight q.

For a critical bosonic string $c = 26$ and $a = 1$, or equivalently $\tilde{c} = 0$ and $h = 0$, and in this case all the states involving C_{-n} are null and so one finds that the count of states is a light-cone count in the sense that there are excitations for only $D-2$ of the D non-zero modes. When the effective bosonic string, or sector, is not critical one finds that the string does not lose 2 degrees of freedom and only some of the states involving C_{-n} are null. When one is dealing with one of the minimal models, which must be the case if $\tilde{c} \leq 1$ and the states obey a no ghost theorem, the explicit form of the characters is known [12] and so equation (2.12) gives the count of states. In this case, one can then consider the question of whether the string theory is modular invariant. The discussion of reference [10] is for the Ising model case, but it is trivial to extend it to the general case. One finds that the cosmological constant is the product of a factor that is associated with the x^μ oscillators, which is modular invariant by itself and a factor associated with the minimal model. For the theory to be modular invariant this latter factor must also be invariant. Consequently, for every modular invariant of the minimal series [13] one finds a corresponding modular invariant string.

3. The physical states of the W_3 string

Before we begin, it will be useful to recall what are the physical states in the ordinary critical bosonic string constructed from the 26 fields x^μ. These states were found, by studying the original dual model [14], to obey the Virasoro constraints $(L_0 - 1)|\psi^x\rangle = 0$, $L_m|\psi^x\rangle = 0$ $m \geq 1$ where $|\psi^x\rangle$ depends only on x^μ. With the discovery of the free gauge covariant action, it became more common to consider the physical states as belonging to the cohomology of the BRST operator Q, subject to a certain ghost number constraint. We recall that the cohomology of Q [4,15] consists of the states $|\psi^x\rangle|\downarrow\rangle$ and $|\psi^x\rangle c_0|\downarrow\rangle$, where $|\downarrow\rangle = c_1|0\rangle$ and $|0\rangle$ is the $SL(2, \mathbf{R})$ invariant vacuum, as well as two further states with zero momentum, namely $|p^\mu = 0\rangle|0\rangle$ and $|p^\mu = 0\rangle c_{-1}c_0c_1|0\rangle$.

For a W_3 string, it is also natural to regard the physical states as being given by the cohomology of Q, subject to a suitable ghost number constraint. By analogy, we expect physical states to be contained in states of the form $|\psi^{x,\varphi}\rangle|\downarrow\rangle$, where $|\downarrow\rangle = c_1e_1e_2|0\rangle$. Applying Q to these states one finds [6] that they belong to the cohomology of Q if

$$\begin{aligned}(L_0^m - 4)|\psi^{x,\varphi}\rangle =&0, \quad W_0^m|\psi^{x,\varphi}\rangle = 0, \\ L_n^m|\psi^{x,\varphi}\rangle =&0, \quad W_n^m|\psi^{x,\varphi}\rangle = 0, \quad n \geq 1.\end{aligned} \tag{3.1}$$

Included amongst the states satisfying these conditions are physical states having the form [7]

$$|\psi^{x,\varphi}\rangle = |\psi^x\rangle|0,\beta\rangle|\downarrow\rangle, \tag{3.2}$$

where $|0,\beta\rangle$ is a state with φ momentum equal to β and no φ oscillators. Such states will satisfy the conditions (3.1) provided that the state $|\psi^x\rangle$, which depends on x^μ alone, satisfies the conditions

$$L_n^x|\psi^x\rangle = 0, \ n \geq 1, \quad (L_0^x - a)|\psi^x\rangle = 0, \tag{3.3}$$

where $a = 1$ for $\beta = 8iQ/7$ and $6iQ/7$, and $a = 15/16$ for $\beta = iQ$. The above values of β are in fact the only ones allowed by the on-shell conditions of equation (3.1). These states exhibit what is a general phenomenon, namely the freezing of the ϕ momentum by the physical state conditions which was first noticed in reference [7]. The reader may observe that if one takes one minus the above effective intercepts then one gets 0 and 1/16 which remind one of the weights that occur in the Ising model. In references [7] and [16] this, and other phenomenological number matching, was shown to extend to a relation between higher W_N strings and minimal models.

Further analysis of low level states was carried out in references [17]and [18], but it was only in reference [19] that a systematic study of the physical states of the type of equation (3.1) at levels up to and including 2 was undertaken. It was found that any state that contained non-zero mode φ oscillators acting on the vacuum was null. By examining all other null states, the count of physical degrees of freedom at these levels was found. It thus became clear that the non-null physical states based on the ghost state $|\downarrow\rangle$ were of the form of equation (3.2) and that the open W_3 string had only one massless particle, a photon, namely the same massless states as the open bosonic string [19].

It will be useful to rewrite the physical states discussed above in terms of vertex operators acting on the vacuum $|0\rangle$ in the ghost sector as well as the usual vacua for the bosonic oscillators. The states of equation(3.2) correspond to the vertex operators

$$V(1,0) = c\partial e\, e\; e^{i\beta(1;0)\varphi} V^x(1) \tag{3.4}$$

$$\bar{V}(1,0) = \; c\partial e\, e\; e^{i\bar{\beta}(1;0)\varphi} V^x(1), \tag{3.5}$$

where, with further developments in mind, we have introduced the notation $\beta(1;n) = (8-8n)\frac{iQ}{7}$ and $\bar{\beta}(1,n) = (6-8n)\frac{iQ}{7}$, and

$$V(15/16,0) = \; c\partial e\; e\; e^{i\beta(15/16,0)\varphi} V^x(15/16) \tag{3.6}$$

where $\beta(15/16,n) = (7-4n)iQ/7$. Here $V^x(a)$ is any vertex operator, constructed from x^μ alone, that has conformal weight a with respect to $T^x(z)$. The simplest is $V^x(a) = e^{ip\cdot x}$, where $\frac{1}{2}p\cdot(p-2i\alpha) = a$.

For the bosonic string, all the physical states, except for those of fixed momentum, are built on a single state with given ghost number, but this is not the case for the W_3 string. States in the cohomology of Q which are not of the above form were first found in the context of the 2 scalar W_3 string in reference [16] and one such state was found in the 3 scalar W_3 string. It was shown, as discussed below, that the consistency of the W_3 scattering [21] and modular invariance [10] also required further states to those listed above and these additional states are given by [10]

$$V(1/2,0) = \left(c\, e - \frac{i}{\sqrt{522}} \partial e\, e \right)\; e^{i\beta(1/2,0)\varphi} V^x(1/2) \tag{3.7}$$

where $\beta(1/2,m) = (4-8m)\frac{iQ}{7}$.

These states were found by a vanishing null state argument which also led to states in the cohomology of Q that were of the form [10]

$$V(15/16,1) = \; \left(c\, e + \frac{i\partial e\, e}{\sqrt{522}} \right)\; e^{i\beta(15/16,1)\varphi} V^x(15/16). \tag{3.8}$$

and

$$\bar{V}(1/2,0) = \Big(-\frac{4}{3\sqrt{58}} b\, c\, \partial e\, e - \frac{4}{3\sqrt{58}} \partial^2 e\, e + \frac{1}{\sqrt{29}} \partial\varphi \partial e\, e + i\sqrt{2} c\, e\, \partial\varphi - \frac{3i}{2} c \partial e \Big)$$

$$\times e^{i\bar{\beta}(1/2,0)\varphi} V^x(1/2), \tag{3.9}$$

where $\bar{\beta}(1/2, m) = (2 - 8m)\frac{iQ}{7}$. This argument also implied the presence of an infinite number of other such states. The above physical states of the W_3 string of equations (3.4), (3.6) and (3.7) are in effect contained in sectors which are effective bosonic strings with intercepts 1,15/16 and 1/2 and central charge $c = 25 + 1/2$. The states of equations (3.5), (3.8)and (3.9) are also of this form with effective intercepts 1, 15/16 and 1/2 respectively and so are in effect only copies of the previous states. Further such copies were found in references [22,26]

We can now analyse the physical content of these states, using the results of the previous section. These physical states are of the form of equation (2.9) with $\tilde{c} = 1/2$ and highest weights h of 0,1/16 and 1/2 and so the count of states of equation (2.12) has characters which are none other than those of the Ising model and further, these states obey a no ghost theorem in that they have positive norm [10]. We note that given $\tilde{c} = \frac{1}{2}$ these are the only values of the intercepts for which this is true. Thus, although the construction of the W_3 string did not seem to involve the Ising model, we find a deep connection between the states of the two theories.

We can now consider whether the above states are sufficient in number for the cosmological constant to be modular invariant. Since the partition function involves the Ising model characters one might suspect that one finds a modular invariant W_3 string for every modular invariant Ising model. In fact there is only one such Ising model and so only one modular invariant W_3 string. We refer the reader to reference [10] for a detailed discussion.

These facts, and the result, discussed below, that the factorisation of the tree level W_3 scattering amplitudes for these states leads to no new states strongly suggests [10] that the cohomology of Q involves the states of equations (3.4),(3.6) and (3.7),copies of them, and possible discrete states.

Considerable further evidence for this conjecture was provided in reference [23] where it was shown how to generate infinite classes of states belonging to the cohomology of Q. We now summarize this work. The existence of copies suggests that there should be a screening charge of the form

$$\int dw\; e^{i\beta\varphi} f(b, c, d, e, \partial\varphi) \tag{3.10}$$

that commute with Q and takes one copy into another. If we can find such charges that do not involve the field x^μ, then the insertion of these charges into a correlation function will not change the effective space-time interpretation of the correlation function. Indeed three such screening charges do exist [23] and are given by

$$\int dz\ e\ e^{i\beta\varphi} \tag{3.11}$$

for $\beta = 6iQ/7$ or $\beta = 8iQ/7$. and, more importantly for our present purposes, the operator

$$S = \oint dz\{d - \frac{5i}{3\sqrt{58}}\partial b - \frac{2}{3.87}\partial b\, b\, e - \frac{4i}{3}\frac{1}{\sqrt{58}} d\, b\, e\} e^{i\beta^s\varphi} \tag{3.12}$$

where $\beta^s = -2iQ/7$. Since S commutes with Q, it follows that whenever the action of S on a physical state is well-defined and non-zero it will produce another physical state.

Let us now consider when two screening operators have a well defined action on a vertex operator with φ momentum β. Since this has not, to our knowledge, been clearly discussed in the literature let us consider the more general case of two screening operators with φ momentum $\beta_i^s, i = 1, 2$. Since the factors in front of the exponentials are made from either ghosts or $\partial\varphi$ they can only contribute integer powers of the coordinates in the operator product expansion, and consequently do not affect whether or not the action of the screening operators is well defined. Consequently, we must focus our attention on the factors

$$\oint_{C_1} dw_1 \oint_{C_2} dw_2\ e^{i\beta_1^s\varphi(w_1)}\ e^{i\beta_2^s\varphi(w_2)}\ e^{\beta\varphi(z)}$$

$$= \oint_{C_1} dw_1 \oint_{C_2} dw_2\ (w_1 - w_2)^{\beta_1^s\beta_2^s}(w_1 - z)^{\beta_1^s\beta}(w_2 - z)^{\beta_2^s\beta}\ e^{i\beta_1^s\varphi(w_1) + i\beta_2^s\varphi(w_2) + i\beta\varphi(z)} \tag{3.13}$$

Since $|w_1 - z| > |w_2 - z|$ we arrange the w_1 and w_2 contours to be around the point z in such a way that the above condition is satisfied.

Let us consider the substitution (w_1, w_2) to (w_1, τ) given by $(w_2 - z) = \tau(w_1 - z)$. The above condition implies that $|\tau| < 1$, but we also demand that $\tau = 1$ at one and only one point in other words the w_1 and w_2 contours touch one another once. This latter condition ensures that the value of the integral is dependent on only the one place where the contours cross the branch cut. We are to regard w_2 as fixed and consider the τ integration. Substituting $dw_2 = d\tau(w_1 - z)$ we find the above integrals become

$$\oint_z dw_1 \oint d\tau \tau^{\beta_1^s \beta}(1-\tau)^{\beta_1^s \beta_2^s}(w_1 - z)^P \\ \exp(i\beta_1^s \varphi(z + (w_1 - z)) + i\beta_2^s \varphi(z + \tau(w_1 - z)) + i\beta\varphi(z)) \tag{3.14}$$

where $P = 1 + \beta_1^s \beta + \beta_1^s \beta_2^s + \beta_2^s \beta$. This integral is well defined if P is an integer.

The generalization to n screening charges with φ momenta β_i^s is straightforward. Their action contains the term

$$\prod_{i=1}^{n} \left(\oint dw_i \, e^{i\beta_i^s(w_i)} \right) e^{i\beta\varphi(z)} = \\ \oint dw_i \prod_{\substack{i<j \\ i,j=1}}^{n} (w_i - w_j)^{\beta_i^s \beta_j^s} \prod_{j=1}^{n} (w_j - z)^{\beta_j^s \beta} \exp i\{\sum_{j=1}^{n} \beta_j^s \varphi(w_j) + \beta\varphi(z)\} \tag{3.15}$$

We now replace w_j, $j = 2, \ldots, n$ by w_1, τ_j $j = 2, ..., n$ using the formula $(w_j - z) = \tau_j(w_1 - z)$. The above expression becomes

$$\oint_z dw_1 \prod_{i=1}^{n-1} \oint d\tau_i \, f(\tau_i) \, (w_1 - z)^{P'} \; \exp i \left\{ \sum_{j=1}^{n} \beta_j^s \varphi(z + \tau_j(w_n - z)) + \beta\varphi(z) \right\} \tag{3.16}$$

where $f(\tau_i)$ is a function of τ_i and

$$P' = (n-1) + \sum_{\substack{i,j=1 \\ i<j}}^{N} \beta_i^s \beta_j^s + \sum_{j=1}^{N} \beta_j^s \beta. \tag{3.17}$$

Thus the integrals are well defined if P' is an integer.

Let us apply this general discussion to the case of interest, namely the action of n screening charges S with momentum $\beta^s = -2iQ/7$. Taking $\beta = isQ/7$ with s an integer, their action is well defined if

$$\frac{n}{4}[-(n-1) + s] \in \mathbb{Z}, \tag{3.18}$$

and then the momentum of the resulting vertex is

$$\frac{iQ}{7}(s - 2n). \tag{3.19}$$

For the action of screening charges on $V(15/16,0)$ we have $s = 7$. We then find that n must be even and that the vertices so constructed have momenta $\beta(15/16,m) = \frac{iQ}{7}(7-4m)$, $m \in \mathbb{Z}_+$. For the action on $V(1,0)$, for which $s = 8$, we find that $n = 4m$ or $4m+1$ with $m \in \mathbb{Z}_+$, which leads to the vertices with momenta $\beta(1,m) = (8-8m)\frac{iQ}{7}$, $m \in \mathbb{Z}_+$, and $\bar{\beta}(1,m) = (6-8m)\frac{iQ}{7}$, $m \in \mathbb{Z}_+$.

Finally, the action of n screening charges on $V(1/2,0)$ is well defined if $n = 4m$ or $4m+1$ with $m \in \mathbb{Z}_+$, which leads to vertices with momenta $\beta(1/2,m) = \frac{iQ}{7}(4-8m)$ and $\bar{\beta}(1/2,m) = \frac{iQ}{7}(2-8m)$.

It can, and does happen, that applying S's alone leads to a vanishing result. This can be avoided, however, by the judicious use of the picture changing operator P which is of the form $P = [a \cdot x + \varphi, Q]$; this operator was first used in the context of W_3 strings in reference [22].

The general pattern is as follows; in the intercept 15/16 sector we have the vertices

$$V(15/16,m) = (S^2P)^m \, V(15/16,0) \tag{3.20}$$

while for the intercept $a = 1$ and $1/2$ vertices, we have the vertices

$$V(a,m) \quad \text{and} \quad \bar{V}(a,m) \tag{3.21}$$

These latter vertices are defined by the relations

$$\bar{V}(a,m) = SPV(a,m), \tag{3.22}$$

$$V(a,m) = S^3P\bar{V}(a,m-1). \tag{3.23}$$

To summarise, we have found [23] that given the basic vertices $V(a,0)$ for $a =$ 1, 1/2 and 15/16, we can use S and P to create the BRST invariant vertices $V(a,m)$; for $a = 1, 1/2$ we obtain in addition the vertices $\bar{V}(a,m)$. We can also obtain further vertices by the action of P on these. Since the physical states for a given intercept have a spectrum generating algebra involving the operators $C_n^i, i = 1, \ldots D-2$ and C_n the states in the cohomology of Q are generated by C_n^i, C_n, S and P.

A number of examples of this procedure were given in reference [23], but here, we give only one example relating the first two copies in the intercept 1/2 sector. A short calculation shows that although $SV(1/2,0)$ is well defined it vanishes. We therefore introduce further powers of the ghost e into the vertex using the picture changing operator. We then have

$$\begin{aligned} PV(1/2,0) =& \big(5\,\partial^2 e\, e - \frac{24Q}{7}\partial\varphi\partial e\, e \\ & - \frac{19i}{3\sqrt{58}}\partial^2 e\,\partial e\, e\big) e^{i\beta(1/2,0)\varphi} V^x(1/2), \end{aligned} \tag{3.24}$$

and acting on this with the screening operator we find the vertex $\bar{V}(1/2,0)$,

$$SPV(1/2,0) \propto \bar{V}(1/2,0). \tag{3.25}$$

4. W_3 String Scattering

The scattering, at tree level, of W_3 strings was first found using the group theoretic method [24]. It was found that these scattering amplitudes contained within them the Ising model correlation functions, in particular the scattering amplitude of N W_3 string states is given by [21]

$$\int \prod_i{}' dz_i\, V f(z_i). \tag{4.1}$$

Here V is the usual scattering vertex in the presence of a background charge, and f, the measure, is an Ising model correlation function that depends on the intercepts of the external states. To be specific, if the N external states have effective intercepts a_i, $\quad i = 1, ..., N$, which can take only the values 1, 15/16 or 1/2, then $f = \langle \prod_{i=1}^{N} \varphi_i(z_i) \rangle$ where φ_i is the Ising field of weight $h_i = 1 - a_i$.

The essential steps, using the group theoretic method to calculate a scattering amplitude are first the computation of the vertex V using overlap relations, and then the determination of the measure f by demanding that null states decouple. Using this technique it was possible to work with the reduced subspace of the full W_3 Fock space in which the physical states sit. It is important to understand that the properties of the vertices and null states used in this reduced Hilbert space are those inherited from the full Fock space of the W_3 string. We found that the decoupling of the null states of the W_3 string implied that f obeyed the differential equations satisfied by the Ising model. The reader is referred to reference [21] for the details of this derivation.

It is straightforward to evaluate the W_3 string scattering given by equation (3.1) whenever the Ising model correlation functions are known. The vertex V can be found in reference [21] for any states of the W_3 string, but for tachyons it has the simple expression

$$\prod_{i<j}(z^i - z^j)^{2\alpha' p_i \cdot p_j} \tag{4.2}$$

For four tachyon scattering, with the choice of Koba-Nielsen coordinates $z_1 = \infty$, $z_2 = 1$, $z_3 = x$ and $z_4 = 0$, this reduces to

$$x^{-\alpha' s - a_3 - a_4}(1-x)^{-\alpha' t - a_2 - a_3}, \tag{4.3}$$

where $s = -(p_1 + p_2)\cdot(p_1 + p_2 - 2i\alpha)$, $t = -(p_2 + p_3)\cdot(p_2 + p_3 - 2i\alpha)$ and where a factor of $(z_1)^{-2a_1}$, which is cancelled by other such factors in the amplitude, has been removed.

Let us give one example of the evaluation of such amplitudes, namely the case of four intercept 15/16 tachyonic W_3 strings, whose subsequent study will be instructive.

$$F(s,t) = \int_0^1 dx x^{-\alpha' s - 15/8}(1-x)^{-\alpha' t - 15/8} f(x) z_1^{1/8} \tag{4.4}.$$

The function $f(x)$ is the Ising correlation function for 4 weight 1/16 fields which is of the form [25]

$$f(x) = [(z_1 - z_3)(z_2 - z_4)]^{-1/8} Y(x) \tag{4.5},$$

where

$$x = \frac{(z_1 - z_2)(z_3 - z_4)}{(z_1 - z_3)(z_2 - z_4)} \tag{4.6}$$

and

$$Y(x) = \frac{1}{[x(1-x)]^{1/8}}(a\cos\theta + b\sin\theta), \tag{4.7}$$

with $x = \sin^2\theta$. The two constants a and b correspond to the two solutions to the second order differential equation that this correlation function satisfies due to the existence of null states in the Ising model. Taking the above choice of z's we find that

$$F(s,t) = \int_0^1 dx x^{-\alpha' s - 2}(1-x)^{-\alpha' t - 2}(a\cos\theta + b\sin\theta). \tag{4.8}$$

In the above we have introduced the slope α', which is usually taken to be 1/2 for the open string.

The values of the constants a and b are to be chosen by the physical requirement of crossing. The open string amplitude $T^{(4)}(p_i)$ is as usual the sum of three terms

$$T^{(4)}(p_i) = F(s,t) + F(t,u) + F(u,s).$$

Crossing for four identical particles means that $T^{(4)}(p_i)$ should be symmetric under the exchange of any two legs or equivalently momenta. This means, for example, that it should be symmetric under $s \leftrightarrow t$, which follows provided that F itself is a symmetric function of its arguments. This property is in turn guaranteed if the integrand is symmetric under $x \leftrightarrow 1-x$, provided, at the same time, we interchange s and t, or p_2 and p_4. The transformation $x \to 1-x$ can be written as $\theta \to \pi/2 - \theta$, whereupon it is obvious that we should take

$$\begin{aligned} F(s,t) &= \int_0^1 dx x^{-\alpha' s-2}(1-x)^{-\alpha' t-2}\cos(\theta/2 - \pi/8) \\ &= \frac{1}{\sqrt{2}}\int_0^1 dx x^{-\alpha' s-2}(1-x)^{-\alpha' t-2}\{\cos\pi/8\sqrt{1+\sqrt{1+x}} \\ &+ \sin\pi/8\sqrt{1-\sqrt{1-x}}\}. \end{aligned} \tag{4.9}$$

Having found the expression for four tachyon scattering we can examine the particles exchanged in a given channel. It is clear from its integral reresentation that the amplitude obeys duality, that is $F(s,t)$ can be expressed as a sum of poles in either the s-channel or the t-channel. Let us consider the s-channel. Expanding the factor $(1-x)^{-\alpha' t-2}$ we find that

$$\begin{aligned} F(s,t) &= \sum_{n=0}^{\infty}\sum_{p=0}^{\infty}\cos\frac{\pi}{8}\sqrt{2}a_p\frac{(\alpha' t+2)(\alpha' t+3)\ldots,\alpha' t+n+1)}{(-\alpha' s-1+p+n)} \\ &+ \sum_{n=0}^{\infty}\sum_{p=0}^{\infty}\sin\frac{\pi}{8}\frac{b_p}{\sqrt{2}}\frac{(\alpha' t+2)(\alpha' t+3)\ldots,\alpha' t+n+1)}{(-\alpha' s-1+p+n+1/2)}, \end{aligned} \tag{4.10}$$

where

$$\sqrt{1+\sqrt{1+x}} = \sqrt{2}\sum_{p=0}^{\infty} a_p x^p \tag{4.11}$$

and

$$\sqrt{1-\sqrt{1-x}} = \sqrt{\frac{x}{2}} \sum_{p=0}^{\infty} b_p x^p. \tag{4.12}$$

The states exchanged in the first and second terms have masses which satisfy $\alpha' m^2 = p + n - 1$ and $\alpha' m^2 = p + n - 1/2$ respectively. While the former states are contained in the intercept 1 sector the latter are the intercept 1/2 sector. Consequently, we find, as discussed above, that factorisation of the W_3 string scattering requires the existence of intercept 1/2 states in its spectrum in addition to those of intercept 1 and 15/16 sectors. The fact that the W_3 scattering involves Ising correlations includes the 3 point function. The form of the 3 point functions of the Ising model, as for any conformal model, are well known to be equivalent to the fusion rules. Consequently, the fusion rule $\sigma\sigma = 1 + \epsilon$, where σ and ϵ are the Ising fields of weight 1/16 and 1/2 respectively implies the existence of intecept 1/2 states in the W_3 string.

Following the general tree-level calculations of W_3 string scattering amplitudes given in [21], covariant approaches to W_3 string scattering have been considered. The first such attempt [22,26] involved calculations of correlation functions using only vertex and picture-changing operators. The resulting scattering amplitudes disagreed with those of reference [21] and did not share the connection with Ising model correlation functions found in [21]; in addition they violated general principles of S-matrix theory and string theory. The problem was most readily apparent when considering the scattering of 4 intercept 15/16 strings considered above. It is easy to see that the vertices with this intercept can never be combined in a correlator so as to balance the φ background charge $2iQ$. As a consequence, the authors of refences [22,26] concluded that this amplitude and many like it vanished. The intercept 15/16 amplitude given above, did however factorise properly [21] and it is clear that this amplitude cannot vanish and be consistent with factorisation i.e duality [29]. At a more elementary level it is clear that the vanishing of such a four point function is inconsistent with the optical theorem and its generalisations[27,28]

The use of only vertex and picture-changing operators to calculate W_3 covariant scattering amplitudes was a straightforward generalization of previous covariant string scattering calculations for the bosonic and superstrings [9]; in view of the expected role of W-moduli it is perhaps not surprising that this simple generalization is not adequate for W-strings.

In reference [23], a general covariant procedure for calculating W_3-scattering was given. It involved using not only the vertices $V(a,0)$ and the picture changing operator P, but also the screening charge S in the computation of scattering amplitudes. This method led to amplitudes in agreement with those found earlier [21] and which satisfied the general principles expected. We now summarize this approach: the W_3 string scattering amplitudes are to be constructed from the building blocks

$$V(a,0)\ ;\ a=1,15/16,1/2\ ;\ S\ ,\ P \tag{4.13}$$

and the operation

$$\int dz \oint_z dv\, b(v). \tag{4.14}$$

The latter is a standard operation used for the $b-c$ ghost system. We must, however, assemble the building blocks so that the φ momentum sums to $2iQ$. This tells us the required number N_s of screening charges. We must also have, after carrying out all the operator product expansions, 3 c ghosts and 5 e ghosts, otherwise the correlator will vanish. As usual, we require for an N string amplitude $N-3$ of the operators of equation (4.14). The blocks $V(a,0)$ with $a=1,15/16$, $V(1/2,0)$, S and P have ghosts number 3, 2, -1 and 1 respectively. The ghost number requirement gives us the number N_p of picture changing operators P. To be precise if we have a scattering of N_1 intercept 1, $N_{15/16}$ intercept 15/16 and $N_{1/2}$ intercept 1/2 strings, then φ momentum conservation demands that

$$8N_1+7N_{15/16}+4N_{1/2}-2N_P=14 \tag{4.15}$$

while the ghost number count yields the relations

$$3N_1+3N_{15/16}+2N_{1/2}-N_S+N_P-(N_1+N_{15/16}+N_{1/2}-3)=8 \tag{4.16}$$

These equations imply that

$$\begin{aligned} N_S&=4\ N_1+\frac{7}{2}N_{15/16}+2N_{1/2}-7\\ N_P&=2\ N_1+\frac{3}{2}N_{15/16}+N_{1/2}-2 \end{aligned} \tag{4.17}$$

To be concrete let us consider the scattering of 4 intercept 15/16 tachyonic strings. The above equations tell us that we require 7 factors of S and 4 factors of P. One way to distribute these factors in the correlator is as follows

$$\begin{aligned}
&\langle 0|PV(15/16,0)(z_1)V(15/16,1)(z_2) \\
&\quad \oint dz_3 \oint_{z_3} dv\, b(v) V(15/16,1)(z_3)V(15/16,1)(z_4)S|0\rangle \\
&= \oint dz_3 \int dw\, \langle 0| \left(c\,\partial^2 e\, \partial e\, e\; e^{i\beta(15/16,0)\varphi} V^x(15/16)\right)(z_1) \\
&\quad \left(c\,e\; e^{i\beta(15/16,1)\varphi} V^x(15/16)\right)(z_2) \left(e\; e^{i\beta(15/16,1)\varphi} V^x(15/16)\right)(z_3) \\
&\quad \left(c\,e\; e^{i\beta(15/16,1)\varphi} V^x(15/16)\right)(z_4) \\
&\quad \left((d - \frac{5i}{3\sqrt{58}}\partial b - \frac{2}{3.87}\partial b\, b\, e - \frac{4i}{3}\frac{1}{\sqrt{58}} d\, b\, e) e^{i\beta^s\varphi}\right)(w)|0\rangle \\
&= -\oint dz_3 \int dw \langle 0| \left(\prod_{\substack{i=1\\ i\neq 3}}^{4} c(z_i)\right) \left(\partial^2 e\, \partial e\, e\right)(z_1) e(z_2) e(z_3) e(z_4) d(w) \\
&\quad e^{i\beta(15/16,0)\varphi}(z_1) \prod_{i=2}^{4} e^{i\beta(15/16,1)\varphi}(z_i) e^{i\beta^s\varphi}(w) \prod_{i=1}^{4} V^x(15/16)(z_i)|0\rangle
\end{aligned} \tag{4.18}$$

By bosonizing the ghosts we find the c ghosts give a factor

$$(z_1 - z_2)(z_1 - z_3)(z_2 - z_3) \tag{4.19}$$

and the e-d ghosts a factor

$$(z_1 - w) \prod_{i=2}^{4} (z_1 - z_i)^3 (z_i - w)^{-1} \prod_{\substack{i,j=2\\ i<j}}^{4} (z_i - z_j). \tag{4.20}$$

Evaluating the exponential factors in the usual way we find that the amplitude for four intercept 15/16 tachyonic states is proportional to

$$\begin{aligned}
&\int dw \int dx\, x(1-x)^{-1/8} [w(1-w)(x-w)]^{-\frac{1}{4}} (1-x)^{p_2\cdot p_3} x^{p_3\cdot p_4} \\
&= \int dw \int dx\; x^{s/2-2}(1-x)^{t/2-2}[(1-w)(x-w)w]^{-\frac{1}{4}}
\end{aligned} \tag{4.21}$$

where we have chosen $z_1 = \infty$, $z_2 = 1$, $z_3 = x$ and $z_4 = 0$. This expression can be shown to agree with that given in equation(4.8), thus agreeing with the result of reference [21].

Let us now consider the general scattering amplitude and how we must distribute all the factors of P and S, whose number is specified in equation (4.17), among the vertices so as to gain a non-zero result. To be concrete, let us consider the scattering of $N_{\frac{15}{16}} = 2n \geq 6$ intercept 15/16 states, in which case $N_S = 7(n-1)$ and $N_P = 3n-2$. One way to do this is to take $n-3$ of the vertices $V(15/16,2) = (S^2P)^2V(15/16,0)$ and $n+3$ of the vertices $V(15/16,1) = S^2PV(15/16,0)$. This leaves over one P factor and $n-1$ S factors and so leads to the correlator

$$\begin{aligned} &\langle 0| \prod_{i=4}^{2n} \left\{ \oint dz_i \oint_{z_i} dv_i \, b(v_i) \right\} \prod_{i=1}^{n+2} V(15/16,1)(z_i) \\ &P\,V(15/16,1)(z_{n+3}) \prod_{j=n4}^{2n} V(15/16,2)(z_j) S^{n-1}|0\rangle \end{aligned} \tag{4.22}$$

We have chosen to assign the final P factor to the $n+3$ vertex, but any other vertex is just as good. We could also use two of the final screening charges on the $PV(15/16,1)$ vertex to yield the correlator

$$\begin{aligned} &\langle 0| \prod_{i=4}^{2n} \left\{ \oint dz_i \oint_{z_i} dv_i \, b(v_i) \right\} \prod_{i=1}^{n+2} V(15/16,1)(z_i) \\ &\prod_{j=n+3}^{2n} V(15/16,2)(z_j) S^{n-3}|0\rangle \end{aligned} \tag{4.23}$$

Clearly there are many ways to write such a correlation using also the vertices $V(15/16,m), m > 2$.

For the scattering of $N_{1/2} = 2n$ intercept 1/2 states, $N_S = 4n-7$ and $N_P = 2(n-1)$, we can assign all the factors of S to the vertices, for example by the choice of $n-2$ of the vertices $V(1/2,1)$, one of $\bar{V}(1/2,0)$ and $n+1$ of the vertices $V(1/2,0)$ and then place on one of these vertices the required extra P. The correlator is then

$$\begin{aligned} &\langle 0| \prod_{i=4}^{2n} \left\{ \int dz_i \oint dv_i \, b(v_i) \right\} \prod_{i=1}^{n+1} V(1/2,0)(z_i) \\ &\prod_{j=n+2}^{2n-1} V(1/2,1)(z_j)\; P\bar{V}(1/2,0)(z_{2n})|0\rangle \end{aligned} \tag{4.24}$$

Finally for N_1 intercept 1 states, $N_S = 4N_1 - 7$ and $N_P = 2N_1 - 2$, which we can assign as $N_1 - 2$ vertices $V(1,1)$, one of $V(1,0)$ and one of $\bar{V}(1,0)$, with one additional factor of P to give the correlator

$$\langle 0| \prod_{i=4}^{N_1} \int dz_i \oint_{z_i} dv_i\, b(v_i) \prod_{j=1}^{N_1-2} V(1;1)(z_j)$$
$$.P\ V(1,0)(z_{N_1-1})\bar{V}(1,0)(z_{N_1})\ |0\rangle \tag{4.25}$$

The many ways of constructing the above correlators lead to the same results. The correlator is independent of the place where the picture changing operator is applied, since

$$P(z_1) - P(z_2) = [Q, \varphi(z_1) - \varphi(z_2)] = [Q, \int_{z_1}^{z_2} \partial\varphi], \tag{4.26}$$

and this is a BRST trivial operator as it contains $\partial\varphi$ and not φ. Further, we can change on which vertex the screening charges act by deforming the w-contours. There are also different choices for the contours of the residual screening charges and these lead to different results. As explained above, in the context of the 4 intercept 15/16 scattering, we require these different solutions since only a particular combination of the contours gives a crossing symmetric amplitude.

5. Discussion

Let us begin by summarising some of the main results. The physical states of the W_3 string belong to three effective bosonic string sub-sectors of the theory which have intercepts $a = 1$, $1/2$ and $15/16$ and an effective central charge of $25 + \frac{1}{2}$. The count of physical degrees of freedom in the sub-sector with intercept a involves the Ising model character, χ_a where $a = 1 - h$. The cohomology of the BRST charge Q contains an infinite number of copies of the above states at different ghost numbers and φ momenta. There also exists strong evidence that, apart from possible discrete states, these are the only states in the cohomology of Q.

The scattering of W_3 stings, at tree level, has been found to contain, as part of its integrand, a factor that is none other than the correlation functions of the Ising model. This result first emerged from an application [21] of the group theoretic approach to string theory, but it can also be found using a gauge covariant approach [23].

It would be interesting to give a path-integral derivation of these scattering results. This would require a knowledge of W-moduli. Any such derivation would, however, have to reproduce the above results, and this could provide a clue to our understanding of W-moduli. We observe that the number of W-moduli is $2N-5$ for the scattering of N strings, and this number emerges from our results in the guise of $N_S - N_P + N_{1/2}$.

The W_3 string shows much of the same magic as the bosonic and superstrings; it obeys a no ghost theorem, it is modular invariant, and the scattering amplitudes satisfy duality and factorisation. Perhaps the most spectacular property of the W_3 string is its connection with the Ising model. This connection was first noticed on the basis of phenomological number matching [7,16], but it also occurs in the count of physical degrees of freedom [10] and the scattering amplitudes [21,23] and as such it pervade all aspects of the W_3 strings. One can wonder how the W_3 string differs from a non-critical string constructed from the Ising model and D-1 scalar fields x^μ. In the latter theory, the Liouville field will dress the vertices and, as first noticed in reference [7], can be identified with the remaining scalar. If we adopt the currently perceived wisdom [30], then the scattering amplitudes for such a theory will be constructed from such dressed vertices and so will agree with those found for the W_3 string.

Given this correspondence, one is entitled to ask what is the role of the W symmetry when the model can seemingly be formulated without it. Presumably, although one finds the Ising model in the W_3 string one cannot find all possible tensored Ising models by starting in this way, while for higher W_N strings this statement would extend to tensor products of the unitary minimal models. One intriguing possibility is that only those strings constructed from minimal models that can also be found from W, or some other larger symmetry, are really consistent.

It is interesting to note that the non-perturbative results that emerge from the matrix model approach to two dimensional quantum gravity do involve W type constraints on the square root of the partition function [38]. Indeed, the Ising model arises from a two matrix model which involves W_3 constraints and the unitary minimal models which should occur in W_N strings arises from an $N-1$ matrix model that involves W_N constraints. It would be good to understand the connections between these results.

The Ising model is usually realised by a Feigin-Fuchs construction [31] involving only one scalar field, but the W_3 string involves an Ising model constructed from two scalars, the field φ and the bosonised d, e ghosts. After this talk was given, it was shown [32] that this realisation of the Ising model involves parafermions, which play a crucial role in determining the properties of the W_3 string.

Acknowledgement

I wish to thank Mike Freeman, with whom many of the results discussed here were found, as well as Lars Brink, Nathan Berkovits, Bengt Nilsson and Christian Preitschopf for discussions.

References

[1] A. B. Zamolodchikov, Theor. Math. Phys 65 (1989) 1205; V.A. Fateev and S. K. Lukyanov, Int. J. Mod. Phys. A3 (1988) 507.

[2] V. A. Fateev and A. B. Zamolodchikov, Nucl. Phys. B280 [FS18] (1987) 644.

[3] L. J. Romans, Nucl. Phys B352 (1991) 829.

[4] M. Kato and K. Ogawa, Nucl. Phys. B212 (1983) 443; S. Hwang, Phys. Rev. D10 (1983) 2614.

[5] A. Neveu, H. Nicolai and P. West; Phys. Lett. 175B (1986) 307

[6] A. Bilal and J. L. Gervais, Nucl. Phys. B326 (1989) 222. P. Howe and P. West, unpublished

[7] S. Das, A. Dhar and S. Kalyana Rama, Mod. Phys. Lett. A6 (1991); Mod. Phys. Lett. 269B (1991) 167; Int. J. Mod. Phys. A7 (1992) 2295.

[8] J. Thierry-Mieg, Phys. Lett. B197 (1987) 368..

[9] For a review, see D. Lust and S. Theisen, "Lectures on string theory," Springer-Verlag 1989.

[10] P. West, Int. J. Mod. Phys. **A8** (1993) 2875.

[11] E. Del Giudice, P. Di Vecchia and S. Fubini; Ann. Phys. 70 (1972) 378; R. C. Brower and P. Goddard; Nucl. Phys. B40 (1972) 437; R. C. Brower; Phys. Rev. D6 (1972) 1655; P. Goddard and C. Thorn; Phys. Lett. 40B (1972) 235; C. B. Thorn; Nucl. Phys. B248 (1984) 551.

[12] A. Rocha-Caridi; In Vertex Operations in Mathematics and Physics; Lepowsky, J et al (eds), M.S.R.I. publication No.3 p451, Springer-Verlag (1984).

[13] A. Capelli, C. Itzykson and J. B. Zuber; Nucl. Phys. B280 [FS18] (1987) 445.

[14] E. del Giudice and P. Di Vecchia, Nuovo Cimento 5A (1971) 90.

[15] M. D. Freeman and D. I. Olive, Phys. Lett. B 175 (1986) 151.

[16] S. Kalyana Rama, Mod. Phys. Lett **A6** (1991)3531.

[17] C. N. Pope, L. J. Romans, K. S. Stelle; Phys. Lett. 268B (1991) 167, 269B (1991) 287.

[18] C.N. Pope, S. Schrans and K. W. Xu, Texas A & M preprint CTP TAMU-5/92.

[19] H. Lu, B. E. W. Nilsson, C. N. Pope, K. S. Stelle and P. West; Int. J. Mod. Phys. **A8** (1993) 4071.

[20] C. N. Pope, E. Sezgin, K. S. Stelle and X. J. Wang, "Discrete states in the W_3 string," CTP-TAMU-64/92, Imperial/TP/91-92/40.

[21] M. Freeman and P. West; Phys. Lett. **B299** (1993) 30.

[22] H. Lu, C. N. Pope, S. Schrans and X-J. Wang, "The interacting W_3 string," CTP-TAMU-86/92.

[23] M. Freeman and P. West; Int. J. Mod. Phys. **A8** (1993) 4261.

[24] For a review see P. West, Nucl. Phys. B (Proc. Suppl) 5B (1988) 217, Nucl. Phys. B320 (1989) 103; M. Freeman and P. West, Phys. Lett. 205B (1988) 30. A. Neveu and P. West, Comm. Math. Phys. 114 (1988) 613.

[25] A. Luther and I. Peschel, Phys. Rev. B12 (1975) 3908; A. Belavin, A. Polyakov and A. Zamolodchikov, Nucl. Phys. B241 (1984) 333. For a review, see C. Itzykson and M. Drouffe, "Statistical Field Theory," Cambridge University Press.

[26] H. Lu, C. N. Pope, S. Schrans and X-J. Wang, "On the spectrum and scattering of W_3 strings," CTP-TAMU-4/93.

[27] See, for example, C. Itzykson and J-B. Zuber, "Quantum Field Theory," McGraw Hill.

[28] R. Eden, P. Landshoff, D. Olive and J. Polkinghorne, "The Analytic S-matrix," Cambridge University Press.

[29] R. Dolan, D. Horn and C. Schmid, Phys. Rev. Lett. 19 (1967) 402; Phys. Rev. Lett. 166 (1968) 1768.

[30] For a review, see N. Ohta, "Discrete States in Two Dimensional Quantum Gravity", preprint OS-GE-92, and references therein.

[31] F. L. Feigin and D. B. Fuchs, Moscow preprint 1983. Vl. S. Dotsenko and V. A. Fateev, Nucl. Phys. B240 (1984) 312

[32] M. Freeman and P. West "W_3, Phys. Lett 314B (1993) 320.

[33] J. Wess and B. Zumino, Nucl. Phys. B70 (1974)139.

[34] A. Salam and J. Strathdee, Phys. Rev. D11 (1975) 1521, Nucl. Phys. B86 (1975)142.

[35] P. Fayet and J. Illiopoulos Phys. Lett 51B (1974) 461, L. O'Raifeartaigh, Nucl. Phys. B96 (1975) 331, P. Fayet, Phys. Lett. 58B (1975) 67

[36] P. West Nucl. Phys. B106 (1976) 219.

[37] E. Witten, Nucl. Phys. B186 (1981) 150, S. Dimopoulos and H. Georgi, Nucl. Phys. B193 (1981) 150, N. Sakai, Z. Physik C11 (1982) 153, R. Kaul Phys. Lett. B109 (1982) 19.

[38] R. Dijkgraaf, E. Verlinde and H. Verlinde Comm. Math. Phys. 123 (1989) 485, M. Fukuma, H. Kawai and R. Nakayama, Int. J. mod Phys A6 (1991) 1385, J. Goeree, Nucl. Phys. B358 (1991) 737.

Aspects of κ–symmetry

E. Sezgin *

Center for Theoretical Physics, Texas A&M University
College Station, Texas 77843-4242, USA

ABSTRACT

We review various aspects of a fermionic gauge symmetry, known as the κ–symmetry, which plays an important role in formulations of superstrings, supermembranes and higher dimensional extended objects. We also review some aspects of the connection between κ–symmetric theories and their twistor-like formulations.

1. Introduction

κ–symmetry [1] is a remarkable symmetry which plays an important role in the manifestly spacetime supersymmetric formulation of string theories [2]. It is also crucial for the existence of super p–brane theories [3][4][5]. While for $p = 1$ (string) there exists a formulation which does not have manifest spacetime supersymmetry but has a world-sheet local supersymmetry (the NSR formulation), such a formulation is forbiddingly complicated for higher super p–branes.

κ–symmetry is a fermionic symmetry, and in most models where it occurs, interestingly enough, a gauge field is not necessary for its realization. Furthermore, κ–symmetry is typically an on-shell symmetry and consequently, there is no simple way to construct higher order κ–invariants. Whether κ–symmetry is powerful enough to improve dramatically the finiteness behaviour of higher super p–branes is probably one of the most interesting and important open questions. A formidable obstacle to answering this question is the absence of a well defined covariant quantization scheme for κ–symmetry and this is basically due to the fact that the BRST quantization of the symmetry requires an infinite number of ghosts [6].

The aim of this relatively short review is to outline some of the salient features of κ–symmetric field theories, and some of the outstanding open problems. This article is based on a talk given at the conference in honor of Professor Abdus Salam. It is a great

* Supported in part by the National Science Foundation, under grant PHY-9106593.

pleasure to make this contribution. Professor Abdus Salam has made so many important contributions in so many areas that I did not have any difficulty in choosing a subject which owes greatly to his work. As we will see below, κ-symmetric field theories are essentially sigma models formulated in a superspace, and the background fields occurring in these models are described by superfields. While these models have a very simple and elegant geometric form in superspace [7][8], it would be a nightmare to describe them in component formalism, in presence of a curved background. The notions of superspace and superfields, which play such a significant role in these theories, were invented by Salam and Strathdee in 1974 [9]. The concluding remark in their paper is rather interesting and it reflects their admirable approach to doing physics and their humility even when they present an important result [9]:

"The approach discussed in this paper [the superspace approach] *may not provide the most serviceable one available but, with the present hazy understanding of this curious and potentially important symmetry* [supersymmetry], *it seems worthwhile to examine every avenue".*

2. Massless Superparticle in Curved Superspace

As mentioned above, κ-symmetric actions are basically sigma models in which the target space is a superspace. We take it to be a Salam-Strathdee type superspace which can be viewed as the supercoset G/H where G is the super Poincaré group and H is the Lorentz group in d dimensions [9]. It would be very interesting to consider more general supermanifolds. Focusing on the usual scenario, we consider a target superspace with coordinates $Z^M = (X^m, \theta^\mu)$. It is useful to define the pull-back supervielbein as

$$E_\tau^A = \partial_\tau Z^M E_M{}^A \ , \tag{2.1}$$

where ∂_τ denotes differentiation with respect to the worldline time variable. The tangent space index splits as $A = (a, \alpha)$, where $a = 0, 1, ..., d-1$ labels a Lorentz vector and α labels the fundamental representation of the Lorentz group $\times$ the automorphism group of the super Poincaré algebra in d dimensions. The simplest action for massless superparticle is given by

$$S = \tfrac{1}{2} \int d\tau e^{-1} \phi E_\tau^a E_\tau^a \ , \tag{2.2}$$

where e is the einbein field, $\phi(Z)$ is the dilaton superfield and it is understood that the contraction of the Lorentz indices is with the Lorentz metric η_{ab}. Note that this action describes the coupling of a massless particle to the target space fields that are contained in the superfields $E_M{}^A$ and ϕ. (In flat target space, this action reduces to the Brink-Schwarz superparticle action [10]). The remarkable property of this action is that, with suitable constraints imposed on the target superspace torsion, it is invariant under the so-called κ-symmetry transformations which take the form [1][8]

$$\delta Z^M = \kappa_\alpha E_\tau^a \Gamma_a^{\alpha\beta} E_\beta{}^M \ , \qquad e^{-1}\delta e = \kappa_\alpha S^\alpha \ , \tag{2.3}$$

where S^α is a function of the superfields to be determined by the κ-symmetry of the action and $E_A{}^M$ is the inverse supervielbein satisfying $E_A{}^M E_M{}^B = \delta_A^B$. The torsion tensor is defined by $T^A = dE^A + E^B \wedge \omega_B{}^A = \frac{1}{2} E^B \wedge E^C T_{CB}{}^A$ where the basis one form is $E^A = dZ^M E_M{}^A$ and the spin connection one form is $\omega_A{}^B = dZ^M \omega_{MA}{}^B$. In standard superspace the tangent space group is Lorentzian so that $\omega_a{}^\alpha = 0$ and $\omega_\alpha{}^a = 0$. Keeping this in mind, and making use of the following lemma

$$\delta E_\tau^A = \partial_\tau (\delta Z^M E_M{}^A) + \delta Z^M E_M{}^B E_\tau^C (-T_{CB}{}^A + \omega_C{}^A{}_B - (-1)^{BC} \omega_B{}^A{}_C) \ , \tag{2.4}$$

we find that the invariance of the action (2.2) under the κ-symmetry transformations (2.3) requires the following torsion constraints

$$T_{\alpha\beta}{}^c = 2(\Gamma^c)_{\alpha\beta} \ , \qquad T_{\alpha(bc)} = u^\beta{}_{(b}\Gamma_{c)\beta\alpha} + \eta_{bc} v_\alpha \ , \tag{2.5}$$

and fixes the function S^α to be

$$S^\alpha = -4E_\tau^\alpha + E_\tau^a (2u^\alpha{}_a + 2\Gamma_a^{\alpha\beta} v_\beta + \Gamma_a^{\alpha\beta} \phi^{-1} D_\beta \phi) \ , \tag{2.6}$$

where $u^\alpha{}_a$ is an arbitrary Γ-*traceless* vector-spinor superfield † and v^α is an arbitrary spinor superfield.

The action is also invariant under the worldline reparametrizations, and provided that $u^\beta{}_a = 0$ and $\phi v_\alpha + \frac{1}{2} D_\alpha \phi = 0$, also under the local λ-symmetry transformations

$$\delta Z^M = \lambda E_\tau^\alpha E_\alpha{}^M \ , \qquad \delta e = 0 \ , \tag{2.7}$$

where $\eta(\tau)$ and $\lambda(\tau)$ are arbitrary parameters. (For flat target space version of (2.7), see ref. [2]). Furthermore, provided that the target superspace admits a *superconformal Killing vector* k^M, the action is also invariant under the following target space rigid superconformal transformations [11]

$$\delta Z^M = k^M \ , \qquad \delta e = 2e(U + \tfrac{1}{2} k^M \partial_M \phi) \ , \tag{2.8}$$

where U is a scalar superfield which occurs in the superconformal Killing equation $\mathcal{L}_k E_M{}^a = U E_M{}^a + E_M{}^b L_b{}^a$, where the second term is a Lorentz transformation. Superconformal

† The freedom for having a vector-spinor in the κ-symmetry imposed constraint on $T_{\alpha(ab)}$ is usually overlooked, but as one allows field redefinitions afterwards to arrive at a standard set of constraints, this omission becomes inconsequential.

Killing vectors can be found at least in certain spacetimes of dimension $d \leq 7$. In higher dimensions one can always find the super Killing vectors which obey the super Poincaré algebra. The action will be invariant under such transformations, and hence its manifest spacetime supersymmetry.

The algebra of the symmetry transformations described above closes on-shell, i.e. modulo the equations of motion which take the form

$$\begin{aligned}
&E_\tau^a E_\tau^a = 0 \,, \\
&E_\tau^a \Gamma^a_{\alpha\beta} S^\beta = 0 \,, \\
&e\phi^{-1}\partial_\tau\left(e^{-1}\phi E_\tau^a\right) + E_\tau^b E_\tau{}^c\left(-T_{abc} + \omega_{bca}\right) + E_\tau^b E_\tau^\alpha\left(T_{\alpha ab} - \omega_{\alpha ab}\right) = 0 \,.
\end{aligned} \tag{2.9}$$

Instead of deriving the algebra of the κ–transformations directly in the Lagrangian formalism, following [12], we shall consider the algebra of first class constraints which generate the κ–transformations. To this end let us define the conjugate momenta

$$\begin{aligned}
P_A &= E_A{}^M \frac{\delta S}{\delta \partial_\tau Z^M} \,, \\
&= e^{-1}\phi E_\tau^a \delta_A^a \,.
\end{aligned} \tag{2.10}$$

The constraints which follow from the form of the Lagrangian are $P_a P^a \approx 0$ and $P_\alpha \approx 0$. However, these do not form a set of first class constraints and therefore do not form a closed algebra. Instead, let us define the combinations [12][13]

$$\mathcal{A} = \tfrac{1}{2}P^a P_a + A^\alpha P_\alpha \,, \qquad \mathcal{B}^\alpha = A^{\alpha\beta} P_\beta \,, \tag{2.11}$$

where A^α and $A^{\alpha\beta}$ are functions of superfields which are to be determined by the closure of the algebra of the above constraints. For definiteness let us consider the *ten dimensional spacetime*. It turns out that $A^\alpha = 0$ and $A^{\alpha\beta} = \Gamma_a^{\alpha\beta} P_a$. Furthermore, in [12] it has been shown that various sets of constraints can be obtained from those required by the closure of the constraint algebra plus the Bianchi identity $\sum_{(ABC)}(D_A T_{BC}{}^D - R_{ABC}{}^D + T_{AB}{}^E T_{EC}{}^D) = 0$, where $R_{ABC}{}^D$ is the Riemann curvature. In addition, if one allows the local Lorentz transformations $\delta E^a = E^b \Lambda_b{}^a$ and transformations of the form $\delta E^\alpha = E^b \Lambda_b{}^\alpha + E^\beta \Lambda_\beta{}^\alpha$, it has been shown that [12] the set of constraints proposed by Nilsson [14]

$$\begin{aligned}
&T_{\alpha\beta}{}^a = 2(\Gamma^a)_{\alpha\beta} \,, \qquad T_{\alpha a}{}^b = v_\alpha \delta_a^b \,, \qquad T_{ab}{}^c = 0 \,, \\
&T_{\alpha\beta}{}^\gamma = (\Gamma^{abcde})_{\alpha\beta}(\Gamma_{abcde})^{\gamma\delta}\, v_\delta \,,
\end{aligned} \tag{2.12}$$

or the set proposed by Witten [8]

$$\begin{aligned}
&T_{\alpha\beta}{}^a = 2(\Gamma^a)_{\alpha\beta} \,, \qquad T_{\alpha a}{}^b = 0 \,, \qquad T_{a(bc)} = 0 \,, \\
&T_{\alpha\beta}{}^\gamma = 0 \,, \qquad T_{a\alpha}{}^\beta = -\tfrac{1}{24}\left(\Gamma_a \Gamma^{bcd}\right)_\alpha{}^\beta T_{bcd} \,,
\end{aligned} \tag{2.13}$$

can be derived. In obtaining (2.13) scale transformations of the form $\delta E^\alpha = -\frac{1}{2}uE^\alpha$ and $\delta E^a = -uE^a$ must also be allowed.

The two sets of constraints (2.12) and (2.13), as well as many other possible choices of constraints are all equivalent via appropriate field redefinitions. It is interesting to see how one such particular set, which differs from (2.12) and (2.13), emerges from the principle of light-like integrability in loop superspace [15]. In fact, one expects a very close connection between light-like integrability and κ–symmetry [8][12][15].

Neither set of the constraints (2.12) and (2.13) are sufficient to imply the $N = 1, d = 10$ supergravity equations of motion. To describe them, one has to introduce a 3-form superfield in Nilsson's case, while in Witten's case, as T_{abc} happens to be totally antisymmetric, one can construct a super 3-form H in terms of T_{abc} and the dilaton superfield ϕ [8][16]. Either set of constraints turn out to arise in superstring theory where the string couples to the two form field B whose field strength is $H = dB$.

The massless superparticle can also be coupled to background Maxwell and Yang-Mills fields. For simplicity let us take the gauge group to be $SO(N)$. Introducing fermionic worldline fields $\psi^I(\tau), I = 1, ..., N$, the action can be written as

$$S = \int d\tau\left(\tfrac{1}{2}e^{-1}\phi E^a_\tau E^a_\tau + \partial_\tau Z^M B_M + \psi^I D_\tau \psi^I\right) , \tag{2.14}$$

where $D_i\psi^I = \partial_\tau\psi^I + \partial_\tau Z^M A^{IJ}_M \psi^J$ and $B_M(Z)$, $A^{IJ}_M(Z)$ are the background Maxwell and Yang-Mills superfields. The κ–symmetry transformations again take the form (2.3) with the additional transformation rule

$$\delta\psi^I = -\delta Z^M A^{IJ}_M \psi^J . \tag{2.15}$$

The κ–symmetry transformations leave the action invariant provided that in addition to the constraints (2.5), the following ones are imposed

$$H_{\alpha\beta} = 0 , \qquad H_{a\alpha} = \phi(\Gamma_a)_{\alpha\beta}h^\beta , \qquad F^{IJ}_{\alpha\beta} = 0 , \qquad F^{IJ}_{a\alpha} = (\Gamma_a)_{\alpha\beta}\chi^{\beta IJ} , \tag{2.16}$$

and that the function S^α is given by

$$S^\alpha = -4E^\alpha_\tau + E^a_\tau\left(2u^\alpha{}_a + 2\Gamma^{\alpha\beta}_a v_\beta + \Gamma^{\alpha\beta}_a \phi^{-1}D_\beta\phi\right) + \left(h^\alpha + \phi^{-1}\chi^\alpha_{IJ}\psi^I\psi^J\right) , \tag{2.17}$$

where h^α, χ^α_{IJ} are arbitrary superfields and $H = dB$, $F = dA + A \wedge A$. It should be noted that while the B–term in (2.14) is not necessary for κ–symmetry of the action, it will become necessary in the case of a *massive* superparticle action, as we will see in Sec. 4.

The action (2.14) is also invariant under the worldline reparametrizations, and provided that A^{IJ}_M and B_M are symmetric tensors with respect to the superconformal Killing vectors, also invariant under the target space rigid superconformal symmetry (2.8). The λ–symmetry (2.7) and (2.15), however, imposes the constraints $h^\alpha = 0$ and $\chi^\alpha_{IJ} = 0$, in addition to the previous ones: $u^\alpha{}_a = 0$, $\phi v^\alpha + \frac{1}{2}D_\alpha\phi = 0$.

Considering again the ten dimensional spacetime, the constraints (2.16) describe super Maxwell and super Yang-Mills equations of motion. The coupled supergravity plus super Maxwell/super Yang-Mills system does not arise from the constraints implied by the κ-symmetry of the superparticle. Superstring theory however does describe such a system as we shall see later. In passing we note that coupling of Yang-Mills fields to the massless superparticle can be achieved also by using bosonic coordinates [12] instead of the fermionic ones used above.

Flat Target Superspace

It is useful to consider the flat target superspace limit of the actions described above. For simplicity let us consider the action (2.2) with the dilaton field set equal to a constant, e.g. $\phi = 1$. In a Salam-Strathdee superspace the spin connection is vanishing but the torsion does not completely vanish. Its only nonvanishing component is $T_{\alpha\beta}{}^{a} = 2(\Gamma^a)_{\alpha\beta}$. Thus from the definition of the torsion one finds the components of the supervielbein in flat superspace to be $E_m{}^a = \delta_m^a$, $E_m{}^\alpha = 0$, $E_\mu{}^a = (\Gamma^a\theta)_\mu$, $E_\mu{}^\alpha = \delta_\mu^\alpha$. Substituting this into (2.2) gives

$$S = \tfrac{1}{2}\int d\tau e^{-1}\left(\partial_\tau X^a - \bar\theta\Gamma^a\partial_\tau\theta\right)\left(\partial_\tau X^b - \bar\theta\Gamma^b\partial_\tau\theta\right)\eta_{ab}\ . \tag{2.18}$$

This is the Brink-Schwarz superparticle action [10]. The κ-transformation rules (2.3) become [1]

$$\delta X^a = \bar\theta\Gamma^a\delta\theta\ , \qquad \delta\theta = \Pi^a\Gamma_a\kappa\ , \qquad e^{-1}\delta e = -4\bar\kappa\partial_\tau\theta\ . \tag{2.19}$$

where $\Pi^a = \left(\partial_\tau X^a - \bar\theta\Gamma^a\partial_\tau\theta\right)$. Since on-shell $\Pi^a\Pi_a = 0$ and that $Tr\ \Pi^a\Gamma_a = 0$, the matrix $\Pi^a\Gamma_a = 0$ has half as many zero eigenvalues, and therefore using κ-symmetry one can gauge away only half as many degrees of freedom θ^α.

This phenomenon of halving the fermionic degrees of freedom is rather similar to a phenomenon discovered long ago by Mack and Salam [17] in their study of conformally invariant field theories formulated on a five dimensional cone. From manifestly $SO(4,2)$ invariant field theories on the cone, by suitably restricting the fields, and in the case of fermions by a gauge symmetry similar to κ-symmetry, they could obtain the 3+1 dimensional description of correct degrees of freedom. For example, let us define the cone by $y^A y_A = 0$, $A = 1, ..., 6$, where y^A are the embedding coordinates of a $4+2$ dimensional plane. Consider the manifestly $SO(4,2)$ invariant field equation for a fermionic field ψ to be of the form:

$$\left(\Gamma^{AB}y_A\partial_B - 2\right)\psi = 0\ , \qquad y^A\partial_A\psi = -2\psi\ , \tag{2.20}$$

where the second equation is the restriction imposed on the field, representing its homogeneity degree on the cone. The above field equation is actually invariant under the following transformations

$$\delta\psi = y^A\Gamma_A\kappa\ , \qquad y^A\partial_A\kappa = -3\kappa\ , \tag{2.21}$$

where, the second equation expresses the condition which has to be imposed on the symmetry parameter κ in the form of homogeneity degree on the cone [18]. Since $y^A y_A = 0$ and $Tr\, y^A \Gamma_A = 0$, indeed the field ψ has has as many degrees of freedom in 3+1 dimension. This phenomenon and the form of the transformation rules is similar to the Siegel's κ–symmetry transformation rules [1]. Given that superparticle actions in $d = 3, 4, 6$ dimensions have also superconformal symmetry, one wonders if there is a deeper connection between the two phenomenon and possibly with the remarkable representations of the superconformal groups known as the supersingletons [19][18].

Quantization

In covariant BRST quantization of the massless superparticle, or indeed any κ–symmetric system, two features arise. Firstly, the κ–transformation rules close on-shell, and second, the κ–symmetry is an infinitely reducible type symmetry, in the terminology of Batalin and Vilkovisky [20]. Both of these features can be handled in the Batalin-Vilkovisky quantization procedure. However, the fact that the system is infinitely reducible means that an infinite tower of ghost fields are needed. This leads to some problems, such as the problem of regularizing infinite sums and the problem of Stueckelberg type residual gauge symmetries of the final BRST invariant action. A period of intense activity in these area occurred in 1989, resulting in a number of papers to which we refer the reader for a detailed discussion of these problems [6][21].

More recently, the idea of trading the κ–symetry for world-line local supersymmetry has been put forward [22] in attempt not only to understand the origin of κ–symmetry as a special supersymmetry transformation, but also to achieve the covariant quantization in a way which may avoid the problems mentioned briefly above, as well as achieving an off-shell formulation which would then enable one to construct higher order invariants. (Some progress has already been made in [23] towards the quantization of these theories). The approach of ref. [22], which is currently gaining popularity, makes use of twistor-like variables, which are essentially commuting fermionic variables arising as superpartners of the fermionic coordinates of the target superspace. We now turn to a brief discussion of this approach.

Twistor-like Formulation

In order to introduce the twistor-like formulations, it is convenient to pass to the first order form of the action, which for the massless superparticle in flat target superspace is given by

$$S = \int d\tau \left[P_a(\partial_\tau X^a - \bar\theta\Gamma^a\partial_\tau\theta) - \tfrac{1}{2}eP_aP^a \right. . \qquad (2.22)$$

This action is invariant under the κ–symmetry transformations

$$\delta X^a = \bar\theta\Gamma^a\delta\theta\ , \qquad \delta\theta = \Gamma^a P_a \kappa\ , \qquad \delta e = -4\bar\kappa\partial_\tau\theta\ , \qquad \delta P_a = 0\ . \qquad (2.23)$$

The constraints which follow from the action (2.22) are the reparametrization constraint $T := P_a P^a \approx 0$, and the fermionic constraint $d_\alpha := P_\alpha - P_a \Gamma^a_{\alpha\beta}\theta^\beta \approx 0$, where P_α is the conjugate momentum associated with θ_α (see eq. (2.10)). The fermionic constraint is a combination of first class constraints generating the κ–symmetry transformations and second class constraints which, of course, have no such an interpretation.

The main idea of the twistor-like formulation is to replace the momentum variable P_m with a suitable combination of *commuting* fermions, which are the twistor-like variables, to reformulate the action (2.3) in a world-line locally supersymmetric fashion [22]. The word *twistor-like* is used to avoid confusion with the supertwistor which consists of a multiplet of fields forming a multiplet of superconformal groups which are known to exists in dimensions $d \leq 6$. In fact such variables have been used previously in a twistor formulation of superparticles and superstrings in $d = 3, 4, 6$ [24]. A similar, but not quite the same, multiplet of variables were introduced in [25] to give a twistor-like formulation of these models in $d = 10$ as well. However, the twistor-like formulation we shall briefly review below, which is due to [22], differs from this formulation, in that it involves a different set of variables yet. The advantage of this formulation is that it allows a natural generalization to curved superspace, as well as higher super p-branes [26][27]. It should be noted that in all these three different formulations, there is a common variable, namely the commuting spinor mentioned earlier which is used to replace the momentum variable P_a.

For superparticles and superstrings the twistorlike formulation works in $d = 3, 4, 6, 10$. To simplify matters, let us focus our attention to the $d = 3$ case. The twistor-like formulation of the massless superparticle in $d = 3$ is given by [22]

$$S = \int d\tau P_a \left(\partial_\tau X^a - \bar\theta \Gamma^a \partial_\tau \theta + \bar\lambda \Gamma^a \lambda\right) , \tag{2.24}$$

where λ is the twistor-like variable, a two component commuting Majorana spinor in $d = 3$. Its equation of motion is $P_a \Gamma^a \lambda = 0$, whose solution can be shown to be

$$P_a = e^{-1} \bar\lambda \Gamma_a \lambda , \tag{2.25}$$

thanks to the identity

$$\Gamma^a_{(\alpha\beta}\Gamma^a_{\gamma)\delta} = 0 . \tag{2.26}$$

Note that $P_a P^a = 0$ as well, due to this identity. In this action κ–symmetry has been replaced by $n = 1$ local world-line supersymmetry generated by $Q := \lambda^\alpha d_\alpha$ [22], were we recall that $d_\alpha = P_\alpha - \left(P_a \Gamma^a \theta\right)_\alpha$.

To close the $n = 1$ supersymmetry off-shell, it is convenient to pass to a superfield formalism which will naturally introduce the necessary auxiliary fields. To this end, let us consider the superline with coordinates (τ, η) and the following superfields

$$P_a(\tau,\eta) = P_a + \eta \rho_a , \qquad X^a(\tau,\eta) = X^a + \eta \chi^a , \qquad \theta(\tau,\eta) = \theta + \eta\lambda , \tag{2.27}$$

where η is the single anticommuting coordinate, and the auxiliary fields ρ and χ have been introduced. Note that the twistor variable λ has been paired with the target space fermionic variable θ. In terms of these *superfields* the off-shell supersymmetric action is [22]

$$S = \int d\tau d\eta P_a \left(DX^a + \bar{\theta}\Gamma^a D\theta \right) , \tag{2.28}$$

where $D = \frac{\partial}{\partial\eta} + \eta\frac{\partial}{\partial\tau}$. Note that this action has the form of a Wess-Zumino term.

To formulate the massless superparticle action in $d > 3$ [28][29]30], and to put the above formulation in a somewhat more geometrical form that will enable us to make the curved superspace and higher p-brane generalizations, it is useful to introduce the following notation. For definiteness let us focus on $d = 10$, in which case the κ symmetry can be replaced by $n = 8$ world-line supersymmetry. The coordinates of $n = 8$ worldline superspace will be denoted by $Z^M = (\tau, \mu)$ and the coordinates of the target superspace by $Z^{\underline{M}} = (\underline{m}, \underline{\mu})$, where $\mu = 1, ...8$, $\underline{m} = 0, 1, ..., 9$, $\underline{\alpha} = 1, ..., 16$. We take the $d = 10$ spinors to be Majorana-Weyl. The world-line supervielbein will be denoted by $E_M{}^A$ and the target space supervielbein by $E_{\underline{M}}{}^{\underline{A}}$, where the tangent space indices split as $A = (0, r)$ and $\underline{A} = (\underline{a}, \underline{\alpha})$, respectively, with $r = 1, ...8$, $\underline{a} = 0, 1, ...9$, $\underline{\alpha} = 1, ..., 16$. Note that the un-underlined indices always refer to the world-line superspace quantities, while their underlined versions always refer to the corresponding target space quantities.

The twistor-like formulation of massless superparticle and superstrings in $d > 3$ with κ-symmetry completely replaced by an n-extended worldline supersymmetry requires the introduction of n twistor-like variables $\lambda_r{}^{\underline{\alpha}}$ satisfying the constraint [29][30][31]

$$\bar{\lambda}_r \Gamma^{\underline{a}} \lambda_s = \tfrac{1}{8}\delta_{rs} \left(\bar{\lambda}_q \Gamma^{\underline{a}} \lambda_q \right) . \tag{2.29}$$

To express this constraint in a geometrical form, it is convenient to introduce the notation

$$E_A{}^{\underline{A}} = E_A{}^M \left(\partial_M Z^{\underline{M}} \right) E_{\underline{M}}{}^{\underline{A}} . \tag{2.30}$$

Let us furthermore make the identifications

$$E_r{}^{\underline{\alpha}}|_{\theta=0} = \lambda_r{}^{\underline{\alpha}} , \qquad E_0{}^{\underline{a}}|_{\theta=0} = \mathcal{E}_0{}^{\underline{a}} , \tag{2.31}$$

where $\mathcal{E}_0{}^{\underline{a}} = \partial_\tau X^{\underline{a}} - \bar{\theta}\Gamma^{\underline{a}}\partial_\tau\theta$, in flat target superspace. Thus, the fermionic coordinates $\theta^{\underline{\alpha}}$ have been elevated to a superfield whose expansion is of the form $\theta^{\underline{\alpha}}(\tau, \theta) = \theta^{\underline{\alpha}}(\tau) + \lambda_r{}^{\underline{\alpha}}(\tau)\theta^r + \cdots$. Furthermore, for definiteness and to simplify matters we shall characterize, both, the worldline and target superspace geometries. Namely, we shall take the worldline superspace to be characterized by the torsion constraints

$$T_{rs}{}^0 = 2\delta_{rs} , \qquad T_{0r}{}^0 = 0 , \qquad T_{s0}{}^r = 0 , \qquad T_{rs}{}^q = 0 , \tag{2.32}$$

and we shall take the target superspace geometry to be characterized by the torsion constraints

$$T_{\alpha\beta}{}^{\underline{c}} = 2\Gamma^{\underline{c}}_{\underline{\alpha\beta}} \,, \qquad T_{\underline{b\alpha}}{}^{\underline{a}} = 0 \,, \qquad T_{\underline{\alpha\beta}}{}^{\underline{\gamma}} = 0 \,. \tag{2.33}$$

Of course, all consequences of these constraints which follow from the Bianchi identities are understood to hold.

The $n = 8$ locally supersymmetric action can now be written as [29][30]

$$S = \int d\tau d^8\theta P_{\underline{a}}{}^{r} E_r{}^{\underline{a}} \,, \tag{2.34}$$

where $P_{\underline{a}}{}^{r}$ is a Lagrange multiplier superfield. The constraints (2.32) leave enough room for worldline diffeomorphisms and local $n = 8$ supersymmetry. (See refs. [29][30] for a detailed description of these transformations).

The field for the Lagrange multiplier is $E_r{}^{\underline{a}} = 0$. Taking the curl of this equation, and recalling (2.31), yields an integrability condition whose $\theta = 0$ component is the twistor constraint (2.29). The classical equivalence of this theory to the usual κ–symmetric one has been shown in [29][30].

Analogous twistor-like formulations have also been proposed for massive superparticle [32][27], superstrings in $d = 3,4,6$ [33], superstrings in $d = 10$ [34][30], supermembranes [26] and all higher super p-branes [27]. The questions of covariant quantization, and higher order invariants have been so far addressed only in limited cases. It has been shown in [23] that the quantization of the superstring in which two of the eight κ–symmetries are replaced by $n = 2$ world-sheet local supersymmetry, a semi-light cone gauge, while not covariant, avoids a problem encountered in the usual light-cone gauge formulation due to the global issues that arise in choosing such a gauge. As for the construction of higher order κ–invariants, results have been obtained for the case of massive superparticle in $d = 2,3$ by using the twistor-like formulation [32]. (See references therein for other approaches to the construction of higher order κ–invariants). We next turn to the description of the κ–symmetric superstring theory in curved superspace.

3. Superstring in Curved Superspace

For definiteness, let us focus our attention on the heterotic string propagating in $d = 10$ supergravity plus $SO(32)$ Yang-Mills background. Let the coordinates of the world-sheet be $\sigma^i, i = 0,1$, and the pulled-back supervielbein

$$E_i^A = \partial_i Z^M E_M^A \,, \tag{3.1}$$

where the tangent space index splits as $A = (a, \alpha)$ with $a = 0,1,...,9$ and $\alpha = 1,...,16$. We shall be dealing with sixteen component *Majorana-Weyl* spinors in $d = 10$. The ingredients for the action are the Kalp-Ramond super 2-form $B = dZ^M \wedge dZ^N B_{NM}$, the Yang-Mills superfield A_M, the dilaton superfield ϕ and the *heterotic fermions* $\psi^I, I = 1, ..., 32$ which are *Majorana-Weyl* spinors on the world-sheet. In terms of these building blocks, the heterotic string action can be written as

$$S = \int d^2\sigma \Big[-\tfrac{1}{2}\phi\sqrt{-g}g^{ij}E_i^a E_j^a + \tfrac{1}{2}\epsilon^{ij}\partial_i Z^M \partial_j Z^N B_{NM} + \tfrac{1}{2}\sqrt{-g}\psi^I\gamma^i D_i\psi^I \Big] , \tag{3.2}$$

where g_{ij} is the world-sheet metric of signature $(-1, +1)$, $g = \det g_{ij}$ and $D_i\psi^I = (\partial_i\delta^{IJ} + \partial_i Z^M A_M^{IJ})\psi^J$. We shall require that the action be invariant under the κ–symmetry transformations of the form

$$\delta Z^M = \kappa_{i\alpha} P_-^{ij} E_j^a \Gamma_a^{\alpha\beta} E_\beta{}^M , \qquad \delta\psi^I = -\delta Z^M A_M^{IJ}\psi^J , \qquad e_{ir}\delta e^{jr} = P_{+im} P_+^{jn} S^{m\alpha}\kappa_{n\alpha} , \tag{3.3}$$

where $\kappa_{i\alpha}$ is the transformation parameter. Note that unlike in the particle case, this parameter is a world-sheet vector in addition to being a target space (Majorana-Weyl) spinor. Note also that only its self-dual projection occurs. The quantity $S^{i\alpha}$ is to be determined by the κ–symmetry of the action. The duality projectors are defined as $P_{\pm ij} = \frac{1}{2}(g_{ij} \pm \sqrt{-g}\epsilon_{ij})$, satisfying the relation $P_+^{ki}P_{+kj} = 0$. Note that the Levi-Civita symbols ϵ_{ij} and ϵ^{ij} are both constants, satisfying $\epsilon^{ij}\epsilon_{jk} = \delta_k^i$.

The κ–symmetry of the action imposes the constraints †

$$T_{\alpha\beta}{}^c = 2(\Gamma^c)_{\alpha\beta} , \qquad T_{\alpha(bc)} = u^\beta{}_{(b}\Gamma_{c)\beta\alpha} + \eta_{bc}v_\alpha , \tag{3.4a}$$
$$H_{\alpha\beta\gamma} = 0 , \qquad H_{a\alpha\beta} = -2\phi(\Gamma_a)_{\alpha\beta} , \qquad H_{ab\alpha} = 2\phi(\Gamma_{ab})_\alpha{}^\beta h_\beta + 2\phi u^\beta{}_{[a}\Gamma_{b]\beta\alpha} , \tag{3.4b}$$
$$F_{\alpha\beta}^{IJ} = 0 , \qquad F_{a\alpha}^{IJ} = (\Gamma_a)_{\alpha\beta}\chi^{\beta IJ} \tag{3.4c}$$
$$h_\alpha = v_\alpha + \tfrac{1}{2}\phi^{-1}D_\alpha\phi . \tag{3.4d}$$

Furthermore, the quantity S_i^α is determined to be

$$S_i^\alpha = -4E_i^\alpha + 2E_i{}^a(-u^\alpha{}_a + \Gamma_a^{\alpha\beta}h_\beta) - \tfrac{1}{2}\phi^{-1}\bar\psi^I\gamma_i\psi^J\chi_{IJ}^\alpha . \tag{3.5}$$

(Concerning the occurrence of $u^\alpha{}_a$ in these formulae, see the footnote below eq. (2.6)). The action (3.2) is also invariant under the world-sheet reparametrization, local Weyl scalings and the spacetime rigid Poincaré supersymmetry transformations.

Assuming the κ–symmetry constraints (3.5), the action (3.2) is also invariant under the following local bosonic λ–symmetry transformations

$$\delta Z^M = P_+^{ij}E_i^\alpha\lambda_j E_\alpha{}^M , \qquad \delta e_{ir} = 0, \tag{3.6}$$

† The study of κ–symmetric string theory in *curved* background was pioneered by Witten [8]. Soon afterwards, various extensions of his results were obtained in refs. [35][36].

provided that $u^{\alpha}{}_a = 0$, $h_\alpha = 0$ and $\chi^{\alpha}_{IJ} = 0$. (For the variation of ψ^I we adopt the same rule as in (2.15)). The flat target space version of this symmetry has been given in ref. [2].

The question arises as to what supergravity theory, if any, do the above constraints describe. In the absence of the Yang-Mills sector, and ignoring the question of κ–symmetry anomalies, it was shown by Witten that the constraints of pure $d = 10, N = 1$ supergravity are consistent with κ–symmetry. Whether they are necessary for κ symmetry is more difficult to establish. To this end, as in the particle case, one may investigate the closure of first class constraints which may be constructed out of the constraints which follow from the Lagrangian. Then allowing field redefinitions (including the scaling of the supervielbein) and taking into account the Bianchi identities, in [12] it has been shown for the case of pure supergravity that Witten or Nilsson constraints, supplemented by additional constraints needed for the description of $N = 1, d = 10$ supergravity, can be derived.

However, pure $N = 1, d = 10$ supergravity has gravitational anomalies. One has to include an $SO(32)$ or $E_8 \times E_8$ Yang-Mills sector, and utilize the Green-Schwarz mechanism to cancel the gravitational, gauge and mixed anomalies. These anomalies reflect themselves in the form of κ–symmetry anomalies, from the world-sheet point of view, which can also be cancelled by a similar mechanism. In trying to understand the consequences of the κ–symmetry constraints (3.4), this anomaly cancellation mechanism must be taken into account. The κ–symmetry anomalies have been discussed in detail in [37], and previously in ref. [38]. To cancel them one adds a counterterm Γ, whose κ–symmetry variation is [37][38]

$$\delta\Gamma = \int d^2\sigma \epsilon^{ij} E_i^A E_j^B \kappa^\alpha \mathcal{K}_{\alpha BA} \ , \tag{3.7}$$

where $\mathcal{K}_{\alpha BA}$ are particular components of a super 3-form $\mathcal{K}$ which takes the form

$$\mathcal{K} = \omega_{3Y} - \omega_{3L} + \cdots \ , \tag{3.8}$$

and ω_{3Y} and ω_{3L} are the Yang-Mills and Lorentz Chern-Simons forms. Then the κ–variation of the classical action plus the anomalous variation $\delta\Gamma$ will yield the same result as in the variation of the classical action alone, except that the curvature $H = dB$ will now be replaced by

$$\mathcal{H} = dB + \mathcal{K} \ . \tag{3.9}$$

If we wish to impose the so-called *standard* set of constraints, i.e. those in (3.4) with $u^{\alpha}{}_a$ and v^α set equal to zero (one could consider any equivalent set), then the above replacement is consistent provided that [37]

$$(d\mathcal{K})_{\alpha\beta\gamma\delta} = 0 \ , \qquad (d\mathcal{K})_{a\alpha\beta\gamma} = 0 \ . \tag{3.10}$$

Such a $\mathcal{K}$ can be found, thanks to the existence of the following relation [39]

$$\mathrm{Tr}(RR) = dX + K \ , \tag{3.11}$$

for some 3-form X and a 4-form K which has the vanishing projections $K_{\alpha\beta\gamma\delta}=0$ and $K_{a\alpha\beta\gamma}=0$. Since $\mathrm{Tr}(FF)$ has the same vanishing projections as well, it follows that a suitable choice for $\mathcal{K}$ is [37]

$$\mathcal{K}=\omega_{3Y}-\omega_{3L}+X\ . \tag{3.12}$$

With this choice, $d\mathcal{K}=\mathrm{Tr}(FF)-K$ indeed has the vanishing projections in the $(\alpha\beta\gamma\delta)$ and $(a\alpha\beta\gamma)$ directions as required.

In summary, the constraints needed to be imposed are the torsion constraints (3.4) with $u^\alpha{}_a=0$, $v^\alpha=0$ and with the replacement $H\to\mathcal{H}$ to be made. When the consequences of the Bianchi identities are taken into account as well, all the superfields get determined in terms of the totally antisymmetric T_{abc}, χ_α and ϕ. In the absence of the Lorentz Chern-Simons term, for example, the remaining projection of $\mathcal{H}$ is found to be: $\mathcal{H}_{abc}=-\frac{3}{2}\phi T_{abc}+\frac{1}{4}(\Gamma_{abc})_{\alpha\beta}\mathrm{Tr}(\chi^\alpha\chi^\beta)$ [16]. In this case the resulting supergravity equations are those of the usual $N=1, D=10$ supergravity plus Yang-Mills system. When the effect of the Lorentz Chern-Simons term is taken into account as outlined above, the resulting equations describe an anomaly free model which can accommodate stringy corrections, in particular an R^4–type correction. For more details, we refer the reader to ref. [39].

So far we have considered the heterotic string in $d=10$. For a discussion of actions for heterotic strings in $d=3,4,6$, see ref. [36], where the occurrence of various supergarvity/matter, as well as *off-shell* supergravity backgrounds is also discussed. A Green-Schwarz type action for the Type IIA superstring has been discussed in ref. [40], and for the Type IIB string in ref. [35]. In the latter two cases, what has been shown is that the superspace constraints of on-shell Type IIA and Type IIB supergravities are ***sufficient*** for the κ–symmetry of the string actions.

4. Massive Superparticle, Supermembranes and Higher Super p–Branes

Just as the action for the massless superparticle is similar to the heterotic string action, the massive superparticle action is similar to the action for supermembranes and higher super p–branes. We shall describe the action for them in a unified manner.

A super p–brane moving in a superspace of d bosonic dimensions will sweep a worldvolume of $p+1$ dimensions. Denoting the coordinates of the world-volume by σ^i and the coordinates of the target superspace by Z^M, we again adopt the definition (3.1), with $i=1,...,p+1$. The tangent space index again splits as $A=(a,\alpha)$ with $a=0,1,...,d-1$ and for definiteness, we take α we label the minimum dimensional spinors possible in d–dimensions. Thus, it can be said that we are considering Type 1 super p–branes, in the sense that we are considering the minimum possible target space supersymmetry. In fact, while the existence of Type 2 super p–branes (i.e. target spaces with higher than minimum supersymmetry) has been proposed, at present no action is known for them, and this is one of the most interesting open problems in the theory of super p–branes.

The action for super p–branes requires a Kalp-Ramond type super $(p+1)$–form whose pull-back is

$$B_{i_1\dots i_{p+1}} = \partial_{i_1} Z^{M_1} \dots \partial_{i_{p+1}} Z^{M_{p+1}}\, B_{M_{p+1}\dots M_1} \; . \tag{4.1}$$

An action for super threebrane in $d=6$ flat target spacetime was first given in ref. [3]. Generalization to all super p–branes in curved target superspace was found in ref. [4]. The action takes the form

$$S = \int d^{p+1}\sigma \Big[-\tfrac{1}{2}\sqrt{-g}g^{ij}\phi E_i^a E_j^a + \epsilon^{i_1\cdots i_{p+1}} B_{i_1\dots i_{p+1}} + \tfrac{1}{2}(p-1)\phi\sqrt{-g} \Big] \; . \tag{4.2}$$

Note the presence of the worldvolume cosmological term for $p \neq 1$. It is essential for the equation of motion for the metric g_{ij} to yield the nondegenerate form

$$g_{ij} = E_i^a E_j^a \; . \tag{4.3}$$

Since this is an algebraic equation it can be used in establishing the κ– symmetry of the action. In this "1.5" formalism we need not vary the metric g_{ij} and the κ–transformation of the only independent variable Z^M is given by

$$\delta Z^M = \kappa^\alpha (1+\Gamma)_\alpha{}^\beta E_\beta{}^M \; , \tag{4.4}$$

where the matrix Γ is given by

$$\Gamma = \frac{\epsilon_p}{(p+1)!\sqrt{-g}} \epsilon^{i_1\cdots i_{p+1}} E_{i_1}^{a_1} \dots E_{i_{p+1}}^{a_{p+1}} \Gamma_{a_1\dots a_{p+1}} \; , \tag{4.5}$$

which satisfies $\Gamma^2 = 1$ provided that we use (4.3) and choose $\epsilon_p = (-1)^{(p+2)(p+3)/4}$. Thus, the matrix $(1+\Gamma)$ is a projection operator on-shell, and the phenomenon of halving the fermionic degrees of freedom occurs for all super p–branes.

The action (4.2) is invariant under the κ–symmetry transformation provided that in addition to the torsion constraints (3.4a), the following following constraints on the super $(p+2)$–form $H = dB$ are satisfied [4]

$$H_{\alpha\beta\gamma A_1\cdots A_{p-1}} = 0 \; , \tag{4.6a}$$

$$H_{\alpha\beta a_p\cdots a_1} = 2\epsilon_p \phi (\Gamma_{a_1\cdots a_p})_{\alpha\beta} \; , \tag{4.6b}$$

$$H_{\alpha a_{p+1}\cdots a_1} = \epsilon_p \phi (\Gamma_{a_1\cdots a_{p+1}})_\alpha{}^\beta h_\beta + \epsilon_p \phi u^\beta{}_{[a_{p+1}} \Gamma_{a_1\cdots a_p]\beta\alpha} \; , \tag{4.6c}$$

$$h_\alpha = v_\alpha + \tfrac{1}{2}\phi^{-1} D_\alpha \phi \; , \tag{4.6d}$$

where v_α and $h^\alpha{}_a$ are arbitrary superfields. (Concerning the occurrence of the vector-spinor $h^\alpha{}_a$ in the constraints, see the footnote below eq. (2.6)). The Bianchi identity $dH = 0$ is satisfied provided that the following Γ–matrix identity is satisfied:

$$\Gamma^a_{(\alpha\beta} \Gamma^{ac_1\cdots c_p}_{\gamma\delta)} = 0 \; . \tag{4.7}$$

This identity which is crucial for the κ–symmetry of the action imposes restrictions on possible dimensions d of spacetime and on possible values of p for a p–dimensional extended object. Assuming Lorentzian signature, these restrictions are [5]: $(p = 1; d = 3,4,6,10)$, $(p = 2; d = 4,5,7,11)$, $(p = 3; d = 6,8)$, $(p = 4; d = 9)$, $(p = 5; d = 10)$. It is a very interesting fact that the existence of the κ–symetry imposes so severe restrictions on both the dimension of spacetime *and* the extension of the fundamental object. We recall that the restriction $d \leq 11$ arises for the existence of supergravities with a single gravitational field, and the restriction $d \leq 6$ arises for the existence of scalar supermultiplets. Since in a gauge in which κ–symmetry is fixed, it is expected that a globally supersymmetric field theory in a $(p+1)$–dimensional worldvolume emerges, the occurrence of the restrictions $d \leq 11$ and $p \leq 5$ can be viewed as a consequence of an interesting fusion of worldvolume and targetspace supersymmetries via the κ–symmetry.

Turning to the constraints (3.4a) and (4.6), only in the case of eleven dimensional it is known rigorously that these constraints (with the dilaton superfield redefined away) do imply the unique $N = 1, d = 11$ supergravity equations of motion [41][4][42]. For the case of super fivebrane in ten dimensions, one expects that the corresponding equations of motion are those of the dual formulation of the $N = 1, d = 10$ pure supergravity, in which instead of a 2-form field, a 6-form field occurs [43]. In the remaining cases one expects that a suitable version of supergravity theories possibly coupled to matter/Yang-Mills supermultiplets will be described by the constraints required for the existence of the super p–brane actions. It would be of considerable interest to work out explicitly the consequences of these constraints and to determine exactly which supergravity theories are described by them. (See ref. [44] for a description of supergravity theories in diverse dimensions).

Heterotic and Type II Super p–Branes

One of the most interesting open problems is to find certain generalizations of the super p–brane actions. The possibility of super p–branes described by non–Poincaré symmetries of non-Lorentzian target spaces has been discussed in ref. [45]. In particular, the case of a $2+2$ worldvolume embedded in $10+2$ dimensional target space is rather interesting, but no action has been found for it. There are two other kinds of generalizations whose existence have been established by indirect means and for which no actions have been written down so far. These are the analogs of the heterotic string, which are called the *heterotic p–branes* (in the sense that they describe the coupling Yang-Mills to the super p–brane) [46][47], and the analogs of the Type IIA and Type IIB superstrings, which are called the *Type II super p–branes* (in the sense that the target space supersymmetry is not the minimal $N = 1$ supersymmetry, but instead an $N = 2$ supersymmetry which could be of $(2,0)$ or $(1,1)$ type) [48][49]. The existence of these types of super p–branes has been deduced from the existence of certain kinds of supersymmetric extended soliton solutions to the anomaly free $N = 1$ supergravity plus Yang-Mills, Type IIA or Type IIB supergravity theories in $d = 10$. These are solutions with $(p+1)$–dimensional Poincaré symmetry and they are asymptotically flat in the internal directions. For example, the super fivebrane solution of ref. [47] is obtained by solving the equations of motion which follow from the following low energy action for the bosonic degrees of freedom of the heterotic string in $d = 10$:

$$S = \frac{1}{\alpha'^4}\int d^{10}x\sqrt{-g}e^{-2\phi}\Big(R + 4(\nabla\phi)^2 - \tfrac{1}{3}H^2 - \tfrac{1}{30}\alpha'\mathrm{Tr}F^2\Big)\ , \tag{4.8}$$

where the notation is self-explanatory. The solution has $(1,0)$ type supersymmetry on the worldvolume, and it takes the form [47]

$$\begin{aligned} &g_{ab} = \eta_{ab}\ , \qquad g_{\mu\nu} = e^{2\phi}\delta_{\mu\nu}\ , \qquad e^{2\phi} = e^{2\phi_0} + 8\alpha'\frac{(x^2+2\rho^2)}{(x^2+\rho^2)^2}\ , \\ &H_{\mu\nu\lambda} = -\epsilon_{\mu\nu\lambda}{}^{\rho}\nabla_\rho\phi\ , \qquad A_\mu = \frac{-2\Sigma_{\mu\nu}x^\nu}{(x^2+\rho^2)}\ , \end{aligned} \tag{4.9}$$

where we have used the notation of ref. [47], according to which, $a = 0, 1, ..., 5$ labels the coordinates of the fivebrane worldvolume, $\mu = 5, ..., 9$ labels the four Euclidean internal coordinates x^μ, while ϕ_0 is the value of the dilaton at spatial infinity, $x^2 = x^\mu x^\nu\delta_{\mu\nu}$, and ρ is the size of the instanton, and $\Sigma_{\mu\nu}$ are the antisymmetric, self-dual 't Hooft matrices which arise in the description of the Yang-Mills instantons in four Euclidean dimensions. The ansatz for A_μ lies in $SU(2)$ subgroup of $E_8 \times E_8$ or $SO(32)$.

This solution breaks the spacetime symmetries, and in particular the $d = 10$ translation group is broken to that of the worldvolume, and *half* of the spacetime symmetries are maintained †. The unbroken symmetries are linearly realized, while the broken symmetries are nonlinearly realized on the worldvolume. The massless Goldstone bosons and fermions arising in this symmetry breaking turn out to be the position coordinates and the heterotic coordinates of the fivebrane. These fields correspond to the zero modes of the wave operator for the fluctuations around the soliton, but they are easier to determine (at least their number) by using index theorems. In ref. [47], it is argued that there are 240 fermionic zero modes (corresponding to 120 degrees of freedom) and 120 bosonic zero modes. These include the 4 translational and 8 supertranslational zero modes.

The effective action describing the dynamics of the massless fields is presumably a $(1,0)$ supersymmetric hyperkahler sigma model in $5+1$ dimensions. However, its exact form has never been spelled out as yet. Such an action would be the gauge fixed version of an action in which the worldvolume symmetries would be realized covariantly, and in particular the worldvolume supersymmetry would manifest itself in the form of a κ–symmetry. This

† There are other interesting solutions which break more than half of the spacetime supersymmetries [50]. Presumably the usual κ–symmetric formulations can not exist for the super p–branes implied by these solutions, but it is tempting to consider that the twistor-like formulations may accommodate them.

covariant form of the action is even more mysterious, and it is one of the major open problems in this subject.

As for the p-brane solitons of the Type IIA and Type IIB supergravities, an extensive review about them can be found in ref. [51]. An interesting new feature emerging is that the zero modes implied by these solutions correspond to vector or antisymmetric tensor supermultiplets of $(1,1)$ or $(2,0)$ supersymmetry on the worldvolume, unlike in the case of Type I super p-branes where the zero modes form only scalar supermultiplets [48]. The values of p, the worldvolume supersymmetry N and the nature of the worldvolume supermultiplet which arises, corresponding to the solutions of Type IIA supergravity in $d=10$ are [48][51]

$$\{p=4, N=2;\ A_\mu, 4\lambda, 5\phi\}\ ,\quad \{p=5, N=(2,0);\ B^-_{\mu\nu}, 4\lambda, 5\phi\}\ ,\quad \{p=6, N=1;\ A_\mu, \lambda, 3\phi\}\ ,$$

while those corresponding to the solutions of Type IIB supergravity in $d=10$ [48][51] are

$$\{p=3, N=2;\ A_\mu, 4\lambda, 5\phi\}\ ,\quad \{p=5, N=2;\ A_\mu, 2\lambda, 4\phi\}\ ,$$

where λ is the fermionic partner of the worldvolume Maxwell field A_μ, and $B^-_{\mu\nu}$ is a worldvolume antisymmetric tensor field with a self-dual field strength.

There is a sense in which some of the super p-brane theories may be dual to each other. A super p-brane in d dimensions may be considered to be dual to a super $(d-p-3)$-brane in d dimensions, when each one exists as a soliton of the other. In a stronger sense, dual theories describe the strong and weak coupling regimes of the same physics, but this is rather difficult to prove rigorously. Among the duality relations probably the most interesting one is the heterotic string-heterotic fivebrane duality [46][47][52]. The two major open questions are how to construct the heterotic fivebrane action and how to covariantly quantize a κ-symmetric theory (or its twistor-like version).

As for the actions for super p-branes *in a physical gauge*, in principle the theory of nonlinear realizations applicable to spacetime supersymmetries can be used for their construction. This procedure has been illustrated for some special cases involving string solitons arising in certain globally supersymmetric field theories [53][3]. However, application of this procedure to higher super p-branes becomes quickly rather cumbersome [54]. Nonetheless, an interesting proposal has emerged in this area, namely, that the effective action for some of the super p-brane actions, in a certain limit, must reduce to certain supersingleton actions which live at the boundary of an appropriate anti de Sitter (AdS) space [55]. (A similar proposal had been made before according to which the action for fluctuations of a super p-brane theory compactified on a $AdS^{p+2}\times S^{d-p-2}$ were to be described by supersingleton theories formulated at the $S^p\times S^1$ boundary of of the AdS space [56][57]).

An alternative approach, which may have the additional advantage of possibly yielding a covariant action, is based on the twistor-like formulation of the super p-branes [26][27]. In this approach, the theory possesses a rigid target space supersymmetry and a local worldvolume supersymmetry. The κ–symmetry emerges as a special worldvolume supersymmetry transformation. The classical equivalence of the twistor–like and the usual κ–symmetric formulation of the heterotic string has been shown to hold [29][30]. A similar conclusion has been reached in ref. [26] for the case of $p = 2$. However, while we agree with the form of the action of ref. [26] and we generalize it to all super p–branes [27], it is by no means clear to us how this theory could be equivalent to the usual κ–symmetric one, even for the case of $p = 2$. When the dust settles on this issue, we will probably learn some intriguing and highly nontrivial aspects of these theories.

There exists the possibility that the formulation of ref. [27], after all, *is* inequivalent to the usual super p–brane action, and that it must be taken in its own right as a candidate for a novel super p–brane theory. In any event, the fact that we are now dealing with worldvolume supersymmetric field theories may give us a handle on the problem of how to construct the so far elusive heterotic and Type II super p–branes actions, as well as the problem of how to covariantly quantize the super p–brane theories.

ACKNOWLEDGEMENTS

I am deeply grateful to Professor Abdus Salam, in whose honor this conference has been held, for more than a decade of continued support, encouragement and inspiration. It is an honor and privilege to know and to have closely worked with a great scientist, a great visionary, a great man.

I am delighted to have been involved with the activities of one of Professor Abdus Salam's most wonderful creations, the International Centre for Theoretical Physics in Trieste, in the capacity of a staff member for many years. I have very fond memories of my Trieste years, working closely with Professor Abdus Salam and benefitting greatly from his profound wisdom. I have always been deeply impressed with his amazing ability to focus on the relevant issues, always going straight to the heart of the matter.

Over the years, I not only came to appreciate the great contributions of Professor Abdus Salam to theoretical physics, but I also witnessed closely his monumental contributions to the development of science in third world countries, through his relentless efforts, and with his continuous idea-generating capacities. I would like to take this opportunity to express my great admiration and respect for him.

REFERENCES

1. W. Siegel, Phys. Lett. **B128** (1983) 397.
2. M.B. Green and J.H. Schwarz, Phys. Lett. **136B** (1984) 367; Nucl. Phys. **B243** (1984) 285.
3. J. Hughes, J. Liu and J. Polchinski, Phys. Lett. **B180** (1986) 370.
4. E. Bergshoeff, E. Sezgin and P.K. Townsend, Phys. Lett. **B189** (1987) 75.
5. A. Achúcarro, J.M. Evans, P.K. Townsend and D.L. Wiltshire, Phys. Lett. **B198** (1987) 441.
6. R. Kallosh, Phys. Lett. **B224** (1989)273; Phys. Lett. **B225** (1989) 49;

   M. Green and C.M. Hull, Phys. Lett. **225B** (1989) 57;

   S.J. Gates, M. Grisaru, U. Lindström. P. van Nieuwenhuizen, M. Roćek, W. Siegel and A. van de Ven, Phys. Lett. **B225** (1989) 44.
7. M. Henneaux and L. Mezincescu, Phys. Lett. **B152** (1985) 340.
8. E. Witten, Nucl. Phys. **B266** (1986) 245.
9. Abdus Salam and J.Strathdee, Nucl. Phys. **B76** (1974) 477.
10. L. Brink and J.H. Schwarz, Phys. Lett. **B100** (1981) 310.
11. J.H. Schwarz, Nucl. Phys. **B185** (1981) 221;

    E.R.C. Abraham, P.S. Howe and P.K. Townsend, Class. Quant. Grav. **6**(1989) 1541.
12. J.A. Shapiro and C.C. Taylor, Phys. Rep. **191** (1990) 221.
13. W. Siegel, Nucl. Phys. **B263** (1985) 285; Phys. Lett. **B203** (1988) 79.
14. B.E.W. Nilsson, Nucl. Phys. **B188** (1981) 176.
15. E. Bergshoeff, P.S. Howe, C.N. Pope, E. Sezgin and E. Sokatchev, Nucl. Phys. **B354** (1991) 113.
16. J.J. Atick, A. Dhar and B. Ratra, Phys. Rev. **D33** (1986) 2824.
17. G. Mack and Abdus Salam, Ann. Phys. **53** (1969) 174.
18. M. Flato and C. Fronsdal, J. Math. Phys. **22** (1981) 1100;

    C. Fronsdal, Phys. Rev. **D26** (1982) 1988.
19. P.A.M. Dirac, J. Math. Phys. **4** (1963) 901.
20. I. Batalin and G. Vilkovisky, Phys. Rev. **D28** (1983) 2567.
21. E. Bergshoeff and R.E. Kallosh, Nucl. Phys. **B333** (1990) 605.
22. D.P. Sorokin, V.I. Tkach and D.V. Volkov, Mod. Phys. Lett. **A4** (1989) 901;

    D.P. Sorokin, V.I. Tkach, D.V. Volkov and A.A. Zheltukhin, Phys. Lett. **B216** (1989) 302.

23. N. Berkovits, Nucl. Phys. **B379** (1992) 96; **B395** (1993) 77.
24. I. Bengston and M. Cederwall, Nucl. Phys. **B302** (1988) 81.
25. N. Berkovits, Phys. Lett. **B247** (1990) 45.
26. P. Pasti and M. Tonin, preprint, DFPD/93/TH/07.
27. E. Bergshoeff and E. Sezgin, preprint, CTP TAMU-28/93, UG-5/93.
28. F. Delduc and E. Sokatchev, Class. Quantum Grav. **9** (1992) 361.
29. A. Galperin and E. Sokatchev, Phys. Rev. **D46** (1992) 714.
30. F. Delduc, A. Galperin, P. Howe and E. Sokatchev, Phys. Rev. **D47** (1993) 578.
31. A.S. Galperin, P.S. Howe and K.S. Stelle, Nucl. Phys. **B368** (1992) 248.
32. E.A. Ivanov and A.A. Kapustnikov, Phys. Lett. **B267** (1991) 175;
    J.P. Gauntlett, Phys. Lett. **B272** (1991) 25.
33. N. Berkovits, Phys. Lett. **B232** (1989) 184;
    F. Delduc, E. Ivanov and E. Sokatchev, Nucl. Phys. **B384** (1992) 334.
34. M. Tonin, Phys. Lett. **B266** (1991) 312; Int. J. Mod. Phys. **A7** (1992) 6013;
    N. Berkovits, Nucl. Phys. **B379** (1992) 96;
    D.P. Sorokin and M. Tonin, preprint, DFPD/93/TH/52.
35. M.T. Grisaru, P.S. Howe, L. Mezincescu, B.E.W. Nilsson and P.K. Townsend, Phys. Lett. **B162** (1985) 116.
36. E. Bergshoeff, E. Sezgin and P.K. Townsend, Phys. Lett. **169B** (1986) 191.
37. M. Tonin, Int. J. Mod. Phys. **3**(1988) 1519; Int. J. Mod. Phys. **4**(1989) 1983; in *Geometrical and Algebraic Aspects of Nonlinear Field Theory*, Eds. S. De Filippo, M. Marinaro, G. Marmo and G. Vilasi (North-Holland, 1989).
38. J.J. Atick, A. Dhar and B. Ratra, Phys. Lett. **B169** (86) 54.
39. L. Bonora, P. Pasti and M. Tonin, Phys. Lett. **B188** (1987) 335;
    L. Bonora, M. Bregola, K. Lechner, P. Pasti and M. Tonin, Nucl. Phys. **B296** (1988) 877.
40. S. Bellucci, S.J. Gates, Jr., B. Badak and S. Vashakidze, Mod. Phys. Lett. **A4** (1989)1985.
41. E. Cremmer and S. Ferrara, Phys. Lett. **B91** (1980) 61;
    L. Brink and P.S. Howe, Phys. Lett. **B91** (1980) 384.
42. M.J. Duff, P.S. Howe, T. Inami and K.S. Stelle, Phys. Lett. **B191** (1987) 70.
43. A.H. Chamseddine, Phys. Rev. **D24** (1981) 3065.
44. Abdus Salam and E. Sezgin, *Supergravities in Diverse Dimensions* (World Scientific, 1989).
45. M. Blencowe and M.J. Duff, Nucl. Phys. **B310** (1988) 387.

46. M.J. Duff, Class. Quant. Grav. **5** (1988) 189.

47. A. Strominger, Nucl. Phys. **B343** (1990) 167. [Erratum: **B353** (1991) 565].

48. C. Callan, J. Harvey and A. Strominger, Nucl. Phys. **B359** (1991) 611; **B367** (1991) 60.

49. M.J. Duff and J.X. Lu, Nucl. Phys. **B390** (1993) 276.

50. J. Harvey and A. Strominger, Phys. Rev. Lett. **66** (1991) 549;
R. Güven, Phys. Lett. **B276** (1992) 49.

51. M.J. Duff and J.X. Lu, preprint, CERN-TH.6675/93, CTP/TAMU-54/92.

52. M.J. Duff and J.X. Lu, Phys. Rev. Lett. **66** (1991) 1402; Nucl. Phys. **B357** (1991) 534.

53. J. Hughes, J. Liu and J. Polchinski, Nucl. Phys. **B278** (1986) 147.

54. E.A. Ivanov and A.A. Kapustnikov, Phys. Lett. **B252** (1990) 212.

55. G.W. Gibbons and P.K. Townsend, preprint, DAMTP/R-93/19.

56. M.P. Blencowe and M.J. Duff, Phys. Lett. **B203** (1988) 229.

57. H. Nicolai, E. Sezgin and Y. Tanii, Nucl. Phys. **B305** (1988) 483.

Conformal Symmetry and Quantum Space Time

Gerhard Mack
II. Institut für Theoretische Physik, Universität Hamburg, Germany

Abstract

We wish to compound a global space time, which is quantum, from local Lorentz frames on which physics is described by a conformal local quantum field theory on a classical space time of some dimension. Quantum field planes furnish a noncommutative differential algebra Ω which substitutes for the commutative algebra of functions and forms on a contractible manifold. The basic data to construct it come from the conformal field theory in the local Lorentz frame.

1 Pleasant memories of Trieste

It is a great pleasure for me to speak at this reunion in honour of Abdus Salam.

I came to Trieste in 1968. Abdus Salam had read my thesis, which dealt with dilatation and conformal symmetry in a Lagrangean frame work. He congratulated me, saying that I had convinced him that conformal symmetry was not a symmetry for the crackpots. (It was suspect because it seemed to interchange spacelike and timelike points, see [18]) He invited me to Trieste and to work with him on conformal symmetry. A joint paper resulted out of this collaboration [1]. But before that I was told that all the world was described by gauge theories, and I had better checked whether the Lagrangians of Yang Mills theories were conformal invariant. It was found that they were.

I got my first introduction to the spirit of the place by Miriam Lewis. She was the head of the publications department. She also published scientific papers once a year, on April 1st, which puzzled many who suspected something very profound. Miriam had the most exquisite english sense of humour. With a perfectly straight face she told me that Abdus Salam had a chain gang which was kept in the cellar, to be left out to get some fresh air once a week. But it was not like that at all. He was like a father to his collaborators. He explained the philosophy and vision behind the center, and how one goes about getting money for such a place. The sincere enthousiasm with which he worked for the scientists in the Third World was overwhelming.

After my first few months there, the center moved from Piazza Oberdan to Miramare. And soon the place filled with unforgettable characters from all over the world. There was Daniel Akyeampong from Ghana, who was always in good spirits. His laughter could be heard through the halls and it brightened our days. There was M.O. Taha from Sudan who never took off his tie. There was Father Ryan from Ireland. He amazed the local girls who met him at the beach. He died later in a tragic mountain accident together with Bruno Renner. There was Nasrallah Nasrallah from Lebanon who chose to be called MichelAngelo, because our teacher of italian could not pronounce his name. There were also Israelis at the time, as Israel was still a member of UNESCO then. That lead to some lively discussions around a ping pong table which I remember.

All the bachelors practically lived in the center. And we became very good friends, in spite of all the differences in cultural background, or just because of it. Wherever I go now in the world, I usually meet somebody who remembers me from Trieste.

I think this kind of understanding, solidarity and friendship between people from different cultures is what the world needs most in these days of global threat to the basis of human life on this planet. Powerful forces are needed to counteract the widespread human tendency to interpret any difference as a difference in value, fit to confirm one's own superiority. Otherwise hate will govern the world - look what is happening in former Yugoslavia.

For this reason I believe that the world needs 1000 such centers, in all fields of human endeavor, but with a clear will to separate knowledge from agitation and propaganda. It needs centers where people from different cultures can meet and be friends. And the International Center of Theoretical Physics in Trieste, which was created by Abdus Salam can serve as a model.

2 Towards quantum space time

2.1 Parallel transport of germs of curves

According to the basic principles of general relativity, there is *no a priori way* of comparing directions of vectors at different points of space time: one needs *dynamically determined parallel transporters*

This also determines "straight lines" = curves with parallel tangent vectors.

Classically, one imagines however, that there is an *a priori definition of straightness in the infinitesimally small*, i.e. an a priori differential structure. When this is given up, one is lead to a string theory.

classical: In the small, all curves (e.g. world lines of particles) are straight. Their germs are tangent vectors.

quantum: Observation at short distances requires input of energy which curves space time. There is then no reason to restrict attention to curves which are straight in the infinitesimally small. [1] Germs of curves will then become vectors in an ∞-dimensional Hilbert space.

Dynamical laws can only involve objects which can be parallel transported.

The space of germs carries a representation of the group of reparametrizations of an interval I with base point $\hat{\tau}$, $\tau \mapsto f(\tau)$ with $F(\tau) = \tau$ for $\tau \notin I \setminus \{\hat{\tau}\}$. Once a parallel transport is fixed, the restriction $f(\hat{\tau}) = \hat{\tau}$ could be lifted because one may combine the action of the diffeomorphism with parallel transport along the curve. In any case, the interval I may be regarded as a subset of the circle S^1, and the reparametrtizatons make a subgroup of $Diff(S^1)$ or of its ∞-sheeted universal covering group.

Thus the action of reparametrizations f on germs is as follows

classical: $v \mapsto \lambda_f v$; $\lambda_f = f'(\hat{\tau}) \in \mathbf{R}$

(1-dimensional representation of the group algebra of $Diff(S^1)$)

[1] One might in fact have to consider "curves" which are not continuous, and conventional string theory does so [34]

quantum: $v \mapsto \lambda_f v;\ \lambda_f \in \mathcal{A} = U(\mathbf{C}Diff(S^1))$
where U is a ∞-dimensional representation of the group algebra of $Diff(S^1)$.

This suggests the following strategy: Start from general relativity and replace $\mathbf{R}$ by a noncommutative algebra $\mathcal{A}$ which includes localized diffeomorphisms of the circle. This will necessitate the use of noncommutative differential geometry. The elements of $\mathcal{A}$ will be called a-*numbers* in the following, since they will substitute for numbers.

Classical general relativity can be regarded as a gauge theory with gauge group the Lorentz group $SO(3,1) = SO(3,1) \otimes \mathbf{R}$. This gauge group will have to be replaced by a gauge symmetry of the form $\mathcal{G} \otimes \mathcal{A}$.

Causality

A notion of positive/negative timelike vs spacelike is basic in relativistic physics. Usually it is defined by reference to the ordering in $\mathbf{R}$. There is a scalar product $<,>$ in the vector space of germs which takes values in $\mathbf{R}$ classically, and v is timelike if $< v, v > > 0$. When $\mathbf{R}$ is replaced by $\mathcal{A}$, we retain an appropriate notion of $>$ because the universal covering $Diff^{\sim}(S^1)$ of the diffeomorphism group of the circle is a group with a global causal ordering:

$$A > 0 \text{ if } A = Ue^{-\tau L_0}U^{-1}$$

with $\tau > 0$ and U any element of the group $Diff^{\sim}(S^1)$ [31]. (Remark: This global causal ordering restricts to the ∞-sheeted covering of $SL(2,\mathbf{R})$) .

Divorce of the Lorentz group

Classical general relativity may be regarded as a gauge theory with gauge group $SO(3,1)$. Note that the local symmetry in global space time is a global symmetry in local Lorentz frames. Of course, $SO(3,1)$ is noncompact. But its finite dimensional representations are tensor products of the "left handed" and "right handed" fundamental representations $(0,\frac{1}{2})$ and $(\frac{1}{2},0)$. These representations are both unitary irreducible representations of $SU(2)$, extended to $SL(2,\mathbf{C})$ by analyticity.

In their geometric interpretation of the standard model, Connes and Lott proposed that space time is double sheeted [5]. It is then natural to postulate that left handed matter fields live on one sheet, and right handed matter fields on the other. In this way the Lorentz group gets divorced into two SU(2) groups. At the level of the fundamental theory there is no need for a group which accomodates representations which are tensor products of representations for left handed and for right handed matter. That comes only at the level of effective theories at length scales longer than the distance between the two sheets of space time.

This suggests that we may replace the Lorentz group by a compact group like SU(2) or possibly a deformation of it, i.e. by some weak quasi Hopf algebra [22].

Since the algebra $\mathcal{A}$ is to become part of the local gauge symmetry, there will be two algebras $\mathcal{A}^{left}$ and $\mathcal{A}^{right}$ attached to to the two sheets of space time.

3 Noncommutative differential geometry

In noncommutative differential geometry, the algebras $\mathcal{F}$ of functions and $\Omega \supset \mathcal{F}$ of forms on a manifold get replaced by possibly noncommutative algebras. Noncommutativity may enter in several stages. In the biggest generality, the notion of point (of an underlying manifold) is given up completely, i.e. the algebra of functions has no ideals. In any case, there will be an exterior derivative d which acts on Ω and which obeys

$$d^2 = 0 \quad , \quad d(\eta\omega) = (d\eta)\omega \pm \eta(d\omega) \quad \text{(Leibniz rule)} \tag{1}$$

with + if ω is an even form, and - otherwise. This requires that Ω is $\mathbf{Z}_2$-graded. (Usually it is assumed that it is $\mathbf{Z}$-graded).

In Connes and Lott's reinterpretation of the standard model, the algebra $\mathcal{F}$ of functions remains commutative.

Müller Hoissen et al. pointed out that the finite difference calculus on a lattice and standard lattice gauge theory may be regarded as examples of a noncommutative differential calculus and geometry [27]. Again, the algebra $\mathcal{F}$ remains a commutative algebra of functions. This example may help to convince the reader that noncommutative differential calculus is nothing exotic. Therefore I mention it briefly.

The algebra $\mathcal{F}$ of functions is generated by coordinate functions x^μ. The algebra of forms is obtained by adjoining differentials $\theta^\mu = dx^\mu$ to $\mathcal{F}$. The differentials anticommute as usual, viz. $dx^\mu \wedge dx^\nu = -dx^\nu \wedge dx^\mu$. Therefore $(dx^\mu)^2 = 0$. But coordinates and differentials do not commute,

$$[x^\mu, dx^\nu] = adx^\nu \quad \text{if } \mu = \nu$$

and $= 0$ otherwise. The parameter $a \in \mathbf{R}$ parametrizes the deformation of the standard differential calculus and will acquire the meaning of lattice spacing.

Consider the 1-dimensional case - generalization is obvious. The commutation relations take the form

$$xdx = dx(x + a) \tag{2}$$

It follows that

$$f(x)dx = dxf(x + a) \tag{3}$$

for all power series $f(x) = \sum c_n x^n$. df is a 1-form, therefore

$$df(x) = dx f'(x) \tag{4}$$

with a function f' which is determined by the requirement that d applied to x is dx, and the Leibnitz rule. It follows that

$$f'(x) = \frac{1}{a}[f(x + a) - f(x)] \tag{5}$$

One may adjoin an operator ∇ to the algebra (right partial derivative) such that $f' = \nabla f$. ∇ obeys relations

$$\nabla x = 1 + (x + a)\nabla \quad , \quad [\nabla, dx] = 0 \tag{6}$$

The Leibniz rule, which is usually thought to be modified on the lattice, gets repaired by the noncommutativity of dx and x. This is true both for d and for ∇.

Still more amusing is the fact that one may retain the usual formulation of continuum gauge theory also on the lattice. Let e_μ be the lattice vector of length a in μ-direction. In Wilsons lattice gauge theory formulation one works with parallel transporters $U(x + e_\mu, x)$ along links from x to $x + e_\mu$. They take their values in the gauge group G and transform under gauge transformations according to

$$U(x + e_\mu, x) \mapsto S(x + e_\mu)U(x + e_\mu, x)S(x)^{-1} .$$

Define a vector potential A as a 1-form by

$$\begin{aligned} A(x) &= dx^{\mu} A_{\mu}(x) \ , &(7)\\ U(x+e_{\mu},x) &= 1 - aA_{\mu}(x) \ . &(8)\end{aligned}$$

Thanks to the noncommutativity of dx^{μ} and $S(x)$, this vector potential transforms in the familiar way under gauge transformations

$$A(x) \mapsto A'(x) = S(x)A(x)S(x)^{-1} + dS(x)A(x)S(x)^{-1} \tag{9}$$

One may also define the field strength F in the standard way, as well as a Hodge *-operation, and "integrals"=lattice sums . The Wilson action S then takes the familiar form

$$\begin{aligned} F &= F_{\mu\nu} dx^{\mu} dx^{\nu} = dA + A^2 \ , &(10)\\ S &= \int tr(*FF^*) &(11)\end{aligned}$$

4 From quantum field theory to quantum space time

The analysis of conformal field theory with the tools of algebraic field theory [7,8,9,12,21] yields some of the commutation relations which one needs for a noncommutative differential calculus, and it yields the "gauge group". Diffeomorphisms of the circle are part of this "gauge group". They enter as multiplication with unitary a-numbers.

Conformal field theory in Minkowski space

Conformal field theory in 2-dimensional Minkowski space can be analytically continued to the tube [18]. If we make the right-left split suggested by Connes and Lott's picture of a two sheeted (quantum) space time we are led to consider chiral conformal field theories which live on the circle (or its covering space). The algebra of observables $\mathcal{A}$ of this theory includes bounded functions of the stress energy tensor, in particular the group algebra of $Diff^{\sim}(\mathbf{S}^1)$. Its generator L_0 serves to define the notion of positive energy representation.

The Hilbert space of states of this chiral theory decomposes into a number of superselection sectors, i.e. subspaces $\mathcal{H}^J$ carrying inequivalent irreducible representations π^J of $\mathcal{A}$ [36]. This includes the vacuum sector $\mathcal{H}^0$ which contains the vacuum $|0>$. In conformal field theory on the circle, $|0>$ is the lowest weight vector of the Virasoro algebra with weight 0. π^0 is called the vacuum representation.

Starting from the information in the observable algebra, one constructs field operators Ψ^J_m which make transitions between different superselection sectors, and a gauge symmetry $\mathcal{G}$ of first kind. $\mathcal{G}$ is a weak quasiassociative quasitriangular *-bialgebra in the most general case. Groups are special cases. In general, $\mathcal{G}$ is not uniquely determined. The properties which it has to fulfill will be mentioned below. Under standard assumptions of algebraic field theory, a suitable $\mathcal{G}$ is guaranteed to exist at least in the rational case, i.e. when the total number of superselection sectors is finite [32]

In algebraic field theory one prefers to work with bounded observables. The algebra $\mathcal{A}$ of observables contains subalgebras $\mathcal{A}(\mathcal{O})$ attached to bounded (and topologically trivial) domains of space time. In chiral conformal theory, $\mathcal{O}$ are intervals on the circle, and $\mathcal{A}(\mathcal{O})$ includes the diffeomorphisms of the circle which act trivially outside $\mathcal{O}$. Observables in subalgebras for relatively spacelike domains commute. On the circle, "relatively spacelike" means disjoint.

Representations of the observable algebra

One starts from general assumptions of algebraic field theory. These assumptions are fulfilled in conformal field theories, provided one completes $\mathcal{A}(\mathcal{O})$ to von Neumann algebras (acting in $\mathcal{H}^0$), and the global algebra $\mathcal{A}$ is constructed from the local algebras $\mathcal{A}(\mathcal{O})$ in an appropriate way. [2,?]. It follows that there exist localizable unital *-endomorphisms

$$\rho^J : \mathcal{A} \mapsto \mathcal{A} \tag{12}$$

such that all positive energy representations can be obtained from the vacuum representation by composition with these morphisms

$$\pi^J = \pi^0 \circ \rho^J \quad \text{i.e.} \quad \pi^J(A) = \pi^0(\rho^J(A)). \tag{13}$$

By definition, unital *-endomorphisms of $\mathcal{A}$ are C-linear maps subject to the conditions

$$\begin{aligned} \rho(AB) &= \rho(A)\rho(B) \ , \\ \rho(A^*) &= \rho(A)^* \ , \\ \rho(\mathbf{1}) &= \mathbf{1} \ . \end{aligned} \tag{14}$$

Localizability means the following. To every bounded space time domain $\mathcal{O}$ there exists a unitary $u \in \mathcal{A}$ such that the morphism

$$\rho^J_u(A) = u^* \rho^J(A) u, \tag{15}$$

is localized in $\mathcal{O}$. This means that

$$\rho^J_u(A) = A \ \text{ if } \ \ A \in \mathcal{A}(\mathcal{O}') \tag{16}$$

and if $\mathcal{O}'$ is relatively spacelike to $\mathcal{O}$.

Using these morphisms, a tensor product of irreducible positive energy representations of $\mathcal{A}$ can be defined as

$$\pi^I \times \pi^J = \pi^0 \circ \rho^I \circ \rho^J \ . \tag{17}$$

These tensor products will decompose into irreducibles according to

$$\pi^I \times \pi^J = \sum_K \pi^K N^{IJ}_K \tag{18}$$

with multiplicities N^{IJ}_K.

Statistics operators

It follows from the standard assumptions of algebraic field theory that the observable algebra determines statitics operators $\epsilon(\rho^J, \rho^K) \in \mathcal{A}$ which enjoy standard properties [16].

One defines spaces of intertwiners between morphisms,

$$T \in \mathcal{T}(\rho_1, \rho_2) \subset \mathcal{A} \quad \text{iff } T\rho_1(A) = \rho_2(A)T \quad \text{for all } \ A \in \mathcal{A} \ .$$

For every pair of localizable endomorphisms $\rho_i, i = 1, 2$ there is a unitary local intertwiner , the

$$\text{statistics operator} \quad \epsilon(\rho_1, \rho_2) \in \mathcal{T}(\rho_1 \circ \rho_2, \rho_2 \circ \rho_1)$$

Explicitly, its intertwining property is

$$\epsilon(\rho_1,\rho_2)\rho_1\circ\rho_2(A) = \rho_2\circ\rho_1(A)\epsilon(\rho_1,\rho_2)$$

The collection of statistics operators in uniquely determined by the following equations (a detailed proof can be found in [16])

$$\begin{aligned}
&\epsilon(\rho_1,\rho_2)\rho_1(T_2)T_1 = T_2\sigma_2(T_1)\epsilon(\sigma_1,\sigma_2) \qquad \text{for all} \qquad T_i\in\mathcal{T}(\sigma_i,\rho_i)\ ,\\
&\epsilon(\rho_1\circ\rho_2,\sigma) = \epsilon(\rho_1,\sigma)\rho_1(\epsilon(\rho_2,\sigma))\ ,\\
&\epsilon(\sigma,\rho_1\circ\rho_2) = \rho_1(\epsilon(\sigma,\rho_2))\epsilon(\sigma,\rho_1)\ ,\\
&\epsilon(\rho_1,\rho_2) = 1 \qquad \text{whenever} \qquad \rho_1 > \rho_2\ .
\end{aligned} \tag{19}$$

In the last row, $\rho_1 > \rho_2$ if ρ_1 and ρ_2 are localized in relatively spacelike domains, and if these domains are ordered, when they can be ordered. This last property will not be used here. On the circle, ordering is left or right, after a reference point on the circle has been chosen. Trivialization for $\rho_1 < \rho_2$ would give rise to the
opposite statistics operator $\epsilon(\rho_2,\rho_1)^*$

Symmetry

A *-bialgebra $\mathcal{G}$ is a complex algebra with a product and a coproduct $\Delta:\mathcal{G}\otimes\mathcal{G}\mapsto\mathcal{G}$, a unit e and a counit $\epsilon:\mathcal{G}\mapsto\mathbf{C}$ subject to standard requirements [22]. Δ is a not necessarily unital *-homomorphism of algebras, i.e. $\Delta(e)\neq e\otimes e$ is allowed. Tensor products of irreducible representations of $\mathcal{G}$ are constructed with the help of the coproduct,

$$\tau^I \boxtimes \tau^J(\xi) = \tau^I\otimes\tau^J(\Delta(\xi)).$$

$\otimes$ is the outer tensor product, it yields a representation of $\mathcal{G}\otimes\mathcal{G}$. $\mathcal{G}$ is called coassociative if the coproduct is coassociative, i.e. if the tensor product is literally associative, not just associative up to equivalence.

We assume that a semisimple *-bialgebra is chosen with the property that its inequivalent irreducible representations τ^I are in bijective correspondence with the inequivalent positive energy representations of $\mathcal{A}$, and the decomposition of their tensor products matches with the decomposition (18)

$$\tau^I\boxtimes\tau^J = \sum_K \tau^K N_K^{IJ} \tag{20}$$

The existence of such a *-bialgebra is guaranteed if there is only a finite number of superselection sectors $\mathcal{H}^J$. The data furnished by the observable algebra permit to construct a reassociator $\varphi \in \mathcal{G} \otimes \mathcal{G} \otimes \mathcal{G}$ and an

R-element $R \in \mathcal{G} \otimes \mathcal{G}$ which make $\mathcal{G}$ into a weak quasiassociative quasitriangular *-bialgebra [23,32].

Field algebra

Starting from the observable algebra $\mathcal{A}$ and its morphisms, one can construct an algebra $\mathbf{\Phi}$ of field operators (field algebra) which acts in the Hilbert space

$$\mathcal{H} = \sum_K \mathcal{H}^K \otimes V^K \ . \tag{21}$$

V^K is the representation space of $\mathcal{G}$ in which acts τ^K.

In algebraic field theory, one prefers to work with bounded field operators Ψ. There will be subalgebras $\mathbf{\Phi}(\mathcal{O})$ of fields which are localized in $\mathcal{O}$. If $\Psi \in \mathbf{\Phi}(\mathcal{O})$ and $A \in \mathcal{A}(\mathcal{O}')$, and if $\mathcal{O}$ and $\mathcal{O}'$ are relatively spacelike, then Ψ and A commute. Fields $\Psi(z)$ at a point z would have to be constructed as suitable limits when $\mathcal{O}$ shrinks to a point. This is a familiar procedure also in loop group theory [28]. For an explicit example, see [3].

Morphisms ρ^J_u of $\mathcal{A}$ which differ by inner automorphisms generate equivalent representations. We choose one representative ρ^J once and for all such that $\pi^J = \pi^0 \circ \rho^J$. In chiral conformal theories it is convenient to choose it so that $|0>$ is a lowest weight vector also for the representation $\pi^0 \circ \rho^J$. (For the chiral conformal Ising model, these morphisms were explicitly described in [21].)

By explicit construction one may exhibit protofields Θ^J_j, ($j = 1...n_J = dimV^J$) with the following properties [23,32]

1. The protofields are operators in $\mathcal{H}$.

2. They are $\mathcal{G} \otimes A$ covariant in the sense that

$$\begin{aligned} \xi A \Theta^J_i &= \Theta^J_j \mu^J_{ji}(\xi A) \ (A \in \mathcal{A}, \ \xi \in \mathcal{G}) \ , & (22) \\ \mu^J_{ji}(\xi A) &= \rho^J(A)(\tau^J_{ji} \otimes id)(\Delta(\xi)) & (23) \end{aligned}$$

3. They obey operator product expansions

$$\Theta^J_j \Theta^K_k = \sum_{Mm} \Theta^M_m B[^M_m \ ^J_j \ ^K_k] \tag{24}$$

The coefficients $B[^M_m \ ^J_j \ ^K_k] \in \mathcal{G} \otimes \mathcal{A}$ in general. If $\mathcal{G}$ is coassociative, they are actually in $\mathcal{A}$. They possess the intertwining property

$$\sum_{j',k'} B[^M_m \ ^J_{j'} \ ^K_{k'}](\mu^K \circ \mu^J)_{k'j',kj}(\xi A) = \sum_{m'} \mu^M_{mm'}(\xi A) B[^M_{m'} \ ^J_j \ ^K_k] \tag{25}$$

$$(\mu^K \circ \mu^J)_{ki,jl}(\xi) = \mu^K_{kj}(\mu^J_{il}(\xi)) \,. \tag{26}$$

4. Local braid relations are satisfied,

$$\Theta^J_j \Theta^K_k \mathcal{R}^{JK+}_{jk,lm} = \Theta^K_l \Theta^J_m \gamma_0 \tag{27}$$

with coefficients $\mathcal{R}^{JI+}_{jk,lm} \in \mathcal{G} \otimes \mathcal{A}$ which enjoy the properties of a statistics operator for $\mathcal{G} \otimes \mathcal{A}$, and γ_0 is a complex number of modulus 1.

The braid relations are consistent with the operator product expansions thanks to the braiding properties of the Clebsch Gordan coefficients.

$$\mathcal{R}^{KL+}_{km,pr} \mu^L_{ml}(B[^K_k \ ^I_i \ ^J_j]) = B[^K_r \ ^I_q \ ^J_s] \mu^J_{sn}(\mathcal{R}^{IL+}_{im,pq}) \mathcal{R}^{JL+}_{jl,mn} \tag{28}$$

μ and $B, \mathcal{R}^+$ are amplimorphisms and intertwiners between them in the sense of Vecsernyes and Szlachanyi, but our notational conventions are different from [30]. Note that the covariance property passes to products in a contravariant way:

$$\zeta \Theta^I \Theta^J = \Theta^I \Theta^J \mu^J \circ \mu^I(\zeta)$$

The explicit formula for $\mathcal{R}^{JI+}$ in terms of the statistics operator for $\mathcal{A}$ and the reassociator φ and R-matrix for $\mathcal{G}$ is as follows.

$$\mathcal{R}^{IJ+} = \epsilon(\rho^I, \rho^J)^* P(\tau^I \otimes \tau^J \otimes id)(\varphi_{213}(R \otimes e)\varphi^{-1}) \tag{29}$$

The standard notation φ_{213} is explained in eq.(49)below, and P ist the permutation matrix $P_{ij,kl} = \delta_{il}\delta_{jk}$.

The braiding properties of Clebsch Gordan coefficients follow from properties of statistics operators, cp. eqs. (19) for $\mathcal{A}$. The same properties hold for the intertwiners for $\mathcal{G}$ [30] which also enter into $\mathcal{R}^+$.

For the construction of the quantum field planes it is important that $\mathcal{R}^{II+}$ admit a decomposition in idempotents $\in M_{N^2}(\mathcal{K})$ with complex coefficients of modulus 1, as in eq.(40). This decomposition follows from similar ones for $P(\tau^I \otimes \tau^I)(R) \in M_N(\mathbf{C})$ and for $\epsilon(\rho^I, \rho^I) \in \mathcal{A}$. They are standard. ($N = dimV^I$).

The operator product expansions for protofields are reminiscent of operator product expansions in topological field theory [6]. The protofields have no space time dependence. But fields localized in domains $\mathcal{O}$ of space time can be constructed from them. Given $\mathcal{O}$, choose any unitary $u \in \mathcal{A}$ such that the morphism ρ^J_u is localized in $\mathcal{O}$. Such u exist because ρ^J is localizable. Define

$$\Psi^J_j(\rho^J_u) = \Theta^J_j u \ . \tag{30}$$

Then

$$A\Psi^J_j(\rho^J_u) = \Psi^J_j(\rho^J_u)\rho^J_u(A) = A \ , \tag{31}$$

if $A \in \mathcal{A}(\mathcal{O}')$ and if $\mathcal{O}'$ is relatively spacelike to $\mathcal{O}$. In other words, $\Psi^J_j(\rho^J_u)$ has the locality properties of a field in $\mathbf{\Phi}(\mathcal{O})$. Fields at a point would have to be constructed as limits. They would have the form

$$\Psi^J_j(z) = \Theta^J_j a^J(z)$$

with $a^J(z)$ affiliated with $\mathcal{A}$. [2]

For simplicity we will assume in the following that all irreducible representations of the symmetry algebra $\mathcal{G}$ are contained in tensor products of a single fundamental representation τ^f. It follows that there is a fundamental protofield multiplet

$$\Theta_i = \Theta^f_i \tag{32}$$

such that the whole field algebra is generated by these Θ_i's together with A.

[2] Actually, the construction of the local field algebras is somewhat subtle. If one includes all the fields with the right locality properties, the algebras $\mathbf{\Phi}(\mathcal{O})$ become too large to have some desirable properties. These subtleties can be dealt with in a way which preserves the Reeh Schlieder property, see [32]

Operator product expansions

Operator product expansions for the protofields imply operator product expansions for the local fields. They are partially summed Wilson operator product expansions: All the derivatives of quasiprimary fields are summed up [11,20], and all descendent fields associated with the same primary field are summed up, resulting in coefficients with values in $\mathcal{A}$ - i.e. functions of the stress energy tensor in minimal conformal theories. The operator product expansions take the form

$$\Psi_j^J(\rho_u^J)\Psi_k^K(\rho_v^K) = \sum_{Mm} \Psi_m^M(\rho^M)B[{}^J_j\ {}^K_k\ {}^M_m]v\rho_v^K(u)\ .. \tag{33}$$

The localization properties of ρ^M can be chosen arbitrarily, but the coefficient B depends on the choice of ρ^M.

Note that an observable $A \in \mathcal{A}(\mathcal{O})$ commutes with every term in the operator product expansion if the localization regions of ρ_u^J and ρ_v^K are relatively spacelike to $\mathcal{O}$.

5 Quantum field planes

We wish to compound a global space time, which is quantum, from local Lorentz frames on which physics is described by a special relativistic local quantum field theory which lives on a classical space time of some dimension. Quantum field planes furnish examples of a noncommutative differential algebra $\Omega \supset \mathcal{F}$ which substitutes for the commutative algebra of forms and functions on a contractible manifold $\mathcal{M} \simeq \mathbf{C}^N$.

$$\Omega = \mathcal{F} \otimes \mathbf{\Phi} \tag{34}$$

as a linear space. $\mathbf{\Phi}$ substitutes for $\wedge T_x^*\mathcal{M}$.

It is convenient to consider right away algebras $\mathcal{F}_\mathcal{K}$ and $\Omega_\mathcal{K}$ which substitute for algebras of $\mathcal{G}$-valued functions and forms. In this section their definition will be given. The consistency proofs which show their existence will be detailed elsewhere [25]. We call these algebras "quantum field planes". They are generalizations of the quantum planes of Manin [26] and of Wess and Zumino [33].

The elements $\theta_a = dz_a$ of the fibre $T_x\mathcal{M}$ of cotangent space get replaced by field operators of a local field theory, and the exterior algebra $\wedge T_x\mathcal{M}$ (or the Clifford algebra) gets replaced by the algebra $\mathbf{\Phi}$ of these field operators. The symmetry of the quantum field theory may be a group or quantum group or, more generally, a weak quasiassociative quasitriangular *-bialgebra $\mathcal{G}$. The exterior derivative d satisfies $d^2 = 0$ and the Leibniz rule; it commutes with $\zeta \in \mathcal{G}$ and with multiplication by a-numbers. These a-numbers are elements of the observable algebra $\mathcal{A} \subset \mathbf{\Phi}$, i.e. $\mathcal{G}$-invariant elements of $\mathbf{\Phi}$. Under more restrictive assumptions, a Hodge *-operation also exists. The global gauge symmetry $\mathcal{G} \otimes \mathcal{A}$ in the local Lorentz frames may be elevated to a local symmetry algebra $\mathcal{F}_\mathcal{K}$ such that

$$D = d + \omega$$

transforms like a covariant derivative ($\omega = \sum_i \omega_i \Theta_i$, $\omega_i \in \mathcal{F}_\mathcal{K}$.) In a chiral conformal field theory $\mathbf{\Phi}$, the gauge symmetry $\mathcal{G} \otimes \mathcal{A}$ includes the diffeomorphisms of the circle.

We proceed to a more formal description.

The term *algebra* shall mean complex algebra with unit element throughout.

Definition 1 (morphism) *A morphism ρ of a *-algebra $\mathcal{A}$ is a unit-preserving *-homomorphism of $\mathcal{A}$. Thus $\rho : \mathcal{A} \mapsto \mathcal{A}$ is a $\mathbf{C}$-linear map satisfying eqs. (14). The product of morphisms ρ and σ is defined in the natural way,*

$$\rho \circ \sigma(a) = \rho(\sigma(a)) \quad (a \in \mathcal{A}). \tag{35}$$

Definition 2 (amplimorphism) *[30] An amplimorphism of dimension N of an algebra $\mathcal{K}$ is a not necessarily unit preserving homomorphism*

$$\mu : \mathcal{K} \mapsto M_N(\mathcal{K}) \tag{36}$$

Products of amplimorphisms μ and ν of dimensions M and N are amplimorphisms of dimension MN and are defined by

$$(\mu \circ \nu)_{ik,jl}(\zeta) = \mu_{ij}(\nu_{kl}(\zeta)) \tag{37}$$

Definition 3 ($\mathcal{K}$-covariant quantum plane) *Let $\mathcal{K}$ be an algebra and μ an amplimorphism of $\mathcal{K}$ of dimension N.*

An N-dimensional quantum plane with symmetry $\mathcal{K}$, fibre $\Phi_{\mathcal{K}}$, fundamental amplimorphism μ and exterior derivative d is a $\mathbf{Z}_2$-graded algebra $\Omega_{\mathcal{K}} = \Omega_{\mathcal{K}}^+ + \Omega_{\mathcal{K}}^-$ with distinguished subalgebras $\Phi_{\mathcal{K}}$ and $\mathcal{F}_{\mathcal{K}} \subset \Omega_{\mathcal{K}}^+$ and properties as follows.

i) $\Phi_{\mathcal{K}}, \mathcal{F}_{\mathcal{K}}$ and $\Omega_{\mathcal{K}}$ are $\mathcal{K}$-bimodules, and $\Omega_{\mathcal{K}} = \mathcal{F}_{\mathcal{K}} \otimes \Phi_{\mathcal{K}}$ as a $\mathcal{K}$-bimodule.

ii) The algebra $\Phi_{\mathcal{K}}$ is generated by $\mathcal{K}$ and generators $\Theta_i (i = 1...N)$. $\mathcal{K}$-covariance (38) and braid relations (39) hold as stated below.

$$\zeta\Theta_j = \Theta_i\mu_{ij}(\zeta) \quad (\forall\zeta \in \mathcal{K}) \tag{38}$$

$$\Theta_i\Theta_j\mathcal{R}^+_{ij,kl} = \Theta_k\Theta_l\gamma_0 , \tag{39}$$

$\mathcal{R}^+ \in M_{N^2}(\mathcal{K})$; it admits a spectral decomposition in orthogonal idempotents $E_\alpha \in M_{N^2}(\mathcal{K})$ with complex eigenvalues γ_α, including $\gamma_0 \neq 0$.

$$\mathcal{R}^+ = \sum_{\alpha=0}^{M} \gamma_\alpha E_\alpha , \tag{40}$$

$$\mu \circ \mu(\zeta)E_\alpha = E_\alpha\mu \circ \mu(\zeta) \quad (\forall\zeta \in \mathcal{K}) . \tag{41}$$

iii) The algebra $\mathcal{F}_{\mathcal{K}}$ is generated by $\mathcal{K}$ and generators $Z_i (i = 1...N)$ subject to the $\mathcal{K}$-covariance and braid relations as follows (we use matrix notation)

$$\zeta Z = Z\mu(\zeta) \quad \forall\zeta \in \mathcal{K} , \tag{42}$$

$$ZZ\prod_{\alpha=1}^{M}(\mathcal{R}^+ - \gamma_\alpha 1) = 0 . \tag{43}$$

iv) The algebra $\Omega_{\mathcal{K}}$ is generated by $\Phi_{\mathcal{K}}$ and $\mathcal{F}_{\mathcal{K}}$ subject to the following additional braid relations among generators

$$Z_i\Theta_j\mathcal{R}^+_{ij,kl} = -\Theta_k Z_l\gamma_0 \tag{44}$$

v) The exterior derivative d obeys $d^2 = 0$, and the following defining relations make an algebra out of $\Omega_{\mathcal{K}} + d\Omega_{\mathcal{K}}$.

$$d\zeta = \zeta d \quad (\forall\zeta \in \mathcal{K}) \tag{45}$$

$$dZ_i = Z_i d + \Theta_i , \tag{46}$$

$$d\Theta_i = -\Theta_i d \tag{47}$$

In general, the entries $\mathcal{R}^{+}_{ij,kl}$ of $\mathcal{R}^{+}$ are elements of $\mathcal{K}$. In the classical examples, they happen to be complex numbers. In this case, the generators Z_i alone, without $\mathcal{K}$, generate a subalgebra $\mathcal{F}$ of $\mathcal{F}_{\mathcal{K}}$. $\mathcal{K}$ can act on $\mathcal{F}$. Therefore one can think of elements of $\mathcal{F}_{\mathcal{K}}$ as acting of $\mathcal{F}$: Z_a is multiplication by $z_a \in \mathcal{F}$ Similarly, Z_i and Θ_i generate a subalgebra Ω of $\Omega_{\mathcal{K}}$.

The amplimorphism μ and the braid matrix $\mathcal{R}^{+}$ in the definition should be thought as part of making a symmetry out of $\mathcal{K}$. The examples differ in what is this symmetry. In quantum field planes there are in addition further relation between $\Theta's$ beyond those stated above.

Example 4 (complex hyperplane $\mathbf{C}^N$) *Its symmetry $\mathcal{K}$ is the group algebra of $GL_N(\mathbf{C})$ or of a subgroup G of it. The fundamental amplimorphism is furnished by the fundamental representation τ of G*

$$\mu_{ij}(\xi) = \tau_{ij}(\xi)\xi \in \mathcal{K}$$

$\mathcal{R}^{+} = P$ is the permutation matrix with eigenvalues $\gamma_0 = -1$ and $\gamma_1 = 1$.

$$P_{ab,cd} = \delta_{ad}\delta_{bc} \ . \tag{48}$$

Thus, $\mathcal{R}^{+}$ is a numerical matrix. $\mathcal{F}$ is the commutative algebra of complex functions (polynomials) on $\mathbf{C}^N$, while $\mathcal{F}_{\mathcal{K}}$ and $\Omega_{\mathcal{K}}$ consist of $\mathcal{G}$-valued functions and forms, respectively. $\mathbf{\Phi}$ is generated by N anticommuting differentials.

If we interchange the eigenvalues γ_0 and γ_1 we get superplanes.

Example 5 (quantum plane with weak quasi quantum group symmetry) *[24,26,33] Let $\mathcal{K}$ be a simple quasitriangular quasicoassociative bi-algebra with unit element e, coproduct Δ , counit ϵ, R-element $R \in \mathcal{K}\otimes\mathcal{K}$ and reassociator $\varphi \in \mathcal{K}\otimes\mathcal{K}\otimes\mathcal{K}$, and let τ be some representation of $\mathcal{K}$, e.g. a fundamental one. We write*

$$\varphi_{213} = \sum_s \varphi_s^2 \otimes \varphi_s^1 \otimes \varphi_s^3 \tag{49}$$

if $\varphi = \sum_s \varphi_s^1 \otimes \varphi_s^2 \otimes \varphi_s^3$ as usual, and φ^{-1} for the quasiinverse of φ, cp. [22] Let P be the permutation matrix (48) as before. Then

$$\begin{aligned} \mu_{ij}(\xi) &= (\tau_{ij}\otimes id)\Delta(\xi) && (50)\\ \mathcal{R}^{+} &= P(\tau\otimes\tau\otimes id)(\varphi_{213}(R\otimes e)\varphi^{-1}) && (51)\\ &= (\tau\otimes\tau\otimes id)(\varphi)P(\tau\otimes\tau\otimes id)(R\otimes e)(\tau\otimes\tau\otimes id)(\varphi^{-1}) && (52) \end{aligned}$$

have the desired properties. If $\Delta(e) = e \otimes e$ (no truncation), $\varphi = e \otimes e \otimes e$ (coassociativity) and if there is an antipode then $\mathcal{K}$ is a Hopf algebra ("quantum group algebra"), and $\mathcal{R}^+$ is a complex matrix. If moreover $N = 2$, we get the familiar algebras $\mathcal{F}$ and Ω of "functions and forms on the quantum plane". The generalization to $N > 2$ was also known.

Example 6 (quantum field plane) *In this case $\mathcal{K} = \mathcal{G} \otimes \mathcal{A}$, where $\mathcal{G}$ is, in the most general case, a quasitriangular quasicoassociative bi- algebra as in the preceding example, and $\mathcal{A}$ is the algebra of observables of a local quantum field theory, i.e. a *-algebra of bounded operators in a Hilbert space. The elements of $\mathcal{A}$ will be called a-numbers. $\mathcal{G}$ and $\mathcal{A}$ should be compatible in the sense that their irreducible representations can be labelled by the same indices $I \in \mathcal{I}$, and their tensor product decompositions match as discussed before eq.(20).*

The amplimorphism $\mu = \mu^f$ is furnished by a N-dimensional representation $\tau = \tau^f$ of $\mathcal{G}$, the coproduct Δ of $\mathcal{G}$, and a morphism $\rho = \rho^f$ of $\mathcal{A}$, by the following formula with $I = f$

$$\mu^I_{ij}(\zeta a) = \rho^I(a)(\tau^I_{ij} \otimes id)\Delta(\zeta) \quad (a \in \mathcal{A}\ , \zeta \in \mathcal{G}). \tag{53}$$

The R-matrix $\mathcal{R}^+$ differs from that in the preceding example by multiplication with the inverse of a unitary a-number $\epsilon(\rho,\rho)$ (= statistics operator for $\mathcal{A}$), which commutes with all a-numbers in $\rho \circ \rho(\mathcal{A})$. It is given by the formula (29), specialized to $I = J = f$, viz.

$$\mathcal{R}^+ = \epsilon(\rho,\rho)^* P(\tau \otimes \tau \otimes id)(\varphi_{213}(R \otimes e)\varphi^{-1}) \tag{54}$$

$\mathbf{\Phi}_{\mathcal{K}}$ is generated by $\mathcal{K}$ and by generators Θ_i ($i = 1...N$) , as required by the definition.

In addition there are further relations in the field algebra $\mathbf{\Phi}_{\mathcal{K}}$ which come from operator product expansions. The algebra $\mathbf{\Phi}_{\mathcal{K}}$ is the $\mathcal{A}$-linear span of protofields Θ^I_i, $I \in \mathcal{I}$ whose covariance property involves μ^I. Their product in $\mathbf{\Phi}_{\mathcal{K}}$ is given by their operator product expansions (24).

It was explained in the last section how a suitable field algebra $\mathbf{\Phi}_{\mathcal{K}}$ is obtained in the frame work of algebraic field theory. The following result shows that the definition of a quantum plane is meaningful.

Theorem 7 *[25] (*Existence of quantum field planes associated with field algebras $\Phi_{\mathcal{K}}$*) Let $\mathcal{A}$ be a *-algebra with morphisms $\rho^I, (I \in \mathcal{I})$ and statistics operators $\epsilon(\rho^J, \rho^K)$. Let $\mathcal{G}$ be a quasiassociative quasitriangular *-bialgebra with representations $\tau^I, (I \in \mathcal{I})$ which is compatible with $\mathcal{A}$ in the sense of eq.(20). Let $\Phi_{\mathcal{K}}$ be a field algebra which contains $\mathcal{K} = \mathcal{G} \otimes \mathcal{A}$ and which has the properties enumerated in the last section ($\mathcal{G} \otimes \mathcal{A}$-covariance, operator product expansions and braid relations). Suppose that all irreducible representations of $\mathcal{G}$ are generated by a single fundamental representation $\tau = \tau^f$.*

If the operator product expansions in $\Phi_{\mathcal{K}}$ are compatible with a $\mathbf{Z}_2$-grading of $\Phi_{\mathcal{K}}$ then the remaining defining relations of a $\mathcal{G} \otimes \mathcal{A}$-covariant quantum plane are consistent with the structure of the field algebra $\Phi_{\mathcal{K}}$ and define a quantum field plane.

The assumption about the single fundamental representation of $\mathcal{G}$ could be given up at the cost of introducing several coordinates Z^f_a and differentials Θ^f_a labelled by f.

We mention without further discussion that partial derivatives $\partial_{\dot{b}}$ can also be constructed such that

$$d = g^{a\dot{b}} \Theta_a \partial_{\dot{b}}$$

with metric tensor $g^{a\dot{b}} \in \mathcal{G} \otimes \mathcal{A}$.

6 Infinitesimal parallel transport

In general relativity, covariant derivatives $\nabla_X \theta$ tell how to parallel transport a cotangent vector θ (or more generally an element $\theta \in \wedge T^*_x\mathcal{M}$) in some direction $X \in T_x\mathcal{M}$. Thus we have a map

$$\nabla : T_x\mathcal{M} \otimes T^*_x\mathcal{M} \mapsto T^*_x\mathcal{M} \tag{55}$$

or, equivalently

$$T^*_x\mathcal{M} \mapsto T^*_x\mathcal{M} \otimes T^*_x\mathcal{M} \tag{56}$$

In string theory or conformal field theory, (co)tangent space is replaced by a Hilbert space of states of germs of curves (representation spaces for $Diff^{\sim}(\mathbf{S}^1)$). Thus we want maps

$$\mathcal{H} \mapsto \mathcal{H} \otimes \mathcal{H} \tag{57}$$

Conformal field theory gives for every Riemann surface with boundary maps between Hilbert spaces $\mathcal{H}$ associated with connected boundary components [29]. For infinitesimal parallel transport we shall presumably only want to consider surfaces without holes, whose boundary consists of three circles. Suitable integration over such surfaces (i.e. their moduli) ought to give the desired map 57. Integrations of such a type have been considered in topological gravity [6].

Covariant exterior derivatives are made from ordinary ones,

$$D = d + \omega \, , \tag{58}$$

where ω is a 1-form with values in the symmetry algebra $\mathcal{G}$. $\mathcal{G}$ is a Lie algebra in the classical case, but more general possibilities can be envisaged [24].

In the last section we constructed "algebras of functions and forms " with values in $\mathcal{G}$ and an exterior derivative acting on them. The fibre $\mathbf{\Phi}$ which substitutes for $\wedge T_x^* \mathcal{M}$ is a field algebra rather than the Hilbert space $\mathcal{H}$, as would be suggested by the above considerations. But this makes no difference. By operator product expansions, the algebra $\mathbf{\Phi}$ is the C-linear span of field operators Ψ_j^J. One demands of a field algebra that it has the Reeh Schlieder property that no local field annihilates the vacuum. As a result there is a bijective correspondence between local fields and a dense set of states in the Hilbert space $\mathcal{H}$.

The construction of the differential algebras deals only with the kinematical aspects of quantum space time. One needs dynamical equations for matter fields and vector potential ω in addition. We don't know them yet. In classical general relativity, the vierbein field behaves like a quadruple of matter fields, and the vector potential is determined by them through algebraic equations. The form of the vertices (57) in conformal field theory suggests that the same should remain true in quantum space time, and it raises the hope that a good choice of dynamical equations could reproduce the expansions of a good string theory when the dynamical equations are solved perturbatively.

7 Beyond string theory

One is lead to a string theory if one starts from a theory of particles in local Lorentz frames. This is not inconsistent with special relativity, since the equivalence principle suggests that there needs to be no interaction.

But one may just as well envisage starting from a 4-dimensional local relativistic quantum field theory in the local Lorentz frames. Everything that has been said about field algebras remains valid. The symmetry $\mathcal{G}$ will be given by a group [9]. The quantum field planes may be constructed just as well in this situation.

References

[1] Abdus Salam, G. Mack, *Finite component field representations of the conformal group* Ann. Phys. (N.Y.) **53** (1969) 174

[2] H. Baumgärtel, *On Haag dual nets over compact spaces* SFB 288 preprint No 61, Berlin April 1993

[3] D. Buchholz, G. Mack, I.T. Todorov, *The current algebra on the circle as a germ of local field theories*, Nucl. Phys B (Proc. Suppl.) **5B** (1988) 20

[4] A. Connes, *Non-commutative differential geometry*, Publ. Math. I.H.E.S. **62** (1985)

[5] A. Connes, J. Lott, *Particle models and noncommutative geometry* Nucl. Phys. **B18** (Proc.Suppl.) (1990) 29 Publ. Math. I.H.E.S. **62** (1985)

[6] R. Dijkgraaf, H. Verlinde, E. Verlinde *Notes on topological field theory and 2-D gravity* Proc. Trieste Spring school 1990 p.91

[7] S. Doplicher, R. Haag, J.E. Roberts, *Fields, observables and gauge transformations I,II*, Commun. Math. Phys. **13** (1969) 1 and **15** (1969) 173

[8] S. Doplicher, R. Haag, J.E. Roberts, *Local observables and particle statistics I,II*, Commun. Math. Phys. **23** (1971) 199 and **35** (1974) 49

[9] S. Doplicher, J.E.Roberts, *C^*-algebras and duality for compact groups: why there is a compact group of internal gauge symmetries in particle physics*, in: VIIIth international congress on mathematical physics, Marseille 1986, M.Mebkhout, R.Sénéor (eds.)

S. Doplicher, J.E. Roberts, *Endomorphisms of C^*–algebras, cross products and duality for compact groups*, Ann. of Math. **130** (1989) 75-119
S. Doplicher, J.E. Roberts, *Wy there is a field algebra with a compact gauge group describing the superselection structure in particle physics*, Commun. Math. Phys. **131** (1990) 51

[10] V.G. Drinfel'd, *Quasi Hopf algebras and Knizhnik Zamolodchikov equations*, in: Problems of modern quantum field theory, Proceedings Alushta 1989, Research reports in physics, Springer Verlag Heidelberg 1989
V.G. Drinfel'd, *Quasi-Hopf algebras*, Leningrad. Math. J. Vol. **1** (1990), No. 6

[11] S. Ferrara, R. Gatto, A.F. Grillo, G. Parisi Nucl. Phys. **B49** (1972) 77; Lett. Nuovo Cimento **4** (1972) 115
S. Ferrara, R. Gatto, A.F. Grillo, Springer tracts in modern physics vol **67** Heidelberg 1973
Ann.Phys.(N.Y.) **76** (1973) 116

[12] K. Fredenhagen, K.H. Rehren, B. Schroer, *Superselection sectors with braid group statistics and exchange algebras, I: general theory*, Commun.Math.Phys. **125** (1989) 201

[13] K. Fredenhagen, *Generalization of the theory of superselection sectors*, in [17]

[14] J. Fröhlich, *Statistics of fields, the Yang Baxter equation and the theory of knots and links*, in: nonperturbative quantum field theory, G.t'Hooft et al.(eds.), Plenum Press 1988

[15] J. Fröhlich, C. King, *Two–dimensional conformal field theory and three–dimensional topology*, Int. J. Mod. Phys. **A4** (1989) 5321
J. Fröhlich, *Quantum statistics and locality*, in: proceedings of the Gibbs Symposium, Yale University, 1989

[16] D. Kastler, M. Mebkhout, K.H. Rehren, *Introduction to the algebraic theory of superselection sectors* in [17].

[17] D. Kastler, *The algebraic theory of superselection sectors. Introduction and recent results*, Proceedings of the Workshop " Algebraic Theory of

Superselection Sectors", Palermo 1989, D. Kastler (ed.), World Scientific (1990)

[18] M. Lüscher, G. Mack, *Global conformal invariance in quantum field theory*, Commun. Math. Phys. **41** (1975) 203

[19] G. Mack, *Introduction to conformal invariant quantum field theories in two dimensions* in: *Nonperturbative quantum field theory*, G.t'Hooft et al. (eds.), Plenum Press New York 1988

[20] G. Mack, *Convergence of operator product expansions on the vacuum in conformal invariant quantum field theory* Commun. Math. Phys. **53** (1977) 155

[21] G. Mack, V. Schomerus, *Conformal field algebras with quantum symmetry from the theory of superselection sectors*, Commun. Math. Phys. **134**(1990) 139

[22] G. Mack, V. Schomerus, *Quasi Hopf Quantum Symmetry in Quantum Theory*, Nucl. Phys. **B370** (1992) 185

[23] G. Mack, V. Schomerus, *Quasi quantum group symmetry and local braid relations in the conformal Ising model*, Phys. Letters **B267** (1991) 207

[24] G. Mack, V. Schomerus, *Action of truncated quantum groups on quasi quantum planes and a quasi-associative differential geometry and calculus*, Commun. Math. Phys. **149** (1992) 513

[25] G. Mack, V. Schomerus. *Models of quantum space time: Quantum field planes* (in preparation)

[26] Yu.I. Manin, *Quantum groups and Non-commutative geometry*, Proc. Int. Congr. Math., Berkeley **1** (1986) 798

[27] F. Müller-Hoissen, *Some Applications of Noncommutative Differential Calculus in Physics* Habilitation thesis Göttingen 1993, and references to work of Dimakis, Müller-Hoissen and Striker given there.

[28] A.Pressley and G. Segal, *Loop groups* Oxford Clarendon 1988

[29] G. Segal *The definition of conformal field theory* in *Differential geometric methods in quantum field theory* Proceedings Como 1987

[30] P. Vecsernyés, *On the Quantum Symmetry of the Chiral Ising model* Princeton preprint PUPT-1406 (June 1993)

[31] V. Schomerus, *Konsequenzen der Lokalität in der 2-dimensionalen konformen Quantenfeldtheorie* Diplomarbeit Universität Hamburg 1989 (unpublished)

[32] V. Schomerus, *Quantum symmetry in quantum theory* Dissertation Hamburg February 1993, preprint DESY 93-018, and to be published.

[33] J. Wess, B. Zumino, *Covariant differential calculus on the quantum hyperplane*, Nucl. Phys. B (Proc. Suppl.) **18B** (1991) 302

[34] H.W. Wiesbrock, *The C^*-algebra of bosonic strings* Commun. Math. Phys. **136** (1991) 369,
The construction of the sh-Lie-algebra of closed bosonic string Commun. Math. Phys **145** (1992) 17

[35] E. Witten, *String field theory and noncommutative geometry* Nucl. Phys. **B268** (1986) 253

[36] G.G. Wick, E.P. Wigner, A.S.Wightman, *Intrinsic parity of elementary particles* Phys. Rev. **88** (1952) 101

[37] S.L. Woronowicz, *Compact matrix pseudogroups*, Commun. Math. Phys. **111** (1987) 613

[38] S.L. Woronowicz, *Twisted SU(2) group: an example of a noncommutative differential calculus*, Publ. Res. Inst. Math. Sciences Kyoto Univ. **23** (1987) 117-181

Particle Physics Models, Grand Unification and Gravity in Non-Commutative Geometry

A. H. Chamseddine[1] * and J. Fröhlich[2] †

[1] Theoretische Physik, Universität Zürich, CH 8001 Zürich Switzerland
[2] Institut des Hautes Etudes Scientifique, F-91440 Bures-Sur-Yvette, France

Abstract

We review the construction of particle physics models in the framework of non-commutative geometry. We first give simple examples, and then progress to outline the Connes-Lott construction of the standard Weinberg-Salam model and our construction of the SO(10) model. We then discuss the analogue of the Einstein-Hilbert action and gravitational matter couplings. Finally we speculate on some experimental signatures of predictions specific to the non-commutative approach.

1. Introduction

The Weinberg-Salam model [1] of electroweak interactions is a milestone in the search for unity of all fundamental interactions. But although this model has passed all experimental tests at present energies, many challenges remain. To name just a few, we have to understand:
a-The role of the Higgs field necessary in the spontaneous breakdown of the $SU(2) \times U(1)$ gauge symmetry.
b-The fermionic mass matrices and family mixing, the gauge coupling constants, the mass and vacuum expectation value (vev) of the Higgs field.
c-Unifying gravity with the strong and electroweak interactions in a renormalizable theory.
There are many attempts to solve these problems using schemes such as grand unification, Kaluza-Klein compactification and string theory, all with and without supersymmetry. The virtues and shortcomings of these lines of research are now well known.

During the past few years, Connes has proposed a construction of particle physics models based on his formulation of non-commutative geometry [2]. This method addresses point a- raised above, in that it predicts the existence of the Higgs field and gives it a geometrical meaning [3]. This article is a short review of Connes' non-commutative construction and intended for particle physicists. The mathematics used here will be the minimum needed. For the more mathematically oriented reader we refer to some of the available reviews [4]. Our plan is as follows. In section 2 we introduce the non-commutative construction and give simple examples. In section 3 we review the derivation of the standard model and in section 4 the grand unified

* Supported by the Swiss National Foundation (SNF)

† Permanent address: Theoretische Physik, ETH, CH 8093 Zürich Switzerland

SO(10) model. In section 5 we describe an analogue of the Einstein-Hilbert action and the gravitational matter couplings, and, under a natural geometrical assumption, obtain some predictions for the top quark mass and the Higgs mass.

2. The non-commutative construction

Connes' non-commutative geometry is very general [2]. A non-commutative geometry is specified by the triple $(\mathcal{A}, h, D)$, where h is a Hilbert space, $\mathcal{A}$ is an involutive algebra of operators on h, and D is an unbounded self-adjoint operator on h. Let $\Omega^\cdot$ be the Z graded differential algebra of universal forms over R or C: $\Omega^\cdot = \oplus_n \Omega^n$, where $\mathcal{A} = \Omega^0$ and Ω^n is the space of n-forms with operations i) $d : \Omega^n \to \Omega^{n+1}$, ii)$m : \Omega^n \otimes \Omega^m \to \Omega^{n+m}$. The algebra of universal forms over $\mathcal{A}$, $\Omega^\cdot(\mathcal{A})$, is generated by f and df, where $f \in \mathcal{A}$. The operator d obeys Leibnitz rule, $d(fg) = (df)g + f(dg)$, where $f, g \in \mathcal{A}$, and $d^2 = 0$. An n-form in $\Omega^n(\mathcal{A})$ is given by $\sum_i a_0^i da_1^i \cdots da_n^i;\quad a_0^i, \cdots a_n^i \in \mathcal{A}$. An involutive representation of $\Omega^\cdot(\mathcal{A})$ on h is provided by the map $\pi : \Omega^\cdot(\mathcal{A}) \to B(h)$ defined by

$$\pi(a_0 da_1 \cdots da_n) = a_0[D, a_1] \cdots [D, a_n], \tag{2.1}$$

where $B(h)$ is the algebra of bounded operators on h. The non-commutativity resides in the fact that ab is not necessarily equal, up to a sign, to ba. Let E be a vector bundle determined by the vector space $\mathcal{E}$ of its sections. We will be mainly interested in the case $\mathcal{E} = \mathcal{A}$. Let ρ be a self-adjoint element in the space $\Omega^1(\mathcal{A})$. It determines a connectiion with curvature $\theta = d\rho + \rho^2 \in \Omega^2(\mathcal{A})$. The Yang-Mills action functional is obtained using the Dixmier trace which permits the definition of integration and volume elements in non-commutative geometry. We set (see [2,3])

$$I_{\rm YM} = {\rm Tr}_w(\theta^2 D^{-4}), \tag{2.2}$$

The same quantity can be defined using the heat kernel expansion (see [5]);i.e.,

$$\lim_{\epsilon\to 0} \frac{{\rm Tr}_H(\theta^2 e^{-\epsilon D^2})}{{\rm Tr}_H(e^{-\epsilon D^2})}. \tag{2.3}$$

We illustrate these notions with two simple examples.

1-Let $\mathcal{A}_1 = C^\infty(M)$, the algebra of functions on a four-dimensional Riemannian manifold M, h the Hilbert space of spinors $L^2(M, \sqrt{g}d^4x)$ and $D_1 = \partial\!\!\!/$, the Dirac operator on h. The one-form $\rho = \sum_i a^i db^i$ has the image under π

$$\pi(\rho) = \sum_i a^i[D, b^i] = \sum_i a^i \partial\!\!\!/ b^i \equiv \gamma^\mu A_\mu. \tag{2.4}$$

Similarly for the two-form $d\rho$ we have

$$\pi(d\rho) = \sum_i [D, a^i][D, b^i] = \sum_i \partial\!\!\!/ a^i \partial\!\!\!/ b^i. \tag{2.5}$$

The curvature $\pi(\theta) = \pi(d\rho) + \pi(\rho)^2$ is then given by

$$\pi(\theta) = \frac{1}{2}\gamma^{\mu\nu} F_{\mu\nu} + X, \tag{2.6}$$

where $\gamma^{\mu\nu} = \frac{1}{2}(\gamma^\mu\gamma^\nu - \gamma^\nu\gamma^\mu)$, $X = g^{\mu\nu}(A_\mu A_\nu + \sum_i \partial_\mu a^i \partial_\nu b^i)$ is an "auxiliary field" and $F_{\mu\nu}$ is the field strength of A_μ. Notice that $\pi(d\rho)_{\rho=0} = g^{\mu\nu}\sum_i \partial_\mu a^i \partial_\nu b^i \neq 0$, is a scalar function. This is the reason behind the presence of the auxiliary field in $\pi(\theta)$. It is possible to work instead with the space $\frac{\Omega^2(\mathcal{A})}{\mathrm{Ker}\pi + d\mathrm{Ker}\pi}$, but we will not do this now. The Yang-Mills action becomes

$$I_{YM} = \int d^4x(-\frac{1}{4}F_{\mu\nu}F^{\mu\nu} + X^2). \tag{2.7}$$

After eliminating the auxiliary field X by its equation of motion, it decouples from the action.

2-For a two point space, we take $\mathcal{A}_2 = C \oplus C$, and $h = C^N \oplus C^N$ and the Dirac operator is $D_2 = \begin{pmatrix} 0 & K \\ K^* & 0 \end{pmatrix}$, where K is an $N \times N$ matrix. The elements $a \in \mathcal{A}_2$ have the representation $a \to \mathrm{diag}(a_1, a_2)$, $a_1, a_2 \in C$. Then

$$\pi(\rho) = \sum_i a^i[D, b^i] = \begin{pmatrix} 0 & K\phi \\ K^*\phi^* & 0 \end{pmatrix}, \tag{2.8}$$

where $\phi = \sum_i a_1^i(b_2^i - b_1^i)$ and $\phi^* = \sum_i a_2^i(b_1^i - b_2^i)$. Then $\pi(d\rho) = \sum_i [D, a^i][D, b^i]$ is equal to

$$\pi(d\rho) = -\begin{pmatrix} KK^*(\phi + \phi^*) & 0 \\ 0 & K^*K(\phi + \phi^*) \end{pmatrix}. \tag{2.9}$$

The Yang-Mills action is easily calculated to be

$$\mathrm{tr}(\theta^2) = 2\mathrm{Tr}(KK^*)^2\left(|\phi - 1|^2 - 1\right)^2, \tag{2.10}$$

It is seen to be of the same form as the Higgs potential for a scalar field ϕ and is positive definite. Notice that $[D, a] = \begin{pmatrix} 0 & K(b_2 - b_1) \\ K^*(b_1 - b_2) & 0 \end{pmatrix}$ is a difference operator in the discrete space.

3. The standard Weinberg-Salam model

With the simple tools introduced in the last section, we now show that it is possible to construct realistic action functionals. Not all models are possible, but for those ones which are, the Higgs structure is fixed. For lack of space we shall only describe the standard Weinberg-Salam model in this section and the grand unification SO(10) model in the next section. Our method is a modified variant of Connes' construction (simplifying some computations [5]).

Combining examples 1 and 2, let the algebra be $\mathcal{A} = \mathcal{A}_1 \otimes \mathcal{A}_2$ acting on the Hilbert space $h = h_1 \otimes h_2$, where $\mathcal{A}_1 = C^\infty(M)$, considered before, and $\mathcal{A}_2 = M_2(C) \oplus M_1(C)$ the algebras of 2×2 and 1×1 matrices. The Hilbert space is that of spinors of the form $L = \begin{pmatrix} l \\ e \end{pmatrix}$ where l is a doublet and e is a singlet. The spinor L satisfies the chirality condition $\gamma_5 \otimes \Gamma_1 L = L$, where $\Gamma_1 = \mathrm{diag}(1_2, -1)$ is the grading operator.

This implies that $l = l_L$ is left-handed and $e = e_R$ is right-handed, and so we can write $l_L = \begin{pmatrix} \nu_L \\ e_L \end{pmatrix}$. The Dirac operator is $D = D_1 \otimes 1 + \Gamma_1 \otimes D_2$, so that

$$D_l = \begin{pmatrix} \partial\!\!\!/ \otimes 1_2 & \gamma_5 M_{12} \otimes k \\ \gamma_5 M_{21} \otimes k^* & \partial\!\!\!/ \end{pmatrix}, \tag{3.1}$$

where $M_{21} = M_{12}^*$ and k is a family mixing matrix. The geometry is that of a four-dimensional manifold M times a discrete space of two points. The column M_{12} in D, the vev of the Higgs field, is taken to be $M_{12} = \mu \begin{pmatrix} 0 \\ 1 \end{pmatrix} \equiv H_0$. The elements $a \in \mathcal{A}$ have the representation $a \to \mathrm{diag}(a_1, a_2)$ where a_1 and a_2 are 2×2 and 1×1 unitary matrix-valued functions, respectively. The self-adjoint one-form ρ has the representation

$$\pi_l(\rho) = \begin{pmatrix} A_1 \otimes 1_3 & \gamma_5 H \otimes k \\ \gamma_5 H^* \otimes k^* & A_2 \otimes 1_3 \end{pmatrix}, \tag{3.2}$$

where $A_1 = \sum_i a_1^i \partial\!\!\!/ b_1^i$, $A_2 = \sum_i a_2^i \partial\!\!\!/ b_2^i$ and $H = H_0 + \sum_i a_1^i H_0 b_2^i$. In a world without quarks, the generalized tracelessness condition $\mathrm{Tr}(\Gamma_1 \pi(\rho)) = 0$ allows the gauge fields to be written in the form $A_1 = -\frac{i}{2} g_2 \sigma^a A_a + i g_1 B$, $A_2 = 2 i g_1 B$ where g_1, g_2 are the U(1) and SU(2) gauge couplings. The leptonic action $< L, (D + \rho) L >$ gives the correct lepton couplings to the gauge and Higgs fields. However, to be realistic, the quarks and the SU(3) gauge group must be introduced. This can be achieved by taking a bimodule structure relating two algebras $\mathcal{A}$ and $\mathcal{B}$, where the algebra $\mathcal{B}$ is taken to be $M_1(C) \oplus M_3(C)$, commuting with the action of $\mathcal{A}$, and the mass matrices in the Dirac operator are taken to be zero when acting on elements of $\mathcal{B}$. Then the one-form η in $\Omega^1(\mathcal{B})$ has the simple form $\pi_l(\eta) = B_1 \mathrm{diag}(1_2, 1)$, where B_1 is a $U(1)$ gauge field associated with $M_1(C)$. The quark Hilbert space is that of the spinor $Q = \begin{pmatrix} u_L \\ d_L \\ d_R \\ u_R \end{pmatrix}$. The representation of $a \in \mathcal{A}$ is: $a \to \mathrm{diag}(a_1, a_2, \overline{a_2})$ where a_1 is a 2×2 matrix-valued function and a_2 is a complex-valued function. The Dirac operator acting on the quark Hilbert space is

$$D_q = \begin{pmatrix} \gamma^\mu(\partial_\mu + \ldots) \otimes 1_2 \otimes 1_3 & \gamma_5 \otimes M_{12} \otimes k' & \gamma_5 \otimes \tilde{M}_{12} \otimes k'' \\ \gamma_5 \otimes M_{12}^* \otimes k'^* & \gamma^\mu(\partial_\mu + \ldots) \otimes 1_3 & 0 \\ \gamma_5 \otimes \tilde{M}_{12}^* \otimes k''^* & 0 & \gamma^\mu(\partial_\mu + \ldots) \otimes 1_3 \end{pmatrix} \otimes 1_3, \tag{3.3}$$

where k' and k'' are 3×3 family mixing matrices, and $\tilde{M}_{12} = \mu \begin{pmatrix} 1 \\ 0 \end{pmatrix}$. Then the one-form in $\Omega^1(\mathcal{A})$ has the representation

$$\pi_q(\rho) = \begin{pmatrix} A_1 \otimes 1_3 & \gamma_5 H \otimes k' & \gamma_5 \tilde{H} \otimes k'' \\ \gamma_5 H^* \otimes k'^* & A_2 \otimes 1_3 & 0 \\ \gamma_5 \tilde{H}^* \otimes k''^* & 0 & \overline{A_2} \otimes 1_3 \end{pmatrix}, \tag{3.4}$$

where $\tilde{H}_a = \epsilon_{ab} H^b$. On the algebra $\mathcal{B}$ the Dirac operator has zero mass matrices, and the one form η in $\Omega^1(\mathcal{B})$ has the representation $\pi_q(\eta) = B_2 \mathrm{diag}(1_2, 1, 1)$ where

B_2 is the gauge field associated with $M_3(C)$. Imposing the unimodularity condition on the algebras $\mathcal{A}$ and $\mathcal{B}$ relates the U(1) factors in both algebras [3]: $\mathrm{tr}(A_1) = 0$, $A_2 = B_1 = -\mathrm{tr}B_2 = \frac{i}{2}g_1 B$. We can then write

$$\begin{aligned} A_1 &= -\frac{i}{2}g_2 A^a \sigma_a \\ B_2 &= -\frac{i}{6}g_1 B - \frac{i}{2}g_3 V^i \lambda_i \end{aligned}$$

where g_3 is the SU(3) guge coupling constant, and σ^a and λ^i are the Pauli and Gell-Mann matrices, respectively. The fermionic action for the leptons is

$$< L, (D+\rho+\eta)L >= \int d^4x\sqrt{g}\Big(\overline{L}(D_l + \pi_l(\rho) + \pi_l(\eta))L\Big), \tag{3.5}$$

and, for the quarks it is

$$< Q, (D+\rho+\eta)Q >= \int d^4x\sqrt{g}\Big(\overline{Q}(D_q + \pi_q(\rho) + \pi_q(\eta))Q\Big), \tag{3.6}$$

and these can be easily checked to reproduce the standard model lepton and quark interactions with the correct hypercharge assignments.

The bosonic actions are the square of the curvature in the lepton and quark spaces, and are given, respectively, by

$$\begin{aligned} I_l &= \mathrm{Tr}(C_l(\theta_\rho + \theta_\eta)^2 D_l^{-4}) \\ I_q &= \mathrm{Tr}(C_q(\theta_\rho + \theta_\eta)^2 D_q^{-4}) \end{aligned} \tag{3.7}$$

To compute the bosonic action, we use a general formula, derived in [5], based on a Dirac operator where the discrete space has N points:

$$D = \begin{pmatrix} \partial\!\!\!/ \otimes 1 \otimes 1 & \gamma_5 \otimes M_{12} \otimes K_{12} & \dots & \gamma_5 \otimes M_{1N} \otimes K_{1N} \\ \gamma_5 \otimes M_{21} \otimes K_{21} & \partial\!\!\!/ \otimes 1 \otimes 1 & \dots & \gamma_5 \otimes M_{2N} \\ \vdots & \vdots & \ddots & \vdots \\ \gamma_5 \otimes M_{N1} \otimes K_{N1} & \gamma_5 \otimes M_{N2} \otimes K_{N2} & \dots & \partial\!\!\!/ \otimes 1 \end{pmatrix}, \tag{3.8}$$

where the K_{mn} are 3×3 matrices commuting with the a_i and b_i. The Yang-Mills action associated with this operator is

$$\begin{aligned} I_b = \sum_{m=1}^{N} Tr\Big(&\frac{1}{2}F_{\mu\nu}^{m}F^{\mu\nu m} - \Big|\sum_{p\neq m}|K_{mp}|^2|\phi_{mp} + M_{mp}|^2 - (Y_m + X'_{mm})\Big|^2 \\ &+ \sum_{p\neq m}|K_{mp}|^2\Big|\partial_\mu(\phi_{mp} + M_{mp}) + A_{\mu m}(\phi_{mp} + M_{mp}) - (\phi_{mp} + M_{mp})A_{\mu p}\Big|^2 \\ &- \sum_{n\neq m}\sum_{p\neq m,n}\Big|K_{mp}K_{pn}\big((\phi_{mp} + M_{mp})(\phi_{pn} + M_{pn}) - M_{mp}M_{pn}\big) - X_{mn}\Big)\Big|^2\Big), \end{aligned} \tag{3.9}$$

where the A^m are the gauge fields in the $m-m$ entry of $\pi(\rho)$ and ϕ_{mn} are the scalar fields in the $m-n$ entry of $\pi(\rho)$. The X_{mn}, X'_{mn} and Y_m are fields whose unconstrained elements are auxiliary fields that can be eliminated from the action. Their expressions in terms of the a^i and b^i are

$$X_{mn} = \sum_i a^i_m \sum_{p\neq m,n} K_{mp}K_{pn}(M_{mp}M_{pn}b^i_n - b^i_m M_{mp}M_{pn}), \qquad m \neq n, \tag{3.10}$$

$$X'_{mm} = \sum_i a^i_m \partial^2 b^i_m + (\partial^\mu A^m_\mu + A^{\mu m} A^m_\mu), \tag{3.11}$$

$$Y_m = \sum_{p\neq m}\sum_i a^i_m |K_{mp}|^2 |M_{mp}|^2 b^i_m. \tag{3.12}$$

Using Eqs (3.9)-(3.12) for the leptons and quarks seperately, yields an action containing the kinetic terms for the U(1), SU(2) and SU(3) gauge fields, as well as the kinetic energy and potential of the Higgs field. The most complicated step is the elimination of the auxiliary fields, but this only changes the coefficients of the Higgs potential, not its form. By writing $C_l = \mathrm{diag}(c_1, c_1, c_2)$ and $C_q = \mathrm{diag}(c_3, c_3, c_4, c_4)$, the bosonic action depends on the constants $c_1, c_2, c_3, c_4, g_1, g_2, g_3$ as well as on the Yukawa couplings. Normalizing the kinetic energies of the $SU(3)$, $SU(2)$ and $U(1)$ gauge fields fixes three of the constants $c_1, \ldots, c_4$ in terms of g_1, g_2, g_3. In the special case when $c_1 = c_2 = c_3 = c_4$, one gets a constraint on the gauge coupling constants as well as fixed values for the Higgs mass and top quark mass [3]. These relations cannot be maintained after quantization, as can be seen from the renormalization group equations for the coupling constants and the masses [6]. We shall not assume any such relations among the $c's$. The Higgs sector is then parametrized in terms of two parameters λ and m which are functions of of the $c's$, k, s and H_0. The bosonic part of the standard model becomes

$$\begin{aligned} L_b = &-\frac{1}{4}\left(F^3_{\mu\nu}F^{\mu\nu 3} + F^2_{\mu\nu}F^{\mu\nu 2} + F^1_{\mu\nu}F^{\mu\nu 1}\right) \\ &+ D_\mu(H + M_{12})^* D_\nu(H + M_{12}) g^{\mu\nu} \\ &- \frac{\lambda}{24}\left| |H + M_{12}|^2 - |M_{12}|^2 \right|^2 \end{aligned} \tag{3.13}$$

The cosmological constant comes out to be zero, naturally, at the classical level.

4. SO(10) unification model.

The way the strong interactions are introduced in the standard model suggests that a more unified picture would be preferable. The starting point is the Hilbert space of spinors and the Dirac operator acting on this space. The arrangement of fermions determines the structure of the discrete space. We place the fermions in the 16_s spinor representation of $SO(10)$ [7]. This is a 32-component spinor subject to the space-time and SO(10) chirality

$$\begin{aligned} (\gamma_5)^\beta_\alpha \psi_{\beta\hat\alpha} &= \psi_{\alpha\hat\alpha} \\ (\Gamma_{11})^{\hat\beta}_{\hat\alpha} \psi_{\alpha\hat\beta} &= \psi_{\alpha\hat\alpha}. \end{aligned} \tag{4.1}$$

where $\Gamma_{11} = -i\Gamma_0\Gamma_1 \cdots \Gamma_9$. This reduces the independent spinor components to two for the space-time indices, and to sixteen for the $SO(10)$ indices. The general fermionic action is given by

$$\overline{\psi^p_{\alpha\hat\alpha}} (\partial\!\!\!/ + A^{IJ}\Gamma_{IJ})^{\beta\hat\beta}_{\alpha\hat\alpha} \psi^p_{\beta\hat\beta} + \psi^{Tp}_{\alpha\hat\alpha} C^{\alpha\beta} H^{pq}_{\hat\alpha\hat\beta} \psi^q_{\beta\hat\beta}, \tag{4.2}$$

where C is the charge conjugation matrix, $p, q = 1, 2, 3$ are family indices, and H is some appropriate combination of Higgs fields breaking the subgroup $SU(2) \times U(1)$ of $SO(10)$ at low energies. An exception of a Higgs field that breaks the symmetry at high energies and yet couples to fermions is the one that gives a Majorana mass to the right handed neutrinos . The other Higgs fields needed to break the $SO(10)$ symmetry at high energies should not couple to the fermions so as not to give the quarks and leptons super heavy masses. The simplest picture corresponds to the spinor $\Psi = \begin{pmatrix} P_+\psi \\ P_+\psi \\ P_-\psi^c \end{pmatrix}$ where $\psi^c \equiv BC\overline{\psi}^T$, B is the SO(10) conjugation matrix satisfying $B^{-1}\Gamma_I B = -\Gamma_I^T$ and $P_\pm = \frac{1}{2}(1 \pm \Gamma_{11})$. However, it turns out that the model associated with this arrangement, although elegant, is not realistic, because the Cabbibo angle vanishes [8]. The correct model is the one with the spinor

$$\Psi = \begin{pmatrix} P_+\psi \\ P_+\psi \\ P_-\psi^c \\ P_-\psi^c \\ \lambda \\ \lambda^c \end{pmatrix}, \tag{4.3}$$

where λ is a singlet fermion that will couple to the right-handed neutrino in the 16_s. The algebra $\mathcal{A}$ is equal to $\mathcal{A}_1 \otimes \mathcal{A}_2$ where $\mathcal{A}_1 = C^\infty(M)$, and

$$\mathcal{A}_2 \equiv P_+\mathrm{Cliff}\big(SO(10)\big)P_+ \oplus R. \tag{4.4}$$

The involutive map π is taken to be

$$\pi(a) = \pi_0(a) \oplus \pi_0(a) \oplus \overline{\pi}_0(a) \oplus \overline{\pi}_0(a) \oplus \overline{\pi}_0(a) \oplus \pi_1(a) \oplus \pi_1(a), \tag{4.5}$$

acting on the Hilbert space $\tilde{h} = h_1 \otimes (h_2^{(1)} \oplus \cdots \oplus h_2^{(6)})$ where $h_2^{(i)} \cong h_2, \quad i = 1, \cdots 4$, h_2 is the 32 dimensional Hilbert space on which $\mathcal{A}_2$ acts, and $h_2^{(i)} \cong C, \quad i = 5, 6$. Let h be the subspace of $\tilde{h}$ which is the image of the orthogonal projection onto elements of the form (4.3). On $\tilde{h}$ the self-adjoint Dirac operator has the form (3.8), for N=6. From Eq (4.5) we have the permutation symmetry $1 \leftrightarrow 2$, $3 \leftrightarrow 4$, $5 \leftrightarrow 6$, and the conjugation symmetry $1 \leftrightarrow 3$, and the one-form $\pi(\rho)$ reads

$$\pi(\rho) = \begin{pmatrix} A & \gamma_5\mathcal{M}K_{12} & \gamma_5\mathcal{N}K_{13} & \gamma_5\mathcal{N}K_{14} & \gamma_5 HK_{15} & \gamma_5 HK_{16} \\ \gamma_5\mathcal{M}K_{12} & A & \gamma_5\mathcal{N}K_{23} & \gamma_5\mathcal{N}K_{24} & \gamma_5 HK_{25} & \gamma_5 HK_{26} \\ \gamma_5\mathcal{N}^*K_{31} & \gamma_5\mathcal{N}^*K_{32} & B\overline{A}B^{-1} & \gamma_5\mathcal{M}'K_{34} & \gamma_5 H'K_{35} & \gamma_5 H'K_{36} \\ \gamma_5\mathcal{N}^*K_{41} & \gamma_5\mathcal{N}^*K_{42} & \gamma_5\mathcal{M}'K_{43} & B\overline{A}B^{-1} & \gamma_5 H'K_{45} & \gamma_5 H'K_{46} \\ \gamma_5 H^*K_{51} & \gamma_5 H^*K_{52} & \gamma_5 H'^*K_{53} & \gamma_5 H'^*K_{54} & 0 & 0 \\ \gamma_5 H^*K_{61} & \gamma_5 H^*K_{62} & \gamma_5 H'^*K_{63} & \gamma_5 H'^*K_{64} & 0 & 0 \end{pmatrix}, \tag{4.6}$$

where the new functions A, $\mathcal{M}$, $\mathcal{N}$ and H are given in terms of the a^i and b^i by

$$
\begin{aligned}
A &= P_+(\sum_i a^i \partial\!\!\!/ b^i)P_+ \\
\mathcal{M} + \mathcal{M}_0 &= P_+(\sum_i a^i \mathcal{M}_0 b^i)P_+ \\
\mathcal{N} + \mathcal{N}_0 &= P_+(\sum_i a^i \mathcal{N}_0 B\overline{b^i}B^{-1})P_- \\
H + H_0 &= P_+(\sum_i a^i H_0 b'^i)
\end{aligned}
\tag{4.7}
$$

and $\mathcal{M}' = B\overline{\mathcal{M}}B^{-1}$, $H' = B\overline{H}$. We can expand these fields in terms of the $SO(10)$ Clifford algebra as follows:

$$
\begin{aligned}
A &= P_+\big(ia + a^{IJ}\Gamma_{IJ} + ia^{IJKL}\Gamma_{IJKL}\big)P_+ \\
\mathcal{M} &= P_+\big(m + im^{IJ}\Gamma_{IJ} + m^{IJKL}\Gamma_{IJKL}\big)P_+ \\
\mathcal{N} &= P_+\big(n^I\Gamma_I + n^{IJK}\Gamma_{IJK} + n^{IJKLM}\Gamma_{IJKLM}\big)P_-.
\end{aligned}
\tag{4.8}
$$

The self-adjointness condition on $\pi(\rho)$ implies, after using the hermiticity of the Γ_I matrices, that all the fields a and m appearing in the expansion of $A, \mathcal{M}$ are real, because both are self-adjoint, while those in $\mathcal{N}$ are complex. Imposing the reality condition on the coefficients of the Clifford algebra expansion of the gauge field A forces $a = 0 = a^{IJKL}$, reducing the gauge group from U(8) to SO(10). The symmetry breaking pattern that breaks the gauge group $SO(10)$ must be coded into the Dirac operator D. The Higgs fields at our disposal are $\mathcal{M}$, $\mathcal{N}$ and H. In terms of $SO(10)$ representations these are $\underline{1}$, $\underline{45}$, $\underline{210}$ in $\mathcal{M}$, complex $\underline{10}$, $\underline{120}$ and $\underline{126}$ in $\mathcal{N}$ and $\underline{16}_s$ in H. To be explicit we shall work in a specific Γ matrix representation. The 32×32 Γ matrices are represented in terms of tensor products of five sets of Pauli matrices $\sigma_i, \tau_i, \eta_i, \rho_i, \kappa_i$ where $i = 1,2,3$. The Γ matrices are given by

$$
\begin{aligned}
&\Gamma_i = \kappa_1\rho_3\eta_i, \qquad \Gamma_{i+3} = \kappa_1\rho_1\sigma_i \\
&\Gamma_{i+6} = \kappa_1\rho_2\tau_i, \qquad \Gamma_0 = \kappa_2, \qquad \Gamma_{11} = \kappa_3
\end{aligned}
\tag{4.9}
$$

where $i = 1,2,3$, and where we have omitted the tensor product symbols. In this basis, an $SO(10)$ chiral spinor will take the form $\psi_+ = \begin{pmatrix} \chi_+ \\ 0 \end{pmatrix}$ where χ is a $\underline{16}_s$. The $SO(10)$ conjugation matrix is defined by $B \equiv -\Gamma_1\Gamma_3\Gamma_4\Gamma_6\Gamma_8$ which, in the basis of equation (4.9), becomes

$$
B = \kappa_1\rho_2\eta_2\tau_2\sigma_2 \equiv \kappa_1 b \tag{4.10}
$$

where the matrix $b = \rho_2\eta_2\tau_2\sigma_2$ is the conjugation matrix in the space of the sixteen component spinors. The action of B on a chiral spinor is then $B\psi_+ = \begin{pmatrix} 0 \\ b\chi_+ \end{pmatrix}$. The advantage of this system of matrices is that $bC\overline{\chi_+}^T$, have the same form as χ_+ but is right-handed not left-handed. To correctly associate the components of χ_+ with quarks and leptons, we consider the action of the charge operator [7] on χ_+:

$$
Q = -\frac{1}{6}(\sigma_3 + \tau_3 + \rho_3\tau_3\sigma_3) + \frac{1}{2}\eta_3 \tag{4.11}
$$

which gives

$$Q\chi_+ = \mathrm{diag}(0, \frac{2}{3}, \frac{2}{3}, \frac{2}{3}, -1, -\frac{1}{3}, -\frac{1}{3}, -\frac{1}{3}, \frac{1}{3}, \frac{1}{3}, \frac{1}{3}, 1, -\frac{2}{3}, -\frac{2}{3}, -\frac{2}{3}, 0)\chi_+ \tag{4.12}$$

Thus the components of the left handed spinor χ_+ are written as the column

$$\begin{aligned}\chi_+ = (&n_L, u_L^1, u_L^2, u_L^3, e_L, d_L^1, d_L^2, d_L^3,\\ &-(d_R^3)^c, (d_R^2)^c, (d_R^1)^c, -(e_R)^c, (u_R^3)^c, -(u_R^2)^c, -(u_R^1)^c, (n_R)^c)\end{aligned}$$

where the c in this equation stands for the usual charge conjugation, eg. $d^c = C\overline{d}^T$. The upper and lower components in χ are mirrors, with the signs chosen so that the spinor $bC\overline{\chi_+}^T$ has exactly the same form as χ_+, but with the left-handed and right handed signs, L and R, interchanged. We now specify the vevs $\mathcal{M}_0$, $\mathcal{N}_0$ and H_0. The group $SO(10)$ is broken at high energies by $\mathcal{M}$ which contains the representations $\underline{45}$ and $\underline{210}$. By taking the vev of the $\underline{210}$ to be $\mathcal{M}^{0123} = O(M_G)$, the $SO(10)$ symmetry is broken to $SO(4) \times SO(6)$ which is isomorphic to $SU(4)_c \times SU(2)_L \times SU(2)_R$. The $SU(4)_c$ is further broken to $SU(3)_c \times U(1)_c$ by the vev of the $\underline{45}$. Therefore we write [8]

$$\begin{aligned}P_+\mathcal{M}_0P_+ &= P_+\Big(M_G\Gamma_{0123} - iM_1(\Gamma_{45} + \Gamma_{78} + \Gamma_{69})\Big)P_+\\ &= \frac{1}{2}(1+\kappa_3)\Big(-M_G\rho_3 + M_1(\sigma_3 + \tau_3 + \rho_3\tau_3\sigma_3)\Big)\end{aligned} \tag{4.13}$$

Therefore $\mathcal{M}_0$ breaks $SO(10)$ to $SU(3)_c \times U(1)_c \times SU(2)_L \times SU(2)_R$ which is also of rank five. The rank is reduced by giving a vev to the components of $\underline{126}$ that couple to the right-handed neutrino.Therefore the vev of $\mathcal{N}_0$ must contain the term

$$M_2(\frac{1}{2^5})(\kappa_1 + i\kappa_2)(\rho_1 + i\rho_2)(\eta_1 + i\eta_2)(\tau_1 + i\tau_2)(\sigma_1 + i\sigma_2) \tag{4.14}$$

The vev of $\mathcal{N}_0$ break $U(1)_c \times SU(2)_R$ to $U(1)_Y$, and the surviving group would be the familiar $SU(3)_c \times SU(2)_L \times U(1)_Y$. This breaking is also obtained for an H_0 whose vev is $H_0 = M_3 \begin{pmatrix} 0 \\ \vdots \\ 0 \\ 1 \end{pmatrix}$. As we shall explain shortly, M_1, M_2 and M_3 must be related for the model to be consistent. The only generators that leave $\mathcal{M}_0$, the part of $\mathcal{N}_0$ given by 4.14 and H_0 invariant are those of the standard model. The eight $SU(3)$ generators are given by $(1-\rho_3\tau_3)\sigma_i$, $(1-\rho_3\sigma_3)\tau_i$, $\rho_3(\tau_1\sigma_1 + \tau_2\sigma_2)$ and $\rho_3(\tau_2\sigma_1 - \tau_1\sigma_2)$. The $SU(2)_L$ generators are $\frac{1}{2}(1 \pm \kappa_3\rho_3)\eta^i$. Finally the $U(1)_Y$ generator is related to the charge operator Q by $Q = \frac{1}{2}Y + T_L^3$, where the action of the $SU(2)_L$ isospin T_L^3 on χ_+ is given by $T_L^3 = \frac{1}{2}(1+\rho_3)\eta_3$.

For the last stage of symmetry breaking of $SU(2)_L \times U(1)_Y$ we can use the field $\mathcal{N}$ which contains the compex representations $\underline{10}$, $\underline{120}$ and $\underline{126}$. The most general vev that preserves the group $SU(3)_c \times U(1)_Q$ is

$$\begin{aligned}P_+\mathcal{N}_0P_-\kappa_1 = \frac{1}{2}(1+\kappa_3)\Big(&s + p\rho_3\eta_3 + a\rho_3 + a'\eta_3\\ &+ (b' + b\rho_3\eta_3 + e\eta_3 + f\rho_3)(\sigma_3 + \tau_3 + \rho_3\tau_3\sigma_3)\\ &+ M_2(\frac{1}{2^5})(\rho_1 + i\rho_2)(\eta_1 + i\eta_2)(\tau_1 + i\tau_2)(\sigma_1 + i\sigma_2)\Big),\end{aligned} \tag{4.15}$$

where all terms containing η_3 break $SU(2)_L \times U(1)_Y$ and s, p, a, a', b, b', e, f are $O(M_W)$. The fermionic action is simply given by

$$
\begin{aligned}
I_{f-\text{mass}} &= \langle \Psi, (D+\rho)\Psi \rangle \\
&= -\int d^4x \Big(\big((s+p+3(e+f))K_{(pq)} + (a+a'+3(b+b'))K_{[pq]}\big)\overline{N^p_R}N^q_L \\
&\quad + \big((s+p-(e+f))K_{(pq)} + (a+a'-(b+b'))K_{[pq]}\big)\overline{u^p_R}u^q_L \\
&\quad + \big((s-p-3(e-f))K_{(pq)} + (a-a'-3(b-b'))K_{[pq]}\big)\overline{e^p_R}e^q_L \\
&\quad + \big((s-p+e-f)K_{(pq)} + (a-a'+b-b')K_{[pq]}\big)\overline{d^p_R}d^q_L \\
&\quad + \big(\sqrt{2}M_3 K'_{pq}\overline{N^p_R}\lambda^q_L + M_2 K''_{(pq)}(N^{pc}_R)^T C^{-1} N^{qc}_R\big) + \text{h.c}\Big),
\end{aligned} \tag{4.16}
$$

where we have denoted the family mixing matrices K_{13}, K_{15} and K_{56} by K, K', K'', respectively. The symmetric and antisymmetric parts of K_{pq} are denoted by $K_{(pq)}$ and $K_{[pq]}$, respectively. Since we have three neutral fields, N_L, N^c_R and λ_L, and their mass eigenstates are mixed, the mass matrix must be diagonalised. Ignoring the mixing due to the generation matrices, the mass matix of the neutral fields is of the form

$$
\begin{array}{c} \\ N_L \\ N^c_R \\ \lambda_L \end{array}
\begin{array}{c} \begin{array}{ccc} N_L & N^c_R & \lambda_L \end{array} \\ \begin{pmatrix} 0 & m & 0 \\ m & M_2 & M_3 \\ 0 & M_3 & 0 \end{pmatrix} \end{array}, \tag{4.17}
$$

and we shall assume a mass hierarchy $m \ll M_2, M_3$, and $M_2 \sim M_3$. Diagonalisation of the matrix (4.17) produces two massive fields whose masses are of order M_2, and the third will be a massless left-handed neutrino.

The bosonic action can be read from Eq (3.9), for N=6. The only complicated step is the elimination of the auxiliary fields, and one finds that the vev's used cannot be arbitrary but must be related for the potential to survive and the model to be consistent. These relations are $M_G = M_1$ and $M_1 M_2 = -\frac{K_{15}\overline{K_{15}}}{2K_{12}K_{13}} M_3^2$. The bosonic action is

$$
\begin{aligned}
&-4g^2 F^{IJ}_{\mu\nu}F^{\mu\nu IJ} + 2|K_{12}|^2 \mathrm{Tr}\Big(\big(D_\mu(\mathcal{M}+\mathcal{M}_0)\big)^2\Big) + 8|K_{13}|^2 \mathrm{Tr}\Big(|D_\mu(\mathcal{N}+\mathcal{N}_0)|^2\Big) \\
&+ 12|K_{15}|^2 \Big|D_\mu(H+H_0)\Big|^2 - V(\mathcal{M},\mathcal{N},\mathcal{H}),
\end{aligned} \tag{4.18}
$$

where the potential $V(\mathcal{M},\mathcal{N},H)$ is

$$
\begin{aligned}
&(\mathrm{Tr}|K_{12}|^4 - (\mathrm{Tr}|K_{12}|^2)^2)\mathrm{Tr}\Big(|\mathcal{M}+\mathcal{M}_0|^2 - |\mathcal{M}_0|^2\Big)^2 \\
&+ 4\Big|K_{13}K_{12}((\mathcal{M}+\mathcal{M}_0)(\mathcal{N}+\mathcal{N}_0) + (\mathcal{N}+\mathcal{N}_0)B(\overline{\mathcal{M}}+\overline{\mathcal{M}_0})B^{-1}) \\
&\qquad + 2K_{15}\overline{K_{15}}((H+H_0)B(\overline{H}+\overline{H_0}))\Big|^2 \\
&+ 8\Big|K_{12}K_{15}(\mathcal{M}+\mathcal{M}_0)(H+H_0) + 2K_{13}\overline{K_{15}}(\mathcal{N}+\mathcal{N}_0)B(\overline{H}+\overline{H_0}) - u(H+H_0)\Big|^2 \\
&+ 16(\mathrm{Tr}|K_{15}|^4 - (\mathrm{Tr}|K_{15}|^2)^2\Big||H^*+H_0^*|^2 - M_3^2\Big|^2 + 16\mathrm{Tr}|K_{15}|^4\Big||H^*+H_0^*|^2 - M_3^2\Big|^2,
\end{aligned} \tag{4.19}
$$

and $u = 2K_{13}K_{15}\big(s+p-3(b+b')+2(a+a')+M_2\big)-2K_{12}K_{25}M_1$. We deduce that the SO(10) model is an attractive model. Its construction is completely dictated by the arrangement of the fermions, their representations, and the Dirac operator acting on them. The nature of the Higgs fields is completely fixed, and their vev's constrained by the requirement that the potential is non-trivial for the consistency of the theory. The mass matrix of the fermions is realistic.

5. Gravity in non-commutative geometry

A natural question to ask is how to introduce gravity in the framework of non-commutative geometry. An answer to this question requires a generalisation of the basic notions of Riemannian geometry. Connes has proposed to define metric proporties of a non-commutative space corresponding to an involutive unital algebra $\mathcal{A}$ in terms of K-cycles over $\mathcal{A}$ [2-3]. In [9] it was shown that every K-cycle over $\mathcal{A}$ yields a notion of "cotangent bundle" associated to $\mathcal{A}$ and a Riemannian metric on the cotangent bundle. One can also introduce analogues of the spin connection, torsion, Riemann curvature tensor, Ricci tensor, and scalar curvature. This allows one to write the generalized Einstein-Hilbert action. Here we shall only describe the gravity action for a two sheeted space, and refer the reader to [9] and [4] for details. We shall also derive, heuristically, an experimental signature of the effect of the geometry on the standard model, which turns out to be a constraint on the Higgs mass and top quark mass.

Consider a space-time X consisting of two copies of a four-dimensional manifold M: $X = M \times Z_2$. The algebra $\mathcal{A}$ is given by $\mathcal{A} = C^\infty(M)\otimes\mathcal{A}_1 \oplus C^\infty(M)\otimes\mathcal{A}_2$, where $\mathcal{A}_1 = \mathcal{A}_2 = C$. The elements of $\mathcal{A}$ are operators of the form diag$(1\otimes a_1, 1\otimes a_2)$ where a_i, $i = 1,2$ are smooth function on M, and 1 is the identity in the Clifford algebra, Cliff(T^*M), of Dirac matrices over M. We consider even K-cycles (π, H, D, Γ) for $\mathcal{A}$, with $\pi = \pi_1 \oplus \pi_2$, where π_i is a representation of $C^\infty(M)\otimes\mathcal{A}_i$ on a Hilbert space $L^2(S_i, \tau_i dv)$, where S_i is a bundle of spinors on M with values in a finitely generated, projective hermitian left $\mathcal{A}_i$ module E_i, τ_i is a normalized trace on $\mathcal{A}_i$ and dv is the volume element on M. Then h is defined by $h = L^2(S_1, \tau_1, dv) \oplus L^2(S_2, \tau_2, dv)$. The Dirac operator is taken to be

$$D = \begin{pmatrix} \not{\nabla}_M \otimes 1 & \gamma_5 \otimes \phi \\ \gamma_5 \otimes \phi^* & \not{\nabla}_M \otimes 1 \end{pmatrix}, \tag{5.1}$$

where $\not{\nabla}_M$ is the standard covariant Dirac operator on M. The Z_2 grading on M is given by $\Gamma = \begin{pmatrix} \gamma_5 & 0 \\ 0 & -\gamma_5 \end{pmatrix}$. The "cotangent bundle" $\Omega^1_D(\mathcal{A}) = \mathcal{O}$⇕]}⊣$^\infty(\mathcal{A})/\ker\pi$ is a free left and right $\mathcal{A}$ module, with a basis $\{e^N\}^5_{N=1}$ given by

$$e^a = \begin{pmatrix} \gamma^a & 0 \\ 0 & \gamma^a \end{pmatrix}, \qquad e^5 = \begin{pmatrix} 0 & \gamma_5 \\ -\gamma_5 & 0 \end{pmatrix}, \qquad a = 1,2,3,4. \tag{5.2}$$

The hermitian structure on $\Omega^1_D(\mathcal{A})$ is given by the trace of 8×8 matrices, normalized such that tr$1 = 1$. Hence

$$< e^N, e^M >= \mathrm{tr}(e^N (e^M)^*) = \delta^{NM}. \tag{5.3}$$

For a one-form $\rho = \sum_i a_i db_i$ in $\Omega^1_D(\mathcal{A})$, $\pi(\rho)$ is parametrized by

$$\pi(\rho) = \begin{pmatrix} \gamma^\mu \rho_{1\mu} & \gamma_5 \phi \rho_5 \\ -\gamma_5 \phi \bar{\rho}_5 & \gamma^\mu \rho_{2\mu} \end{pmatrix}, \tag{5.4}$$

where $\rho_{1\mu} = \sum_i a_{i1}\partial_\mu b_{i1}$, $\rho_5 = \sum_i a_{i1}(b_{i2} - b_{i1})$, and similarly for $\rho_{2\mu}$ and $\tilde{\rho_5}$. Evaluating $\pi(d\rho) = \sum_i [D, a_i][D, b_i]$, we obtain

$$\pi(d\rho) = -\begin{pmatrix} g^{\mu\nu}\partial_\mu a_{i1}\partial_\nu b_{i1} & 0 \\ 0 & g^{\mu\nu}\partial_\mu a_{i2}\partial_\nu b_{i2} \end{pmatrix}. \tag{5.5}$$

One sees that, for a suitable choice of a_i, b_i subject to the constraint $\pi(\rho) = 0$, any expression of the form $\mathrm{diag}(X_1, X_2)$ can be obtained, where X_1, X_2 are scalar functions. Therefore, we can express $\pi(d\rho)$ modulo auxiliary fields in terms of its components:

$$\pi(d\rho) = \begin{pmatrix} \gamma^{\mu\nu}\partial_\mu\rho_{1\nu} & \phi\gamma^\mu\gamma_5(\partial_\mu\rho_5 + \rho_{1\mu} - \rho_{2\mu}) \\ -\phi\gamma^\mu\gamma_5(\partial_\mu\tilde{\rho_5} + \rho_{1\mu} - \rho_{2\mu}) & \gamma^{\mu\nu}\partial_\mu\alpha_{2\nu} \end{pmatrix}. \tag{5.6}$$

This is a representative of $\pi(d\rho)$ in $\pi(\Omega^2(\mathcal{A}))/\pi(d\mathrm{Ker}\pi(|_{\Omega^1(\mathcal{A})}))$ orthogonal to the auxiliary fields. Let ∇ be a connection on $\Omega^1_D(\mathcal{A})$ and $\omega^N{}_M \in \Omega^1_D(\mathcal{A})$ defined by $\nabla e^N = -\omega^N{}_M \otimes_\mathcal{A} e^M$. The components of $\pi(\nabla)$ in the basis $\{e^N\}_{N=1}^{N=5}$ are given by

$$\omega^N{}_M = \begin{pmatrix} \gamma^\mu\omega^N_{1\mu M} & \gamma_5\phi l^N{}_M \\ -\gamma_5\phi\tilde{l}^N{}_M & \gamma^\mu\omega^N_{2\mu M} \end{pmatrix}. \tag{5.7}$$

Hermiticity of ∇ then implies that

$$\omega^N_{i\mu M} = -\omega^M_{i\mu N}, \qquad i = 1, 2, \qquad \tilde{l}^N{}_M = -l^M{}_N. \tag{5.8}$$

Let $T^N \in \Omega^2_D(\mathcal{A})$ be the components of the torsion $T(\nabla)$ defined by $T^N = T(\nabla)e^N$. Then

$$T^N = de^N + \omega^N{}_M e^M \tag{5.9}$$

Similarly define $R^N{}_M \in \Omega^2_D(\mathcal{A})$ by $R(\nabla)e^N = R^N{}_M \otimes_\mathcal{A} e^M$ where $R(\nabla)$ is the Riemann curvature of ∇ defined by $R(\nabla) := -\nabla^2$. Then

$$R^N{}_M = d\omega^N{}_M + \omega^N{}_P\omega^P{}_M. \tag{5.10}$$

Imposing the condition that the torsion $T(\nabla)$ vanishes gives

$$\begin{gathered} \omega^a_{1\mu b} = \omega^a_{2\mu b} \equiv \omega^a_{\mu b}, \qquad \omega^a_{1\mu 5} = -\omega^a_{2\mu 5} = \phi l^a{}_b e^b{}_\mu, \\ l^a{}_b = l^b{}_a, \qquad l^5{}_a = -l^a{}_5, \qquad l^5{}_a e^a{}_\mu = -\partial_\mu\phi^{-1}, \end{gathered} \tag{5.11}$$

where $\omega^a_{\mu b}$ is the classical Levi-Civita connection derived from the metric $g_{\mu\nu} = e^a_\mu\delta_{ab}e^b_\nu$ on M. The analogue of the Einstein-Hilbert action is

$$\begin{aligned} I(\nabla) &:= \kappa^{-2} < R^N{}_M e^M, e_N > +\Lambda < 1, 1 > \\ &= \kappa^{-2}\int_M \mathrm{tr}(R^N{}_M e^M (e_N)^*) + \Lambda\int 1, \end{aligned} \tag{5.12}$$

where κ^{-1} is the Planck scale. This action is then calculated to be

$$I(\nabla) = \kappa^{-2} \int_M \Big(2r - 4\phi \nabla_\mu \partial^\mu \phi^{-1} + 4\phi^2 l^a{}_a l^5{}_5 + \phi^2 \big((l^a{}_a)^2 - l^a{}_b l^b{}_a \big) \Big) \sqrt{g} d^4x + 2\Lambda \int_M \sqrt{g} d^4x, \tag{5.13}$$

where r is the scalar curvature of the classical Levi-Civita connection. The fields $l^a{}_b$ and $l^5{}_5$ decouple, and by setting $\phi = e^{-\kappa\sigma}$ one finds

$$I(\nabla) = 2 \int_M (\kappa^{-2} r - 2\partial_\mu \sigma \partial^\mu \sigma + \Lambda) \sqrt{g} d^4x. \tag{5.14}$$

Therefore a theory of gravity on $M \times Z_2$ is equivalent to general relativity on M, with an additional massless scalar field σ that couples to the metric of M. To better understand the role of the field σ we can study the coupling of gravity to the Yang-Mills sector [10]. In the case of the standard model the field $\phi = e^{-\kappa\sigma}$ replaces the electroweak scale. In other words, the vev of the field ϕ determines the electroweak scale. This simple result has some unexpected consequences. To determine the σ dependence in the Yang-Mills action of the standard model, we consider the σ dependence in the Dirac operator. For example, the leptonic Dirac operator is

$$D_l = \begin{pmatrix} \gamma^a e^\mu_a (\partial_\mu + \ldots) \otimes 1_2 \otimes 1_3 & \gamma_5 e^{-\kappa\sigma} \otimes M_{12} \otimes k \\ \gamma_5 e^{-\kappa\sigma} \otimes M^*_{12} \otimes k^* & \gamma^a e^\mu_a (\partial_\mu + \ldots) \otimes 1_3 \end{pmatrix}. \tag{5.15}$$

From this one can easily verify that the bosonic part of the standard model is

$$\begin{aligned} L_b = & -\frac{1}{4} \Big(F^3_{\mu\nu} F^{\mu\nu 3} + F^2_{\mu\nu} F^{\mu\nu 2} + F^1_{\mu\nu} F^{\mu\nu 1} \Big) \\ & + D_\mu (H + M_{12})^* D_\nu (H + M_{12}) g^{\mu\nu} e^{-2\kappa\sigma} \\ & - \frac{\lambda}{24} \Big| |H + M_{12}|^2 - |M_{12}|^2 \Big|^2 e^{-4\kappa\sigma}. \end{aligned} \tag{5.16}$$

The σ dependence in Eq (5.16) is a consequence of the "Weyl invariance" of the action (3.7) under rescaling of the Dirac operator $D \to e^{-w} D$, as this implies $g_{\mu\nu} \to e^{2w} g_{\mu\nu}$ and $\kappa\sigma \to \kappa\sigma + w$. This can be easily seen from the scalings: $\pi(\rho) \to e^{-w}\pi(\rho)$ and $\pi(\theta) \to e^{-2w}\pi(\theta)$. By redefining $H + M_{12} \to e^{\kappa\sigma} H$, the H dependent terms in (5.16) become

$$D_\mu H^* D^\mu H + \kappa \partial_\mu (H^* H) \partial^\mu \sigma + \kappa^2 H^* H \partial_\mu \partial^\mu \sigma - \frac{\lambda}{24} \Big| (H^* H)^2 - \mu^2 e^{-2\kappa\sigma} \Big|^2. \tag{5.17}$$

The potential in Eq (5.17) could be rewritten in the familiar form

$$V_0 = \frac{\lambda}{24} (H^* H)^2 - \frac{1}{2} m^2 (H^* H) + \frac{3}{2\lambda} m^4, \tag{5.18}$$

where we have set $m^2 = \frac{\lambda \mu^2}{6} e^{-2\kappa\sigma}$, so that m is now a field and not just a parameter. The potential V_0 is of the same form as that of the standard model. We assume that, after renormalization, the bosonic action takes the same form as I_l+I_q. In the absence of some understanding of symmetries, it is not possible to prove this assumption at the quanutm level. Let ϕ be the component of the Higgs field that develops a vev. We are then mainly interested in the potential

$$V_0 = \frac{\lambda}{24}\phi^4 - \frac{1}{2}m^2\phi^2 + \frac{3}{2\lambda}m^4. \tag{5.19}$$

Minimizing with respect to ϕ and m yields the same asymmetric phase $\phi^2 = \frac{6}{\lambda}m^2$, and the weak scale, $e^{-\kappa\sigma}$, is undetermined at the classical level. The quantum corrections to the potential are given, in the one-loop approximation, by the effective Coleman-Weinberg [11] potential of the standard model [12]:

$$\begin{aligned} V_1 = \frac{1}{16\pi^2}\Big(&\frac{1}{4}H^2(\ln\frac{H}{M^2}-\frac{3}{2}) + \frac{3}{4}G^2(\ln\frac{G}{M^2}-\frac{3}{2}) + \frac{3}{2}W^2(\ln\frac{W}{M^2}-\frac{5}{6}) \\ &+ \frac{3}{4}Z^2(\ln\frac{Z}{M^2}-\frac{5}{6}) - 3T^2(\ln\frac{T}{M^2}-\frac{3}{2})\Big), \end{aligned} \tag{5.20}$$

where

$$\begin{aligned} &H = -m^2 + \frac{1}{2}\lambda\phi^2, \qquad G = -m^2 + \frac{1}{6}\lambda\phi^2 \\ &W = \frac{1}{4}g_2^2\phi^2, \qquad Z = \frac{1}{4}(g_2^2+g_1^2)\phi^2, \qquad T = \frac{1}{2}h^2\phi^2, \end{aligned} \tag{5.21}$$

and M is the renormalization scale. Minimizing the total potential $V_0 + V_1$ with respect to the fields ϕ and m, after rescaling

$$G = \overline{G}M^2, \qquad H = \overline{H}M^2, \qquad T = \overline{T}M^2,$$

the asymmetric solution is given by the solution to the following two equations:

$$\begin{aligned} 0 &= \overline{G} + \frac{M^2}{32\pi^2\phi^2}(\overline{H}-\overline{G})\Big(\overline{H}(\ln\overline{H}-1) + 3\overline{G}(\ln\overline{G}-1)\Big), \\ 0 &= \overline{G} + \frac{3M^2}{32\pi^2\phi^2}(\overline{H}-\overline{G})\Big(\overline{H}(\ln\overline{H}-1) + \overline{G}(\ln\overline{G}-1)\Big) - \frac{g_2^2+g_1^2}{64\pi^2} \\ &+ \frac{3g_2^4\phi^2}{128\pi^2M^2}\left(\ln\frac{g_2^2\phi^2}{4M^2}-\frac{1}{3}\right) - \frac{3M^2}{4\pi^2\phi^2}\overline{T}^2(\ln\overline{T}-1). \end{aligned} \tag{5.22}$$

At the scale $M = m_Z$, the mass of the Z-particle, the coupling constants g_1, g_2 as well as the vev ϕ are known from experimental data, corrected with the help of the renormalization group equations [12]:

$$g_2 = 0.650, \qquad g_1 = 0.358, \qquad \phi = 246\ Gev. \tag{5.23}$$

The only unknowns in the minimization equations are λ, m and the square of the top quark mass $T = m_t^2$. These equations, being complicated functions of $\overline{H}$ and $\overline{G}$, can only be solved numerically, for various values of $\overline{T}$. The numerical solutions are easily obtained using *Mathematica*. The Higgs mass can be determined from the formula $m_H^2 = \frac{\partial^2 V}{\partial \phi^2}$ which gives

$$\begin{aligned} m_H^2 = M^2 \Big((\overline{H} - \overline{G}) + \frac{9M^2}{16\pi^2\phi^2}(\overline{H} - \overline{G})^2(\ln \overline{H} + \frac{1}{3}\ln \overline{G}) \\ + \frac{3g_2^4\phi^2}{64\pi^2 M^2}\ln\frac{g_2^2\phi^2}{4M^2} - \frac{3M^2}{2\pi^2\phi^2}\overline{T}^2 \ln \overline{T} \Big). \end{aligned} \tag{5.24}$$

We now quote the results: There are only two classes of solutions, for $\overline{G} \ll \overline{H}$ and for $\overline{H} \ll \overline{G}$. In the first case, we find that there are only two narrow bands for the top quark mass where solutions exist. The first band is $0.365 \leq \overline{T} \leq 0.455, \quad \overline{G} \ll \overline{H}$, corresponding to a top quark mass

$$54.90 \text{ Gev} \leq m_t \leq 61.35 \text{ Gev}, \tag{5.25}$$

which is already ruled out experimentally. The second band is very narrow: $2.57 \leq \overline{T} \leq 2.61, \quad \overline{G} \ll \overline{H}$, corresponding to the top quark mass

$$146.23 \text{ Gev} \leq m_t \leq 147.37 \text{ Gev}, \tag{5.26}$$

and a Higgs mass $117.26 \text{ Gev} \leq m_H \leq 142.61$ Gev. Clearly this band of values for the top quark mass lies within the present experimental average of [13]

$$m_t = 149 + \begin{pmatrix} +21 \\ -47 \end{pmatrix} \text{ Gev}. \tag{5.27}$$

The second class of solutions occurs when $1.30 \leq \overline{T} \leq 2.61, \quad \overline{H} \ll \overline{G}$, corresponding to the top quark mass

$$104.07 \text{ Gev} \leq m_t \leq 147.48 \text{ Gev}, \tag{5.28}$$

and a Higgs mass $1208 \text{ Gev} \geq m_H \geq 1197$ Gev. However, since $\overline{H} \ll \overline{G}$, and since the coupling constant $\lambda = O(-100)$, the potential, in this domain, becomes unbounded from below, signaling the break down of the perturbative regime. Requiring stability of the electroweak potential excludes this solution. Therefore the only acceptable solution is (5.26) which is remarkably constrained, considering the wide range of possibilities that one might have, a priori. We note that the field σ becomes massive with the square of the mass given by: $m_\sigma^2 = \frac{\partial^2 V}{\partial \sigma^2}$. This is equal to

$$m_\sigma^2 = \kappa^2 m^2 \Big(2\phi^2 \frac{\overline{H} - 4\overline{G}}{\overline{H} - \overline{G}} + \frac{M^2}{16\pi^2}(\overline{H}(1 - \ln \overline{H}) + 3\overline{G}(1 - \ln \overline{G})) \Big). \tag{5.29}$$

For the physically acceptable solutions we have $\overline{H} = O(1)$, $\overline{G} = O(10^{-4})$ and $m^2 = O(M^2)$. Then we find from Eq (5.29) that

$$m_\sigma^2 = O(\kappa^2 M^4), \tag{5.30}$$

so that $m_\sigma = O(10^{-15})$ Gev, which is unobservable. These predictions have at best a heuristic value, since the problem of fixing the form of the cosmological constant at the one-loop level by imposing natural geometrical constraints is not understood. However, they do suggest that gravitational effects may play a role in understanding masses of fermions and Higgses.

Acknowledgments by A.H.C: I would like to express my indebtedness and gratitude to Abdus Salam for his role as my thesis advisor at Imperial College, during the period 1974-1976, and as my mentor during 1977 when I stayed as a postdoctoral fellow in Trieste. His quest to understand the fundamental forces in nature, including gravity, and his way of thinking have shaped my research since then. I consider myself fortunate to have studied under such a great physicist, teacher and a generous person, who found pleasure in sharing his knowledge with the people around him.

References

[1] S. Glashow *Nucl. Phys.* **22**, 579 (1961);
A. Salam and J. Ward *Phys. Lett* **13**, 168 (1964);
S. Weinberg, *Phys. Rev. Lett.* **19**, 1264 (1967);
A. Salam, in *Elementary Particle Theory* (editor N. Svartholm), Almquist and Forlag, Stockholm.

[2] A. Connes, *Publ. Math. IHES* **62** 44 (1983);
A. Connes, in *the interface of mathematics and particle physics* , Clarendon press, Oxford 1990, Eds D. Quillen, G. Segal and S. Tsou.

[3] A. Connes and J. Lott,*Nucl.Phys.B Proc.Supp.* **18B** 29 (1990), North-Holland, Amsterdam; *Proceedings of 1991 Summer Cargese conference* p.53 editors J. Fröhlich et al (1992),Plenum Pub.;
D. Kastler , Marseille preprints CPT-91/P.2610, CPT-91/P.2611 and CPT-92/P.2814;

For an alternative construction see
R. Coquereaux, G. Esposito-Farése, G. Vaillant, *Nucl. Phys.* **B353** 689 (1991);
M. Dubois-Violette, R. Kerner, J. Madore, *J. Math. Phys.* **31** (1990) 316;
B. Balakrishna, F. Gürsey and K. C. Wali, *Phys. Lett.* **254B** (1991) 430;
H.-G. Ding, H.-Y. Guo, J.-M. Li and Ke Wu, Beijing preprint ASITP-93-23.

[4] A. Connes, *Non-Commutative Geometry*, Academic press, to be published;
D.Kastler and M. Mebkhout, *Lectures on non-commutative geometry and applications to elementary particle physics*, to be published;
A.H. Chamseddine and J. Fröhlich, *Some elements of Connes' non-commutative geometry and space-time physics*, Zurich preprint ZU-TH-17/1993

[5] A.H. Chamseddine, G. Felder and J. Fröhlich, *Phys. Lett.* **296B** (1993) 109; *Nucl. Phys.* **B395** (1993) 672.

[6] E. Alvarez, J.M. Garcia-Bondia and C.P. Martin, *Madrid preprint*, May 1993.

[7] H. Georgi in *Particles and Fields* **1974** p.575, editor C.E. Carlson (AIP, New York, 1975);
H. Fritsch and P. Minkowski, *Ann. Phys.* (N.Y) **93** (1977) 193;
M.S. Chanowitz, J. Ellis and M. Gaillard, *Nucl. Phys.* **B129** (1977) 506.

[8] A.H. Chamseddine and J. Fröhlich, *SO(10) unification in non-commutative geometry", Zurich preprint ZU-TH-10/1993.*

[9] A.H. Chamseddine, G. Felder and J. Fröhlich, Comm.Math.Phys **155** *(1993) 205.*

[10] A.H. Chamseddine and J. Fröhlich, Constraints on the Higgs and top quark masses from effective potential and non-commutative geometry, Zurich preprint ZU-TH-18/1993, to appear in Phys. Lett. **B**.

[11] E. Weinberg and S. Coleman, Phys.Rev. **D7** (1973) 1888.

[12] M. Sher, Phys.Rep. **179** (1989) 274 and references therein;
C. Ford, D.R.T. Jones, Y.W. Stephenson and M.B. Einhorn, Liverpool preprint LTH 288.

[13] U. Amaldi, This Proceedings.

CHROMOGRAVITY EXPLAINS "STRONG GRAVITY"

Yuval Ne'eman*

Wolfson Chair Extraordinary,

Raymond and Beverley Sackler Faculty of Exact Sciences

Tel-Aviv University, Tel-Aviv, Israel 69978

and

Center for Particle Physics*,

University of Texas, Austin, USA 78712

and

Djordje Šijački¶

Institute of Physics, P.O.B. 57, Belgrade

Yugoslavia

* Supported in part by DOE Grant DE-FG05-85-ER40200.

¶ Supported in part by Science Foundation (Belgrade).

1. The Strong Interactions Crisis in the Fifties

The successes of QED in the late forties, coupled with the discovery of the pions, confirming Yukawa's (and Stueckelberg's) prognostication for the Strong Interactions, generated the hope that something similar to QED :could be achieved for that strong nuclear force, using nucleons and pions instead of electrons and photons. Abdus Salam managed to prove that the method of Renormalization could indeed be adapted to the new structure. Unfortunately, in one of the first applications of the method of Dispersion Relations, which Gell-Mann and Goldberger had coopted from optics into Particle Physics, Habershaim calculated the pionnucleon coupling, obtaining $g^2/4\pi = 15$, a value excluding any idea of using perturbation techniques for the new force. Under the charismatic influence of G. Chew, the USA West Coast abandoned Relativistic Quantum Field Theory (RQFT) altogether, vesting its hopes in "S Matrix Theory", the name selected for Dispersion Relations, when reformulated as a fundamental theory, with no RQFT in the background. One of the authors of this article (Yuval Ne'eman) was a student of Salam's at Imperial College in 1958-61 and can testify to the paramount position of RQFT in Salam's thinking. As a matter of fact, the three students who got their Ph.D. in 1961-62 are a testimonial to Salam's full-breadth approach: R. Streater plunged into Axiomatic Field Theory, C. Lovelace worked in Dispersion Relations and Ne'eman was encouraged to follow his inclination and concentrate on symmetries, with RQFT as guiding paradigm. In the mid-sixties, Salam initiated programs centering on nonlinear Lagrangians, on Borel Summability as a replacement of the usual perturbation techniques, etc., all the while sticking by RQFT.

Meanwhile, there was no theory for the Strong Interactions, except for the wealth of results deriving from SU(3) and quarks. These could be put in a compact form in a "phenomenological" RQFT which was shown by Schwinger and Weinberg to reproduce even the results of Current Algebra; the latter had been originally formulated in an independent manner - Gell-Mann's answer to the West Coast taboo on RQFT. The phenomenological model included right from the start gauge vector meson couplings: both original Eightfold Way SU(3) proposals (YN's and Gell-Mann's) had been inspired by the Yang-Mills mechanism and by a 1960 paper based on it, by the late J.J. Sakurai, "Vector Theory of the Strong Interactions" which made a convincing case for the existence of universally coupled vector mesons for Isospin and Hypercharge and for their role.

The other ad hoc treatment was provided by the systematics of Regge trajectories, again abstracting a pattern from phenomenology, as in the case of SU(3). Actually, the dynamical input assumed in "S-Matrix Theory" having been abstracted from generalized Yukawa dynamics, bore no resemblance to the phenomenological picture. The Regge sequences of the Chew-Frautschi plot turned out to be straight lines $J \sim M^2$, with the appearance of going on indefinitely, somewhat like those of a harmonic oscillator.

2. Salam's Strong Gravity

In the hunt for hadron resonance particles, opened around 1960 by the Alvarez bubble chamber group's discovery of the vector mesons, spreading everywhere later, the f^0, a spin 2 state with the internal quantum numbers of the vacuum was found in 1962 [1]. P.G.O. Freund pointed out the similarity with the graviton and the universality features thereby induced, in a Dispersion Relations "pole dominance" context [2]. Several years later, the idea was revived by Cabibbo, Horwitz and Ne'eman [3] ("CHN") and by Delbourgo, Salam and Strathdee [4]. CHN displayed highly successful predictions based on "universal" 1^- and 2^+ Regge couplings (including the Pomeron, with the graviton's quantum numbers) emulating Yang-Mills and Gravity theories. In 1970, Isham, Salam and Strathdee [5], taking their cue from the discussion of ρ-γ mixing in the context of hadron electrodynamics, adapted the idea to "f-g mixing" for hadron gravity, while still motivated by the dominance argument.

In 1973, QCD was adopted as the paradigmal foundation of the Strong Interactions. t'Hooft's success in completing the task of renormalizing the Yang-Mills interaction had opened the way for Asymptotic Freedom, proved by Gross and Wilczek and by Politzer as well as by t'Hooft himself. The proven Asymptotic Freedom of the theory appeared to provide a good explanation of the Quark Model results, by providing for a "very weak" coupling of the quarks within the hadron itself, while nevertheless making the quarks sit in an "impossibly deep" potential - i.e. a possibility of confinement. RQFT had already been rehabilitated and restored to its regal position in 1971, for the Electroweak interaction. S-Matrix theory became a historical relic. Apparently, the error had been in making a RQFT of nucleons and mesons rather than one of quarks and gluons.

At the same time, it was still hard to see how the new theory could reproduce some of the other features of Strong Interactions - especially the spectrum of Regge sequences - which had indeed been successfully reproduced by the 1968-73 candidate theory, namely the String.

Salam was not entirely satisfied with the new paradigm. It was hard to see how the theory could indeed reproduce the successes of the String. Confinement (a "natural" in the String, a theory modeled on harmonic oscillators) was not proven, except on a lattice (this is still the state of affairs in 1993.). Ordinary hadron-hadron forces, which had been understood in terms of various meson exchanges, would just have to be posited - with the assumption that the entire Quark Model could be reproduced by QCD. Salam tried to find an improvement on QCD, such as would complete it as the theory of Strong Interactions (SI). I think Salam's presentation in January 1977, at the Marshak Anniversary Conference [6] contains the most eloquent argumentation for the need to add a gravity-like component to QCD, for a complete description of SI. He suggests equivalent roles for spin 1 (the gluons) and spin 2 (f-gravity) in SI, just as they appeared in the String (J=1 for open,

J=2 for closed strings), both being essential to the renormalizability of the theory. SI are presented as a "link" between unification with the Electroweak interactions, within a GUT mediated by J=1 connections, and unification with Gravity (f-g unification) through common mediation by J=2 connections. Salam especially discussed the possibility of deriving confinement from a quantized soliton-like solution of f-gravity.

One way of implementing this scheme was to gauge the tensor-completion of Lorentz SL(2,C) with color SU(3) - a return to the SU(6)-type structures of 1964-65, now with SL(6,C), the gauging, however, doing away with the various no-go theorems [7-9], all proven for global symmetries. The model has been pursued by Sivaram and Sinha in a series of papers since 1973; we refer the reader to their 1979 review [10].

That the String, born as a theory of the SI (1967-73) could then metamorphose into a theory of Quantum Gravity (unified with all other forces - a TOE) is yet another example of the basic resemblances between hadron dynamics and Gravity.

3. Poincaré and (Metric) Affine Gravity and Strong Interactions

In 1977, Y. Ne'eman started a collaboration with F.W. Hehl and his collaborators at Cologne. The latter had furthered the understanding of Torsion in gravitational theories, including generalizations of Einstein-Cartan in the Kibble-Sciama version, i.e. theories with local but non-propagating torsion. They had then tried to generalize this approach into a Metric-Affine theory.

To start with, they had been wondering about extending the angular momentum tensor current, to include the symmetric (shear) part. Y. Ne'eman realized that what Hehl et al. were after was something that had already been constructed and analyzed in 1965 by him, with Gell-Mann and the late Yossef Dothan [11] (in a paper which also suggested SL(6,C) $\subset$ U(6,6)), for the description of Regge sequences in the hadron spectrum in a spin-unitary spin unification; in the same paper, however, they had suggested applying SL(3,R) for the Regge trajectories themselves and for quadrupolar excitations in nuclei.

In this work on Regge and nuclear bands, Ne'eman and his coauthors had noticed a relationship with gravity - the Regge sequences represent the quasi-conservation of the time-derivatives of gravitational quadrupoles; the nuclear sequences had also involved moments of inertia. Somehow, the SI managed to relate to Gravity in a variety of ways. What Hehl et al. also needed in order to make progress was a way of representing spinors in an Affine picture, i.e. representations of the double-covering of SL(4,R), which were supposed not to exist (and are still said not to exist in many texts on Gravity). In 1978, Ne'eman proved the existence of the double covering and constructed some of its representations [12-14]. Since that time, Mickelsson and Ne'eman have also developed the theory of "world spinors" [15-16], thus going beyond the "anholonomic" (tangent space) SL(4,R) spinors.

The work of Ne'eman with Hehl and his group also included an attempt to study the dynamics deriving from Lagrangians quadratic in curvature and torsion, later adding Non-Metricity, the appropriate field strength relating to shear. In this work [17] they found that the quadratic Poincaré Gauge Theory Lagrangian already produces confinement, at the classical level; like Salam, YN thought of gravity as contributing to the SI.

Of course, the various Superunification schemes based on $8 \geq N > 1$ all contain sectors connecting Gravity and SI, as well as the Electroweak. Note that the quantization of electric charge and the cancellation of Chiral Anomalies between quarks and leptons in each generation still require a GUT, such as the Salam-Pati model etc., i.e beyond the simplest SU(5) (invalidated by the proton decay lifetime lower bound) or SuperGUT, as a unifying and constraining structure. There is thus a high probability that such sectors, relating SI and gravity, will some day be identified. So much more so, should the String return from its present eclipse and redominate the physics sky.

4. Chromogravity

Meanwhile, however, the Standard Model paradigm requires "low-energy" ($<10^3$GeV) hadron physics (=SI) to be fully described by QCD. We believe that recent work by the authors of this paper has managed to isolate and identify a specific category of elements in the Infrared (IR) region of QCD which are precisely responsible for all those aspects of hadron physics which appeared to Salam [6] to require the intervention of gravity. These elements indeed generate an analog to gravity, a J=2 "chromograviton" or "pseudo-graviton" whose action effectively generates Salam's "Strong Gravity" or his "f-gravity", without having to introduce the theory as an additional input. Salam's dynamics should then be regarded as "effective", resulting, however, from elements within QCD itself. We are thus not yet drawing from those bridging elements of the Superunification theory which connect Einstein's and Newton's Gravity to the SI. Everything - including those features of hadron dynamics which were reproduced by the String, another gravitationally structured theory - results from a class of Feynman diagrams in QCD. This is our "Chromogravity" or "Pseudo-Gravity", which we present in broad lines in the next sections [18-20].

QCD is often compared with QED, confinement then appearing as a case of infinite ionization energy. In this analogy, hadrons correspond to neutral atoms or molecules - and the inter-hadron forces to those responsible e.g. for the binding of atoms within molecules. From atomic and molecular physics, we know that these involve three levels: Heteropolar (ionic) molecules, made of ions (e.g. $Na^+ + Cl^-$), atomic (e.g. He + He → He_2) and Van der Waals. The binding of $q+\bar{q}$ into mesons, or of q+q+q into baryons, is very much like the heteropolar case,

Coulomb-like. The making of nuclei from nucleons, on the other hand, is very much like the case of the atomic molecules, involving electron exchange forces. In QCD, the exchange involves mesons, quark-antiquark b$\overline{p}$ nd pairs. It is a short range force, like other exchange forces.

Is there then also an analog to the Van der Waals forces? We know from the complex-J plane analysis that Regge trajectories do not represent Yukawa-pion effects, as had originally been conjectured. The excitations represent $\Delta J = 2$ jumps, adding angular momentum to the $q\overline{q}$ and qqq systems. Moreover, in nuclei there is direct evidence for a longer-range component of QCD, responsible for such excitations as those studied phenomenologically in even-even nuclei: the ground state has J = 2 and 0, the 5+1 =6 states of the U(6) defining representation of the "IBM" (Arima-Iachello) model [21]. It thus appears that the relevant agent has the quantum numbers of a metric $g_{\mu\nu}$ in 4 dimensions: the 10 components of this symmetric tensor have J = 2 + 1 + 0 + 0, but 1 + 0 are cancelled by the Riemannian condition $D_\sigma g_{\mu\nu} = 0$. Returning to our comparison with molecules, we note that the Van der Waals forces are represented by an effective interaction in which two photons are emitted or reabsorbed at each vertex [22]. In QCD, the minimal such contribution would correspond to the exchange of two gluons with J = 1 each.

5. The Emergence of a Gravity-like Metric from QCD

Do we have components of QCD that might have long range or quasi-long range? The only possible candidates would consist in sets of gluons, provided some gauge condition could preserve their masslessness to a limited extent. We are thus again drawn to look at gluon exchanges.

Between hadrons, only colour-neutral exchanges are allowed. The simplest examples are [$B^a{}_\mu$ is a gluon, a,b its SU(3) indices, η_{ab} is a metric (1,..,1) in SU(3), d_{abc} the SU(3) symmetric tensor]

$$G_{\mu\nu} = \eta_{ab}\, B^a{}_\mu\, B^b{}_\nu \tag{5.1}$$

$$G^{(3)}{}_{\mu^1\mu^2\mu^3} = d_{abc}\, B^a{}_{\mu^1}\, B^b{}_{\mu^2}\, B^c{}_{\mu^3} \tag{5.2}$$

and more generally

$$G^{(n)}{}_{\mu^1\mu^2..\mu^n} := d^{(n)}{}_{a^1a^2..a^n}\, B^{a^1}{}_{\mu^1}\, B^{a^2}{}_{\mu^2} \cdots B^{a^n}{}_{\mu^n} \tag{5.3}$$

where

$$d^{(2)}{}_{a^1a^2} = \eta_{a^1a^2} \tag{5.4}$$

$$d^{(n)}{}_{a^1a^2..\,a^n} = d_{a^1a^2b^1}\, \eta^{b^1c^1}\, d_{c^1b^2a^3} \cdots \eta^{b^{n-4}c^{n-4}}\, d_{c^{n-4}b^{n-3}a^{n-2}}\, \eta^{b^{n-3}c^{n-3}}\, d_{c^{n-3}a^{n-1}a^n}. \tag{5.5}$$

Let us now analyze the behaviour of these effective entities under the colour SU(3) local gauge. We have for an infinitesimal variation ($\{\lambda_b\}$ is the appropriate representation of the SU(3) algebra matrix λ_b)

$$\delta_\epsilon B^a{}_\mu = \partial_\mu \epsilon^a + B^b{}_\mu \{\lambda_b\}^a{}_c\ \epsilon^c = \partial_\mu \epsilon^a + i\ f^a{}_{bc}\ B^b\ \epsilon^c \tag{5.6a}$$

where we have used the adjoint representation

$$\{\lambda_b\}^a{}_c = -i f_b{}^a{}_c \tag{5.6b}$$

We separate the constant component $N^a{}_\mu$, corresponding to vanishing field strength,

$$\partial_\mu N^a{}_\nu - \partial_\nu N^a{}_\mu = i f^a{}_{bc}\ N^b{}_\mu\ N^c{}_\nu \tag{5.7a}$$

$$B^a{}_\mu = N^a{}_\mu + A^a{}_\mu \tag{5.7b}$$

from the physical field $A^a{}_\mu$.

We now evaluate the colour SU(3) variation of the metric-like $G_{\mu\nu}$ of equation (5.1),

$$\delta_\epsilon G_{\mu\nu} = \delta_\epsilon \{\eta_{ab}(N^a{}_\mu + A^a{}_\mu)(N^b{}_\nu + A^b{}_\nu)\}$$

$$= \eta_{ab}(\partial_\mu \epsilon^a\ N^b{}_\nu + N^a{}_\mu\ \partial_\nu \epsilon^b + \partial_\mu \epsilon^a\ A^b{}_\nu + A^a{}_\mu\ \partial_\nu \epsilon^b)$$

$$+ i\eta_{ab}\ \{f^a{}_{cd}\ B^c{}_\mu\ \epsilon^d\ B^b{}_\nu\ + f^b{}_{cd}\ B^a{}_\mu\ B^c{}_\nu\ \epsilon^d\} \tag{5.8a}$$

The last bracket vanishes, since it represents the homogeneous SU(3) transformation of the SU(3) scalar expression in (5.1)

$$i f_{bcd}\ (B^b{}_\mu\ B^c{}_\nu + B^c{}_\mu\ B^b{}_\nu)\ \epsilon^d \tag{5.8b}$$

(more technically, due to the total antisymmetry of f_{abc} in a compact group)

With $N^a{}_\mu$, $N^b{}_\nu$, representing constant fields, we rewrite the terms in which they appear in terms of a new infinitesimal variation,

$$\xi_\mu = \eta_{ab}\ \epsilon^a\ N^b{}_\mu \tag{5.9}$$

Integrating by parts the terms in $A^a{}_\mu$, $A^b{}_\nu$ we get

$$\eta_{ab}\ (\epsilon^a\ \partial_\mu A^b{}_\nu + \partial_\nu A^a{}_\mu\ \epsilon^b) \tag{5.8c}$$

an expression whose Fourier transform vanishes for $k \to 0$, i.e. a contribution vanishing in the IR (infrared) sector. As a result

$$\delta_\xi G_{\mu\nu} = \partial_\mu \xi_\nu + \partial_\nu \xi_\mu = \partial_\mu(\xi^\sigma G_{\sigma\nu}) + \partial_\nu(\xi^\sigma G_{\mu\sigma}) \qquad (5.10)$$

where we have changed over to the ξ^σ variable of (5.9), and where we can reidentify δ as a variation under a formal infinitesimal R^4 diffeomorphism. Eq. (5.10) is then indeed the variation of a "world tensor" $G_{\mu\nu}$ under Einstein's covariance group, with

$$x^\sigma \to x^\sigma + \xi^\sigma = x^\sigma + \delta x^\sigma,$$

taking into consideration that ξ^σ has to be defined as a contravariant vector, playing the role of an "effective" infinitesimal diffeomorphism of the physical R^4 coordinate. $G_{\mu\nu}$ of eq. (5.1) is thus a "pseudo--metric" or the "chromometric" - when focusing on the QCD dynamics in the IR sector. A necessary condition for a metric is to be non-degenerate, i.e. invertibility; it is fulfilled by $G_{\mu\nu}$ because of the constant part $N^a{}_\mu$, in (5.1)

$$G_{\mu\nu} = \eta_{ab} N^a{}_\mu N^b{}_\nu + ..$$

which will provide the inverse in $G^{\mu\nu}$.

The μ,ν indices are Lorentz indices, acted upon by the physical Lorentz group. This means that the manifold is Riemannian: only Riemannian manifolds - with or without torsion - have tangent manifolds with orthogonal or pseudo-orthogonal symmetry. Thus

$$D_\sigma G_{\mu\nu} = 0 \qquad (5.11)$$

6. Pseudo-diffeomorphisms

To make sure of the fact that we are indeed reproducing the group Diff(4,R), for which the invariant will be supplied by the above $G_{\mu\nu}$, we evaluate the commutator of two δ variations,

$$[\delta_{\xi^1}, \delta_{\xi^2}] G_{\mu^1\mu^2} = \delta_{\xi^3} G_{\mu^1\mu^2},$$

$$\xi^3{}_\mu := \partial_\nu \xi^1{}_\mu \, \xi^2{}_\nu + \partial_\mu \xi^1{}_\nu \, \xi^2{}_\nu - \partial_\nu \xi^2{}_\mu \, \xi^1{}_\nu - \partial_\mu \xi^2{}_\nu \, \xi^1{}_\nu \qquad (6.1)$$

and verify that it indeed closes on the covariance group's commutation relations.

We now return to the n-gluon SU(3) scalars $G^{(n)}{}_{\mu^1\mu^2..\mu^n}$ of eq. (5.3).

We construct an infinite-dimensional vector space

$$V := V[G^{(2)}{}_{\mu^1\mu^2}, G^{(3)}{}_{\mu^1\mu^2\mu^3}, \ldots, G^{(n)}{}_{\mu^1\mu^2\ldots\mu^n}] \qquad (6.2)$$

We now define an infinite set of field dependent algebraic operators

$$L(0)^{\rho}{}_{\nu^1} := d^{(2)}{}_{a^1a^2} \quad B^{a^1}{}_{\nu^1} \quad \frac{\delta}{\delta(\eta_{a^2 b} B^b{}_\rho)}$$

(where $d^{(2)}{}_{a^1a^2} := \eta_{a^1a^2}$),

$$L(1)^{\rho}{}_{\nu^1\nu^2} := d^{(3)}{}_{a^1a^2a^3} \quad B^{a^1}{}_{\nu^1} \; B^{a^2}{}_{\nu^2} \quad \frac{\delta}{\delta(\eta_{a^3 b} B^b{}_\rho)}$$

....

$$L(n)^{\rho}{}_{\nu^1\nu^2..\nu^{n+1}} := d^{(n+2)}{}_{a^1a^2..a^{n+1}} B^{a^1}{}_{\nu^1} B^{a^2}{}_{\nu^2}..B^{a^{n+1}}{}_{\nu^{n+1}} \frac{\delta}{\delta(\eta_{a^{n+2} b} B^b{}_\rho)} \tag{6.3}$$

whose action on the n-gluon SU(3) scalars is

$$L(0)^{\rho}{}_{\nu^1} G(2)_{\mu^1\mu^2} = \delta^\rho{}_{\mu^1} G(2)_{\nu^1\mu^2} + \delta^\rho{}_{\mu^2} G(2)_{\mu^1\nu^1},$$

...

$$L(0)^{\rho}{}_{\nu^1} G^{(m)}{}_{\mu^1\mu^2..\mu^m} = \delta^\rho{}_{\mu^1} G^{(m)}{}_{\nu^1\mu^2..\mu^m} + \delta^\rho{}_{\mu^2} G^{(m)}{}_{\mu^1\nu^1\mu^3..\mu^m} + .. + \delta^\rho{}_{\mu^m} G^{(m)}{}_{\mu^1\mu^2..\mu^{m-1}\nu^1}$$

...

$$L(n)^{\rho}{}_{\nu^1\nu^2..\nu^{n+1}} G^{(m)}{}_{\mu^1\mu^2..\mu^m} = \delta^\rho{}_{\mu^1} G^{(n+m)}{}_{\nu^1\nu^2..\nu^{n+1}\mu^2\mu^3..\mu^m} +.. + \delta^\rho{}_{\mu^2} G^{(n+m)}{}_{\mu^1\nu^1\nu^2..\nu^{n+1}\mu^3..\mu^m} +..+ \delta^\rho{}_{\mu^m} G^{(n+m)}{}_{\nu^1\nu^2..\nu^{n+1}\mu^1\mu^2..\mu^{m-1}}. \tag{6.4}$$

We can now evaluate the commutator,

$$[L(m),L(n)] \subset L(m+n) \tag{6.5}$$

this is in fact the infinite Ogievetsky algebra of Diff(4,R) [23] provided the SU(3) symmetric tensors $d^{(n+2)}$ are unique. If they are not (for each grading m), we get an extension of diff(4,R).

As a result, we note the fact that the set of all no-colour gluon exchanges forms an infinite representation of the group $\mathrm{Diff}(4,R)_{chrom}$, a dynamical active Spectrum Generating Group and passive symmetry, analogous to Einstein's. The pseudo-metric $G_{\mu\nu}$ of eq. (5.1) provides for the construction of invariants, as in "true" gravity. Note that the fact that a gravity-like theory can arise in this "effective" manner should be no surprise: in recent times, Adler, Zee and Novozhilov have proposed models in which "true" gravity itself also emerges from such an "effective" mechanism [24-26].

7. Conclusions

In several articles, we have shown how this "chromo-gravity" explains the emergence of Regge trajectories, including the $M^2 \sim J$ linear behaviour [19]. A calculation of the Casimir operator of $\overline{SA}(4,R)$, the relevant particle Hilbert space group [27] showed that there are no kinematical constraints on the manifield representations classifying the hadron spectrum [28,29].

We have also been able to reproduce [20] the emergence of J=2+0 degenerate states in even-even nuclei as the fundamental excitation of proton-proton and neutron-neutron pairs, i.e. the ground state and postulated symmetry of the Arima-Iachello "IBM" nuclear model [21]. For hadrons, we have also shown [18] that chromo-gravity predicts colour confinement ($1/p^4$ propagators) and explains the SL(4,R) classification of hadron resonances [28,29] with its good fit and its dynamical origin, also related to confinement. Salam's 1971 intuition is entirely vindicated in that it is a gravity-like component which reproduces these specific features.

We wish Abdus Salam good health and a continuation of his creative work. One of us (Yuval Ne'eman) would particularly like to thank him for his open door at Imperial College, which made it easier for Ne'eman to shortcut and though entering physics at a relatively late stage in life, go straight to the heart of the matter.

8. References

1. W. Selove, V. Hagopian, H. Brody, A. Baker and E. Leboy, Phys. Rev. Lett. 9 (1962) 272.

2. P.G.O. Freund, Phys. Lett. 2 (1962) 136.

3. N. Cabibbo, L. Horwitz and Y. Ne'eman, Phys. Lett. 22 (1966) 336.

   N. Cabibbo, L. Horwitz, J.J.J. Kokkedee and Y. Ne'eman, Nuovo Cimento 45 (1966) 275.

   L. Horwitz and Y. Ne'eman, Phys. Lett. B26 (1967) 88.

4. R. Delbourgo, A. Salam and J. Strathdee, Nu. Cimento 49 (1967) 593.

5. C.J. Isham, A. Salam and J. Strathdee, Phys. Rev. D3 (1971) 867. A. Salam, C. Isham and J. Strathdee, Phys. Rev. D3 (1971) 1805.

6. A. Salam, in Five Decades of Weak Interactions, Ann. New York Academy of Sciences 294 (1977) 12.

7. C.J. Isham, A. Salam and J. Strathdee, Lett. Nuo. Cimento, Ser.3, 1 (1972) 969.

8. C.J. Isham, A. Salam and J. Strathdee, Phys. Rev. D8 (1973) 2600.

9. C.J. Isham, A. Salam and J. Strathdee, Phys. Rev. D9 (1974) 1702.

10. C. Sivaram and K. Sinha, Phys. Reports, 51 (1979) 11.

11. Y. Dothan, M. Gell-Mann and Y. Ne'eman, Phys. Lett. 17 (1965) 148.

12. Y. Ne'eman, Ann. Inst. H. Poincaré 28 (1978) 369.

13. Y. Ne'eman, Proc. Nat. Ac. Sc. USA 74 (1977) 4157.

14. Y. Ne'eman, Proc. VIII Intern. Conf. Gen. Rel. and Grav., M.A. McKiernan ed., (1977) 262.

15. Y. Ne'eman, "World Spinors in Riemannian Gravity", in Fibre Bundle Workshop, R. Sen ed., Ben Gurion Univ., Beersheva (1985).

16. Y. Ne'eman, in Spinors in Physics and Geometry, G. Furlan and A. Trautman, eds., World Scientific Pub., Singapore (1988) 313.

17. F.W. Hehl, Y. Ne'eman, J. Nitzsch and P. Von der Heyde, Phys. Lett. B78 (1978) 102.

18. Dj. Šijački and Y. Ne'eman, Phys. Lett. B247 (1990) 571.

19. Y. Ne'eman and Dj. Šijački, Phys. Lett. B276 (1992) 173.

20. Dj. Šijački and Y. Ne'eman, Phys. Lett. B250 (1990) 1.

21. A. Arima and F. Iachello, Phys. Rev. Lett. 35 (1975) 1069; Ann. Phys.(NY) 99 (1976) 253 and 111 (1978) 201, 123 (1979) 468.

22 G. Feinberg and J. Sucher, Phys. Rev. A2 (1970) 2395.

23. V.I. Ogievetsky, Lett. Nuovo Cimento 8 (1973) 988.

24. S.L. Adler, Rev. Mod. Phys. 54 (1982) 729.

25. A. Zee, Phys. Rev. D23 (1981) 858.

26. Yu.V. Novozhilov and D.V. Vassilevich, Lett. Math. Phys. 33 (1991) 253.

27. J. Lemke, Y. Ne'eman and J. Pecina-Cruz, J. Math. Phys. 33 (1992) 2656.

28. Y. Ne'eman and Dj. Šijački, Phys. Rev. D37 (1988) 3267.

29. Dj. Šijački and Y. Ne'eman, Phys. Rev. D47 (1993) 4133.

Twenty Years of the Weyl Anomaly

M. J. Duff ‡

Center for Theoretical Physics
Physics Department
Texas A & M University
College Station, Texas 77843

ABSTRACT

In 1973 two Salam protégés (Derek Capper and the author) discovered that the conformal invariance under Weyl rescalings of the metric tensor $g_{\mu\nu}(x) \rightarrow \Omega^2(x)g_{\mu\nu}(x)$ displayed by classical massless field systems in interaction with gravity no longer survives in the quantum theory. Since then these Weyl anomalies have found a variety of applications in black hole physics, cosmology, string theory and statistical mechanics. We give a nostalgic review.

‡ Research supported in part by NSF Grant PHY-9106593.

When all else fails, you can always tell the truth.

Abdus Salam

1 Trieste and Oxford

Twenty years ago, Derek Capper and I had embarked on our very first post-docs here in Trieste. We were two Salam students fresh from Imperial College filled with ideas about quantizing the gravitational field: a subject which at the time was pursued only by mad dogs and Englishmen. (My thesis title: *Problems in the Classical and Quantum Theories of Gravitation* was greeted with hoots of derision when I announced it at the Cargese Summer School en route to Trieste. The work originated with a bet between Abdus Salam and Hermann Bondi about whether you could generate the Schwarzschild solution using Feynman diagrams. You can (and I did) but I never found out if Bondi ever paid up.)

Inspired by Salam, Capper and I decided to use the recently discovered *dimensional regularization*[1] to calculate corrections to the graviton propagator from closed loops of massless particles: vectors [1] and spinors [2], the former in collaboration with Leopold Halpern. This involved the self-energy

[1]Dimensional regularization had just been invented by another Salam student and contempory of ours, Jonathan Ashmore [3], and independently by Bollini and Giambiagi [4] and by 't Hooft and Veltman [5]. I briefly shared a London house with Jonathan Ashmore and Fritjof Capra. Both were later to leave physics, as indeed was Derek Capper. Ashmore went into biology, Capper into computer science and Capra into eastern mysticism (a decision in which, as far as I am aware, Abdus Salam played no part).

insertion

$$\Pi_{\mu\nu\rho\sigma}(p) = \int d^n x e^{ipx} < T_{\mu\nu}(x) T_{\rho\sigma}(0) > |_{g_{\mu\nu}=\delta_{\mu\nu}} \tag{1}$$

where n is the spacetime dimension and $T_{\mu\nu}(x)$ the energy-momentum tensor of the massless particles. One of our goals was to verify that dimensional regularization correctly preserved the Ward identity

$$p^{\mu}\Pi_{\mu\nu\rho\sigma}(p) = 0 \tag{2}$$

that follows as a consequence of general covariance. If we denote by ϵ the deviation from the physical spacetime dimension one is interested in, and expand about $\epsilon = 0$, we obtain

$$\Pi_{\mu\nu\rho\sigma} = \frac{1}{\epsilon}\Pi_{\mu\nu\rho\sigma}(pole) + \Pi_{\mu\nu\rho\sigma}(finite) \tag{3}$$

Capper and I were able to verify that the pole term correctly obeyed the Ward identity

$$p^{\mu}\Pi_{\mu\nu\rho\sigma}(pole) = 0 \tag{4}$$

and that the infinity could then be removed by a generally covariant counterterm. We checked that the finite term also obeyed the identity

$$p^{\mu}\Pi_{\mu\nu\rho\sigma}(finite) = 0 \tag{5}$$

and hence that there were no diffeomorphism anomalies.[2]

---

[2] Had we looked at closed loops of Weyl fermions or self-dual antisymmetric tensors in 2 *mod* 4 dimensions, and had we been clever enough, we would have noticed that this Ward identity breaks down. But we didn't and we weren't, so this had to wait another ten years for the paper by Alvarez-Gaumé and Witten [6].

We were also aware that since the massless particle systems in question were invariant under the Weyl transformations [3] of the metric

$$g_{\mu\nu}(x) \to \Omega^2(x) g_{\mu\nu}(x) \tag{6}$$

together with appropriate rescalings of the matter fields, this implied that the stress tensors in (1) should be traceless and hence that the self energy should also obey the trace identity

$$\Pi^{\mu}{}_{\mu\rho\sigma}(p) = 0 \tag{7}$$

We verified that the pole term was OK:

$$\Pi^{\mu}{}_{\mu\rho\sigma}(pole) = 0 \tag{8}$$

consistent with our observation that the counterterms were not only generally covariant but Weyl invariant as well. For some reason, however, I did not get around to checking the finite term until Christmas of 73 by which time I was back in England on my second postdoc, in Oxford, where Dennis Sciama was gathering together a group of quantum gravity enthusiasts. To my surprise, I found that

$$\Pi^{\mu}{}_{\mu\rho\sigma}(finite) \neq 0 \tag{9}$$

I contacted Derek and he confirmed that we hadn't goofed. The Weyl invariance (6) displayed by classical massless field systems in interaction with gravity, first proposed by Hermann Weyl in 1918 [7, 8, 9], no longer survives

[3]For fermions this is true $\forall n$; for vectors only for $n = 4$

in the quantum theory! We rushed off a paper [10] to *Nuovo Cimento* (How times have changed!).

I was also able to announce the result at The First Oxford Quantum Gravity Conference, organised by Isham, Penrose and Sciama, and held at the Rutherford Laboratory in February 74 [11]. Unfortunately, the announcement was somewhat overshadowed because Stephen Hawking chose the same conference to reveal to an unsuspecting world his result [12] that black holes evaporate! Ironically, Christensen and Fulling [13] were subsequently to show that in two spacetime dimensions the Hawking effect is due entirely to the trace anomaly. Two dimensional black holes, and the effects of the Weyl anomaly in particular [14, 15, 16], are currently enjoying a revival of interest.

2 Anomaly? What anomaly?

Some cynic once said that in order for physicists to accept a new idea, they must first pass through the following three stages:

(1) It's wrong

(2) It's trivial

(3) I thought of it first.

In the case of the Weyl anomaly, however, our experience was that (1) and (2) got interchanged. Being in Oxford, one of the first people we tried to impress with our new anomaly was J. C. Taylor who merely remarked "Isn't that rather obvious?". In a sense, of course, he was absolutely right. He presumably had in mind the well-known result that theories which are Weyl invariant in n-dimensional curved space are automatically invariant under the conformal group $SO(2,n)$ in the flat space limit, which implies in particular that the dilatation current $D^\mu \equiv x^\nu T^\mu{}_\nu$ is conserved:

$$\partial_\mu D^\mu = T^\mu{}_\mu + x^\nu \partial_\mu T^\mu{}_\nu = T^\mu{}_\mu = 0 \tag{10}$$

Moreover, one already knew from the work of Coleman and Jackiw [17] that such flat-space symmetries suffered from anomalies. Of course, the Weyl invariance (6), in contrast to the conformal group, is a *local* symmetry for which there is therefore no Noether current. Nevertheless, perhaps one should not be too surprised to discover that there is a curved space generalization in the sense of a non-vanishing trace for the stress tensor. Consequently, Capper and I were totally unprepared for the actual response of the rest of the physics community: NO-ONE BELIEVED US! To be fair, we may have put some people off the scent by making the correct, but largely irrelevant, remark that at one loop the anomalies in the *two-point* function could be removed for $n \neq 2$ by adding finite local counterterms [10]. So we wrote another paper [18] stressing that the anomalies were real and could not be ignored, but to no avail. To rub salt in the wounds, among those dismissing our result as spurious were physicists for whom we youngsters had the greatest respect.

First the Americans:

...the finite W_{reg} that is left behind by Schwinger's method, after the infinities have been split off, is both coordinate invariant and conformally invariant, DeWitt [19]

Something is wrong, Christensen [20]

The form of the conformal anomaly in the trace of the stress tensor proposed by a number of authors violates axiom 5, Wald [21]

Thus we find no evidence of the conformal trace anomalies reported by a number of other authors, Adler, Lieberman and Ng [22],

then the Europeans:

Conformal anomalies in a conformally invariant theory do not arise..., Englert, Gastmans and Truffin [23]

There are important, physically relevant differences: most noticeable, normalized energy-momentum tensors do not possess a 'trace anomaly', Brown and Ottewill [24],

especially the Russians:

The main assumption of our work is that a regularization scheme exists which preserves all the formal symmetry properties (including the Weyl symmetry)...Therefore we hope dimensional regularization will give no anomalies.., Kallosh [25]

It turns out that conformal anomalies, discovered in gravitating systems, are not true anomalies, since in modified regularizations they do not arise....

The above point of view on conformal anomalies is shared by Englert et al., Fradkin and Vilkovisky [26]

The presence or absence of the conformal anomaly depends on the choice made between two existing classes of covariant regularizations, Vilkovisky [27]

and, to be democratic, let us not forget the Greeks:

The above method of regularization and renormalization preserves the Ward identities...and trace anomalies do not arise, Antoniadis and Tsamis [28]

There exists a regularization scheme which preserves both general coordinate invariance and local conformal invariance (Englert et al)..., Antoniadis, Iliopoulos and Tomaras [29]

In fact one needs a regularization scheme which preserves the general coordinate and Weyl invariance...such a scheme exists (Englert [4] *et al)*, Antoniadis, Kounnas and Nanopoulos [30].

---

[4]If one starts with a classically non-Weyl invariant theory (e.g. pure Einstein gravity) and artificially makes it Weyl invariant by introducing via a change of variables $g'_{\mu\nu}(x) = e^{2\sigma(x)}g_{\mu\nu}(x)$ an unphysical scalar *spurion* $\sigma(x)$, then unitarity guarantees that no anomalies can arise because this artificial Weyl invariance of the quantum theory, $g'_{\mu\nu}(x) \to \Omega^2(x)g'_{\mu\nu}(x)$ with $e^{2\sigma(x)} \to \Omega^2(x)e^{2\sigma(x)}$, is needed to undo the field redefinition and remove the spurious degree of freedom. Professor Englert informed me in Trieste that this is what the authors of [23] had in mind when they said that anomalies do not arise. Let us all agree therefore that many of the apparent contradictions are due to this misunderstanding.

3 London

As chance would have it, my third postdoc brought me to King's College, London, at the same time as Steve Christensen, Paul Davies, Stanley Deser, Chris Isham and Steve Fulling. Bill Unruh was also a visitor. It was destined therefore to become be a hot-bed of controversy and activity in Weyl anomalies. Provoked by Christensen and Fulling, who had not yet been converted, Deser, Isham and I decided to write down the most general form of the trace of the energy-momentum tensor in various dimensions [31]. By general covariance and dimensional analysis, it must take the following form: For n=2,

$$g^{\alpha\beta} < T_{\alpha\beta} >= aR, \tag{11}$$

where a is a constant. For n=4,

$$g^{\alpha\beta} < T_{\alpha\beta} >= \alpha R^2 + \beta R_{\mu\nu}R^{\mu\nu} + \gamma R_{\mu\nu\rho\sigma}R^{\mu\nu\rho\sigma} + \delta\Box R + cF_{\mu\nu}{}^a F^{\mu\nu a}, \tag{12}$$

where $\alpha, \beta, \gamma, \delta$ and c are constants. (In (12) we have allowed for the possibility of an an external gauge field in addition to the gravitational field.) For $n = 6$, $g^{\alpha\beta} < T_{\alpha\beta} >$ would have to be cubic in curvature and so on. (At one-loop, and ignoring boundary terms, there is no anomaly for n odd). I showed these expressions to Steve Christensen, with whom I was sharing an office, and asked him if he had seen anything like this before. He immediately became very excited and told me that these were precisely the Schwinger-Dewitt b_n coefficients. These are the t-independent terms that appear in the asymptotic expansion of the heat kernel of the appropriate differential operators Δ:

$$Tr\ e^{-\Delta t} \sim \sum_{k=0}^{\infty} B_k t^{(k-n)/2} \qquad t \to 0^+ \tag{13}$$

where

$$B_k = \int d^n x\ b_k \tag{14}$$

For example, if Δ is the conformal Laplacian

$$\Delta = -\Box + \frac{R}{6} \tag{15}$$

then b_4 is given by

$$b_4 = \frac{1}{180(4\pi)^2}[-R_{\mu\nu}R^{\mu\nu} + R_{\mu\nu\rho\sigma}R^{\mu\nu\rho\sigma} + \Box R] \tag{16}$$

This was the road to Damascus for Steve as far as Weyl anomalies were concerned and, like many a recent convert, he went on to become their most ardent advocate [5]. This was also the beginning of a very fruitful collaboration between the two of us. The significance of my paper with Deser and Isham was that, with the exception of the $\Box R$ term in (12), none of the above anomalies could be removed by the addition of finite local counterterms (hence the title *Non-local Conformal Anomalies*) and thus this laid to rest any lingering

[5]A delightful set of reminiscences on the Weyl anomaly by Steve Christensen can also be found in Bryce DeWitt's festschrift [32]. During their stay at King's, he and Fulling shared a flat in the London borough of Ealing (home of the famous *Ealing Comedy* movies). According to Christensen, the connection between trace anomalies and the Hawking effect occurred, Archimedes-like, to Fulling while taking a bath. He did not run through the streets of Ealing shouting "Eureka", but did run upstairs to the payphone to tell Bill Unruh.

doubts about the inevitability of Weyl anomalies (or, at least, it should have done). By this time, or shortly afterwards, the Hawking radiation experts at King's and elsewhere were arriving at the same conclusion[13, 33, 34, 35, 36] as in fact was Hawking himself [37]. There followed a series of papers calculating the numerical coefficients in (11) and (12) and confirming that these were indeed just the b_n coefficients [38, 39, 40, 41, 42]. These results were further generalized to self-interacting theories [43, 44, 45, 46, 47].

One day about this time I answered the phone in my office only to hear those five words most designed to instill fear and trembling into the heart of a young postdoc: "Hi. This is Steven Weinberg". Pondering on the non-renormalizability problem, Weinberg had become interested in quantum gravity in $2 + \epsilon$ dimensions [48, 49]. Inspired by workers in statistical mechanics, who frequently work with non-renormalizable field theories but who nevertheless manage to extract sensible predictions, Weinberg wondered whether this might be true for gravity: was the theory "asymptotically safe"? The answer seemed to rely on the *sign* of the two-dimensional trace anomaly i.e. on the constant a in (11). Accordingly, Weinberg set n=2 in the n-dimensional calculations of [2] and concluded that fermions had the wrong sign:

$$a = \frac{1}{24\pi} \tag{17}$$

He repeated our calculation for scalars himself and found the same sign and magnitude (consistent with the observation that in two-dimensions there is a bose-fermi equivalence, and consistent with the black hole calculations [33, 13]). In the case of vector bosons, however, he found from [1] that there was a sign flip. His question was simple but crucial: did I agree with him or could there be an overall sign error? Not wanting to be the victim of Weinberg's wrath should I get it wrong, I spent several frantic days and sleepless nights checking and rechecking the calculations. Those who have ever chased a minus sign and those who know Steve Weinberg will appreciate my discomfort! In fact I agreed. Unfortunately, asymptotic safety became asymptotically unpopular but my contact with Weinberg later led to a very fruitful semester in Austin, and to my continuing affection for the state of Texas.

The scalar terms of order $n/2$ in the curvature which appear in the n-dimensional gravitational trace anomaly are reminiscent of the pseudoscalar terms of order $n/2$ in the curvature which appear in the n-dimensional gravitational axial anomaly, as calculated by Delbourgo and Salam [50]. I was musing on this shortly after moving across town to Queen Mary College, when I saw a paper by Eguchi and Freund [51] on the then new and exciting topic of gravitational instantons. They considered the two topological invariants, the Pontryagin number, P, and the Euler number, χ, and posed the question: To what anomalies do P and χ contribute? In the case of the Pontryagin number, they were able to answer this question by relating P to the integrated axial anomaly; in the case of the Euler number, however, they found no anomaly. I therefore wrote a short note [52] relating χ to the integrated trace anomaly. This result was later to prove important in the two-dimensional context of string theory, where

$$\chi = \frac{1}{4\pi}\int d^2x\sqrt{g}R \tag{18}$$

and hence from (11)

$$\frac{1}{4\pi}\int d^2x\sqrt{g}g^{\alpha\beta} < T_{\alpha\beta} >= a\chi \tag{19}$$

Unfortunately, the referee's vision did not extend that far and the paper was rejected. Rather than resubmit it, I decided to incorporate the results into a larger paper [53] which re-examined the Weyl anomaly in the light of its applications to the Hawking effect, to gravitational instantons, to asymptotic freedom and Weinberg's asymptotic safety. In the process, I discovered that the constants α, β, γ and δ are not all independent but obey the constraints

$$4\alpha + \beta = \alpha - \gamma = -\delta \tag{20}$$

In other words, the gravitational contribution to the anomaly depends on only two constants (call them b and b') so that (12) may be written

$$g^{\alpha\beta} < T_{\alpha\beta} >= b(F + \frac{2}{3}\Box R) + b'G + cH \tag{21}$$

where

$$F = R_{\mu\nu\rho\sigma}R^{\mu\nu\rho\sigma} - 2R_{\mu\nu}R^{\mu\nu} + \frac{1}{3}R^2, \tag{22}$$

$$G = R_{\mu\nu\rho\sigma}R^{\mu\nu\rho\sigma} - 4R_{\mu\nu}R^{\mu\nu} + R^2 \tag{23}$$

and

$$H = F_{\mu\nu}{}^{a} F^{\mu\nu a} \tag{24}$$

In four (but only four) dimensions

$$F = C^{\mu\nu\rho\sigma} C_{\mu\nu\rho\sigma} \tag{25}$$

where $C_{\mu\nu\rho\sigma}$ is the Weyl tensor, and G is proportional to the Euler number density

$$G = {}^{*} R_{\mu\nu\rho\sigma} {}^{*} R^{\mu\nu\rho\sigma} \tag{26}$$

where $*$ denotes the dual. Note the absence of an R^2 term in (21). This result was later rederived using the Wess-Zumino consistency conditions [54, 55, 56, 57, 58]. Furthermore, the constants a, b, b' and c are those which determine the counterterms

$$\Delta L = \frac{a}{\epsilon} \sqrt{g} R \qquad n = 2 \tag{27}$$

$$\Delta L = \frac{1}{\epsilon} \sqrt{g} (bF + b'G + cH) \qquad n = 4 \tag{28}$$

(and hence the renormalization group β functions) at the one-loop level. The Euler number counterterms are frequently ignored on the grounds that they are total divergences, but will nevertheless contribute in spacetimes of non-trivial topology. We emphasize that the above results are valid only for theories which are classically conformally invariant (e.g. Maxwell/Yang-Mills for n=4 only, and conformal scalars and massless fermions for both n=2 and n=4). For other theories (e.g. Maxwell/Yang-Mills for n=2, pure quantum gravity for n=4, or any theory with mass terms) the "anomalies" will still survive, but will be accompanied by contributions to $g^{\alpha\beta} < T_{\alpha\beta} >$ expected anyway through the lack of conformal invariance. Since the anomaly arises because the operations of regularizing and taking the trace do not commute, the anomaly in a theory which is not classically Weyl invariant may be defined as:

$$Anomaly \ = g^{\alpha\beta} < T_{\alpha\beta} >_{reg} - < g^{\alpha\beta} T_{\alpha\beta} >_{reg} \tag{29}$$

Of course, the second term happens to vanish when the classical invariance is present.

Note that in an expansion about flat space with $g_{\mu\nu} = \delta_{\mu\nu} + h_{\mu\nu}$, R is $O(h)$, so it is sufficient to calculate the two-point function as in [10] to fix the a coefficient of R in (11). For n=4, $\Box R$ is $O(h)$ while F and G are $O(h^2)$. Nevertheless, because of the constraint (20), a calculation of the two-point function is again sufficient to fix the b coefficient of $C^{\mu\nu\rho\sigma}C_{\mu\nu\rho\sigma}$ in (21), notwithstanding the ability to remove the $\Box R$ piece by a finite local counterterm, and notwithstanding some contrary claims in the literature. Indeed, this is how the coefficients of the Weyl invariant $C^{\mu\nu\rho\sigma}C_{\mu\nu\rho\sigma}$ counterterms were first calculated [1, 10].

Explicit calculations [1, 10, 2, 59, 13, 38, 36, 41, 39, 53] yield

$$b = \frac{1}{120(4\pi)^2}[N_S + 6N_F + 12N_V] \tag{30}$$

$$b' = -\frac{1}{360(4\pi)^2}[N_S + 11N_F + 62N_V] \tag{31}$$

where N_S, N_F and N_V are respectively the numbers of scalars, spin 1/2 Dirac fermions and vectors in the theory. Note that the contributions to the b coefficient are all positive, as they must be by spectral representation positivity arguments [2, 59, 58] on the vacuum polarization proper self-energy part in four dimensions.

I also noted that the non-local effective action responsible for the anomalies would contain a term

$$S_{eff} \sim \int d^n x \sqrt{g} R \, \Box^{(n-4)/2} R \tag{32}$$

By setting $n = 2$ one obtains what what later to be known as the *Polyakov action.*

Lack of space prevents me from including the applications of the trace anomaly to cosmology, supergravity, string theory and statistical mechanics. A complete version of this talk will be published elsewhere [60].

4 Acknowledgements

I would like to thank my anomalous collaborators Derek Capper, Steve Christensen, Stanley Deser, Chris Isham, and Peter van Nieuwenhuizen.

Most of all, however, my thanks go to my former supervisor Abdus Salam who first kindled my interest in quantum gravity and who continues to provide inspiration to us all as a scientist and as a human being. When the deeds of great men are recalled, one often hears the cliché "He did not suffer fools gladly", but my memories of Salam at Imperial College were quite the reverse. People from all over the world would arrive and knock on his door to expound their latest theories, some of them quite bizarre. Yet Salam would treat them all with the same courtesy and respect. Perhaps it was because his own ideas always bordered on the outlandish that he was so tolerant of eccentricity in others; he could recognize pearls of wisdom where the rest of us saw only irritating grains of sand. As but one example of a crazy Salam idea, I distinctly remember him remarking on the apparent similarity between the mass and angular momentum relation of a Regge trajectory and that of an extreme black hole. Nowadays, of course, string theorists will juxtapose black holes and Regge slopes without batting an eyelid, but to suggest this back in the late 1960's was considered preposterous by minds lesser than Salam's.[6]

Theoretical physicists are, by and large, an honest bunch: occasions when scientific facts are actually deliberately falsified are almost unheard of. Nevertheless, we are still human and consequently want to present our results in the best possible light when writing them up for publication. I recall a young student approaching Abdus Salam for advice on this ethical dilemma: "Professor Salam, these calculations confirm most of the arguments I have been making so far. Unfortunately, there are also these other calculations which do not quite seem to fit the picture. Should I also draw the reader's

[6]Historical footnote: at the time Salam had to change the gravitational constant to match the hadronic scale, an idea which spawned his *strong gravity*; today the fashion is the reverse and we change the Regge slope to match the Planck scale!

attention to these at the risk of spoiling the effect or should I wait? After all, they will probably turn out to be irrelevant." In a response which should be immortalized in *The Oxford Dictionary of Quotations*, Salam replied: "When all else fails, you can always tell the truth".

Amen.

References

[1] D. M. Capper, M. J. Duff and L. Halpern, *Photon Corrections to the Graviton Propagator*, Phys. Rev **D10** (1974) 461.

[2] D. M. Capper and M. J. Duff, *Neutrino Corrections to the Graviton Propagator*, Nucl. Phys. **B82** (1974) 147.

[3] J. F. Ashmore, *A Method of Gauge-Invariant Regularization*, Nuovo Cimento Lett. **4** (1972) 289.

[4] C. G. Bollini and J. J. Giambiagi, *Dimensional Renormalization: The number of Dimensions as a Regulating Parameter*, Nuovo Cimento **12B** (1972) 20.

[5] G. 't Hooft and M. Veltman, *Regularization and Renormalization of Gauge Fields*, Nucl. Phys. **B44** (1972) 189.

[6] L. Alvarez-Gaumé and E. Witten, *Gravitational Anomalies*, Nucl. Phys. **B234** (1983) 264.

[7] H. Weyl, Sitz. Ber. Preuss. Ak. Wiss.(1918) 465.

[8] H. Weyl, Math. Zietschr. **2** (1918) 384.

[9] H. Weyl, Ann. Der Phys. **59** (1918) 101.

[10] D. M. Capper and M. J. Duff, *Trace Anomalies in Dimensional Regularization*, Nuovo Cimento **23A** (1974) 173.

[11] M. J. Duff, *Covariant Quantization of Gravity* in *Quantum Gravity: An Oxford Symposium*, Eds. C. J. Isham, R. Penrose and D. Sciama (O.U.P.,1975).

[12] S. W. Hawking, *Particle Creation by Black Holes* in *Quantum Gravity: An Oxford Symposium*, Eds. C. J. Isham, R. Penrose and D. Sciama (O.U.P.,1975).

[13] S. M. Christensen and S. A. Fulling, *Trace Anomalies and the Hawking Effect*, Phys. Rev. **D15** (1977) 2088.

[14] N. Sanchez, *Semiclassical Quantum Gravity and Liouville Theory: a Complete Solution to the Back Recation Problem in Two Dimensions*, Nucl. Phys. **B266** (1986) 487.

[15] J. A. Harvey and A. Strominger, *Quantum aspects of Black Holes*, Chicago preprint EFI-92-41.

[16] T. Jacobson and G. Kang, *Conformal Invariance of Black Hole Temperature*, NSF-ITP-93-91.

[17] S. Coleman and R. Jackiw, *Why Dilatation Generators Do Not Generate Dilatations*, Ann. Phys. **67** (1971) 552.

[18] D. M. Capper and M. J. Duff, *Conformal Anomalies and the Renormalizability Problem in Quantum Gravity*, Phys. Lett. **53A** (1975) 361.

[19] B. S. DeWitt, *Quantum Field Theory in Curved Spacetime*, Phys. Rep. **19C** (1975) 297.

[20] S. M. Christensen, *Vacuum Expectation Value of Stress Tensor in an Arbitrary Curved Background: The Covariant point-splitting method*, Phys. Rev. **D 14** (1976) 2490.

[21] R. M. Wald, *The Back Reaction Effect in Particle Creation in Curved Spacetime*, Commun. Math. Phys. **54** (1977) 1.

[22] S. L. Adler, J. Lieberman and Y. J. Ng, *Regularization of the Stress-Energy Tensor for Vector and Scalar Fields Propagating in a General Background Metric*, Ann. Phys. **106** (1977) 279.

[23] F. Englert, C. Truffin and R. Gastmans, *Conformal Invariance in Quantum Gravity*, Nucl. Phys. **B117** (1976) 407.

[24] M. R. Brown and A. C. Ottewill, *The Energy-Momentum Operator in Curved Space-time*, Proc. Roy. Soc. **A389** (1983) 379.

[25] R. Kallosh, *On the Renormalization Problem of Quantum Gravity*, Phys. Lett. **55B** (1975) 321.

[26] E.S. Fradkin and G. A. Vilkovisky, *Conformal Off-Mass-Shell Extension and Elimination of Conformal Anomalies in Quantum Gravity*, Phys. Lett. **73B** (1978) 209.

[27] G. Vilkovisky, *The Gospel According to DeWitt* in *Quantum Theory of Gravity: Essays in Honor of the 60th Birthday of Bryce S. DeWitt*, Ed. S. M. Christensen, (Adam Hilger, 1984).

[28] I. Antoniadis and N. C. Tsamis, *On the Cosmological Problem*, Phys. Lett. **144B** (1984) 55.

[29] I. Antoniadis, J. Iliopoulos and T. N. Tomaras, *On the Stability of Background Solutions in Conformal Gravity*, Nucl. Phys. **B261** (1985) 157.

[30] I. Antoniadis, C. Kounnas and D. V. Nanopoulos, *Non-compact Symmetries and the Cosmological Constant*, Phys. Lett. **162B** (1985) 309.

[31] S. Deser, M. J. Duff and C. J. Isham, *Non-local Conformal Anomalies*, Nucl. Phys. **B111** (1976) 45.

[32] S. M. Christensen, *The World of the Schwinger-DeWitt Algorithm and the Magical a_2 Coefficient*, in *Quantum Theory of Gravity: Essays in Honor of the 60th Birthday of Bryce S. DeWitt*, Ed. S. M. Christensen, (Adam Hilger, 1984).

[33] P. C. W. Davies, S.A. Fulling and W. G. Unruh, *Energy-Momentum Tensor near an Evaporating Black Hole*, Phys. Rev. **D13** (1976) 2720.

[34] J. S. Dowker and R. Critchley, *Effective Lagrangian and Energy-Momentum Tensor in De Sitter Space*, Phys. Rev. **D13** (1976) 3224.

[35] P. C. W. Davies and S. A. Fulling, *Quantum Vacuum Energy in Two Dimensional Spacetimes*, Proc. Roy. Soc. **A354** (1977) 59.

[36] T. S. Bunch and P. C. W. Davies, *Stress Tensor and Conformal Anomalies for Massless Fields in a Robertson-Walker Universe*, Proc. Roy. Soc. London, **A360** (1978) 117.

[37] S. W. Hawking, *Zeta Function Regularization of Path Integrals in Curved Spacetime*, Commun. Math. Phys. **55** (1977) 133.

[38] L.S. Brown, *Stress Tensor Trace Anomaly in a Gravitational Metric: Scalar Field*, Phys. Rev. **D15** (1977) 1469.

[39] L.S. Brown and J. P. Cassidy, *Stress Tensor Trace Anomaly in a Gravitational Metric: General Theory, Maxwell Field*, Phys. Rev. **D15** (1977) 2810.

[40] A. Duncan, *Conformal Anomalies in Curved Spacetime*, Phys. Lett. **66B** (1977) 170.

[41] J. S. Dowker and R. Critchley, *The Stress-tensor Conformal Anomaly for Scalar and Spinor Fields*, Phys. Rev. **D16** (1977) 3390.

[42] H. S. Tsao, *Conformal Anomalies in a General Background Metric*, Phys. Lett. **68B** (1977) 79.

[43] I. Drummond and S. J. Hathrell, *Renormalization of the Gravitational Trace Anomaly in QED*, Phys. Rev. **D21** (1980) 2226.

[44] G. M. Shore, *Conformal Anomaly for Massless Quantum Electrodynamics in Spherical Spacetime*, Phys. Rev. **D21** (1980) 2226.

[45] G. M. Shore, *Radiatively Induced Spontaneous Symmetry Breaking and Phase Transitions in Curved Space-Time*, Ann. Phys. **128** (1980) 376.

[46] S. J. Hathrell, *Trace Anomalies and $\lambda\phi^4$ Theory in Curved Space*, Ann. Phys. **139** (1982) 136.

[47] S. J. Hathrell, *Trace Anomalies and QED in Curved Space*, Ann. Phys. **142** (1982) 34.

[48] S. Weinberg, *Ultraviolet Divergences in Quantum Gravity* in *General Relativity: an Einstein Centenary Survey*, Eds. S. W. Hawking and W. Israel, (C. U. P., 1979).

[49] S. M. Christensen and M. J. Duff, *Quantum Gravity in 2+ϵ Dimensions*, Phys. Lett. **79B** (1978) 213.

[50] R. Delbourgo and A. Salam, *The Gravitational Correction to PCAC*, Phys. Lett. **40B** (1972) 381.

[51] T. Eguchi and P. G. O. Freund, *Quantum Gravity and World Topology*, Phys. Rev. Lett. **37** (1976) 1251.

[52] M. J. Duff, *Comment on Quantum Gravity and World Topology*, QMC-76-29 (1976).

[53] M. J. Duff, *Observations on Conformal Anomalies*, Nucl. Phys. **B125** (1977) 334.

[54] L. Bonora, P. Cotta-Ramusino and C. Reina, *Conformal Anomaly and Cohomology*, Phys. Lett. **B126** (1983) 305.

[55] P. Pasti, M. Bregola and L. Bonora, *Weyl Cohomology* , Class. Quantum Grav. **3** (1986) 635.

[56] A. Cappelli and A. Coste, *On the Stress Tensor of Conformal Theories in Higher Dimensions*, Nucl. Phys. **B314** (1989) 707.

[57] H. Osborn, *Weyl Consistency Conditions and a Local Renormalization Group Equation for General Renormalizable Field Theories*, Nucl. Phys. **B363** (1991) 486.

[58] A. Cappelli, D. Friedan and J. Latorre, *C-Theorem and Spectral Representation*, Nucl. Phys. **B352** (1991) 616.

[59] S. Deser and P. van Nieuwenhuizen, *One Loop Divergences of Quantized Einstein-Maxwell Fields*, Phys. Rev. **D10** (1974) 401.

[60] M. J. Duff, *Twenty Years of the Weyl Anomaly*, CTP-TAMU-06/93, hep-th/9308075.

The Dawning of Gauge Theory

L. O'Raifeartaigh
Dublin Institute for Advanced Studies, 10 Burlington Road Dublin 4

Foreword.

It is a great honour and pleasure for me to be invited to give a talk at the Fest in honour of Professor Abdus Salam. Unlike many of the speakers I did not spend any appreciable time at Imperial College or the ICTP and so I did not have the opportunity of having a short-range interaction with him. But his long-range influence on me has been tremendous and I find it a measure of his power that I could experience such a strong action-at-a-distance. I should also pay tribute to his personal charm, which has been manifest on all occasions on which I have had the pleasure of meeting him. Although Professor Salam has been active in many areas he will perhaps best be rememembered for his work on non-abelian gauge-theories, first for promoting them in the late fifties and early sixties, when it was neither popular nor profitable to do so, and then for bringing this work to such a successful conclusion in 1968. For this reason I should like to devote my talk to non-abelian gauge theories and, because, I had the good luck to acquire recently some little-known literature on the early development of these theories, I have thought that a historical talk on the early emergence of these theories might be of some interest. The same thought has clearly occurred to Tom Kibble who has scheduled a talk with almost the same title. But the two talks will actually be complementary, mine covering the early period 1916-1956 and his covering the subsequent developments in which Professor Salam was so deeply involved.

Introduction.

Future historians of Physics will certainly remember the twentieth century as the century in which Quantum Mechanics and Relativity made their appearance. But they will also remember it as the century in which it was realized that the known fundamental forces (gravitation, electromagnetism, weak and strong nuclear interactions) were governed by a common principle, namely the gauge principle. Indeed, it could be that the discovery of the gauge-principle was the most fundamental one, on the grounds that Quantum Mechanics and Relativity involved changes only in our conception of phase-space, leaving the form of the force quite arbitrary, whereas the gauge-principle determined the fundamental forces. Furthermore it gave the forces a geometrical significance. The first discovery of the gauge-invariance was in Einstein's theory of gravitation, where it took the form of coordinate-invariance, but its transferal to the the other interactions was a slow and tortuous process which took about sixty years. It is about the first forty years of this process that I wish to speak.

The concept of what we now know as a gauge-potential first surfaced in the theory of magnetism, when it was realized that magnetic field should be expressed not as the gradient of a scalar but as the curl of a vector, $\vec{B} = \vec{\nabla} \times \vec{A}$. The concept of the magnetic vector-potential A, and the fact that it was defined only up to the gradient of a scalar, was well-known by the middle of the 19th century (motivating Stoke's theorem for example) but the name vector-potential seems to have been introduced later, though not later than 1892 [1]. Maxwell used the term electromagnetic momentum, presumably because of the substitution $\vec{p} \to \vec{p} + e\vec{A}$. The vector potential was used extensively, of course, in the work of Lorentz and his contemporaries and it became an even more useful concept with the advent of special relativity, when it could be extended to a four-vector $\mathbf{A}$ and thus include the electric scalar potential ϕ. However, even then, the significance of gauge-invariance does not seem to have been fully appreciated. For example, Mie's 1912 theory of electromagnetism [2] was not gauge-invariant, but was not criticized it on those grounds and Einstein's development of his gravitional theory was delayed because, by failing to take the gauge-freedom (in this context coordinate-freedom) fully into account, he thought for some time that his field equations were underdetermined.

Einstein's Gravitational Theory and its Repercussions.

Einstein soon realized the true situation, of course, and proceeded to construct his famous 1916 theory of Gravitation. From our point of view this theory constituted not only a breakthrough of the first magnitude in gravitational theory but it laid the foundations for the later gauge theories. Indeed it triggered such a development not only in Physics but in mathematics, starting with Levi-Civita's [3] invention of parallel transport in 1917 and ending with the modern theory of fibre-bundles [4] in 1950. Perhaps the most important aspect of Einstein's theory from the point of view of gauge-theories was that it was the first theory to derive an interaction from a geometrical principle 9indeed from any principle other than a minimal principle). Fittingly enough, the next important step in the progress to other gauge-theories was an attempted extension [5] of Einstein's theory by Weyl in 1918. What Weyl proposed was that general coordinate invariance should be extended to include invariance with respect to transformations of the metric tensor of the form $g_{\mu\nu}(x) \rightarrow e^{\int dx.A(x)} g_{\mu\nu}(x)$. Incidentally, it was in this paper that the word gauge(*Eich*)-transformation first made its appearance and had the everyday meaning of a change of length or a change in calibration. Einstein (in a postscript at the end of Weyl's paper) was quick to point out that the proposed extension was not acceptable, because the non-integrability of the scale-factor implied that the time-scales of clocks would depend on their history. Nevertheless, the idea of scale transformations laid the foundation for later developments. The main development was the realization that it was the sphere of application of Weyl's idea, and not the idea itself, that was wrong, and that the correct sphere of application was as a phase-transformation of the quantum-mechanical wave-function. In other words, $g(x) \rightarrow e^{\int dx.A(x)} g(x)$ should be replaced by $\psi(x) \rightarrow e^{\int dx.A(x)} \psi(x)$. This seems to have been first suggested in a paper [6] of Schrödinger in 1922 and played an important role in his later invention of wave-mechanics [7] (although he does not refer to it in the 1926 papers). However, the 1922 suggestion was quite obscure since it had to be formulated without a wave-function (actually in terms of Bohr quantization rules). With the advent of wave-mechanics the statement became much clearer because then it could be formulated as the statement (due Fock [8] and London [9] in 1927) that in wave-mechanics the change $\vec{p} \rightarrow \vec{p} + e\vec{A}$ of classical electrodynamics becomes $\vec{\partial} \rightarrow \vec{\partial} + \frac{ie}{\hbar}\vec{A}$ and hence a gauge-transformation of $\vec{A}$ could be viewed as a phase-transformation $\psi(x) \rightarrow e^{\frac{ie}{\hbar}\alpha(x)}\psi(x)$ of the wave-function. Whatever the origin, the idea came to full fruit in the now-famous paper [10] by Weyl in 1929, in which it was proposed for the first time that the standard theory of electromagnetism could be *derived* from the gauge-principle. Unfortunately, Weyl continued to call the transformations $\psi(x) \rightarrow e^{\frac{ie}{\hbar}\alpha(x)}\psi(x)$ gauge-transformations, although they no longer had anything to do with length or calibration. But, apart from this minor flaw, the 1929 paper was fundamental in so many ways that it must rank among ther most important contributions of the century.

Weyl's 1929 Paper

In this paper [10], which was neglected for many years but has now become famous, Weyl laid the foundation not only for modern gauge-theory but for many other later developments. Among these were

The use of parallel transport in General Relativity

The use of the Vierbein (which he called Achsenkreuz) for gravitation itself and for gravitating spinors

The derivation of Energy, Momentum and Angular Momentum conservation from Euclidean invariance

The introduction of the 2-component Weyl spinor

The first discussion of the possibility of P and T violation

The first mention of a possible relation between the P,T and and charge-conjugation

The use of covariant differentiation for electromagnetism

The derivation of charge conservation from gauge-invariance

The introduction of the non-integrable phase-factor $\exp(i\frac{e}{\hbar}\int A_\mu dx^\mu)$.

Most important of all, the introduction of the electromagnetic gauge-p

Not all of these developments were completely new. Some had been introduced by Weyl himself in earlier work, the derivation of conservation laws from symmetry principles was originally due to Noether [11], and the use of the Vierbein for gravitating spinors had been proposed by Wigner [12] shortly before. But in the 1929 paper these and the new concepts were presented in a manner that was consistent and comprehensive. Einstein's comment on Weyl's earlier book 'Raum-Zeit-Materie' [13], namely that it was *a master-symphony, in which each detail had its part to play in the whole, and the total was grandios,* was equally applicable here. Weyl himself laid particular stress on the derivation of conservation laws from symmetry principles (Noether's theorem). Indeed he claimed that it was the derivation of charge conservation from gauge-invariance that convinced him of the correctness of the gauge-principle.

As is now well-known, the non-integrable phase-factor plays a crucial role in the Aharonov-Bohm effect [14] and in the Yang-Wu [15] formulation of electrodynamics, and has been generalized to non-abelian theories in the form of the Wilson loop [16]. An interesting comment, due to Yang [17], is that from the non-integrable phase-fac point of view Einstein's objection to Weyl's gravitational theory is just the objection that there is no Aharonov-Bohm effect for clocks in a gravitational field!

It is amusing to recall the contrasting reactions of Einstein and Pauli to this paper. Einstein, as before, was very enthusiastic. But Pauli, on first seeing a preliminary version in U.S. National Academy Proceedings, made some quite scathing remarks and only partially withdrew them on reading the final version. (An amusing account of the two letters which he wrote to Weyl can be found in ref.18). He seems to particularly upset by the two-component spinor, on the grounds of parity-violation. But, behind the facade, he really appreciated Weyl's approach, as was shown by subsequent events (see below) and by his Handbuch article on relativity [2].

In spite of the depth and originality of Weyl's 1929 paper it appears to have had little impact during the following two decades, except for the fact that the concepts of Vierbein and parallel-transpor came into common use in General Relativity theory and that it had some influence on the attempts by Einstein and Schrödinger to construct a unified theory of Gravitation and Electromagnetism. It was only toward the end of the forties that the ideas were again taken up, this time by Yang.

Yang-Mills Theory.

Today it is well-known now that Yang's considerations of Weyl's gauge-principle lead to what is now called Yang-Mills, or non-abelian, gauge theory in 1954. What is perhaps not so well-known is that Yang had already begun to consider the problem when he was a graduate student, and that similiar ideas were also being considered independently by other people. What I wish to do here is sketch the development by Yang and Mills and mention parallel approaches which were carried out Klein, Pauli, Shaw and Utiyama. I should hasten to add that there is no question of priority here since Yang and Mills were the first to formulate the theory theory in a clear and logical manner and to publish in a regular journal.

The other approaches were much more speculative (except possibly for Utiyama, whose results were obtained slightly later, though not as late as 1956, as is generally thought). The fact that others were on the same track no more detracts from the Y-M result than the fact that Pauling and others were on the track of DNA detracts from the Crick-Watson result. Rather does it emphasize the fundamentality of the Y-M idea by showing how it could emerge from a number of quite different considerations.

As mentioned above, Yang first considered the idea of extending the local gauge-invariance of Weyl to other symmetries, in particular isospin symmetry, while still a graduate student. He writes [19]: *While a graduate student in Kunming and in Chicago, I had thoroughly studied Pauli's review articles on field theory. I was very much impressed with the idea that charge conservation was related to the invariance of the theory under phase changes, an idea, I later found out, due originally to H. Weyl. I was even more impressed with the fact that gauge invariance* determined *all the electromagnetic interactions. While in Chicago I tried to generalize this to isotopic spin interactions by the procedure later written up in* [20]*, equations (1) and (2)* (covariant derivative). *Starting from these it was easy to get equation (3)* (transformation law for the gauge-potential) *Then I tried to define the field-strengths* $F_{\mu\nu}$ *by* $F_{\mu\nu} = \partial_\mu B_\nu - \partial_\nu B_\mu$ *which was a 'natural' generalization of electromagnetism. This led to a mess, and I had to give up. But the basic idea remained attractive, and I came back to it several times in the next few years, always getting stuck at the same point...As more and more mesons were discovered and all kinds of interactions were being considered, the necessity to have a* principle *for writing down interactions became more obvious to me. So while at Brookhaven* (in the summer of 1953) *I returned once more to the idea of generalizing gauge invariance. My office mate was R.L.Mills who was about to finish his Ph.D. at Columbia with N.Kroll. We worked on the problem and eventually produced* [20]*. We also wrote an Abstract for the April 1954 meeting of the AMS in Washington, which became* [21]*. Different motivations were emphasized in the two papers. The formal aspect of our work did not take long and was essentially finished by February 1954. But we found that we were unable to conclude what the mass of the gauge-particle should be. We toyed with the dimensional argument that, for a pure gauge theory, there is no quantity with the dimension of a mass to start with, and therefore the gauge particle must be massless. But we quickly rejected this line of reasoning.* The first public presentation of their results seems to have been during a seminar given at the Institute in Princeton at the end of

February. In this seminar Pauli (who, as we shall see later, had himself been working on the problem) caused some disruption by refusing to accept Yang's statement that he was not sure about gauge-field masses. (In the light of our present knowledge concerning spontaneous breakdown and confinement, Pauli's demand for immediate information was quite a tall order!). In any case Yang and Mills soon realized that the mass and renormalization problems for this theory were not going to be solved overnight, and sent the manuscript for publication in June 1954. The rest you know.

Klein's Approach.

The contribution of Oskar Klein is a very isolated one. It is not connected with any previous work except Kaluza-Klein theory, it was produced only as a talk at the 1938 Kasimierz Conference on New Theories in Physics [22], it was apparently appreciated by only one member, Moeller, of an illustrious audience that included Bohr, Brillouin, Eddington, Gamow, Goudsmit, Kramers, von Neumann, Rosenfeld and Wigner and it was not even quoted, much less developed, in any subsequent paper. Klein was unlucky enough to be too far ahead of his time and to have done his work just prior to World War II. What he did was to make the natural generalization of the existing Kaluza-Klein theory, which had *one* extra dimension (a circle) to the corresponding theory with *two* extra dimensions (a 2-sphere). Since $SO(3)$ is the symmetry group of the 2-sphere and Klein included the nucleons as a multiplet, it is not surprising that he arrived at an $SU(2)$ gauge-theory. Because of the Kaluza-Klein derivation the field strengths came out automatically and he wrote them in the form

$$A_{rs} = \partial_r A_s - \partial_s A_r + \frac{ie}{hc}(B_r \tilde{B}_s - B_s \tilde{B}_r) \qquad B_{rs} = \delta_r A_s - \delta_s A_r, \qquad (30)(31)$$

and similiarly for $\tilde{B}_{rs}$, where δ_r is the EM convariant derivative. He then postulated a Lagrangian and derived the field equations

$$\epsilon_r \partial_r - \frac{ie}{2hc}\epsilon_r \begin{pmatrix} 0 & \tilde{B}_r \\ B_r & 2A_r \end{pmatrix} \begin{pmatrix} n \\ p \end{pmatrix} = \frac{Mc}{h} \begin{pmatrix} n \\ p \end{pmatrix} \qquad (22)$$

where ϵ_r are the Dirac matrices, for the nucleons, together with the field equations

$$\delta_l B_{kl} - \frac{2ie}{hc} A_{kl} B_l + \frac{\mu^2 c^2}{h^2} B_k = j_l, \qquad (40)$$

and similarly for $\tilde{B}_l$, where j_l is the matter-current, for the gauge fields. I have used Klein's own equation numbering in order to give some idea of the sequence and the number of intermediate steps.

Klein's problem was how to identify the gauge-fields physically and how to give them masses. The mass problem he solved by simply inserting masses for the B-fields in the Lagrangian (in gross violation of gauge-invariance, of course). He then identified the A-field with the electromagnetic field and suggested that the B-fields should mediate the nuclear forces. But it was not clear which of the nuclear forces they should mediate, which is not surprising if we recall that the same problem was around twenty years later, when the nuclear interactions were much better understood. Recall that in 1938 Yukawa's meson had only recently been proposed, and it was not clear whether it should be scalar or vector! Indeed, the only property of the nuclear forces that had been established beyond doubt was the charge independence of the strong interactions. This property prompted Moeller to ask the single question that followed the talk, namely, how the fact that Klein's two new mesons would have to be electrically charged could be reconciled with charge-independence. Astonishingly, Klein responded by pointing out that this difficulty could be overcome by making the following simple change in his array of vector mesons

$$\begin{pmatrix} A_k & \tilde{B}_k \\ B_k & A_k \end{pmatrix} \quad \rightarrow \quad \begin{pmatrix} A_k - C_k & \tilde{B}_k \\ B_k & A_k + C_k \end{pmatrix} \tag{2}$$

If one takes into account that the array (2) is just diag(1,-1) times the usual present-day array one sees that Klein there and then made the generalization $SU(2) \rightarrow SU(2) \times U(1)$! It is interesting to note that the change (2) is just the generalization that was made twenty years later for the *weak* interactions by Glashow when he generalized Schwinger's 1957 $SU(2)$ model to the standard model. Finally, Klein proceeded to second-quantize the system in the manner used at the time for QED. The full Klein paper, the Moeller discussion and a commentary by David Gross can be found in ref.22.

Pauli's Approach.

Pauli's approach stemmed from a discussion following a talk by Pais at a Lorentz Conference [23] in Leiden in 1953. During this discussion Pauli said: *I am very much in favour of the general principle to bring empirical conservation laws and invariance properties in connection with the mathematical groups of transformations of the laws of Nature. If besides the conservation of energy-momentum and of charge the conservation of the property defined as number of nucleons and charge-independence of the nuclear forces are well established they have indeed, as Pais tried now to express mathematically, also to be connected with group theoretical proerties of the laws of nature...I would like to ask in this connection whether the transformation group* (isospin) *with constant phases can be amplified in a way analogous to the gauge group for electromagnetic potentials in such a way that the meson-nucleon interaction is connected with the amplified group. The main problem here seems to be the proper incorporation of the coupling constant into the group.* To which Pais replied that *an amplified group is very much on my mind as a means of specifying the coupling in more detail. I have no results so far.* Pauli and Pais were clearly unaware of Klein's contribution, and, stimulated by this discussion Pauli independently generalized Kaluza-Klein theory from the circle to the 2-sphere. The results, similiar to Klein's, were sent to Pais in letters dated July and December 1953, in which Pauli says he wrote the theory down *just to see how it looks.* He considered his most important result to be the expression for the gauge-field, which, having been induced from the Kaluza-Klein theory, was automatically the correct one. Pauli did not, however, explicitly use this expression in a kinetic term in a Lagrangian, nor did he attempt to abstract his theory from the Kaluza-Klein context. But he was very much aware that the classical gauge-fields masses would have to be zero, which explains his reaction to Yang's talk the following February.

Shaw's Approach.

One of the most remarkable and least known approaches to non-abelian gauge-theory was that of Ronald Shaw, who in 1954 was studying for his doctorate [24] under Professor Salam at Cambridge. Quite unaware of the work of Yang and Mills, Shaw introduced $SU(2)$ Yang-Mills theory in his thesis, starting from a completely different angle, namely from a preprint of Schwinger's in which the usual $U(1)$ formulation of electromagnetism was replaced by an $SO(2)$ formulation. According to Shaw, he saw *in a flash* that the $SO(2)$ formulation could be generalized to $SU(2)$. Even the construction of the kinetic terms does not seem to have given him any difficulty. It may be of interest reproduce the relevant two pages of his thesis (in slightly abbreviated form and with the original equation-numbering):

Now under a general transformation (13) ($\psi' = (1 + \frac{1}{2}ic_i\tau_i)\psi$)with $\vec{c}$ a function of position, L_0 is no longer invariant but becomes

$$L_0' = L_0 + \frac{1}{2}\partial^\alpha \vec{c}.\bar{\psi}\gamma_\alpha \vec{\tau}\psi. \tag{16}$$

The additional term can be compensated by introducing an interaction

$$L_1 = -\frac{iq}{2}\vec{B}^\alpha.\bar{\psi}\gamma_\alpha\vec{\tau}\psi \qquad \text{where} \qquad q\vec{B}^\alpha = q(\vec{B}^\alpha - \vec{c}\times\vec{B}^\alpha) + \partial^\alpha\vec{c} \qquad (17)(18)$$

In order to obtain a theory invariant under general isotopic spin transformations it only remains to construct an invariant free field Lagrangian for the $\vec{B}$-field. It is natural in anology with with electromagnetic case, to define

$$\vec{f}^{\alpha\beta} = \partial^\alpha\vec{b}^\beta - \partial^\beta\vec{b}^\alpha \qquad L_2 = -\frac{1}{4}\vec{f}^{\alpha\beta}.\vec{f}_{\alpha\beta}. \qquad (19)(20)$$

However (20) is not invariant under the transformation (18).... A properly invariant Lagrangian can be obtained by defining

$$\vec{F}^{\alpha\beta} = \vec{f}^{\alpha\beta} - q(\vec{B}^\alpha\times\vec{B}^\beta - \vec{B}^\beta\times\vec{B}^\alpha \qquad L_2 = -\frac{1}{4}\vec{F}^{\alpha\beta}.\vec{F}_{\alpha\beta} \qquad (22)(23)$$

L_2 is now invariant under (18) as $\vec{F}^{\alpha\beta}$ transforms solely as a vector in E_3."

This work was done in the Spring of 1954, around the time that Yang was giving his seminar in Princeton, and thus it came a little later than the work of Yang and Mills. But, of course, it was completely independent. The thesis was not submitted until 1955 because it had to be completed by further work on the Poincare and Lorentz groups. (The latter work appeared finally as Part I and the above-quoted extract in Part II, which adds to the confusion in dates!). As Professor Salam was Shaw's thesis adviser it is sometimes maintained that he should have urged Shaw to publish his result at once. But, in Shaw's own words: *Sometime later (still early in 1954?) I showed it to Salam, but I do not think he appreciated it at the time. I do not at all blame him for this–I probably told him about it in a very dismissive way, since the relevant particles, surely m=0, I thought, did not exist. Much later (end of 1954?) he told me that Yang and Mills had had the same idea and told me to write mine up (which I did not).*

Utiyama's Approach.

Utiyama's published contribution [25] is quite well-known but as it was first published in 1956, quoted Yang and Mills and dealt with all simple compact groups and with Gravitation, it is usually regarded as a generalization of the Yang-Mills paper which was written as a sequel to that paper. That this is not the case we now know from some later statements of Utiyama himself. These statements appear in two books which, as far as I can ascertain, are available only in Japanese and Japanese/Russian respectively. Two of my colleagues have been kind enough to translate the relevant relevant parts of eack book (the translations agree in detail where they overlap). So I can let Utiyama speak for himself.

Book 1 [26] *Brief summary of Chap.10, entitled Memoir of Remorse*

1954, January: In the end of January, I (R. Utiyama) received a letter from the Institute for Advanced Study (IAS), Princeton, inviting me to stay at the IAS from September, 1954. I decided to prepare some completed work prior to going to the IAS so that I could publish papers even if I could not work successfuly there. In fact, I had already the idea of the general gauge theory, including the basic structure of the theory, in my mind before I received the letter from the IAS. So what I should do was simply to spell it out in the form of a paper I completed the work at the end of March 1954.

1954, May or June: I gave a talk on the work at a small workshop held at the then Research Institute for Fundamental Physics (now Yukawa Institute for Theoretical Physics.) *Kyoto University. I got a negative response.*

1954, September: I arrived at Princeton. I met Prof. Minoru Kobayashi (then Professor at Kyoto University) who was my supervisor at Osaka University and had been staying at the IAS from one year earlier as a visiting fellow. When I explained to him my idea of general gauge theory, he told me that recently Prof. Yang announced a theory which is quite similar to mine. Some days later, I found a preprint by Yang in my mailbox. I took a look of it and found many formulae already familiar to me — I immediately realised that he had found the same theory that I had developed. I was shocked too deeply to examine Yang's paper closely and to compare it with my work with care.

1955, March: As my appointment was prolonged for one more year, I felt that I should work on a subject which was more important than those I had engaged so far at the IAS. So I decided to return to the general gauge theory and took a closer look on Yang's paper which had been published in 1954. At this moment I realised for the first time that there was a significant difference between Yang's theory and mine. The difference was that Yang merely found an example of non-Abelian gauge theory whereas I had developed a general idea of gauge theory which could contain gravity as well as electromagnetic theory. Thus I decided to publish my work by translating it into English, and adding an extra section where Yang's theory is discussed as a example of my general gauge theory. I did so partly because his paper had already been published, and partly because I thought that it would not be fair if I did not refer to his paper after I read it, even if I developed the idea of the general gauge theory before knowing his work. My paper was accepted for publication in Phys. Rev. in July 1955 and published early in 1956

Remark: I admit that it was Yang and Mills who first found a new type of (non-Abelian) gauge theory (though not gravity). However, it is very much disappointing for me to see that non-Abelian gauge theory is called by the name 'Yang-Mills theory' only, without a slightest mention to me who developed the general framework of the gauge theory. I very much regret not having submitted my paper to some Japanese journal in March, 1954 when I completed the work.

The most interesting aspect of Utiyama's approach is that it was motivated by gravitational theory and that (partly as a result of this) it was formulated for arbitrary simple Lie groups, not just $U(1)$ and $SU(2)$ or even compact simple groups. In fact, Utiyama had hoped to use this approach to unify the interactions. As we now know (partly thanks to Utiyama himself), strict unification of gravitation with the other interactions along these lines is impossible since gravitation cannot be formulated as a set of local field equations for the gauge-potentials, but only for the metric (or the Vierbein) from which the potentials are derived. Nevertheless, Utiyama was the first to suggest that the broader analogy between gravitation and electromagnetism which had so fascinated Weyl, Einstein and Schrödinger [27] could be extended to include the nuclear interactions. As he says himself in the second book [28]:

If I am allowed to refer to my work on the general gauge theory, I would like to stress that it is my paper which first showed clearly that the theory of gravity could fall into the framework of the gauge theory, which would be a first step toward the the grand unified theory in the modern viewpoint. Even more importantly, my paper pointed out that fields carrying a fundamental force — either gravity or electromagnetism — must in fact be those termed as 'connections' in mathematics, which are now called 'gauge fields'. That the notion of connections is indispensable in establishing a theory of interactions was the basic assertion that I wanted to make in my paper.

Summary

As everybody knows the Yang-Mills paper did not immediately herald a new dawn for the nuclear interactions. Indeed the hurdles still to be overcome, the infra-red, unitarity and renormalization problems on the theoretical side and the zero-mass, neutral current and gauge-field identifications on the experimental side, were so formidable that for more than a decade the theory remained in a backwater. Only with the advent of the concept of spontaneous symmetry breaking and adequate renormalization techniques, such as dimensional regularization, did the theory become feasible, and only with the the experimental discovery of neutral currents and the theoretical discovery of asymptotic freedom did it become completely acceptable. But that is another story, much of which will be told by Tom Kibble.

Meanwhile I should like to end with an amusing anecdote arising out of Weyl's two-component neutrino paper. At Salam's doctoral examination, Peierls had asked him if he knew of any principle that would account for the masslessness of the neutrino in the same way that gauge-invariance accounts for the masslessness of the photon. In 1955, after hearing a talk by Yang on the possibility of parity-violation, Salam realized that Weyl's two-component neutrino would account for both parity violation and the masslessness of the neutrino. But Peierls, who in 1955 still believed firmly in parity-conservation, did not take kindly to the idea. So Salam, unaware of Pauli's long-standing antipathy to the two-component neutrino, communicated the idea to Pauli, through Villars. The reply from Pauli was brief and to the point: *Give my regards to my friend Salam and tell him to think of something better.* Which, as we all know, he did!

Acknowledgements

I am very grateful to Professor C.N. Yang for a discussion on these matters and for some very relevant material, and to Professor Ronald Shaw for sending me a copy of his thesis and comments. I am indebted to Professor Norbert Straumann for a copies of the Pauli-Pais correspondence and for his own contribution to the Weyl Centenary Volume. Finally, I should like to thank Professor Z. Ajduk for extracts from the Proceedings of the Kasimierz Conferences, Professor Misha Saveliev and Dr. Izumi Tsutsui for obtaining and translating the Uchiyama books and Professor A. Wightman for some pertinant historical comments.

References

[1] A. Pais, *Inward Bound* (Clarendon Press, Oxford, 1986).

[2] W. Pauli, *Theory of Relativity* (Pergamon Press, 1958).

[3] T. Levi-Civita, *Rend. Circ. Mat.* Palermo **42** (1917).

[4] C. Ehresmann, *Colloque de Topologie*, Brussels, 1950 p.29-55, S. Kobayashi and K. Nomizu, *Foundations of Differential Geometry*, Interscience, I, 1963; II 1969.

[5] H. Weyl, S.B. Preuss. Akad. Wiss (1918) 465, Gesammelte Abhandlungen, Springer 1968 II p.1 p.29.

[6] E. Schroëdinger, *Zeit. Phys.*, (1922) . [7] V. Raman and P. Forman, *Hist. Studies Phys. Sci.* **1** (1969) 291.

[8] V. Fock, *Z. Physk.* **39** (1927) 226.

[9] F. London, *Z. Physk.* **42** (1927) 375.

[10] H. Weyl, *Zeitschrift fur Physik* **56**(1929) 330.

[11] E. Noether, *Nachr. d. Kgl. Ges. d. Wiss.* Gottingen (1918) 235.

[12] E. Wigner, *Zeit. F. Phys.* **53** (1929) 592 . [13] Space-Time-Matter, Methuen, London 1922 . [14] M. Peshkin and Y. Tonomura, *The Aharonov-Bohm Effect* (Springer, 1991).

[15] T.T. Wu and C.N. Yang, *Phys.Rev.* **D12**(1975) 3845.

[16] See for example T-P Cheng and L-F Li, *Gauge Theory of Elementary Particle Physics* (Clarendon Press, Oxford, 1984).

[17] C. N. Yang, in *Hermann Weyl 1885-1985* ed. K. Chandrasekharan (Springer, 1986).

[18] C. N. Yang, *Selected Papers, with Commentary* (Freeman 1983).

[19] N. Straumann, DMV-Seminar Geschichte der Mathematik H. Weyl's Raum-Zeit-Materie (1992) . [20] C. N. Yang and R. L. Mills, *Phys. Rev.* **96** (1954) 191.

[21] C. N. Yang and R. L. Mills, *Phys. Rev.* **95** (1954) 631 . [22] New Theories in Physics, Proc. XI Warsaw Symposium on Elementary Particle Physics, held at Kasimierz, Poland, May 1988 (eds. Z. Ajduk, S. Pokorski and A. Trautman, World Scientific, 1989) . [23] A. Pais, *Physica* XIX (1953) 869.

[24] R. Shaw, Thesis, Cambridge University, 1955 (unpubl.).

[25] R. Utiyama, *Phys. Rev.* **101** (1956) 1597. (rec. July 7th 1955).

[26] Book 1: Butsurigaku wa dokomade susundaka (How far has Physics progressed?) by Ryoyu Uchiyama (Iwanami Shoten, Japan, 1983).

[27] E. Schroedinger, *Space-Time Structure* (Cambridge University Press, 1953).

[28] Book 2: Ippan gauge ba ron josetsu (Introduction to the General Gauge Field Theory), Ryoyu Uchiyama, (Iwanami Shoten, Japan, 1987).

GENESIS OF UNIFIED GAUGE THEORIES — PERSONAL RECOLLECTIONS FROM IMPERIAL

T.W.B. KIBBLE

Blackett Laboratory, Imperial College, London SW7 2BZ, UK

ABSTRACT

I review the highlights of the historical development of unified gauge theories during the sixties from the perspective of my own recollections as a member of the Imperial College group led by Professor Abdus Salam.

1. Introduction

I had the great good fortune to join the Imperial College Theoretical Physics Group in 1959, just two years after it was set up by Professor Abdus Salam, accompanied by his erstwhile PhD supervisor, Paul Matthews. It was a time of tremendous excitement and rapid progress, and Imperial was I believe one of the best places to be.

What I would like to do today is to recall some of the flavour of those times, to try to tell you what it was like. I want to review some of the highlights of the sixties from the perspective of my personal recollections, and to explain how the problems seemed at the time. You will not learn any new physics, but perhaps the story of our various misconceptions may serve as a cautionary tale.

Let me emphasize that this is not intended as a definitive historical account of the development of gauge theories. I am not equipped for such a task; I am not a historian of science and I have not tried to conduct a comprehensive survey of the literature. What I am aiming for is a much more personal story. I want to try to convey how it felt to be a member of that community and to participate in one of the most exciting developments in theoretical physics in our lifetimes. If in doing so I concentrate on those parts of the story with which I was most closely involved, I hope I may be forgiven for that degree of bias.

2. Imperial College in 1960

What was Imperial like around 1960? When I arrived, as Bob Delbourgo mentioned, we were still part of the Mathematics Department, occupying one corridor of the old Huxley Building which had once been part of a museum and is now again a wing of the Victoria and Albert Museum — a purpose to which it is much better suited! Our rooms were of very strange dimensions, often high and narrow, because they had been carved out of much larger areas. My office had only the top half of a window. The acoustics were such that we could hear every word of the lectures in the adjacent lecture theatre, and no doubt the students could hear everything we said.

All the PhD students were crowded into one big office, which was also where we met for tea and coffee. There were three permanent faculty: Salam and Matthews and John C. Taylor. I joined as a lecturer the following year.

Let me say at the start that it is a great sorrow that Paul Matthews could not have been here today, because of his tragic death five years ago. For many years, Imperial *was* Salam and Matthews. They made a superb team, exactly complementing each other's strengths and abilities. Paul would have loved to have been here — and if he had been alive I am sure it would have been he rather than I who would be giving this talk.

One of the great things about our group in those early days was that we always had lots of visitors, both long- and short-term; our guests included Murray Gell-Mann, Ken Johnson, John Ward, Lowell Brown, Gordon Feldman and Steven Weinberg. Overseeing us all was Professor Salam's formidable secretary Bridget, of whom we were all terrified — all, that is, except Professor Salam! Personally, I always typed my own papers rather than face having to ask her to do it — and remember, there were no word processors then!

About a year after I arrived we moved into the Physics Department's new building. The then Head of Department was another formidable and colourful character, Patrick (or P.M.S.) Blackett. Generally speaking, Blackett, having been brought up in the Cavendish Laboratory tradition under Lord Rutherford, was rather scornful of the value of theoretical physicists, but he knew a good thing when he saw one and persuaded Salam to join the rapidly expanding Physics Department, in what is now the Blackett Laboratory. There we had much better rooms and facilities.

3. Physics at Imperial

What about Physics?

Many people have forgotten — and perhaps the younger among you never knew — that in 1960 field theory was widely regarded as very passé. Of course it had had its triumphs: renormalization theory had made sense of the old problem of ultraviolet divergences, and QED had been magnificently vindicated, with the predictions of the Lamb shift and the magnetic moment of the electron — it is still among the most impressively verified of all physical theories.

But field theory didn't seem to work for anything else, particularly not for the strong interactions. By 1960 the dominant ideology, especially in California from where I had just returned, was S-matrix theory. It was a very attractive philosophy. People always like to get something for nothing, even theoretical physicists — and it was amazing in fact how far one could go on the basis of a few very simple assumptions: relativistic invariance, unitarity, crossing symmetry and causality. The most radical version of the ideology, the self-consistent bootstrap, had as its aim to get all of strong-interaction physics from just these principles and nothing else. There were to be no elementary particles; everything was a bound state of everything else.

Of course, people did study field theory, often rather apologetically; they looked to field theory to provide a concrete model of how scattering amplitudes should behave, particularly to understand their complex analytic structure.

There were, however, a few places in the world where field theory was still studied unashamedly *qua* field theory. Imperial College was one of them. Harvard was certainly another; it is very noticeable how many of our visitors over the next few years were Julian Schwinger's students from Harvard. No doubt there were a few other places too!

Even we were not immune from the charms of S-matrix theory. Much of my early work was on dispersion relations, as indeed was much of Salam's. Salam and Matthews wrote an influential series of papers on dispersion relations for K-meson–nucleon scattering. But at Imperial it was never a major theme. There were two dominant themes: symmetries and gauge theories. Both in a way had their origins in the tremendously successful concept of isospin (then still usually called isotopic spin, or isobaric spin by purists).

In the case of symmetries that origin is fairly obvious. The isospin symmetry between protons and neutrons had shown how two apparently disparate particles might be regarded as different states of a single fundamental entity, the nucleon. The symmetry was generalized to include Yukawa's mesons in an important paper by Nick Kemmer in 1938 [1] — which is incidentally perhaps one of the first papers to suggest the need for a neutral current. Kemmer was a very influential figure in British theoretical physics in the immediate post-war period. He was Paul Matthews' supervisor in Cambridge and he was my Head of Department when I was a student in Edinburgh, succeeding Max Born in the Tait Chair of Mathematical Physics in 1953.

In the forties and fifties, as new particles proliferated, it was natural to try to bring some order into this chaos by enlarging the symmetry group beyond SU(2), especially after the discovery of the new quantum number, strangeness.

One of the first students to work at Imperial on extended symmetries was Yuval Ne'eman. From his work on SU(3), together with the independent work of Murray Gell-Mann, stemmed the eightfold way, with its triumphant vindication in the discovery of the Ω^-, completing the spin-$\frac{3}{2}$ decuplet. In the early sixties, Salam had students working on every conceivable group — on SO(4), on G_2, and so on.

Salam himself made many important contributions then and later to unravelling these symmetries, but this was not I believe his first love. His real goal was always to find the ultimate theory that would describe the weak, electromagnetic and strong interactions, and even gravity — what we would now call a theory of everything. Discovering the symmetries provided important clues to this theory, and was therefore more of a means than an end in itself.

From an early stage, certainly well before I joined Imperial, Salam was convinced that the ultimate theory would be a gauge theory.

I myself caught that enthusiasm when I arrived at Imperial. I began working on gravity as a gauge theory, following in the pioneering steps of Utiyama, whose work in the fifties [2] was mentioned the other day by Lochlainn O'Raifeartaigh [3]. I showed [4] that in that approach one rather naturally arrives at a generalization of the Einstein theory, involving non-zero torsion (as suggested long before by Cartan), interacting with the spin density.

4. Yang-Mills theory

So I come to the start of my real story: the development of unified gauge theory as I saw it.

O'Raifeartaigh gave us a fascinating account of the early history, but from my point of view the first chapter in the story is where his ended, with the epoch-making paper of Yang and Mills [5], published in 1954.

In retrospect there may be others who deserve some of the credit — Weyl, Klein, Shaw, Utiyama — but it was undeniably the paper of Yang and Mills that caught people's imagination. They articulated very clearly what has come to be known as the "gauge principle" — sometimes paraphrased as "Nature abhors a rigid symmetry".

I remarked earlier that both our major themes — symmetries and gauge theories — grew from the success of isospin symmetry. The work of Yang and Mills was specifically in that context. They argued that a rigid, global isospin symmetry is incompatible with the notion of locality inherent in relativistic field theory. Their point was that once the concept of isospin symmetry has been accepted (and neglecting for the moment the small electromagnetic symmetry-breaking effect), it becomes a matter of arbitrary convention which component is identified with the proton and which with the neutron; it is a question of how one chooses one's axes in isospin space. But it then seems odd that making the choice here and now should automatically fix it throughout all of space and at all times. So they asked the question: can we make isospin into a *local* symmetry?

Of course, we all know the answer now, but it may still be instructive to recall the argument, very briefly.

We start with a Lagrangian $\mathcal{L}$ invariant under the rigid isospin transformation

$$\delta\psi(x) = -i\delta\vec{\omega}\cdot\vec{T}\psi(x), \tag{1}$$

where $\vec{T} = \{T_i\}$ are the isospin generators, satisfying

$$[T_i, T_j] = i\epsilon_{ijk}T_k. \tag{2}$$

The question is: can it be made invariant also under the corresponding local transformation

$$\delta\psi(x) = -i\delta\vec{\omega}(x)\cdot\vec{T}\psi(x)? \tag{3}$$

The original Lagrangian is not invariant, because the derivative of ψ transforms inhomogeneously. It can be made invariant by replacing $\partial_\mu\psi$ by a *covariant* derivative,

$$D_\mu\psi = \partial_\mu\psi - ig\vec{A}_\mu\cdot\vec{T}\psi, \tag{4}$$

where $\vec{A}_\mu = \{A_{\mu i}\}$ is a triplet of gauge fields transforming according to

$$\delta\vec{A}_\mu = \delta\vec{\omega} \times \vec{A}_\mu - \frac{1}{g}\partial_\mu\delta\vec{\omega}, \tag{5}$$

so that

$$\delta(D_\mu\psi) = -i\delta\vec{\omega}\cdot\vec{T}D_\mu\psi. \tag{6}$$

The Lagrangian for the gauge field must also be invariant, and may be taken to be

$$\mathcal{L}_g = -\tfrac{1}{4}\vec{F}_{\mu\nu}\cdot\vec{F}^{\mu\nu}, \tag{7}$$

where

$$\vec{F}_{\mu\nu} = \partial_\mu\vec{A}_\nu - \partial_\nu\vec{A}_\mu + g\vec{A}_\mu \times \vec{A}_\nu. \tag{8}$$

The great point of all this is that it provides a reason for interactions to exist, and also prescribes a very specific form for the interaction. Of course the argument is not in the nature of a deductive proof — no such proof exists — but it is a very strong heuristic argument from symmetry to interaction.

The gauge principle provided a beautiful description of the electromagnetic interactions. So it was very natural after the work of Yang and Mills that people began to look for gauge theories of the strong and weak interactions.

The emphasis in the early papers was on finding a gauge theory of strong interactions; that is what Yang and Mills themselves were after. But it was hard to make progress because no one knew how to calculate anything. Given the large value of the coupling constant, one would not expect perturbation theory to work; of course no one knew then about asymptotic freedom.

So gradually the weak interactions began to seem a better bet. There were certainly tantalizing hints of a structure very similar to QED.

5. The *V–A* interaction

The phenomenologically successful theory of beta decay was Fermi's theory involving a direct four-fermion interaction. This was obviously non-renormalizable. It was natural to suppose that it was really an effective interaction due to the exchange of a massive boson, so that in place of the diagram of Figure 1 one should really have that of Figure 2.

Progress was held up for quite a while because of experimental results that seemed to show that the interaction was predominantly scalar and tensor rather than vector and axialvector. The real breakthrough came with another contribution of Yang's — the suggestion by Lee and Yang [6] that parity is not conserved in weak interactions. As you all know, that led directly to the $V - A$ theory of Marshak and Sudarshan [7] and Feynman and Gell-Mann [8].

As O'Raifeartaigh mentioned, Salam was one of the first people [9] to point out the connection between γ_5 symmetry and the masslessness of the neutrino. This is a very interesting and significant paper.

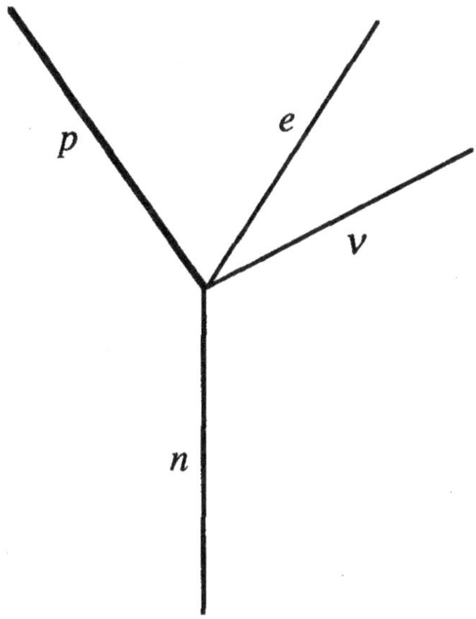

Figure 1: The Fermi interaction

The discovery of the $V-A$ structure of weak interactions in 1957–58 implied that they could be seen as mediated by vector bosons, $W^{\pm}$, thus naturally encouraging the idea that they were described by a gauge theory.

But there still seemed to be an insuperable difficulty: the large mass of the W bosons. If the interaction were of the same strength as electromagnetism, the W would have a mass of about 40 GeV. But of course putting in a mass term in the Lagrangian would destroy the gaüge invaiance. Moreover, the propagator of a massive vector field,

$$\left(g_{\mu\nu} - \frac{k_\mu k_\nu}{m^2}\right) \frac{-i}{k^2 - m^2 + i0} \tag{9}$$

is very badly behaved at infinity and leads to a non-renormalizable theory.

As early as 1958, Salam and Ward [10] proposed a unified gauge theory of weak and electromagnetic interactions, involving a triplet of vector mesons with W^0 identified with the photon. They placed the electron, neutrino and positron too in a triplet, and assumed the interaction

$$\mathcal{L}_{\text{int}} = g\bar{\vec{\psi}} \times \vec{\psi}\cdot\vec{W}, \tag{10}$$

where

$$\vec{\psi} = \begin{pmatrix} e^+ \\ \nu \\ e^- \end{pmatrix}, \qquad \vec{W} = \begin{pmatrix} W^+ \\ \gamma \\ W^- \end{pmatrix}. \tag{11}$$

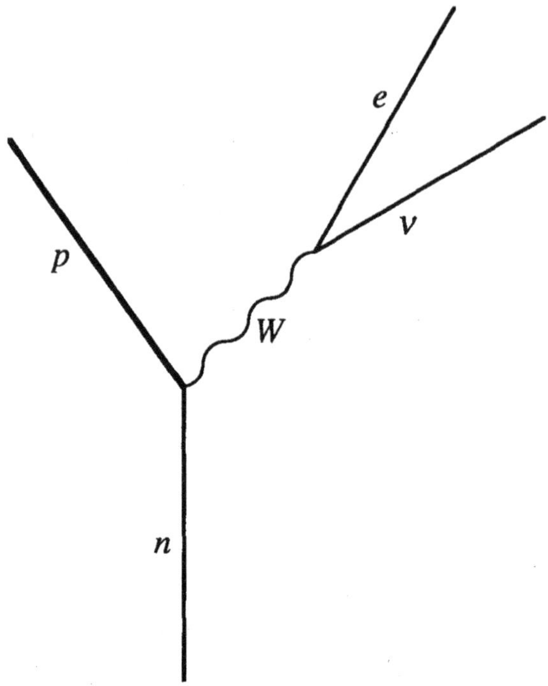

Figure 2: The intermediate vector boson interaction

This was a very ingenious theory, but of course they could only obtain directly the parity-conserving part of the weak interaction. They generated the parity violation by imposing on the resulting theory Salam's γ_5 invariance. And they had to put in the W mass by hand.

Two years later [11] they actually proposed a unified theory of weak, electromagnetic *and strong* interactions, based on the gauge group SO(8) — a paper well ahead of its time, foreshadowing later ideas of grand unification.

But these theories did not really work; nor did similar ones proposed by Glashow and others. The major obstacle remained the vector meson mass. This was essential to make the interaction weak and short-range, but apparently incompatible with both gauge invariance and renormalizability. The only way anyone knew to make a vector-meson theory renormalizable was to make it a gauge theory with zero-mass gauge bosons.

6. Folk theorems

As often happens, progress was delayed by a "folk theorem". Theoretical physicists have a tendency to develop such things — "theorems" that everyone believes and quotes but that turn out not to be true, or at least not applicable to the cases of interest.

One such folk theorem was the assertion that the photon mass is zero *because* of gauge invariance. The theorem is highly plausible. The bare mass is obviously zero, and within perturbation theory the zero mass is apparently protected by gauge invariance. Indeed the zero photon mass was seen as one of the predictive successes of the gauge principle.

It was Julian Schwinger who pointed out in 1961 that this theorem might be false [12]. This was another of the key papers in the development of the subject — though in a way it propagated a myth of its own. For Schwinger said that "gauge invariance of a vector field does not necessarily imply zero mass for an associated particle *if the current vector coupling is sufficiently strong*" (my emphasis). He was of course thinking primarily of a theory of strong interactions. But at first sight his (perfectly correct) statement seemed to imply that the idea would not work for weak interactions. Of course we now understand that non-perturbative symmetry breaking can occur even in weak-coupling theories, but that was not so obvious at the time.

There was another, related problem too — and another folk theorem to contend with. People began edging towards the idea of spontaneous symmetry breaking as the origin of the vector meson masses. But then there was the Goldstone theorem. In one form, this states that in a relativistic theory, if the Lagrangian but not the vacuum state is invariant under a continuous symmetry transformation, there must exist massless spin-zero particles. But of course such particles were ruled out phenomenologically.

Because of the importance of this issue, when Steven Weinberg came to Imperial College on sabbatical in 1961–62, he and Salam spent a great deal of time going over the proof of the theorem, collaborating at long range with Jeffrey Goldstone. This resulted in a paper that gave a very careful and thorough proof of the theorem [13].

To see just how inescapable the difficulty seemed at the time, I think it is worthwhile recalling the argument in its simplest form.

Let j^μ be the relevant current, satisfying

$$\partial_\mu j^\mu = 0, \tag{12}$$

so that the corresponding charge,

$$Q = \int d^3\mathbf{x}\, j^0(t, \mathbf{x}), \tag{13}$$

is conserved. Suppose there is a field $\phi(x)$ whose vacuum expectation value is not invariant under the symmetry transformation, *i.e.*,

$$i\,\langle 0|[Q, \phi(0)]|0\rangle \neq 0. \tag{14}$$

Define

$$f^\mu(k) = i \int d^4x\, e^{ik\cdot x}\, \langle 0|[j^\mu(x), \phi(0)]|0\rangle\,. \tag{15}$$

From relativistic invariance, since f^μ is a function of a single four-vector k^μ, we conclude that it must have the form

$$f^\mu(k) = k^\mu[g(k^2) + \epsilon(k^0)h(k^2)]. \tag{16}$$

Current conservation, Eq. (12), then implies

$$0 = k_\mu f^\mu = k^2 g(k^2) + \epsilon(k^0)k^2 h(k^2). \tag{17}$$

So g and h can only be proportional to the mass-shell delta function, $\delta(k^2)$. On the other hand the symmetry-breaking condition, Eq. (14), implies

$$\int \frac{dk^0}{2\pi} f(k^0, \mathbf{0}) \neq 0. \tag{18}$$

Thus h cannot vanish. This implies that there are zero-mass states that couple to the vacuum through the current in question (and through ϕ).

In condensed-matter physics, counter-examples to the Goldstone theorem were known, in the case of long-range forces. But relativistic invariance seemed to rule out such a mechanism. For quite a while this seemed a very severe obstacle. Zero-mass particles coupled in the appropriate way could not possibly have existed and escaped detection.

7. Evading the Goldstone theorem

The next step in the story, at least from my perspective, was an important paper by Phil Anderson, published in April 1963 [14]. The declared aim of that paper was to exhibit an example of Schwinger's suggestion that in suitable cases a gauge field can acquire a mass. One example he gave was the plasmon: the fact that in a high-density plasma the photon acquires a non-zero "mass", namely the plasma frequency,

$$m_\gamma = \omega_{\text{plasma}} = \sqrt{\frac{n_e e^2}{m_e}}. \tag{19}$$

But this paper actually did much more than exhibit a special case of Schwinger's mechanism. Anderson also pointed out, using the example of superconductivity, how in some cases Goldstone bosons "become tangled up with Yang-Mills gauge bosons and, thus, do not in any true sense have zero mass". He concluded that "the Goldstone zero-mass difficulty is not a serious one, because we can probably cancel it off against an equal Yang-Mills zero-mass problem". This is just what is now known as the Higgs mechanism.

This should have cleared everything up. Reading Anderson's paper now, it seems fairly straightforward, but that is not how it seemed at the time. I have to admit that one of the reasons I started working on this problem was my total failure to understand that paper. If my memory serves me correctly, I took it to Gerry Guralnik, and possibly also

Dick Hagen, both of whom were on sabbatical at Imperial that year, saying "Can you make any sense of this?"

Basically, what we found hard to understand was just *how* Anderson had managed to evade the Goldstone theorem. It took us quite a long time to appreciate that the theorem actually doesn't apply to gauge symmetries. By the time we reached that conclusion, others had reached it too. The result was published independently by Englert and Brout [15] and by Peter Higgs [16]; our paper followed a little later [17].

Of course the reason the Goldstone theorem does not apply to gauge theories is by now very well known. If one uses a physical gauge such as the Coulomb gauge, then manifest relativistic invariance is lost. On the other hand, in a covariant gauge, the theorem does apply, but the predicted massless states are unphysical, pure gauge modes.

So 1963–64 saw a great step forward. At least in principle the problem of the origin of mass was solved. Further work on more explicit models was done in the next few years, by Peter Higgs [18] and by myself [19], among others.

8. Dénouement

But there was still another major hurdle to overcome. The key remaining difficulty was to reconcile the requirement of unification with the fact that the weak intereactions are parity-violating, while the electromagnetic interactions are not. In a gauge theory, the $W^{\pm}$ had to have a neutral partner, W^0. At least in the simplest models, $W^{\pm}$ would acquire a mass but W^0 would not. That was fine if W^0 was the photon; but that idea didn't work because if W^0 was partnered with $W^{\pm}$ it too would have parity-violating interactions.

Once again, there seemed no way out — either one had an unobserved massless gauge boson or else a parity-violating photon. Neither was acceptable. It took another three years for people to realise that the resolution of this dilemma was, as often happens, to accept neither horn, but rather a synthesis of the two. The essential key of course was to extend the gauge group from SU(2) to SU(2) × U(1).

Actually the solution, or something very like it, was already there in the literature, in the form of a 1961 paper by Sheldon Glashow [20], which proposed an SU(2) × U(1) model with mixing between the neutral particles, but of course without the key concept of spontaneous symmetry breaking and the Higgs mechanism.

At Imperial, Salam kept plugging away at the problem, especially in collaboration with John Ward. I had quite a few discussions with him about symmetry breaking during that period.

As fate would have it, in a sense I missed the dénouement of this story, but I would still like to tell you about my reaction to it.

In the summer of 1967, I went to Brookhaven and then on to a year's sabbatical at the University of Rochester. In the autumn of that year, Salam gave a series of lectures at Imperial in which he expounded the SU(2) × U(1) theory, though I only heard about them later — remember there was no e-mail then, and we still used transatlantic telephones, at least from the British end, only in dire emergencies! The work he talked about was presented at a Nobel Symposium early the following year [21]. Meanwhile, the same model had been found independently by Steven Weinberg [22].

It is easy with hindsight see the discovery of the spontaneously broken SU(2) × U(1) theory as a crucial turning point. But it is salutory to recall how it seemed at the time.

When Weinberg's paper appeared in Phys. Rev. Letters, Marshak (who of course knew of my interest in such things) asked me to give a talk to his weekly discussion group. The entire high energy group used to meet regularly once a week in his office — fortunately a very big office — for an informal and wide-ranging seminar. Various group members would be asked to talk about exciting recent papers or current problems.

So I gave a talk about Weinberg's paper, mentioning that Salam and Ward had been working on very similar ideas. I talked about the various problems one faced in constructing a unified theory of weak and electromagnetic interactions and pointed out how ingeniously the model avoided them. The most important thing, I felt, was a sort of existence proof: it proved that this *could* be done. I described it, I believe, as a really wonderful toy model — but, of course, I said, one cannot seriously believe this has anything to do with the real world!

In retrospect, I was certainly myopic. But in a sense, my reasons were absolutely right. The whole thing seemed much too *ad hoc* and ugly, with its curious built-in asymmetry between the left- and right-handed fermions and its large number of independent adjustable parameters. If one thinks of it as part of the final, fundamental theory, it *is* ugly; surely, one feels, the Creator was having an off day! But that is not the right way to look at it. Seen merely as one step towards a still undiscovered final theory, the intricate way SU(2) × U(1) fits together does have a remarkable beauty.

9. Postscript

That was by no means the end of the story. So far as electroweak unification was concerned, there were several important steps still to come: the GIM mechanism in 1970 [23], 't Hooft's proof of renormalizability in 1971 [24], the discovery of neutral currents in 1973 [25], and of the W and Z bosons in 1983 [26, 27].

And of course Salam went on to even grander unification, for example in his work with Pati on lepton-hadron unification [28] and with Strathdee on supersymmetry and superfields [29] and on Kaluza-Klein theory [30]; I could mention many others.

But although it is not the end of the story — indeed, will there be an end to the story? — it is a good place for me to stop.

Before I do so, however, I would like to repeat what I said at the beginning. It was a tremendous privilege to be part of the team led by Professor Abdus Salam, and to participate in this voyage of discovery. I cannot imagine a place I would rather have been during those critical and exciting years. Many of us here, myself included, owe Professor Salam a great debt for the stimulus, support and friendship he has given us over many years. I hope that this meeting has helped in some way to express that gratitude as well as our warm good wishes for the future.

Acknowledgements

I am grateful to Seifallah Randjbar-Daemi and his fellow organizers of the *Salamfest* for arranging a very exciting meeting and for inviting me to give this talk.

References

[1] N. Kemmer, *Phys. Rev.* **52** (1937) 906.
[2] R. Utiyama, *Phys. Rev.* **101** (1956) 1597.
[3] L. O'Raifeartaigh, these proceedings.
[4] T.W.B. Kibble, *J. Math. Phys.* **2** (1961) 212.
[5] C.N. Yang & R.L. Mills, *Phys. Rev.* **96** (1954) 191.
[6] T.D. Lee & C.N. Yang, *Phys. Rev.* **104** (1956) 254.
[7] R.E. Marshak & E.C.G. Sudarshan, *Proceedings of the Padua-Venice Conference on Mesons and Recently Discovered Particles* (1957); *Phys. Rev.* **109** (1958) 1860.
[8] R.P. Feynman & M. Gell-Mann, *Phys. Rev.* **109** (1958) 193.
[9] A. Salam, *Il Nuovo Cimento* **5** (1957) 299.
[10] A. Salam & J.C. Ward, *Il Nuovo Cimento* **11** (1959) 568.
[11] A. Salam & J.C. Ward, *Il Nuovo Cimento* **19** (1961) 165.
[12] J. Schwinger, *Phys. Rev.* **125** (1962) 397.
[13] J. Goldstone, A. Salam & S. Weinberg, *Phys. Rev.* **127** (1962) 965.
[14] P.W. Anderson, *Phys. Rev.* **130** (1963) 439.
[15] F. Englert & R. Brout, *Phys. Rev. Letters* **13** (1964) 321.
[16] P.W. Higgs, *Physics Letters* **12** (1964) 132.
[17] G.S. Guralnik, C.R. Hagen & T.W.B. Kibble, *Phys. Rev. Letters* **13** (1964) 585.
[18] P.W. Higgs, *Phys. Rev.* **145** (1966) 1156.
[19] T.W.B. Kibble, *Phys. Rev.* **155** (1967) 1554.
[20] S.L. Glashow, *Nuc. Phys.* **22** (1961) 579.
[21] A. Salam, in *Elementary Particle Theory*, ed. N. Svartholm (Almqvist & Wiksell, Stockholm, 1968).
[22] S. Weinberg, *Phys. Rev. Letters* **19** (1967) 1264.
[23] S.L. Glashow, J. Iliopoulos & L. Maiani, *Phys. Rev.* **D2** (1970) 1285.
[24] G. 't Hooft, *Nuc. Phys.* **B35** (1971) 167.
[25] F.J. Hasert *et al.*, *Physics Letters* **46B** (1973) 138.
[26] G. Arnison *et al.*, *Physics Letters* **122B** (1983) 102.
[27] G. Arnison *et al.*, *Physics Letters* **126B** (1983) 398.
[28] J.C. Pati & A. Salam, *Phys. Rev.* **D8** (1973) 1240.
[29] A. Salam & J. Strathdee, *Nuc. Phys.* **B76** (1974) 477.
[30] A. Salam & J. Strathdee, *Ann. Phys.* **141** (1982) 316.

LIST OF SPEAKERS

ALI, Ahmed
DESY
Notkestrasse 85
2000 Hamburg 52
Germany

AMALDI, Ugo
Present:
DELPHI Collaboration
CERN
CH-1211 Geneva 23
Switzerland
Permanent: Polo Univ. di Como

AMATI, Daniele
SISSA
Via Beirut 2-4
34014 Trieste
Italy

BAHCALL, John
School of Natural Sciences
Institute for Advanced Study
Olden Lane
Princeton, N.J. 08540
U.S.A.

BELLOTTI, Enrico
Laboratorio Nazionale del Gran Sasso
I.N.F.N.
S.S.17 bis km. 18 + 910
67010 Assergi (Aq.)
Italy

CABIBBO, Nicola
Dipartimento di Fisica
II Università di Roma "Tor Vergata"
Via delle Ricerca Scientifica 1
00133 Roma
Italy

CHAMSEDDINE, Ali
Institut für Theoretische Physik
Universität Zürich
Schonbergasse 9
CH 8001 Zürich
Switzerland

DELBOURGO, Robert
Department of Physics
University of Tasmania
G.P.O. Box 252 C
Hobart - Tasmania 7001
Australia

DUFF, Michael
Department of Physics
Texas A & M University
College Station, Texas 77843
U.S.A.

ELLIS, John
TH Division
CERN
CH-1211 Geneva 23
Switzerland

FELTESSE, Joel
DESY - Dpt. H1K
Notkestrasse 85
2000 Hamburg 52
Germany

GREEN, Michael
Department of Physics
Queen Mary & Westfield College
University of London
Mile End Road
London, E1 4NS
U.K.

GROSS, David
Joseph Henry Laboratory
Princeton University
Jadwin Hall
P.O. Box 708
Princeton, N.J. 08544
U.S.A.

KIBBLE, Thomas W.B.
Blackett Laboratory
Imperial College
Prince Consort Road
London, SW7 2BZ
U.K.

LAUGHLIN, Robert Betts
Department of Physics
Stanford University
Stanford, CA 94305-4060
U.S.A.

RUBAKOV, Valeri A.
Institute for Nuclear Research
Academy of Sciences
60th October Anniversary Prospect 7A
117312 Moscow
Russia

MACK, Gerhard
II Institut für Theoretische Physik
Universität Hamburg
Luruper Chaussee 149
2000 Hamburg 50
Germany

SCHRAMM, David N.
Department of Astronomy and
Astrophysics
The University of Chicago
5640 South Ellis Avenue
Chicago, IL 60637
U.S.A.

MUELLER, Alfred
Physics Department
Columbia University
538 West 120th Street
New York, N.Y. 10027
U.S.A.

SCHRIEFFER, J. Robert
Department of Physics
Florida State University
Tallahassee, FL 32306
U.S.A.

O'RAIFEARTAIGH, Lochlainn
School of Theoretical Physics
Dublin Institute for Advanced Studies
10 Burlington Road
Dublin 4
Ireland

SCHRÖDER, Henning
DESY
Argus-F15
Notkestrasse 85
2000 Hamburg 52
Germany

PATI, Jogesh
Department of Physics and Astrophysics
University of Maryland
College Park, MD 20742-4111
U.S.A.

SCIAMA, Dennis W.
SISSA
Via Beirut 2-4
34014 Trieste
Italy

PECCEI, Roberto
Department of Physics
University of California at Los Angeles
405 Hilgard Avenue
Los Angeles, CA 90024
U.S.A.

SEZGIN, Ergin
Department of Physics
Texas A & M University
College Station, Texas 77843
U.S.A.

RANDJBAR-DAEMI, Seifallah
High Energy Section
ICTP
P.O. Box 586
Trieste
Italy

SHAFI, Qaisar
Bartol Research Institute
University of Delaware
217 Sharp Laboratory
Newark, Delaware 19716
U.S.A.

VAFA, Cumrun
Lyman Laboratory of Physics
Harvard University
Cambridge, MA 02138
U.S.A.

von KLITZING, Klaus
Max-Planck-Institut für
Festkorperforschung
Heisenbergstrasse 1
Postfach 80 06 65
7000 Stuttgart
Germany

WEST, Peter
Department of Mathematics
King's College London
Strand
London, WC2R 2LS
U.K.

WINSTEIN, Bruce
The Enrico Fermi Institute
University of Chicago
5640 South Ellis Avenue
Chicago, IL 60637
U.S.A.

YANG, Chen Ning
Institute for Theoretical Physics
SUNY at Stony Brook
Stony Brook, N.Y. 11794-3840
U.S.A.

LIST OF PARTICIPANTS

ABDALLA, Maria C.B.
Departamento de Fisica Matematica
Instituto de Fisica
Universidade de Sao Paulo
01498 Sao Paulo, S.P.
Brazil

AHMED, Elsayed M.
Maths Department
University of U.A.E.
P.O. Box 15551
Al-Ain
United Arab Emirates

AKDENIZ, K. Gediz
Balkan Physical Union
University of Istanbul
Vezneciler 34459
Istanbul
Turkey

AKHMEDOV, Eugueni
High Energy Section
International Centre for Theoretical Physics
Trieste
Italy

ARDALAN, Farhad
Physics Department
Sharif University of Technology
Tehran
Iran

AREF'EVA, Irina
Dept. of Quantum Field Theory
Steklov Mathematical Institute
Vavilova St. 42
117 966 Moscow, GSP 1
Russia

ARGYRES, Ernestos N.
Institute of Nuclear Physics
"Demokritos"
P.O. Box 60228
Aghia Paraskevi
153 10 Athens
Greece

AYDIN, Z. Zekeriya
Department of Physics
Fen. Fakültesi
Ankara University
06100 Tandogan - Ankara
Turkey

BAAQUIE, Belal E.
Department of Physics
National University of Singapore
Lower Kent Ridge Road
Singapore 0511

BAETA, Robert D.
Department of Physics
University of Ghana
P.O. Box 63
Legon - Accra
Ghana

BAGHDADI, Mohamed
Lab. de Physique Theorique
Université Mohammed V
Avenue Ibn Battota
B.P. 1014
Rabat
Morocco

BAHADIROGLU, A. Gülay
Physics Department
University of Istanbul
Vezneciler
Istanbul
Turkey

BAHARI, Mohammad Reza
Executive Editor
Iranian Journal of Physics
c/o Iran University Press
85 Park Avenue
Tehran
Iran

BAHRAN, Moustafa
Dept. of Physics and Astronomy
University of Oklahoma
440 W. Brooks
Norman, OK 73019-0225
U.S.A.

BALLOCCHI, Giuseppe
PPE Division
CERN
CH-1211 Geneva 23
Switzerland

BARBIELLINI, Guido
INFN - Sezione di Trieste
Università di Trieste
Trieste
Italy

BASSANI, Franco
Scuola Normale Superiore
Piazza dei Cavalieri 7
56100 Pisa
Italy

BEG, Ismat
Department of Mathematics
Quaid-i-Azam University
P.O. Box 1090
Islamabad
Pakistan

BEREZHIANI, Zurab
INFN
Dipartimento di Fisica
Università di Ferrara
Ferrara
Italy

BONINI, Marisa
INFN di Parma
Università di Parma
Viale delle Scienze
43100 Parma
Italy

CASTELLI, Edoardo
INFN - Sezione di Trieste
Università di Trieste
Trieste
Italy

D'ATTANASIO, Marco
Dipartimento di Fisica
Università di Parma
Via delle Scienze
43100 Parma
Italy

DIAMANTINI, M. Cristina
Dipartimento di Fisica
Università di Perugia
Via A. Pascoli
06100 Perugia
Italy

DU, Dongsheng
Institute of High Energy Physics
Academia Sinica
P.O. Box 918(4)
Beijing 100039
People's Republic of China

DUBROVIN, Boris
SISSA
Via Beirut
Trieste
Italy

FANCHIOTTI, Sergio
TH Division
CERN
CH-1211 Geneva 23
Switzerland

FRAMPTON, Paul
Dept. of Physics and Astronomy
University of Chapel Hill
CB# 3255 Phillips Hall
Chapel Hill, NC 27599-3255
U.S.A.

FRASER, Gordon
CERN Courier
CERN
CH-1211 Geneva 23
Switzerland

FRONSDAL, Christian
Physics Department
U C L A
Los Angeles, CA 90024
U.S.A.

HOODBHOY, Pervez
Department of Physics
Quaid-i-Azam University
Islamabad
Pakistan

FU, Xin Yong
Department of Physics
Jiaotong University
Shanghai
People's Republic of China

HORVAT, Dubravko
Department of Physics
University of Zagreb
P.O. Box 170
41001 Zagreb
Croatia

GALIZA, Pitagoras
International Centre for Science
Via Grignano 9
Trieste
Italy

HORVATH, Jorge E.
Instituto Astronomica e Geofisico
Universidade de Sao Paulo
Ac. M. Stefano 4200
04301-904 Sao Paulo
Brazil

GOLSHANI, Mehdi
Institute for Studies in Theoretical Physics & Mathematics
Sharif University of Technology
P.O. Box 19395-1795
Tehran
Iran

JEZABEK, Marek
Inst. für Theoretische Teilchenphysik
Universität Karlsruhe
Kaiserstrasse 12
7500 Karlsruhe 1
Germany

GUBERINA, Branko
Dept. of Theoretical Physics
Ruder Boskovic Institute
Bijenicka 54
P.O. Box 1016
41000 Zagreb
Croatia

KARIM, Munawar
Department of Physics
St. John Fisher College
Rochester, N.Y. 14618
U.S.A.

HALMAGEAN, Eugenia
R&D Institute for Semiconductor Devices
Str. Erou Iancu Nicolae Nr. 32B
72996 Bucharest
Romania

KHADKIKAR, Surendra B.
Physical Research Laboratory
Navrangpura
Ahmedabad 380009
India

HEYROVSKY, David
Dept. of Theoretical Physics
Faculty of Maths and Physics
Charles University
v. Holesovickach 2
180 00 Praha 8
Czech Republic

KOCA, Mehmet
Physics Department
University of Cukurova
P.O. Box 171
Adana
Turkey

LAZARIDES, George
Department of Physics
Aristotelian University of Thessaloniki
Thessaloniki 54006
Greece

LEONTARIS, George K.
Physics Department
Ioannina University
451 10 Ioannina
Greece

MARX, George
Department of Atomic Physics
Eötvös University
Puskin utca 5
Budapest 1088
Hungary

MEJDANI, Rexhap
Department of Physics
Faculty of Natural Sciences
University of Tirana
Tirana
Albania

MOORE, Anthony J.
World Scientific Publishing Co.
73 Lynton Mead
Totteridge
London, N20 8DH
U.K.

MURTAZA, Ghulam
Department of Physics
Quaid-i-Azam University
Islamabad
Pakistan

NAMAZIE, M.A.M.
Dipartimento di Fisica
II Università di Roma "Tor Vergata"
Via della Ricerca Scientifica 1
00133 Roma
Italy

NARANCIC, Zoran
Department of Physics
University of Zagreb
P.O. Box 170
41001 Zagreb
Croatia

NICODEMI, Francesco
Dipartimento di Fisica
Università di Napoli
Mostra d'Oltremare, Pad. 19-20
80125 Napoli
Italy

NUYTS, Jean
Departement de Physique
Université de Mons-Hainaut
Place du Parc 20
7000 Mons
Belgium

PANAGIOTAKOPOULOS
Department of Theoretical Physics
Aristotle University of Thessaloniki
54006 Thessaloniki
Greece

PIETRUSZKO, Stanislaw
Institute of Microelectronics and Optoelectronics
Warsaw University of Tech.
Kosykowa 75
00-662 Warsaw
Poland

RAHMAN, Naseem
Department of Chemistry
University of Trieste
Trieste
Italy

RAITHER, Siegbert
Division of Basic Sciences
U N E S C O
7 place de Fontenoy
75700 Paris
France

RAVNDAL, Finn
Institute of Physics
University of Oslo
P.O. Box 1048
0316 Oslo
Norway

REMIDDI, Ettore
Dipartimento di Fisica
Università di Bologna
Via Irnerio 46
40126 Bologna
Italy

RESTUCCIA, Alvaro E.
Departamento de Fisica
Universidad Simon Bolivar
Apartado 89000
Caracas 1080
Venezuela

RIAZUDDIN
Department of Physics
King Fahd University of Petroleum and Minerals
UPM #654
Dhahran 31261
Saudi Arabia

RIVERS, Raymond J.
Blackett Laboratory
Imperial College
Prince Consort Road
London, SW7 2BZ
U.K.

ROSU, Haret N.
International Centre for Theoretical Physics
Trieste
Italy

SAKELLARIADOU, Maria
Lab. de Modèles de Physique Mathématique
Faculté des Sciences et Tech.
Parc. de Grandmont
37200 Tours
France

SARADZHEV, Fuad M.
Institute of Physics
Academy of Sciences
Narimanova pr. 33
37043 Baku
Azerbaijan

SARKAR, Uptal
Theory Group
Physical Research Laboratory
Ahmedabad 380009
India

SHEN, Xiang
TH Division
CERN
CH-1211 Geneva 23
Switzerland

SIVARAM, Chandra
Indian Institute of Astrophysics
Bangalore
India

SONG, Hi Sung
Center for Theoretical Physics
Seoul National University
Seoul 151-742
Korea

SRIVASTAVA, PREM
Centro Brasileiro de Pesquisas Fisicas
Rua Xavier Sigaud 150
22290 Rio de Janeiro
Brazil

STELLE, K.
Blackett Laboratory
Imperial College
Prince Consort Road
London, SW7 2BZ
U.K.

STROCCHI, Franco
SISSA
Via Beirut 2-4
34014 Trieste
Italy

TABANDEH, Ladan
INFN Microprocessor Laboratory
ICTP
Trieste
Italy

TABANDEH, Mohmoud
INFN Microprocessor Laboratory
ICTP
Trieste
Italy

TOMASEK, Roman
Nuclear Centre
Charles University
V. Holesovickach 2
180 00 Praha 2
Czech Federal Republic

TONIN, Mario
Istituto di Fisica
Università di Padova
Via F. Marzolo 8
35131 Padova
Italy

TRUGENBERGER, Carlo A.
TH Division
CERN
CH-1211 Geneva 23
Switzerland

VANDERHAEGHEN, Marc
Institute for Nuclear Physics
Proeftuinstraat 86
9000 Gent
Belgium

VENTURI, Giovanni
Dipartimento di Fisica
Università di Bologna
Via Irnerio 46
40126 Bologna
Italy

VIOLINI, Galileo
Centro Internacional de Fisica
Apartado Aereo 49490
Bogota
Colombia

ZAMIRALOV, Valeri
Institute of Nuclear Physics
Moscow State University
Moscow 119 899
Russia

ZEMBA, Guillermo
TH Division
CERN
CH-1211 Geneva 23
Switzerland

ZOVKO, Nicola
Ruderkovic Institute
P.O. Box 1016
Zagreb
Croatia

www.ingramcontent.com/pod-product-compliance
Lightning Source LLC
LaVergne TN
LVHW010539300926
849162LV00002B/177
* 9 7 8 9 8 1 0 2 1 4 2 1 0 *